Odoo Development Cookbook

Build effective business applications using the latest features in Odoo 17

Husen Daudi

Jay Vora

Parth Gajjar

Alexandre Fayolle

Holger Brunn

Daniel Reis

Odoo Development Cookbook

Group Product Manager: Aaron Tanna
Publishing Product Manager: Kushal Dave
Book Project Manager: Prajakta Naik
Senior Editor: Kinnari Chohan
Technical Editor: Jubit Pincy
Copy Editor: Safis Editing
Proofreader: Nithya Sadanandan
Indexer: Pratik Shirodkar
Production Designer: Joshua Misquitta
Senior DevRel Marketing Executive: Deepak Kumar
DevRel Marketing Coordinator: Mayank Singh

First published: April 2016
Second edition: January 2018
Third Edition: April 2019
Fourth Edition: December 2020
Fifth Edition: May 2024
Production reference: 2280524

Published by Packt Publishing Ltd.
Grosvenor House
11 St Paul's Square
Birmingham
B3 1RB, UK.

ISBN 978-1-80512-427-6
www.packtpub.com

Contributors

About the authors

Husen Daudi, a software developer with a Master's Degree from Gujarat University, India, is also a Six Sigma Black Belt consultant. He co-founded Serpent Consulting Services Pvt. Ltd., a prominent Open Source ERP Service provider with over 100 IT specialists serving clients in more than 170 countries. With extensive experience in ERP implementation since 2007, he brings a unique approach to his work. Husen has played a pivotal role in developing and maintaining various ERP implementations in both public and private sectors. Outside of work, he is a hobbyist painter and cherishes spending time with his sons, Mufaddal and Yusuf.

Jay Vora, a software engineer with a bachelor's degree from Gujarat University, India, is known for his thoughtful leadership, passion for development, and enthusiasm for technology. With over a decade of experience in ERPs since 2007, he co-founded Serpent Consulting Services Pvt. Ltd., a leading provider of Odoo services. The company boasts a team of over 100 IT specialists serving clients across 170 countries. Jay is known for his sociable nature and active participation in various Odoo forums and social platforms. In addition to his technical pursuits, he is also a poet, writer, and avid blogger on topics ranging from motivation and cricket to ERP-related subjects.

Parth Gajjar is an Odoo expert with a deep understanding of the Odoo framework. He started his career at Odoo and spent 7 years in the R&D department at Odoo India. During his time at Odoo, he worked on several key features, including a marketing automation app, mobile application, report engine, domain builder, and more. He also worked as a code reviewer and helped manage the code quality of the new features. Later, he started his own venture named Droggol and now provides various development services related to Odoo. He loves working on Odoo and solving real-world business problems with different technologies. He often gives technical training to Odoo developers.

I would like to thank my parents and family members for all of the support they have given throughout the writing of this book.

Alexandre Fayolle started working with Linux and free software in the mid-1990s and quickly became interested in the Python programming language. In 2012, he joined Camptocamp to share his expertise on Python, PostgreSQL, and Linux with the team implementing Odoo. He currently manages projects for Camptocamp and is strongly involved in the Odoo Community Association. In his spare time, he likes to play jazz on the vibraphone.

Holger Brunn has been a fervent open source advocate since he came into contact with the open source market sometime in the nineties. He has programmed for ERP and similar systems in different positions since 2001. For the last 10 years, he has dedicated his time to TinyERP, which became OpenERP and evolved into Odoo. Currently, he works at Therp BV in the Netherlands as a developer and is an active member of the Odoo Community Association.

Daniel Reis has had a long career in the IT industry, largely as a consultant implementing business applications in a variety of sectors, and today works for Securitas, a multinational security services provider. He has been working with Odoo (formerly OpenERP) since 2010, is an active contributor to the Odoo Community Association projects, is currently a member of the board of the Odoo Community Association, and collaborates with ThinkOpen Solutions, a leading Portuguese Odoo integrator.

Thank you note

We at SerpentCS (`https://www.serpentcs.com/`) would like to thank everyone who participated actively:

- *Ammar Officewala*
- *Chirag Patel*
- *Deepak Ahir*
- *Maitree Abhishek Pandya*
- *Murtuza Saleh*
- *Nikul Chaudhary*
- *Parvez Qureshi*
- *Prince Patel*
- *Rajan Soni*
- *Ritesh Bambhaniya*
- *Vacha Harshil Bhatt*

About the reviewer

Maxime Chambreuil serves as the Managing Director for Latin America at Open Source Integrators, a firm specializing in expert consulting and implementation services for open source solutions. With nearly two decades of experience in ERP and CRM consulting, Maxime brings a robust background in Information Systems Engineering, Business Management, and Free Software to his role

Maxime holds certifications as a Scrum Master and a Red Hat Partner Platform Certified Salesperson. He is also one of the founders and a former Vice-President of the Odoo Community Association, a non-profit organization dedicated to promoting and supporting Odoo and its collaborative development. Maxime has contributed significantly to the development of Odoo modules for Management Systems, Field Service Management, and various localizations, as well as to translations and bug reports.

Maxime's primary mission is to assist businesses in optimizing their processes, automating workflows, and transitioning to open source solutions.

Table of Contents

3

Creating Odoo Add-On Modules 49

4

Application Models 73

5

Basic Server-Side Development 111

6

Managing Module Data 143

7

Debugging Modules 159

8

Advanced Server-Side Development Techniques 177

9

Backend Views 207

10

Security Access 261

11

Internationalization 285

12

Automation, Workflows, Emails, and Printing 303

14

15

Web Client Development 411

16

The Odoo Web Library (OWL) 447

17

In-App Purchasing with Odoo 471

18

Automated Test Cases 477

19

Managing, Deploying, and Testing with Odoo.sh 507

20

Remote Procedure Calls in Odoo 535

22

Point of Sale 587

23

Managing Emails in Odoo 607

24

Managing the IoT Box 633

25

Web Studio 659

Index 719

Other Books You May Enjoy 736

Preface

While you're reading this, you've already been introduced to Odoo, one of the fastest-growing open-source ERP Business suites. Odoo is a full-featured open-source platform that assists in building solutions for various industries. If you're a developer, you're sitting on a goldmine. If you're an end user, you're gifted with an amazing tool to simplify your business processes, covering everything from pre-sales to sales, inventory, and accounting.

In addition to the extensive list of applications available in Odoo, it's like a nice dough (mmm, does it remind you of mouthwatering pizza?) that can be molded according to your requirements. Technically, it's a very flexible ORM-controller (Object Relational Mapping) driven application development framework built with extensibility in mind. Following the rule of inheritance, features/extensions, and modifications can be implemented as modules categorized as Apps. With ORM mentioned here, Odoo exhibits a monolithic architecture.

The Odoo 17 Development Cookbook provides a solid platform for developers, whether they're beginners or proficient. The code snippets cover most questions and use cases, and the explained fields assist in developing modules accurately while maintaining code quality and usability. As a bonus, *Chapter 25* is a special one that aids developers and non-developers in generating quick prototypes.

This book is written and supported by the entire Serpent Consulting Services Pvt Ltd team, with everyone contributing their time and effort to make the dream a reality.

Who this book is for

This book caters to developers at all levels, requiring a minimum understanding of object-oriented programming, with Python being a mandatory skill. Even newcomers to Python programming can find this book suitable. It's written with the intention of accommodating developers with minimal programming knowledge but a strong desire to learn.

The preferred development editors are PyCharm, Eclipse, or Sublime, but the majority of developers are expected to run Odoo on an Ubuntu/Debian-based operating system. The code examples are intentionally kept simple and clear, accompanied by thorough explanations to facilitate understanding. Newcomers will grasp the concepts from the basics, ensuring an enjoyable learning journey.

Experienced developers already familiar with Odoo should also find value in this book. It not only enhances their existing knowledge but also offers an easy way to stay updated on the latest Odoo versions, with significant changes highlighted.

Ultimately, this book aims to serve as a solid reference for daily use by both newcomers and experienced developers alike. Additionally, the documentation of differences between various Odoo versions will be a valuable resource for developers working with different versions simultaneously or porting modules.

What this book covers

Chapter 1, Installing the Odoo Development Environment, explains how to create a development environment for Odoo, start Odoo, create a configuration file, and activate Odoo's developer tools.

Chapter 2, Managing Odoo Server Instances, provides useful tips for working with add-ons installed from GitHub and organizing the source code of your instance.

Chapter 3, Creating Odoo Add-On Modules, explains the structure of an Odoo add-on module and gives a step-by-step guide for creating a simple module from scratch.

Chapter 4, Application Models, focuses on the Odoo model structure, and explains all types of fields with their attributes. It also covers techniques to extend existing database structures via extended modules.

Chapter 5, Basic Server-Side Development, explains various framework methods to perform CRUD operations in Odoo. This chapter also includes different ways to inherit and extend existing methods.

Chapter 6, Managing Module Data, shows how to ship data along with the code of your module. It also explains how to write a migration script when a data model provided by an add-on is modified in a new release.

Chapter 7, Debugging Modules, proposes some strategies for server-side debugging and an introduction to the Python debugger. It also covers techniques to run Odoo in developer mode.

Chapter 8, Advanced Server-Side Development Techniques, covers more advanced topics of the ORM framework. It is useful for developing wizards, SQL views, installation hooks, on-change methods, and more. This chapter also explains how to execute raw SQL queries in the database.

Chapter 9, Backend Views, explains how to write business views for your data models and how to call server-side methods from these views. It covers the usual views (list view, form view, and search view), as well as some complex views (kanban, graph, calendar, pivot, and so on).

Chapter 10, Security Access, explains how to control who has access to what in your Odoo instance by creating security groups, writing access control lists to define what operations are available to each group on a given model, and, if necessary, by writing record-level rules.

Chapter 11, Internationalization, shows how language translation works in Odoo. It shows how to install multiple languages and how to import/export translated terms.

Chapter 12, Automation, Workflows, Emails, and Printing, illustrates the different tools available in Odoo to implement business processes for your records. It also shows how server actions and automated rules can be used to support business rules. This also covers the QWeb report to generate dynamic PDF documents.

Chapter 13, Web Server Development, covers the core of the Odoo web server. It shows how to create custom URL routes to serve data on a given URL, and also shows how to control access to these URLs.

Chapter 14, CMS Website Development, shows how to manage a website with Odoo. It also shows how to create and modify beautiful web pages and QWeb templates. This chapter also includes how to create dynamic building blocks with options. It includes some dedicated recipes for managing SEO, user forms, UTM tracking, sitemaps, and fetching visitor location information. This chapter also highlights the latest concept of a multiwebsite in Odoo.

Chapter 15, Web Client Development, dives into the JavaScript part of Odoo. It covers how to create a new field widget and make RPC calls to the server. This also includes how to create a brand-new view from scratch. You will also learn how to create onboarding tours.

Chapter 16, The Odoo Web Library (OWL), gives introductions to the new client-side framework called OWL. It covers the life cycle of the OWL component. It also covers recipes to create a field widget from scratch.

Chapter 17, In-App Purchasing with Odoo, covers everything related to the latest concept of IAP in Odoo. In this chapter, you will learn how to create client and service modules for IAP. You will also learn how to create an IAP account and draw IAP credits from the end user.

Chapter 18, Automated Test Cases, includes how to write and execute automated test cases. This includes both server-side and client-side test cases. This chapter also covers tour test cases and setting up headless Chrome to get videos for failed test cases.

Chapter 19, Managing, Deploying, and Testing with Odoo.sh, explains how to manage, deploy, and test Odoo instances with the PaaS platform, Odoo.sh. It covers how you can manage different types of instances, such as production, staging, and development. This chapter also covers various configuration options for Odoo.sh.

Chapter 20, Remote Procedure Calls in Odoo, covers different ways to connect Odoo instances from external applications. This chapter teaches you how to connect to and access the data from an Odoo instance through XML-RPC, JSON-RPC, and the odoorpc library.

Chapter 21, Performance Optimization, explains the different concepts and patterns used to gain performance improvements in Odoo. This chapter includes the concept of prefetching, ORM-cache, and profiling the code to detect performance issues.

Chapter 22, Point of Sale, covers customization in a PoS application. This includes customization of the user interface, adding a new action button, modifying business flow, and extending customer recipes.

Chapter 23, Managing Emails in Odoo, explains how to manage email and chatter in Odoo. It starts by configuring mail servers and then moves to the mailing API of the Odoo framework. This chapter also covers the Jinja2 and QWeb mail templates, chatters on the form view, field logs, and activities.

Chapter 24, Managing the IoT Box, gives you the highlights of the latest hardware of IoT Box. This chapter covers how to configure, access, and debug IoT Box. It also includes a recipe to integrate IoT Box with your custom add-ons.

Chapter 25, delves into an alternative approach to module development. While it's not typically the best recommendation for implementation, analysts can swiftly create probable designs, prototypes, reports, or views using the techniques outlined in this module.

To get the most out of this book

Our primary and most valuable advice is simply 'Practice'! Each chapter provides detailed insights into the development aspect, so it's crucial to apply what you've learned.

To fully benefit from this book, we recommend supplementing your reading with additional resources on the Python programming language, the Ubuntu/Debian Linux operating system, and the PostgreSQL database.

The book includes installation instructions for Odoo, so all you need is Ubuntu 20.04 or later, or any other Linux-based OS. For other operating systems, you can utilize it through a virtual machine. If you're using Windows, you can also install Ubuntu as a subsystem:

Software/hardware covered in the book	Operating system requirements
Odoo 17 + Python 3.6 and above	Ubuntu 20.04 and above

This book is intended for developers who have basic knowledge of the Python programming language, as the Odoo backend runs on Python. In Odoo, data files are created with XML, so basic knowledge of XML is required.

This book also covers the backend JavaScript framework, PoS applications, and the website builder, which requires basic knowledge of JavaScript, jQuery, and Bootstrap 4. The Community Edition of Odoo is open source and freely available, but a few features, including IoT, cohort, and the dashboard, are available only in the Enterprise Edition, so to follow along with that recipe, you will need the Enterprise Edition.

To follow *Chapter 24*, *Managing the IoT Box*, you will require the Raspberry Pi 3 Model B+, which is available at `https://www.raspberrypi.org/products/raspberry-pi-3-model-b-plus/`.

If you are using the digital version of this book, we advise you to type the code yourself or access the code from the book's GitHub repository (a link is available in the next section). Doing so will help you avoid any potential errors related to the copying and pasting of code.

Download the example code files

You can download the example code files for this book from GitHub at `https://github.com/PacktPublishing/Odoo-17-Development-Cookbook-Fifth-Edition/tree/main`. If there's an update to the code, it will be updated in the GitHub repository.

We also have other code bundles from our rich catalog of books and videos available at `https://github.com/PacktPublishing/`. Check them out!

Conventions used

There are a number of text conventions used throughout this book.

`Code in text`: Indicates code words in text, database table names, folder names, filenames, file extensions, pathnames, dummy URLs, user input, and Twitter handles. Here is an example: "Given that `book` is a browse record, we can simply recycle the first example's function by passing `book.id` as a `book_id` parameter to give out the same content."

A block of code is set as follows:

```
@http.route('/my_library/books/json', type='json', auth='none')
def books_json(self):
records = request.env['library.book'].sudo().search([]) return
records.read(['name'])
```

Any command-line input or output is written as follows:

```
$ ./odoo-bin -d mydb --i18n-export=mail.po --modules=mail
$ mv mail.po ./addons/mail/i18n/mail.pot
```

Bold: Indicates a new term, an important word, or words that you see onscreen. For instance, words in menus or dialog boxes appear in **bold**. Here is an example: "Another important usage is providing demonstration data, which is loaded when the database is created with the **Load demonstration data** checkbox checked."

> **Tips or import ant notes**
> Appear like this.

Sections

In this book, you will find several headings that appear frequently (Getting ready, How to do it..., How it works..., There's more..., and See also).

To give clear instructions on how to complete a recipe, use these sections as follows:

Getting ready

This section tells you what to expect in the recipe and describes how to set up any software or any preliminary settings required for the recipe.

How to do it...

This section contains the steps required to follow the recipe.

How it works...

This section usually consists of a detailed explanation of what happened in the previous section.

There's more...

This section consists of additional information about the recipe in order to make you more knowledgeable about the recipe.

See also

This section provides helpful links to other useful information for the recipe.

Get in touch

Feedback from our readers is always welcome.

General feedback: If you have questions about any aspect of this book, email us at `customercare@packtpub.com` and mention the book title in the subject of your message.

Errata: Although we have taken every care to ensure the accuracy of our content, mistakes do happen. If you have found a mistake in this book, we would be grateful if you would report this to us. Please visit `www.packtpub.com/support/errata` and fill in the form.

Piracy: If you come across any illegal copies of our works in any form on the internet, we would be grateful if you would provide us with the location address or website name. Please contact us at `copyright@packt.com` with a link to the material.

If you are interested in becoming an author: If there is a topic that you have expertise in and you are interested in either writing or contributing to a book, please visit `authors.packtpub.com`.

Share your thoughts

Once you've read *Odoo Development Cookbook*, we'd love to hear your thoughts! Scan the QR code below to go straight to the Amazon review page for this book and share your feedback.

`https://packt.link/r/1805124277`

Your review is important to us and the tech community and will help us make sure we're delivering excellent quality content.

Download a free PDF copy of this book

Thanks for purchasing this book!

Do you like to read on the go but are unable to carry your print books everywhere?

Is your eBook purchase not compatible with the device of your choice?

Don't worry, now with every Packt book you get a DRM-free PDF version of that book at no cost.

Read anywhere, any place, on any device. Search, copy, and paste code from your favorite technical books directly into your application.

The perks don't stop there, you can get exclusive access to discounts, newsletters, and great free content in your inbox daily

Follow these simple steps to get the benefits:

1. Scan the QR code or visit the link below

`https://packt.link/free-ebook/9781805124276`

2. Submit your proof of purchase
3. That's it! We'll send your free PDF and other benefits to your email directly

1

Installing the Odoo Development Environment

To begin our Odoo development journey, we must set up our development environment by installing Odoo using source code that we can use to enhance, debug, and improve our development skills. There are several ways to set up an Odoo development environment, but this chapter proposes the best of them. You will find several other tutorials on the web explaining the other approaches. Keep in mind that this chapter is about setting up a development environment that has different requirements from a production environment; production has different parameters we must set in the config file, as per the volume of data and number of users in the system. We will cover configuration file parameters and their usage in this chapter.

If you are new to Odoo development, you must know about certain aspects of the Odoo ecosystem. The first recipe will give you a brief introduction to the Odoo ecosystem, after which we will install Odoo for development purposes.

In this chapter, we will cover the following recipes:

- Understanding the Odoo ecosystem
- Installing Odoo from the source
- Managing Odoo server databases
- Storing the instance configuration in a file
- Activating Odoo developer tools
- Updating the add-on modules list

Technical requirements

All the code that's used in this chapter can be downloaded from this book's GitHub repository at `https://github.com/PacktPublishing/Odoo-17-Development-Cookbook-Fifth-Edition/tree/main/Chapter01`.

Understanding the Odoo ecosystem

Odoo provides developers with out-of-the-box modularity and its powerful framework helps them build projects quickly. There are various characters in the Odoo ecosystem that you should be familiar with before embarking on your journey of becoming a successful Odoo developer.

Let's assume you have a system with 4 CPU cores, 8 GB of RAM, and 30 concurrent Odoo users.

To determine the number of workers needed, divide the number of users by 6. In this case, 30 users divided by 6 equals 5, which is the theoretical number of workers required.

To calculate the theoretical maximum number of workers, multiply the number of CPU cores by 2 and add 1. For 4 CPU cores, (4 * 2) + 1 equals 9, which is the theoretical maximum number of workers.

Based on these calculations, you can use 5 workers for the Odoo users and an additional worker for the cron worker, making a total of 6 workers.

To estimate the RAM consumption, use the following formula:

```
RAM = Number of workers * ((0.8 * 150) + (0.2 * 1024))
```

In this case, 6 workers multiplied by `((0.8 * 150) + (0.2 * 1024))` equals approximately 2 GB of RAM.

Therefore, based on these calculations, the Odoo installation will require around 2 GB of RAM.

Odoo editions

Odoo comes in two different editions. The first one is **Community Edition**, which is open source, while the second one is **Enterprise Edition**, which has licensing fees. Unlike other software vendors, Odoo Enterprise Edition is just a pack of extra applications that adds extra features or new apps to the Community Edition. The Enterprise Edition runs on top of the Community Edition. The Community Edition comes under the **Lesser General Public License v3.0** (**LGPLv3**) and comes with all of the basic **enterprise resource planning** (**ERP**) applications, such as sales, **customer relationship management** (**CRM**), invoicing, purchases, and a website builder. Alternatively, the Enterprise Edition comes with the Odoo Enterprise Edition license, which is a proprietary license. Odoo Enterprise Edition has several advanced features, such as full accounting, studio, **Voice over Internet Protocol** (**VoIP**), mobile responsive design, e-sign, marketing automation, delivery and banking integrations, **Internet of Things** (**IoT**), and more. The Enterprise Edition also provides you with unlimited *bugfix support*. The following diagram shows that the Enterprise Edition depends on the Community Edition, which is why you need the latter to use the former:

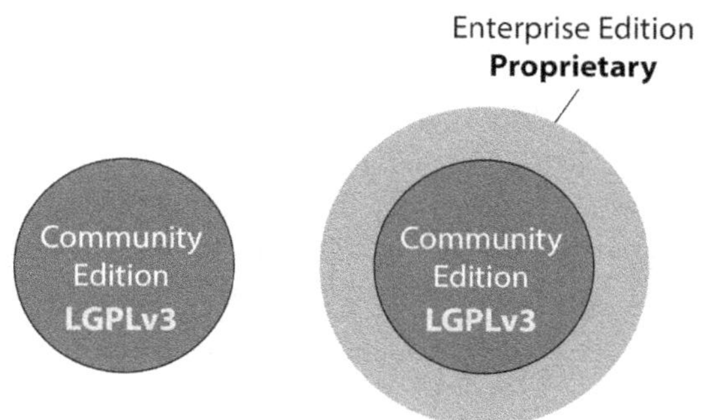

Figure 1.1 – Differences between the Community Edition and Enterprise Edition

You can find a full comparison of both editions here: `https://www.odoo.com/page/editions`.

> **Note**
>
> Odoo has the largest number of community developers among all open source ERPs on the market with 20K+ forks on GitHub, hence why you will find a large number of third-party apps (modules) on the app store. Some of the free apps use an **Affero General Public License version 3** (**AGPLv3**). You cannot use the proprietary license on your app if your application has dependencies on such apps. Apps with an Odoo proprietary license can only be developed on modules that have LGPL or other proprietary licenses.

Git repositories

The entire code base of Odoo is hosted on GitHub. You can post bugs/issues for stable versions here. You can also propose a new feature by submitting a **pull request** (**PR**). There are several repositories in Odoo. See the following table for more information:

Repositories	**Purpose**
`https://github.com/odoo/odoo`	This is the Community Edition of Odoo. It's available publicly.
`https://github.com/odoo/enterprise`	This is the Enterprise Edition of Odoo. It's available to official Odoo partners only.
`https://github.com/odoo-dev/odoo`	This is an ongoing development repository. It's available publicly.

Table 1.1 – Odoo git repositories

Every year, Odoo releases a major release, which is a long-term support version that will last for 3 years, and a few minor versions. Minor versions are mostly used in Odoo's online **Software-as-a-Service (SaaS)** offering, meaning that Odoo SaaS users get early access to these features. Major version branches have names such as 17.0, 16.0, 15.0, 14.0, 13.0, and 12.0, while minor version branches have names such as saas-17.1 and saas-17.2 on GitHub. These minor versions are mostly used for Odoo's SaaS platform. The `master` branch is under development and is unstable, so it is advisable not to use this for production since it might break your database.

Runbot

Runbot is Odoo's automated testing environment. Whenever there is a new commit in Odoo's GitHub branch, Runbot pulls those latest changes and creates the builds for the last four commits. Here, you can test all stable and in-development branches. You can even play with the Enterprise Edition and its development branches.

Every build has a different background color, which indicates the status of the test cases. A green background color means that all of the test cases run successfully and you can test that branch, while a red background color means that some test cases have failed on this branch and some features might be broken on that build. You can view the logs for all test cases, which show exactly what happens during installation. Every build has two databases. The `all` database has all of the modules installed on it, while the `base` database only has base Odoo modules installed. Every build is installed with basic demo data, so you can test it quickly without extra configurations.

> **Note**
>
> You can access Runbot at `http://runbot.odoo.com/runbot`.

The following credentials can be used to access any Runbot build:

- **Login ID**: admin **Password**: admin
- **Login ID**: demo **Password**: demo
- **Login ID**: portal **Password**: portal

> **Note**
>
> This is a public testing environment, so other users might be using/testing the same branch that you are testing.

Odoo app store

Odoo launched the app store a few years back, and it was an instant hit. At the time of writing, over 39,000+ different apps are hosted there. You will find lots of free and paid applications for different versions here. This includes specific solutions for different business verticals, such as education, food industries, and medicine. It also includes apps that extend or add new features to existing Odoo applications. The app store also provides numerous beautiful themes for the Odoo website builder. In *Chapter 3, Creating Odoo Add-On Modules*, you will learn how to set pricing and currency for your custom module.

You can access the Odoo app store by going to `https://www.odoo.com/apps`.

You can access Odoo's themes by going to `https://www.odoo.com/apps/themes`.

> **Note**
>
> Odoo has open sourced several themes after version 13 and now works with an advanced JavaScript called *OWL*. We will cover this in *Chapter 16*. Note that these were paid themes in previous versions. This means that, in Odoo versions 15 and 16, you can download and use those beautiful themes at no extra cost.

Odoo Community Association

Odoo Community Association (**OCA**) is a non-profit organization that develops/manages community-based Odoo modules. All OCA modules are open source and maintained by Odoo community members. OCA's GitHub account contains multiple repositories for different Odoo applications. Apart from Odoo modules, it also contains various tools, a migration library, accounting localizations, and so on.

Here is the URL for OCA's official GitHub account: `https://github.com/OCA`.

Official Odoo help forum

Odoo has a very powerful framework, and tons of things can be achieved just by using/activating options or by following specific patterns. Consequently, if you run into some technical issues or if you are not sure about some complex cases, then you can post your query on Odoo's official help forum. Lots of developers are active on this forum, including some official Odoo employees.

You can search for questions or post your new questions by going to `https://www.odoo.com/forum/help-1`.

Odoo's eLearning platform

Recently, Odoo has launched a new eLearning platform. This platform provides lots of videos that explain how to use different Odoo applications. At the time of writing, this platform does not have technical videos, just functional ones.

Here is the URL for Odoo's eLearning platform: `https://www.odoo.com/slides`.

Installing Odoo from the source

It is highly recommended that you use the **Linux Ubuntu** operating system to install Odoo since this is the operating system that Odoo uses for all its tests, debugging, and installations of Odoo Enterprise. Additionally, most Odoo developers use GNU/Linux distributions, so they are much more likely to get support from the Odoo community for operating system-level issues that occur in **GNU/Linux** than *Windows* or *macOS*.

It is also recommended to develop Odoo add-on modules using the same environment (the same distribution and the same version) as the one that will be used in production. This will avoid nasty surprises, such as discovering on the day of deployment that a library has a different version than expected, with a slightly different and incompatible behavior. If your workstation is using a different operating system, a good approach is to set up a **virtual machine** (**VM**) on your workstation and install a GNU/Linux distribution in the VM.

> **Note**
>
> Ubuntu is available as an app in the **Microsoft Store**, so you can use that if you don't want to switch to Ubuntu.

For this book, we will be using Ubuntu Server 22.04 LTS, but you can use any other Debian GNU/Linux operating system. Whatever Linux distribution you choose, you should have some notion of how to use it from the command line, and knowing about system administration will certainly not do any harm.

Getting ready

We are assuming that you have Ubuntu 22.04 up and running and that you have an account with root access or that `sudo` has been configured. In the following sections, we will install Odoo's dependencies and download Odoo's source code from GitHub.

> **Note**
>
> Some of the configurations require a system login username, so we will use `$(whoami)` whenever a login username is required in a command line. This is a shell command that will substitute your login in the command you are typing.

Some operations will be easier if you have a GitHub account. If you don't have one already, go to `https://github.com` and create one.

How to do it...

To install Odoo from the source, perform the following steps:

1. Run the following commands to install the main dependencies:

```
$ sudo apt-get update
$ sudo apt install openssh-server fail2ban python3-pip python3-dev libxml2-dev libxslt1-dev zlib1g-dev libsasl2-dev libldap2-dev build-essential libssl-dev libffi-dev libmysqlclient-dev libpq-dev libjpeg8-dev liblcms2-dev libblas-dev libatlas-base-dev git curl python3-venv python3.10-venv fontconfig libxrender1 xfonts-75dpi xfonts-base -y
```

2. Download and install **wkhtmltopdf**:

```
$ wget https://github.com/wkhtmltopdf/packaging/releases/download/0.12.6.1-2/wkhtmltox_0.12.6.1-2.jammy_amd64.deb
$ sudo dpkg -i wkhtmltox_0.12.6.1-2.jammy_amd64.deb
```

If you encounter any errors after running the previous command, force install the dependencies with the following command:

```
$ sudo apt-get install -f
```

3. Now, install the PostgreSQL database:

```
$ sudo apt install postgresql -y
```

4. Configure PostgreSQL:

```
$ sudo -i -u postgres createuser -s  $(whoami)
$ sudo su postgres
$ psql
alter user $(whoami) with password 'your_password';
\q
exit
```

5. Configure `git`:

```
$ git config --global user.name "Your Name"
$ git config --global user.email youremail@example.com
```

6. Clone the Odoo code base:

```
$ mkdir ~/odoo-dev
$ cd ~/odoo-dev
$ git clone -b 17.0 --single-branch --depth 1 https://github.com/odoo/odoo.git
```

7. Create an `odoo-17.0` virtual environment and activate it:

```
$ python3 -m venv ~/venv-odoo-17.0
$ source ~/venv-odoo-17.0/bin/activate
```

8. Install the Python dependencies of Odoo in `venv`:

```
$ cd ~/odoo-dev/odoo/
$ pip3 install -r requirements.txt
```

9. Create and start your first Odoo instances:

```
$ createdb odoo-test
$ python3 odoo-bin -d odoo-test -i base --addons-path=addons
--db-filter=odoo-test$
```

10. Point your browser to `http://localhost:8069` and authenticate it by using the `admin` account and using `admin` as the password.

> **Note**
>
> If you need RTL support, please install `node` and `rtlcss` by running `sudo apt-get install nodejs npm -y sudo npm install -g rtlcss`.

How it works...

In *step 1*, we installed several core dependencies. These dependencies include various tools, such as **git**, **pip3**, **wget**, Python setup tools, and more. These core tools will help us install other Odoo dependencies using simple commands.

In *step 2*, we downloaded and installed the `wkhtmltopdf` package, which is used in Odoo to print PDF documents such as sale orders, invoices, and other reports. Odoo 17.0 needs version 0.12.6.1 of `wkhtmltopdf`, and that exact version might be not included in the current Linux distributions. Fortunately for us, the maintainers of `wkhtmltopdf` provide pre-built packages for various distributions at `http://wkhtmltopdf.org/downloads.html` and we have downloaded and installed it from that URL.

After this, we configured PostgreSQL, which is used for Odoo's database management.

PostgreSQL configuration

In *step 3*, we installed the PostgreSQL database.

In *step 4*, we created a new database user with the login name of our system. `$(whoami)` is used to fetch your login name, and the `-s` option is used to give superuser rights. Let's see why we need these configurations.

Odoo uses the `psycopg2` Python library to connect with a PostgreSQL database. To access a PostgreSQL database with the `psycopg2` library, Odoo uses the following values by default:

- By default, `psycopg2` tries to connect to a database with the same username as the current user on local connections, which enables password-less authentication (this is good for the development environment)
- The local connection uses Unix domain sockets
- The database server listens on port `5432`

That's it! Your PostgreSQL database is now ready to be connected with Odoo.

As this is a development server, we have given `--superuser` rights to the user. It is OK to give the PostgreSQL user more rights as this will be your development instance. For a production instance, you can use the `--createdb` command line instead of `--superuser` to restrict rights. The –`superuser` rights in a production server will give additional leverage to an attacker exploiting a vulnerability in some part of the deployed code.

If you want to use a database user with a different login, you will need to provide a password for the user. This is done by passing the `--pwprompt` flag on the command line when creating the user, in which case the command will prompt you for the password.

If the user has already been created and you want to set a password (or modify a forgotten password), you can use the following command:

```
$ psql -c "alter role $(whoami) with password 'newpassword'"
```

If this command fails with an error message saying that the database does not exist, it is because you did not create a database named after your login name in *step 4* of this recipe. That's fine; just add the `--dbname` option with an existing database name, such as `--dbname template1`.

Git configuration

For the development environment, we are using Odoo sourced from GitHub. With `git`, you can easily switch between different Odoo versions. Note that you can fetch the latest changes with the `git pull` command.

In *step 5*, we configured our `git` user.

In *step 6*, we downloaded the source code from Odoo's official GitHub repository. We used the `git clone` command to download Odoo's source code. We used a single branch as we only wanted a branch for the 17.0 version. We also used `--depth 1` to avoid downloading the full commit history of the branch. These options will download the source code very quickly, but if you want, you can omit those options.

Odoo developers also propose nightly builds, which are available as tarballs and distribution packages. The main advantage of using `git clone` is that you will be able to update your repository when new bug fixes are committed in the source tree. You will also be able to easily test any proposed fixes and track regressions so that you can make your bug reports more precise and helpful for developers.

> **Note**
>
> If you have access to the Enterprise Edition source code, you can download that in a separate folder under the `~/odoo-dev` directory.

Virtual environments

Python **virtual environments**, or **venvs** for short, are isolated Python workspaces. These are very useful to Python developers because they allow different workspaces with different versions of various Python libraries to be installed, possibly on different Python interpreter versions.

You can create as many environments as you wish using the `python3 -m venv ~/newvenv` command. This will create a `newvenv` directory in the specified location, containing a `bin/` subdirectory and a `lib/python3.10` subdirectory.

In *step 7*, we created a new virtual environment in the `~/venv-odoo-17.0` directory. This will be our isolated Python environment for Odoo, and all of Odoo's Python dependencies will be installed in this environment.

To activate the virtual environment, we need to use the `source` command. We used the `source ~/venv-odoo-17.0/bin/activate` command to activate the virtual environment.

Installing Python packages

Odoo's source code has a list of Python dependencies in `requirements.txt`. In *step 8*, we installed all those requirements via the `pip3 install` command.

That's it. Now, you can run the Odoo instance.

Starting the instance

Now comes the moment you've been waiting for. To start our first instance, in *step 9*, we created a new empty database, used the `odoo-bin` script, and then started the Odoo instance with the following command:

```
python3 odoo-bin -d odoo-test -i base --addons-path=addons
--db-filter=odoo-test$
```

You can also omit python3 by using ./ before odoo-bin as it is an executable Python script:

```
./odoo-bin -d odoo-test -i base --addons-path=addons --db-filter=odoo-
test$
```

With odoo-bin, a script with the following command-line arguments is used:

- -d database_name: Use this database by default.
- --db-filter=database_name$: Only try to connect to databases that match the supplied regular expression. One Odoo installation can serve multiple instances that live in separate databases, and this argument limits the available databases. The trailing $ is important as the regular expression is used in match mode. This enables you to avoid selecting names starting with the specified string.
- --addons-path=directory1,directory2,...: This is a comma-separated list of directories in which Odoo will look for add-ons. This list is scanned at instance creation time to populate the list of available add-on modules in the instance. If you want to use Odoo's Enterprise Edition, then add its directory with this option.
- -i base: This is used to install a base module. This is required when you have created a database via the command line.

If you are using a database user with a database login that is different from your Linux login, you need to pass the following additional arguments:

- --db_host=localhost: Use a TCP connection to the database server
- --db_user=database_username: Use the specified database login
- --db_password=database_password: This is the password for authenticating against the PostgreSQL server

To get an overview of all available options, use the --help argument. We will see more of the odoo-bin script later in this chapter.

When Odoo is started on an empty database, it will create the database structure that's needed to support its operations. It will also scan the add-on path to find the available add-on modules and insert some into the initial records in the database. This includes the admin user with the default admin password, which you will use for authentication.

Pointing your web browser to http://localhost:8069/ leads you to the login page of your newly created instance, as shown in the following screenshot:

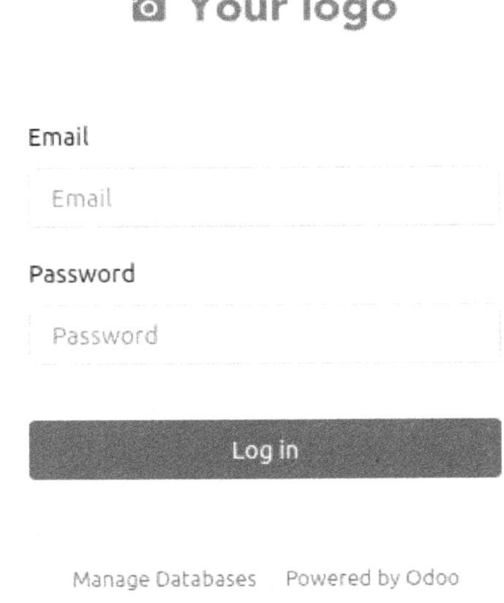

Figure 1.2 – Login screen of the Odoo instance

This is because Odoo includes an HTTP server. By default, it listens on all local network interfaces on TCP port `8069`.

Managing Odoo server databases

When working with Odoo, all the data in your instance is stored in a PostgreSQL database. All the standard database management tools you are used to are available, but Odoo also proposes a web interface for some common operations.

Getting ready

We are assuming that your work environment is set up and that you have an instance running.

How to do it...

The Odoo database management interface provides tools to create, duplicate, remove, back up, and restore a database. There is also a way to change the master password, which is used to protect access to the database management interface.

Accessing the database management interface

To access the database, perform the following steps:

1. Go to the login screen of your instance (if you are authenticated, log out).
2. Click on **Manage Databases**. This will navigate you to `http://localhost:8069/web/database/manager` (you can also point your browser directly to that URL):

Figure 1.3 – Database manager

Setting or changing the master password

If you've set up your instance with default values and haven't modified it yet, as we will explain in the following section, the database management screen will display a warning, telling you that the `master password` instance hasn't been set and will advise you to set one with a direct link:

Figure 1.4 – Master password warning

To set the master password, perform the following steps:

1. Click on the **Set Master Password** button. You will get a dialog box asking you to fill in the **New Master Password** field:

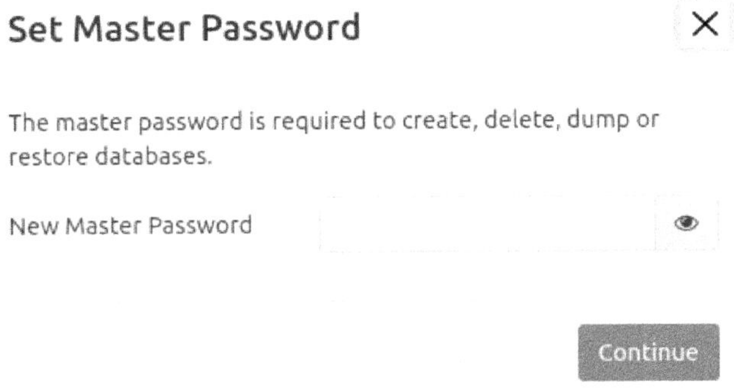

Figure 1.5 – Setting a new master password

2. Type in a complex new password and click **Continue**.

If the master password is already set, click on the **Set Master Password** button at the bottom of the screen to change it. In the dialog box that opens, type the previous master password and the new one, and then click **Continue**:

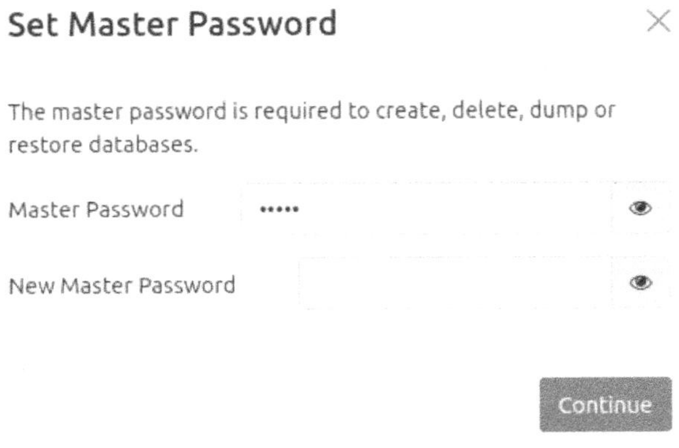

Figure 1.6 – Changing the master password

Note

The master password is the server configuration file under the `admin_passwd` key. If the server is started without a configuration file being specified, a new one will be generated in `~/.odoorc`. Refer to the next recipe for more information about the configuration file.

Creating a new database

This dialog box can be used to create a new database instance that will be handled by the current Odoo server. Follow these steps:

1. On the database management screen, click on the **Create Database** button, which can be found at the bottom of the screen. This will bring up the following dialog:

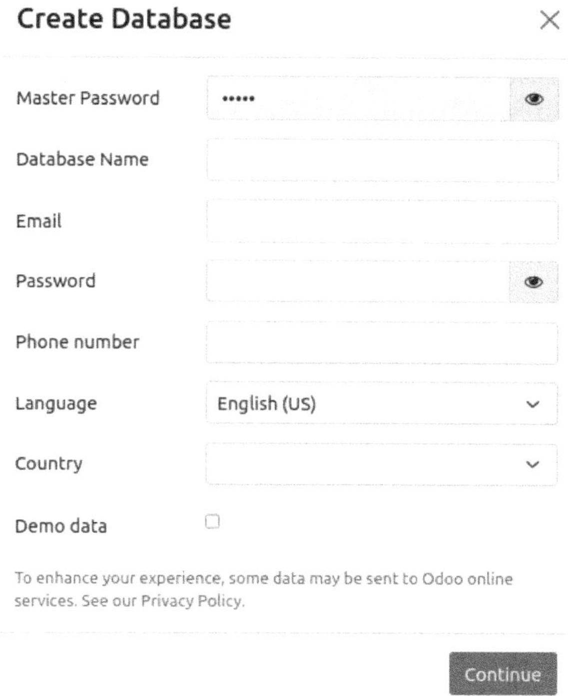

Figure 1.7 – The Create Database dialog

2. Fill in the form, as follows:

   - **Master Password**: This is the master password for this instance.
   - **Database Name**: Input the name of the database you wish to create.
   - **Email**: Add your email address here; this will be your username later.
   - **Password**: Type in the password you want to set for the admin user of the new instance.
   - **Phone Number**: Set your phone number (optional).
   - **Language**: Select the language that you wish to be installed by default in the new database in the drop-down list. Odoo will automatically load the translations for the selected language.

- **Country**: Select the country of the main company in the drop-down list. Selecting this will automatically configure a few things, including the company's currency.
- **Demo data**: Check this box to obtain demonstration data. This is useful for running interactive tests or setting up a demonstration for a customer, but it should not be checked for a database that is designed to contain production data.

> **Note**
> If you wish to use the database to run automated tests for the modules (refer to *Chapter 7, Debugging Modules*), you need to have the demonstration data since the vast majority of the automated tests in Odoo depend on these records to run successfully.

3. Click **Continue** and wait until the new database is initialized. After, you will be redirected to the instance and connected as the administrator.

> **Troubleshooting**
> If you are redirected to a login screen, this is probably because the **--db-filter** option was passed to Odoo and the new database name didn't match the `filter` option. Note that the `odoo-bin start` command does this silently, making only the current database available. To work around this, simply restart Odoo without the `start` command, as shown in the *Installing Odoo from the source* recipe. If you have a configuration file (refer to the *Storing the instance configuration in a file* recipe later in this chapter), check that the `db_filter` option is unset or set to a value matching the new database name.

Duplicating a database

Often, you will have an existing database, and you will want to experiment with it to try a procedure or run a test, but without modifying the existing data. The solution here is simple: duplicate the database and run the test on the copy. Repeat this as many times as required:

1. On the database management screen, click on the **Duplicate Database** link next to the name of the database you wish to clone:

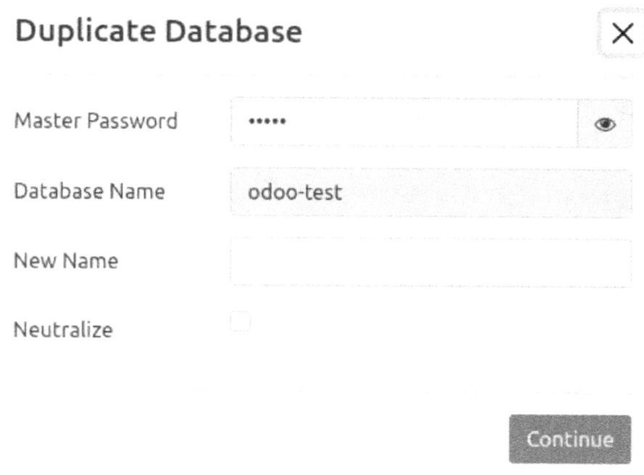

Figure 1.8 – The Duplicate Database dialog

2. Fill in the form, as follows:
   - **Master Password**: This is the master password of the Odoo server
   - **New Name**: The name you want to give to the copy
3. Click **Continue**.
4. You can then click on the name of the newly created database on the database management screen to access the login screen for that database.

Removing a database

When you have finished your tests, you will want to clean up the duplicated databases. To do this, perform the following steps:

1. On the database management screen, you will find the **Delete** button next to the name of the database. Clicking on it will bring up the following dialog:

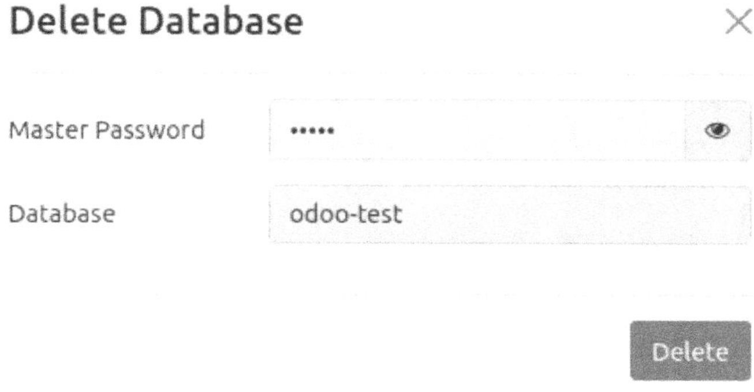

Figure 1.9 – The Delete Database dialog

2. Fill in the form, as well as the **Master Password** field, which is the master password of the Odoo server.
3. Click **Delete**.

> **Caution! Potential data loss!**
> If you selected the wrong database and have no backup, there is no way to recover the lost data.

Backing up a database

To create a backup, perform the following steps:

1. On the database management screen, you will find the **Backup** button next to the database's name. Clicking on it will bring up the following dialog:

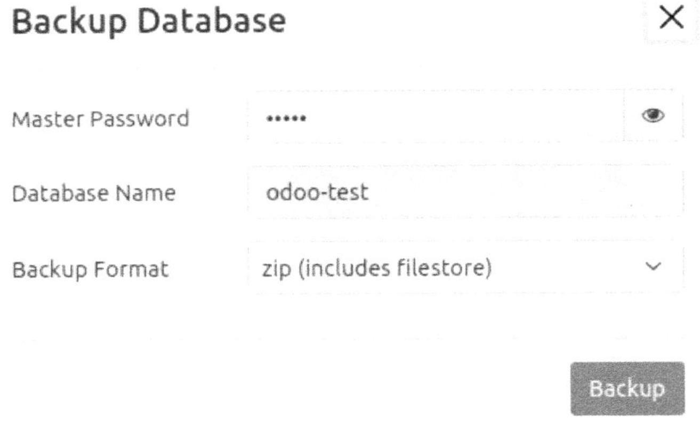

Figure 1.10 – The Backup Database dialog

2. Fill in the form, as follows:
   - **Master Password**: This is the master password of the Odoo server.
   - **Backup Format**: Always use `zip` for a production database since this is the only real full backup format. Only use the `pg_dump` format for a development database when you don't care about the file store.
3. Click **Backup**. The backup file will be downloaded to your browser.

Restoring a database backup

If you need to restore a backup, this is what you need to do:

1. On the database management screen, you will find a **Restore Database** button at the bottom of the screen. Clicking on it will bring up the following dialog:

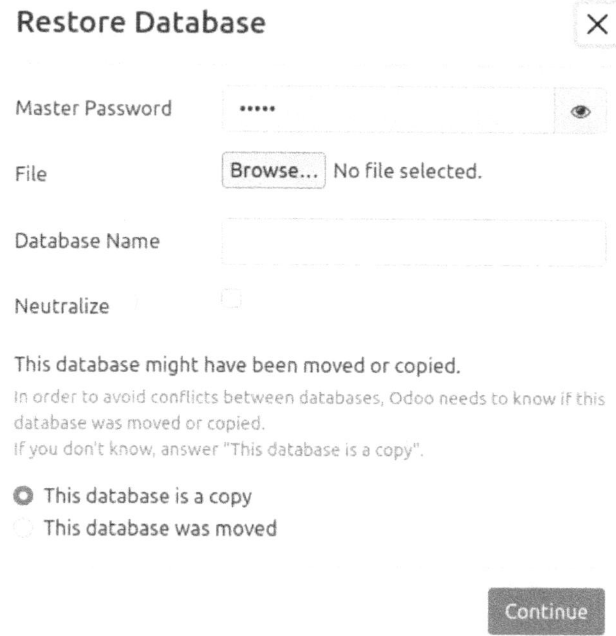

Figure 1.11 – The Restore Database dialog

2. Fill in the form, as follows:
   - **Master Password**: This is the master password of the Odoo server.
   - **File**: This is a previously downloaded Odoo backup.
   - **Database Name**: Provide the name of the database in which the backup will be restored. The database must not exist on the server.
   - **This database might have been moved or copied**: Choose **This database was moved** if the original database was on another server or if it has been deleted from the current server. Otherwise, choose **This database is a copy**, which is the safe default option.
3. Click **Continue**.

> **Note**
>
> It isn't possible to restore a database on top of itself. If you try to do this, you will get an error message (**Database restore error: Database already exists**). You need to remove the database first.

How it works...

These features, apart from the **Change master password** screen, run PostgreSQL administration commands on the server and report back through the web interface.

The master password is a very important piece of information that only lives in the Odoo server configuration file and is never stored in the database. There used to be a default value of `admin`, but using this value is a security liability. In Odoo v9 and later, this is identified as an *unset* master password, and you are urged to change it when accessing the database administration interface. Even if it is stored in the configuration file under the `admin_passwd` entry, this is not the same as the password of the `admin` user; these are two independent passwords. The master password is set for an Odoo server process, which itself can handle multiple database instances, each of which has an independent `admin` user with their own password.

> **Security considerations**
>
> Remember that we are considering a development environment in this chapter. The Odoo database management interface is something that needs to be secured when you are working on a production server as it gives access to a lot of sensitive information, especially if the server hosts Odoo instances for several different clients.

To create a new database, Odoo uses the PostgreSQL `createdb` utility and calls the internal Odoo function to initialize the new database in the same way as when you start Odoo on an empty database.

To duplicate a database, Odoo uses the `--template` option of `createdb`, passing the original database as an argument. This duplicates the structure of the template database in the new database using internal and optimized PostgreSQL routines, which is much faster than creating a backup and restoring it (especially when using the web interface, which requires you to download the backup file and upload it again).

Backup and restore operations use the `pg_dump` and `pg_restore` utilities, respectively. When using the `zip` format, the backup will also include a copy of the file store that contains a copy of the documents when you configure Odoo so that it doesn't keep these in the database; this is the default option in 14.0. Unless you change it, these files reside in `~/.local/share/Odoo/filestore`.

If the backup becomes too large, downloading it may fail. This will be either because the Odoo server itself is unable to handle the large file in memory or because the server is running behind a reverse proxy because there is a limit to the size of HTTP responses that were set in the proxy. Conversely, for the same reasons, you will likely experience issues with the database restore operation. When you start running into these issues, it's time to invest in a more robust external backup solution.

There's more...

Experienced Odoo developers generally don't use the database management interface and perform operations from the command line. To initialize a new database with demo data, for instance, the following single-line command can be used:

```
$ createdb testdb && odoo-bin -d testdb
```

The bonus of using this command line is that you can request add-ons to be installed while you are using it – for instance, `-i sale,purchase,stock`.

To duplicate a database, stop the server and run the following commands:

```
$ createdb -T dbname newdbname
$ cd ~/.local/share/Odoo/filestore # adapt if you have changed the data_dir
$ cp -r dbname newdbname
$ cd -
```

Note that, in the context of development, the file store is often omitted.

> **Note**
>
> The use of `createdb -T` only works if there are no active sessions on the database, which means that you have to shut down your Odoo server before duplicating the database from the command line.

To remove an instance, run the following command:

```
$ dropdb dbname
$ rm -rf ~/.local/share/Odoo/filestore/dbname
```

To create a backup (assuming that the PostgreSQL server is running locally), use the following command:

```
$ pg_dump -Fc -f dbname.dump dbname
$ tar cjf dbname.tgz dbname.dump ~/.local/share/Odoo/filestore/dbname
```

To restore the backup, run the following command:

```
$ tar xf dbname.tgz
$ pg_restore -C -d dbname dbname.dump
```

> **Caution!**
>
> If your Odoo instance uses a different user to connect to the database, you need to pass `-U username` so that the correct user is the owner of the restored database.

Storing the instance configuration in a file

The `odoo-bin` script has dozens of options, and it is tedious to remember them all, as well as how to remember to set them properly when starting the server. Fortunately, it is possible to store them all in a configuration file and only specify the ones you want to alter for development, for example.

How to do it...

For this recipe, perform the following steps:

1. To generate a configuration file for your Odoo instance, run the following command:

   ```
   $ ./odoo-bin --save --config myodoo.cfg --stop-after-init
   ```

2. You can add additional options, and their values will be saved in the generated file. All the unset options will be saved with their default value set. To get a list of possible options, use the following command:

   ```
   $ ./odoo-bin --help | less
   ```

   This will provide you with some help about what the various options perform.

3. To convert from the command-line form into the configuration form, use the long option name, remove the leading dashes, and convert the dashes in the middle into underscores. So, in this case, `--without-demo` will become `without_demo`. This works for most options, but there are a few exceptions, all of which are listed in the following section.

4. Edit the `myodoo.cfg` file (use the table in the following section for some parameters you may want to change). Then, to start the server with the saved options, run the following command:

   ```
   $ ./odoo-bin -c myodoo.cfg
   ```

> **Note**
>
> The `--config` option is commonly abbreviated as `-c`.

How it works...

At startup, Odoo loads its configuration in three passes. First, a set of default values for all options is initialized from the source code. After, the configuration is parsed, and then any value that's defined in the file overrides the defaults. Finally, the command-line options are analyzed, and their values override the configuration that was obtained from the previous pass.

As we mentioned earlier, the names of the configuration variables can be found by looking at the names of the command-line options by removing the leading dashes and converting the middle dashes into underscores. There are a few exceptions to this, notably the following:

Command line	Configuration file
`--db-filter`	`dbfilter`
`--no-http`	`http_enable = True/False`
`--database`	`db_name`
`--dev`	`dev_mode`
`--i18n-import/--i18n-export`	Unavailable

Table 1.1 – Difference in Odoo parameters regarding the command line and the configuration file

Here's a list of options that are commonly set through the configuration file:

Option	Format	Usage
`without_demo`	Comma-separated list of module names	This prevents module demo data from being loaded. Give the value `all` to disable demo data for all modules, or `False` to enable demo data for all modules. To disable demo data for specific modules, provide module names, for example, `sale,purchase,crm`.
`addons_path`	Comma-separated list of paths	This is a list of directory names in which the server will look for add-ons.
`admin_passwd`	Text	This is the master password (take a look at the preceding recipe).
`data_dir`	Path to a directory	This is a directory in which the server will store session information, add-ons downloaded from the internet, and documents if you enable the file store.
`http_port` `longpolling_port`	Port number	These are the ports on which the Odoo server will listen. You will need to specify both to run multiple Odoo servers on the same host; `longpolling_port` is only used if `workers` is not `0`. `http_port` defaults to `8069`, and `longpolling_port` defaults to `8072`.
`logfile`	Path to a file	The file in which Odoo will write its logs.
`log_level`	Log verbosity level	Specifies the level of logging. Accepted values (in increasing order of verbosity) include `critical`, `error`, `warn`, `info`, `debug`, `debug_rpc`, and `debug_rpc_answer`, `debug_sql`.
`workers`	Integer	The number of worker processes. Refer to *Chapter 3*, *Server Deployment*, for more information.
`proxy_mode`	True/False	Activate reverse proxy WSGI wrappers. Only enable this when running behind a trusted web proxy!

Table 1.2 – Odoo parameters and their usage

Here's a list of configuration options related to the database:

Options	Format	Usage
db_host	Hostname	This is the name of the server running the PostgreSQL server. Use False to use local Unix domain sockets, and localhost to use TCP sockets locally.
db_user	Database user login	This is generally empty if db_host is False. This will be the name of the user used to connect to the database.
db_password	Database user password	This is generally empty if db_host is False and when db_user has the same name as the user running the server. Read the main page of pg_hba.conf for more information on this.
db_name	Database name	This is used to set the database name on which some commands operate by default. This does not limit the databases on which the server will act. Refer to the following dbfilter option for this.
db_sslmode	Database SSL mode	This is used to specify the database SSL connection mode.
dbfilter	A regular expression	The expression should match the name of the databases that are considered by the server. If you run the website, it should match a single database, so it will look like ^databasename$. More information on this can be found in *Chapter 3, Server Deployment*.
list_db	True/False	Set to **True** to disable the listing of databases. See *Chapter 3, Server Deployment*, for more information.

Table 1.3 – Odoo parameters and their usage

The configuration file is parsed by Odoo using the Python ConfigParser module. However, the implementation in Odoo 11.0 has changed, and it is no longer possible to use variable interpolation. So, if you are used to defining values for variables from the values of other variables using the %(section.variable)s notation, you will need to change your habits and revert to explicit values.

Some options are not used in config files, but they are widely used during development:

Options	Format	Usage
`-i` or `--init`	Comma-separated list of module names	It will install given modules by default while initializing the database.
`-u` or `--update`	Comma-separated list of module names	It will update given modules when you restart the server. It is mostly used when you modify source code or update the branch from git.
`--dev`	`all`, `reload`, `qweb`, `werkzeug`, and `xml`	This enables developer mode and the auto-reload feature.

Table 1.4 – Odoo parameters and their usage

Activating Odoo developer tools

When using Odoo as a developer, you need to know how to activate **developer mode** in the web interface so that you can access the technical settings menu and developer information. Enabling debug mode will expose several advanced configuration options and fields. These options and fields are hidden in Odoo for better usability because they are not used daily.

How to do it...

To activate developer mode in the web interface, perform the following steps:

1. Connect to your instance and authenticate as `admin`.
2. Go to the **Settings** menu.
3. Scroll to the bottom and locate the **Developer Tools** section:

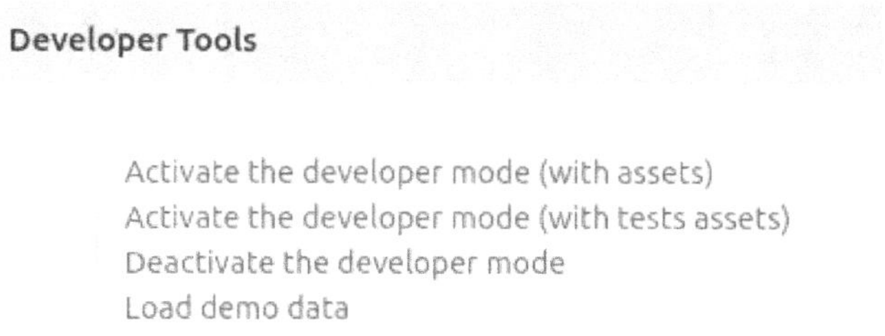

Figure 1.12 – Links to activate different developer modes

4. Click **Activate the developer mode**.
5. Wait for the UI to reload.

> **Alternative way**
>
> It is also possible to activate developer mode by editing the URL. Before the # sign, insert `?debug=1`. For example, if your current URL is `http://localhost:8069/web#menu_id=102&action=94` and you want to enable developer mode, then you need to change that URL to `http://localhost:8069/web?debug=1#menu_id=102&action=94`. Furthermore, if you want to use debug mode with assets, then change the URL to `http://localhost:8069/web?debug=assets#menu_id=102&action=94`.

To exit developer mode, you can perform any one of the following operations:

- Edit the URL and write `?debug=0` in the query string
- Use **Deactivate the developer mode** from the same place in the **Settings** menu
- Click on the bug icon in the top menu and click on the **Leave Developer Tools** option

Lots of developers are using browser extensions to toggle debug mode. By doing this, you can toggle debug mode quickly without accessing the **Settings** menu. These extensions are available for Firefox and Chrome. The following screenshot shows one such plugin you can use and find in the Chrome store:

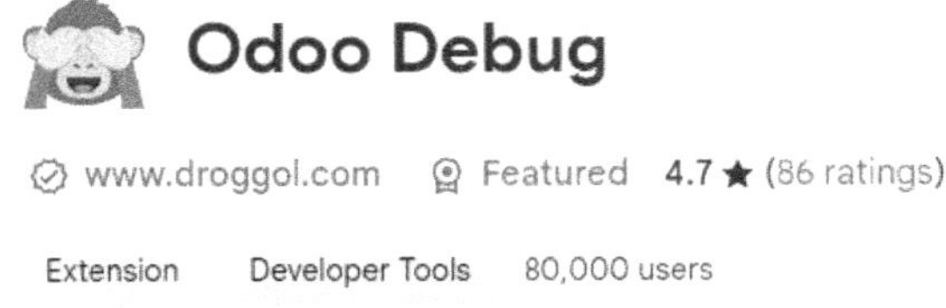

Figure 1.13 – Browser extension for debug mode

> **Note**
>
> The behavior of the debug mode has changed since Odoo v13. Since v13, the status of debug mode is stored in the session, implying that even if you have removed `?debug` from the URL, debug mode will still be active.

How it works...

In developer mode, two things happen:

- You get tooltips when hovering over a field in a form view or over a column in a list view. These provide technical information about the field (internal name, type, and so on).
- A drop-down menu with a bug icon is displayed next to the user's menu in the top-right corner, giving you access to technical information about the model being displayed, the various related view definitions, the workflow, custom filter management, and so on.

There is a variant of developer mode called **Developer mode (with assets)**. This mode behaves like the normal developer mode, but the JavaScript and CSS code that's sent to the browser is not minified, which means that the web development tools of your browser can easily be used to debug the JavaScript code (more on this in *Chapter 15, Web Client Development*).

> **Caution!**
>
> Test your add-ons both with and without developer mode since the unminified versions of the JavaScript libraries can hide bugs that only bite you in the minified version.

Updating the add-on modules list

When a new add-on module is added, you need to run the **Update Module List** wizard to get your new application in the app list. In this recipe, you will learn how to update the app list.

Getting ready

Start your instance and connect to it using your **Administrator** account. After doing this, activate developer mode (if you don't know how to activate developer mode, refer to the *Activating Odoo developer tools* recipe).

How to do it...

To update the list of available add-on modules in your instance, you need to perform the following steps:

1. Open the **Apps** menu.
2. Click **Update Apps List**:

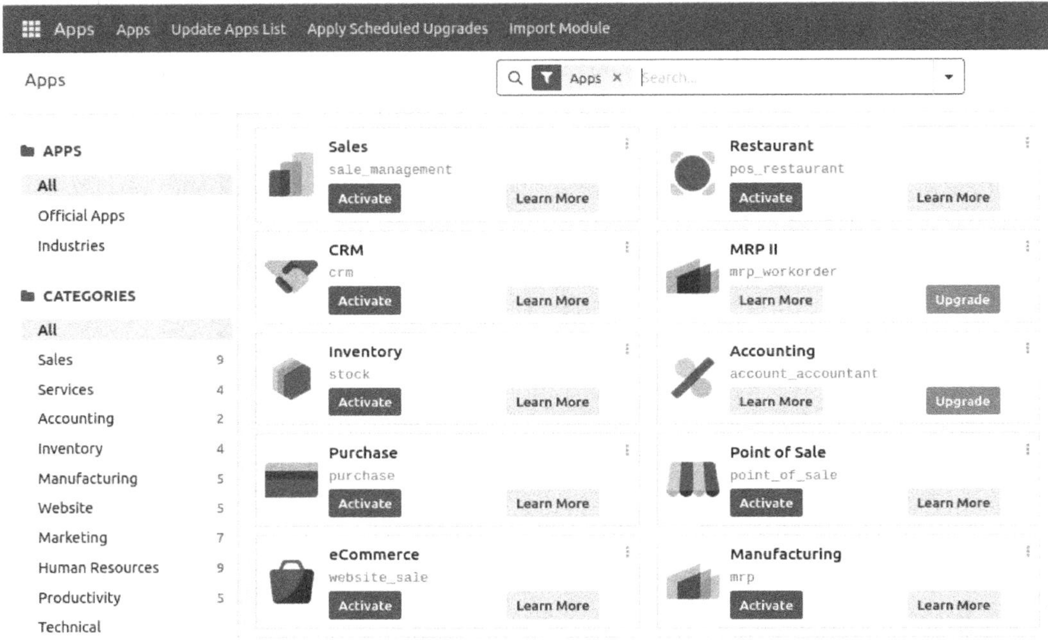

Figure 1.14 – Update Apps List

3. In the dialog that appears, click **Update**:

Module Update

Click on Update below to start the process...

Figure 1.15 – Dialog to update the apps list

4. At the end of the update, you can click on the **Apps** entry to see the updated list of available add-on modules. You will need to remove the default filter on **Apps** in the search box to see all of them.

How it works...

When the **Update** button is clicked, Odoo will read the add-on path configuration variable. For each directory in the list, it will look for immediate subdirectories containing an add-on manifest file, which is a file named `__manifest__.py` that's stored in the add-on module directory. Odoo reads the manifest, expecting to find a Python dictionary. Unless the manifest contains a key `installable` instance set to `False`, the add-on module metadata is recorded in the database. If the module is already present, the information is updated. If not, a new record is created. If a previously available add-on module is not found, the record is not deleted from the list.

> **Note**
> An updated apps list is only required if you add the new add-on path after initializing the database. If you add the new add-on path to the configuration file before initializing the database, then there will be no need to update the module list manually.

To summarize what we have learned so far, after installing, you can start the Odoo server by using the following command line (if you are using a virtual environment, then you need to activate it first):

```
python3 odoo-bin -d odoo-test -i base --addons-path=addons
--db-filter=odoo-test
```

Once you've run the module, you can access Odoo from `http://localhost:8069`.

You can also use a configuration file to run Odoo, as follows:

```
./odoo-bin -c myodoo.cfg
```

Once you start the Odoo server, you can install/update modules from the **App** menu.

2
Managing Odoo Server Instances

In *Chapter 1*, *Installing the Odoo Development Environment*, we looked at how to set up an Odoo instance using only the standard core add-ons that are shipped with the source. As a standard practice to customize Odoo default features, we create a separate module and keep it in a different repository so that you can later upgrade Odoo default and your own repository to keep it clean. This chapter focuses on adding non-core or custom add-ons to an Odoo instance. In Odoo, you can load add-ons from multiple directories. In addition, it is recommended that you load your third-party add-ons or your own custom add-ons from separate folders to avoid conflicts with Odoo core modules. Even Odoo Enterprise Edition is a type of add-ons directory, and you need to load this just like a normal add-ons directory.

In this chapter, we will cover the following recipes:

- Configuring the add-ons path
- Standardizing your instance directory layout
- Installing and upgrading local add-on modules
- Installing add-on modules from GitHub
- Applying changes to add-ons
- Applying and trying proposed **pull requests** (**PRs**)

About the terminology

In this book, we will use the terms **add-on**, **module**, **app**, and **add-on module** interchangeably. All of them refer to the Odoo app or extension app that can be installed in Odoo from the user interface.

Configuring the add-ons path

With the help of the `addons_path` parameter, you can load your own add-on modules into Odoo. When Odoo initializes a new database, it will search for add-on modules within directories that have been provided in the `addons_path` configuration parameter. Odoo will search in these directories for the potential add-on module.

Directories listed in `addons_path` are expected to contain subdirectories, each of which is an add-on module. Following the initialization of the database, you will be able to install modules that are given in these directories.

Getting ready

This recipe assumes that you have an instance ready with a configuration file generated, as described in the *Storing the instance configuration in a file* recipe in *Chapter 1, Installing the Odoo Development Environment*. Note that the source code of Odoo is available in `~/odoo-dev/odoo`, and the configuration file in `~/odoo-dev/odoo/myodoo.cfg`.

How to do it...

To add the `~/odoo-dev/local-addons` directory to the `addons_path` parameter of the instance, perform the following steps:

1. Edit the configuration file for your instance; that is, `~/odoo-dev/myodoo.cfg`.
2. Locate the line starting with `addons_path=`. By default, this should look like the following:

```
addons_path = ~/odoo-dev/odoo/odoo/addons,~/odoo-dev/odoo/addons
```

3. Modify the line by appending a comma, followed by the name of the directory you want to add to `addons_path`, as shown in the following code:

```
addons_path = ~odoo-dev/odoo/odoo/addons,~odoo-dev/odoo/
addons,~/odoo-dev/local-addons
```

4. Restart your instance from the terminal:

```
$ ~/odoo-dev/odoo/odoo-bin -c myodoo.cfg
```

How it works...

When Odoo is restarted, the configuration file is read. The value of the `addons_path` variable is expected to be a comma-separated list of directories. Relative paths are accepted, but they are relative to the current working directory and therefore should be avoided in the configuration file.

At this point, we have only listed the add-ons directory in Odoo, but no add-on modules are present in `~/odoo-dev/local-addons`. And even if you add a new add-on module to this directory, Odoo does not show this module in the user interface. For this, you need to perform an extra operation, as explained in the previous chapter's *Updating the add-on modules list* recipe.

> **Note**
>
> The reason behind this is that when you initialize a new database, Odoo automatically lists your custom modules in available modules, but if you add new modules following database initialization, you then need to manually update the list of available modules, as shown in the *Updating the add-on modules list* recipe in *Chapter 1, Installing the Odoo Development Environment*.

There's more...

When you call the `odoo-bin` script for the first time to initialize a new database, you can pass the `--addons-path` command-line argument with a comma-separated list of directories. This will initialize the list of available add-on modules with all of the add-ons found in the supplied add-ons path. When you do this, you have to explicitly include the base add-ons directory (`odoo/odoo/addons`) as well as the core add-ons directory (`odoo/addons`). A small difference with the preceding recipe is that the local add-ons must not be empty; they must contain at least one subdirectory, which has the minimal structure of an add-on module.

In *Chapter 3, Creating Odoo Add-on Modules*, we will look at how to write your own modules. In the meantime, here's a quick hack to produce something that will make Odoo happy:

```
$ mkdir -p ~/odoo-dev/local-addons/dummy
$ touch ~/odoo-dev/local-addons/dummy/__init__.py
$ echo '{"name": "dummy", "installable": False}' > \
~/odoo-dev/local-addons/dummy/__manifest__.py
```

You can use the `--save` option to save the path to the configuration file:

```
$ odoo/odoo-bin -d odoo-test \
--addons-path="odoo/odoo/addons,odoo/addons,~/odoo-dev/local-addons" \
--save -c ~/odoo-dev/myodoo.cfg --stop-after-init
```

In this case, using relative paths is OK since they will be converted into absolute paths in the configuration file.

> **Note**
>
> Since Odoo only checks directories in the add-ons path for the presence of add-ons when the path is set from the command line, not when the path is loaded from a configuration file, the dummy module is no longer necessary. You may, therefore, remove it (or keep it until you're sure that you won't need to create a new configuration file).

Standardizing your instance directory layout

We recommend that your development and production environments all use a similar directory layout. This standardization will prove helpful when you have to perform maintenance operations, and it will also ease your day-to-day work.

This recipe creates a directory structure that groups files with similar life cycles or similar purposes in standardized subdirectories.

> **Note**
>
> This recipe is only useful if you want to manage similar folder structure development and production environments. If you do not want this, you can skip this recipe.
>
> Also, it is not compulsory to observe the same folder structure as in this recipe. Feel free to alter this structure to suit your needs.

We generate a clean directory structure with clearly labeled directories and dedicated roles. We are using different directories to store the following:

- Code maintained by other people (in `src/`)
- Local-specific code
- The file store of the instance

How to do it...

To create the proposed instance layout, you need to perform the following steps:

1. Create one directory per instance:

   ```
   $ mkdir ~/odoo-dev/projectname
   $ cd ~/odoo-dev/projectname
   ```

2. Create a Python `virtualenv` object in a subdirectory called `env/`:

   ```
   $ python3 -m venv env
   ```

3. Create some subdirectories, as follows:

   ```
   $ mkdir src local bin filestore logs
   ```

4. The functions of the subdirectories are as follows:

   - `src/`: This contains the clone of Odoo itself, as well as various third-party add-on projects (we have added Odoo source code to the next step in this recipe)

- `local/`: This is used to save your instance-specific add-ons
- `bin/`: This includes various helper executable shell scripts
- `filestore/`: This is used as a file store
- `logs/` (*optional*): This is used to store the server log files

5. Clone Odoo and install the requirements (refer to *Chapter 1, Installing the Odoo Development Environment*, for details on this):

```
$ git clone -b 17.0 --single-branch --depth 1 https://github.
com/odoo/odoo.git src/odoo
$ env/bin/pip3 install -r src/odoo/requirements.txt
```

6. Save the following shell script as `bin/odoo`:

```
#!/bin/sh ROOT=$(dirname $0)/..
PYTHON=$ROOT/env/bin/python3 ODOO=$ROOT/src/odoo/odoo-bin
$PYTHON $ODOO -c $ROOT/projectname.cfg "$@" exit $?
```

7. Make the script executable:

```
$ chmod +x bin/odoo
```

8. Create an empty dummy local module:

```
$ mkdir -p local/dummy
$ touch local/dummy/__init__.py
$ echo '{"name": "dummy", "installable": False}' >\ local/
dummy/__manifest__.py
```

9. Generate a configuration file for your instance:

```
$ bin/odoo --stop-after-init --save \
--addons-path src/odoo/odoo/addons,src/odoo/addons,local \
--data-dir filestore
```

10. Add a `.gitignore` file, which is used to tell GitHub to exclude given directories so that Git will ignore these directories when you commit the code; for example, `filestore/`, `env/`, `logs/`, and `src/`:

```
# dotfiles, with exceptions:
.*
!.gitignore
# python compiled files
*.py[co]
# emacs backup files
*~
```

```
# not tracked subdirectories
/env/
/src/
/filestore/
/logs/
```

11. Create a Git repository for this instance and add the files you've added to Git:

```
$ git init
$ git add .
$ git commit -m "initial version of projectname"
```

How it works...

By having one `virtualenv` environment per project, we are sure that the project's dependencies will not interfere with the dependencies of other projects that may be running a different version of Odoo or will use different third-party add-on modules that require different versions of Python dependencies. This comes at the cost of a little disk space.

In a similar way, by using separate clones of Odoo and third-party add-on modules for our different projects, we are able to let each of these evolve independently and only install updates on instances that need them, hence reducing the risk of introducing regressions.

The `bin/odoo` script allows us to run the server without having to remember the various paths or activate the `virtualenv` environment. This also sets the configuration file for us. You can add extra scripts in there to help you in your day-to-day work. For instance, you can add a script to check out the different third-party projects that you need to run your instance.

Regarding the configuration file, we have only demonstrated the bare minimum options to set up here, but you can obviously set up more, such as the database name, the database filter, or the port on which the project listens. Refer to *Chapter 1, Installing the Odoo Development Environment*, for more information on this topic.

Finally, by managing all of this in a Git repository, it becomes quite easy to replicate the setup on a different computer and share the development among a team.

> **Speedup tip**
>
> To facilitate project creation, you can create a template repository containing the empty structure and fork that repository for each new project. This will save you from retyping the `bin/odoo` script, the `.gitignore` file, and any other template file you need (**continuous integration** (**CI**) configuration, `README.md`, changelog, and so on).

There's more...

The development of complex modules requires various configuration options, which leads to updating the configuration file whenever you want to try any configuration option. Updating the configuration file frequently can be a headache, and to avoid this, an alternative way is to pass all configuration options from the command line, as follows:

1. Activate `virtualenv` manually:

```
$ source env/bin/activate
```

2. Go to the Odoo source directory:

```
$ cd src/odoo
```

3. Run the server:

```
./odoo-bin --addons-path=addons,../../local -d test-16 -i
account,sale,purchase --log-level=debug
```

In *step 3*, we passed a few configuration options directly from the command line. The first is `--addons-path`, which loads Odoo's core add-ons directory, `addons`, and your add-ons directory, `local`, in which you will put your own add-on modules. The `-d` option will use the `test-16` database or create a new database if it isn't present. The `-i` option will install the `account`, `sale`, and `purchase` modules. Next, we passed the `log-level` option and increased the log level to `debug` so that it would display more information in the log.

> **Note**
>
> By using the command line, you can quickly change the configuration options. You can also see live logs in the terminal. For all available options, refer to *Chapter 1, Installing the Odoo Development Environment*, or use the `--help` command to view a list of all options and a description of each option.

Installing and upgrading local add-on modules

The core functionality of Odoo comes from its add-on modules. You have a wealth of add-ons available as part of Odoo itself, as well as add-on modules that you can download from the app store or that have been written by yourself.

In this recipe, we will demonstrate how to install and upgrade add-on modules through the web interface and from the command line.

The main benefits of using the command line for these operations include being able to act on more than one add-on at a time and having a clear view of the server logs as the installation or update progresses, which is very useful when in development mode or when scripting the installation of an instance.

Getting ready

Make sure that you have a running Odoo instance with its database initialized and the add-ons path properly set. In this recipe, we will install/upgrade a few add-on modules.

How to do it...

There are two possible methods to install or update add-ons—you can use the web interface or the command line.

From the web interface

To install a new add-on module in your database using the web interface, perform the following steps:

1. Connect to the instance using the **Administrator** account and open the **Apps** menu:

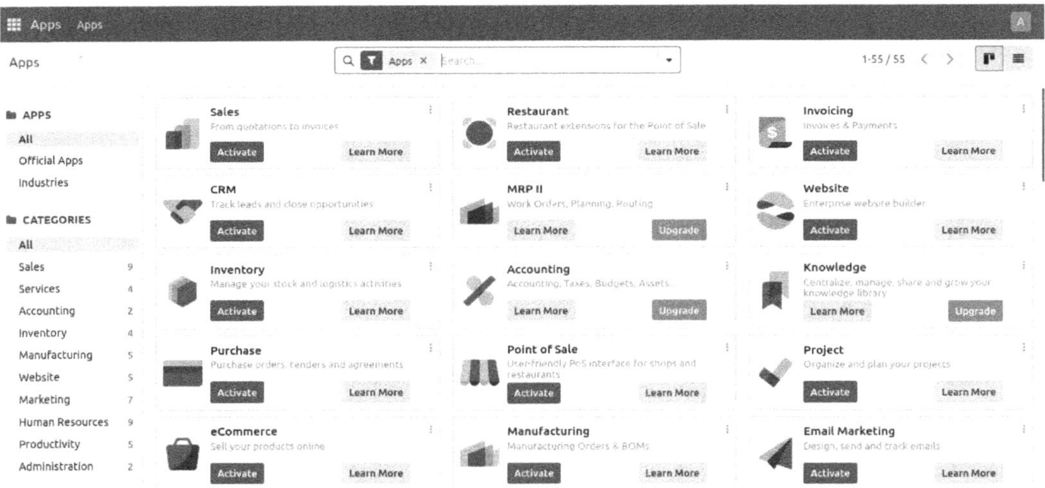

Figure 2.1 – List of Odoo apps

2. Use the search box to locate the add-on you want to install. Here are a few instructions to help you with this task:

   i. Activate the **Not Installed** filter.

   ii. If you're looking for a specific functionality add-on rather than a broad functionality add-on, remove the **Apps** filter.

   iii. Type a part of the module name in the search box and use this as a **Module** filter.

   iv. You may find that using the list view gives something more readable.

3. Click on the **Install** button under the module name on the card.

Note that some Odoo add-on modules have external Python dependencies. If Python dependencies are not installed in your system, then Odoo will abort the installation, and it will show the following dialog:

Invalid Operation

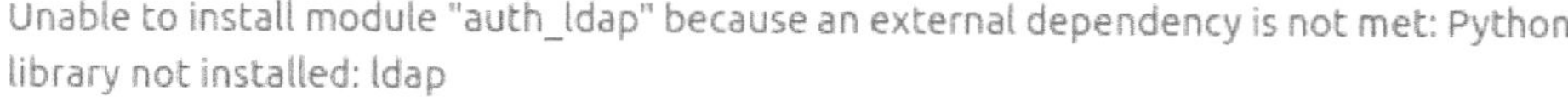

Unable to install module "auth_ldap" because an external dependency is not met: Python library not installed: ldap

Figure 2.2 – Warning for external library dependency

To fix this, just install the relevant Python dependencies on your system.

To update a pre-installed module in your database, perform the following steps:

1. Connect to the instance using the **Administrator** account.
2. Open the **Apps** menu.
3. Click on **Apps**:

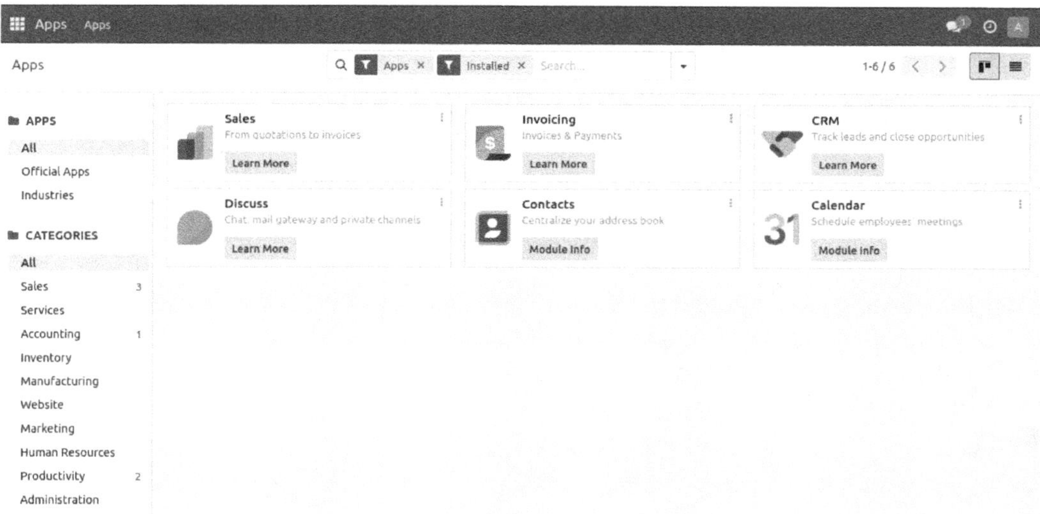

Figure 2.3 – Odoo apps list

4. Use the search box to locate the add-on you want to install. Here are a few tips:

   i. Activate the **Installed** filter.

   ii. If you're looking for a specific functionality add-on rather than a broad functionality add-on, remove the **Apps** filter.

   iii. Type a part of the add-on module name into the search box and then press *Enter* to use this as a **Module** filter. For example, type `crm` and press *Enter* to search CRM apps.

   iv. You may find that using the list view gives you something more readable.

5. Click on the three dots in the top-right corner of the card and click on the **Upgrade** option:

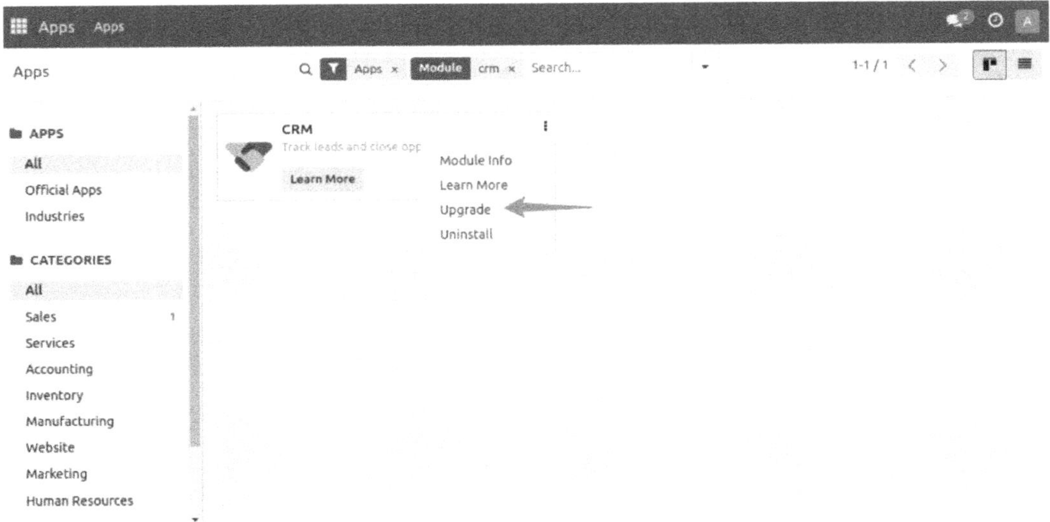

Figure 2.4 – Drop-down link for upgrading the module

Activate **Developer** mode to see the technical name of the module. See *Chapter 1, Installing the Odoo Development Environment*, if you don't know how to activate developer mode:

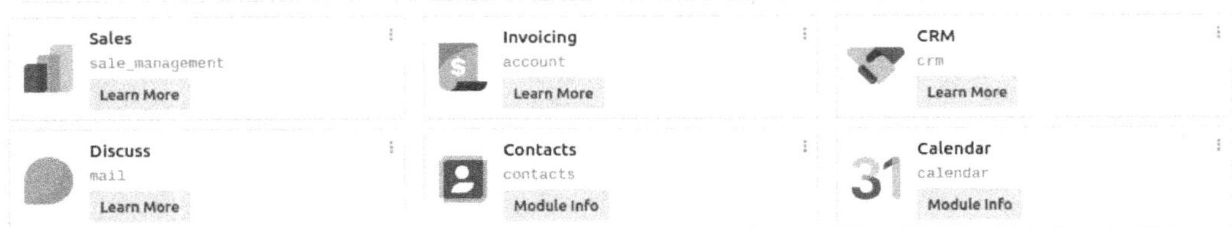

Figure 2.5 – Application's technical names

After activating developer mode, it will show the module's technical name in red. If you are using Odoo Community Edition, you will see some extra apps with the **Upgrade** button. Those apps are Odoo Enterprise Edition apps, and in order to install/use them, you need to purchase a license.

From the command line

To install new add-ons in your database, perform the following steps:

1. Find the names of the add-ons. This is the name of the directory containing the `__manifest__.py` file, without the leading path.
2. Stop the instance. If you are working on a production database, make a backup.
3. Run the following command:

```
$ odoo/odoo-bin -c instance.cfg -d dbname -i addon1,addon2 \
--stop-after-init
```

You may omit `-d dbname` if this is set in your configuration file.

1. Restart the instance.

To update an already installed add-on module in your database, perform the following steps:

1. Find the name of the add-on module to update; this is the name of the directory containing the `__manifest__.py` file, without the leading path.
2. Stop the instance. If you are working on a production database, make a backup.
3. Run the following command:

```
$ odoo/odoo-bin -c instance.cfg -d dbname -u addon1 \
--stop-after-init
```

You may omit `-d dbname` if this is set in your configuration file.

1. Restart the instance.

How it works...

The add-on module installation and update are two closely related processes, but there are some important differences, as highlighted in the following two sections.

Add-on installation

When you install an add-on, Odoo checks its list of available add-ons for an uninstalled add-on with the supplied name. It also checks for the dependencies of that add-on and, if there are any, it will recursively install them before installing the add-on.

The installation process of a single module consists of the following steps:

1. If there are any dependencies, run the `preinit` add-on hook.
2. Load the model definitions from the Python source code and update the database structure, if necessary (refer to *Chapter 4, Application Models*, for details).
3. Load the data files of the add-on and update the database contents, if necessary (refer to *Chapter 6, Managing Module Data*, for details).
4. Install the add-on demo data if demo data has been enabled in the instance.
5. If there are any dependencies, run the add-on `postinit` hook.
6. Run a validation of the view definitions of the add-on.
7. If demo data is enabled and a test is enabled, run the tests of the add-on (refer to *Chapter 18, Automated Test Cases*, for details).
8. Update the module state in the database.
9. Update the translations in the database from the add-on's translations (refer to *Chapter 11, Internationalization*, for details).

Note

The `preinit` and `postinit` hooks are defined in the `__manifest__.py` file using the `pre_init_hook` and `post_init_hook` keys, respectively. These hooks are used to invoke Python functions before and after the installation of an add-on module. To learn more about `init` hooks, refer to *Chapter 3, Creating Odoo Add-on Modules*.

Add-on update

When you update an add-on, Odoo checks in its list of available add-on modules for an installed add-on with the given name. It also checks for the reverse dependencies of that add-on (these are add-ons that depend on the updated add-on). If any, it will recursively update them, too.

The update process of a single add-on module consists of the following steps:

1. Run the add-on module's pre-migration steps, if any (refer to *Chapter 6, Managing Module Data*, for details).
2. Load the model definitions from the Python source code and update the database structure if necessary (refer to *Chapter 4, Application Models*, for details).
3. Load the data files of the add-on and update the database's contents if necessary (refer to *Chapter 6, Managing Module Data*, for details).
4. Update the add-on's demo data if demo data is enabled in the instance.
5. If your module has any migration methods, run the add-on post-migration steps (refer to *Chapter 6, Managing Module Data*, for details).
6. Run a validation of the view definitions of the add-on.

7. If demo data is enabled and a test is enabled, run the tests of the add-on (refer to *Chapter 18, Automated Test Cases*, for details).
8. Update the module state in the database.
9. Update the translations in the database from the add-on's translations (refer to *Chapter 11, Internationalization*, for details).

> **Note**
>
> Note that updating an add-on module that is not installed does nothing at all. However, installing an add-on module that is already installed reinstalls the add-on, which can have some unintended effects on some data files that contain data that is supposed to be updated by the user and not updated during the normal module update process (refer to the *Using the noupdate and forcecreate flags* recipe in *Chapter 6, Managing Module Data*). There is no risk of error from the user interface, but this can happen from the command line.

There's more...

Be careful with dependency handling. Consider an instance where you want to have the `sale`, `sale_stock`, and `sale_specific` add-ons installed, with `sale_specific` depending on `sale_stock`, and `sale_stock` depending on `sale`. To install all three, you only need to install `sale_specific`, as it will recursively install the `sale_stock` and `sale` dependencies. To update all three, you need to update `sale` as this will recursively update the reverse dependencies, `sale_stock` and `sale_specific`.

Another tricky part with managing dependencies is when you add a dependency to an add-on that already has a version installed. Let's understand this by continuing with the previous example. Imagine that you add a dependency on `stock_dropshipping` in `sale_specific`. Updating the `sale_specific` add-on will not automatically install the new dependency, and neither will requesting the installation of `sale_specific`. In this situation, you can get very nasty error messages because the Python code of the add-on is not successfully loaded, but the data of the add-on and the models' tables in the database are present. To resolve this, you need to stop the instance and manually install the new dependency.

Installing add-on modules from GitHub

GitHub is a great source of third-party add-ons. A lot of Odoo partners use GitHub to share the add-ons they maintain internally, and the **Odoo Community Association** (**OCA**) collectively maintains several hundred add-ons on GitHub. Before you start writing your own add-on, ensure you check that nothing already exists that you can use as is or as a starting point.

This recipe will show you how to clone the `partner-contact` project of the OCA from GitHub and make the add-on modules it contains available in your instance.

Getting ready

Suppose you want to add new fields to the customer (partner) form. By default, the Odoo customer model doesn't have a `gender` field. If you want to add a `gender` field, you need to create a new module. Fortunately, someone on a mailing list tells you about the `partner_contact_gender` add-on module, which is maintained by the OCA as part of the `partner-contact` project.

The paths that are used in this recipe reflect the layout that was proposed in the *Standardizing your instance directory layout* recipe.

How to do it...

To install `partner_contact_gender`, perform the following steps:

1. Go to your project's directory:

   ```
   $ cd ~/odoo-dev/my-odoo/src
   ```

2. Clone the `17.0` branch of the `partner-contact` project in the `src/` directory:

   ```
   $ git clone --branch 17.0 \
   https://github.com/OCA/partner-contact.git src/partner-contact
   ```

3. Change the add-ons path to include that directory and update the add-ons list of your instance (refer to the *Configuring the add-ons path* recipe in this chapter and the *Updating the add-on modules list* recipe in the previous chapter). The `add-ons_path` line of `instance.cfg` should look like this:

   ```
   addons_path = ~/odoo-dev/my-odoo/src/odoo/odoo/addons, \
   ~/odoo-dev/my-odoo/src/odoo/addons, \
   ~/odoo-dev/my-odoo/src/partner-contact, \
   ~/odoo-dev/local-addons
   ```

4. Install the `partner_contact_gender` add-on (if you don't know how to install the module, take a look at the previous recipe, *Installing and upgrading local add-on modules*).

How it works...

All OCA code repositories have their add-ons contained in separate subdirectories, which is coherent per what is expected by Odoo regarding the directories in the add-ons path. Consequently, just cloning the repository somewhere and adding that location in the add-ons path is enough.

There's more...

Some maintainers follow a different approach and have one add-on module per repository, living at the root of the repository. In that case, you need to create a new directory, which you will add to the add-ons path, and clone all of the add-ons from the maintainer you need in this directory. Remember to update the add-on modules list each time you add a new repository clone.

Applying changes to add-ons

Most add-ons that are available on GitHub are subject to change and do not follow the rules that Odoo enforces for its stable release. They may receive bug fixes or enhancements, including issues or feature requests that you have submitted, and these changes may introduce database schema changes or updates in data files and views. This recipe explains how to install the updated versions.

Getting ready

Suppose you reported an issue with `partner_contact_gender` and received a notification that the issue was solved in the last revision of the `17.0` branch of the `partner-contact` project. In this case, you would want to update your instance with this latest version.

How to do it...

To apply a source modification to your add-on from GitHub, you need to perform the following steps:

1. Stop the instance using that add-on.
2. Make a backup if it is a production instance (refer to the *Managing Odoo server databases* recipe in *Chapter 1*, *Installing the Odoo Development Environment*).
3. Go to the directory where `partner-contact` was cloned:

   ```
   $ cd ~/odoo-dev/my-odoo/src/partner-contact
   ```

4. Create a local tag for the project so that you can revert to that version in case things break:

   ```
   $ git checkout 17.0
   $ git tag 17.0-before-update-$(date --iso)
   ```

5. Get the latest version of the source code:

   ```
   $ git pull --ff-only
   ```

6. Update the `partner_address_street3` add-on in your databases (refer to the *Installing and upgrading local add-on modules* recipe).
7. Restart the instance.

How it works...

Usually, the developer of the add-on module occasionally releases the newest version of the add-on. This update typically contains bug fixes and new features. Here, we will get a new version of the add-on and update it in our instances.

If `git pull --ff-only` fails, you can revert to the previous version using the following command:

```
$ git reset --hard 17.0-before-update-$(date --iso)
```

Then, you can try `git pull` (without `--ff-only`), which will cause a merge, but this means that you have local changes on the add-on.

See also

If the update step breaks, refer to the *Updating Odoo from source* recipe in *Chapter 1, Installing the Odoo Development Environment*, for recovery instructions. Remember to always test an update on a copy of a database production first.

Applying and trying proposed PRs

In the GitHub world, a PR is a request that's made by a developer so that the maintainers of a project can include some new developments. Such a PR may contain a bug fix or a new feature. These requests are reviewed and tested before being pulled into the `main` branch.

This recipe explains how to apply a PR to your Odoo project in order to test an improvement or a bug fix.

Getting ready

As in the previous recipe, suppose you reported an issue with `partner_address_street3` and received a notification that the issue was solved in a PR, which hasn't been merged in the `17.0` branch of the project. The developer asks you to validate the fix in PR *#123*. You need to update a test instance with this branch.

You should not try out such branches directly on a production database, so first create a test environment with a copy of the production database (refer to *Chapter 1, Installing the Odoo Development Environment*).

How to do it...

To apply and try out a GitHub PR for an add-on, you need to perform the following steps:

1. Stop the instance.
2. Go to the directory where `partner-contact` was cloned:

   ```
   $ cd ~/odoo-dev/my-odoo/src/partner-contact
   ```

3. Create a local tag for the project so that you can revert to that version in case things break:

```
$ git checkout 17.0
$ git tag 17.0-before-update-$(date --iso)
```

4. Pull the branch of the PR. The easiest way to do this is by using the number of the PR, which should have been communicated to you by the developer. In our example, this is PR number `123`:

```
$ git pull origin pull/123/head
```

5. Update the `partner_contact_gender1` add-on module in your database and restart the instance (refer to the *Installing and upgrading local add-on modules* recipe if you don't know how to update the module).
6. Test the update—try to reproduce your issue, or try out the feature you wanted.

If this doesn't work, comment on the PR page of GitHub, explaining what you did and what didn't work so that the developer can update the PR.

If it works, say so on the PR page too; this is an essential part of the PR validation process, and it will speed up merging in the `main` branch.

How it works...

We are using a GitHub feature that enables PRs to be pulled by number using the `pull/nnnn/head` branch name, where `nnnn` is the number of the PR. The `git pull` command will merge the remote branch in ours, applying the changes in our code base. After this, we update the add-on module, test it, and report back to the author of the change with regard to any failures or successes.

There's more...

You can repeat *step 4* of this recipe for different PRs in the same repository if you want to test them simultaneously. If you are really happy with the result, you can create a branch to keep a reference to the result of the applied changes:

```
$ git checkout -b 17.0-custom
```

Using a different branch will help you remember that you are not using the version from GitHub, but a custom one.

> **Note**
>
> The `git branch` command can be used to list all local branches you have in your repository.

From then on, if you need to apply the latest revision of the `17.0` branch from GitHub, you will need to pull it without using `--ff-only`:

```
$ git pull origin 17.0
```

3

Creating Odoo Add-On Modules

Now that we have a development environment and know how to manage Odoo server instances and databases, we will learn how to create Odoo add-on modules.

Our main goal in this chapter is to understand how an add-on module is structured and the typical incremental workflow to add components to it. The various components mentioned in this chapter's recipe names will be covered extensively in subsequent chapters.

An Odoo module can contain several elements:

- **Business objects**:
  - Declared as Python classes, these resources are automatically persisted by Odoo based on their configuration
- **Object views**:
  - A definition of business objects' UI display
- **Data files (XML or CSV files declaring the model metadata)**:
  - Views or reports
  - Configuration data (module parametrization and security rules)
  - Demonstration data and more
- **Web controllers**:
  - Handle requests from web browsers, static web data images, or CSS or JavaScript files used by the web interface or website

In this chapter, we will cover the following recipes:

- Creating and installing a new add-on module
- Completing the add-on module manifest
- Organizing the add-on module file structure
- Adding models
- Adding menu items and views
- Adding access security
- Using the `scaffold` command to create a module

Technical requirements

For this chapter, you are expected to have Odoo installed, and you are also expected to have followed the recipes in *Chapter 1*, *Installing the Odoo Development Environment*. You are also expected to be comfortable in discovering and installing extra add-on modules, as described in *Chapter 2*, *Managing Odoo Server Instances*.

All the code used in this chapter can be downloaded from the GitHub repository at `https://github.com/PacktPublishing/Odoo-17-Development-Cookbook-Fifth-Edition/tree/main/Chapter03`.

What is an Odoo add-on module?

Except for the framework code, all of the code bases of Odoo are packed in the form of modules. These modules can be installed or uninstalled at any time from the database. There are two main purposes for these modules. You can either add new apps/business logic, or you can modify an existing application. Put simply, in Odoo, everything starts and ends with modules.

Odoo offers various business solutions such as Sales, Purchase, POS, Accounting, Manufacturing, Project, and Inventory. Creating a new module involves adding new features to a business or upgrading the existing ones.

The latest version of Odoo introduces numerous new modules in both the Community and Enterprise editions. These include Meeting Rooms, To-Do, and several WhatsApp-related integration modules.

In addition, this version comes packed with exciting new features such as a redesigned user interface, improved search functionality, and new features for CRM, manufacturing, and e-commerce. The new version also includes several other improvements, such as enhanced performance, improved security, and more integrations.

Odoo is used by companies of all sizes; each company has a different business flow and requirements. To deal with this issue, Odoo splits the features of the application into different modules. These modules can be loaded into the database on demand. Basically, the administrator can enable/disable these features at any time. Consequently, the same software can be adjusted for different requirements. Check out the following screenshot of Odoo modules; the first module in the column is the main application, and others are designed to add extra features to that app. To get a modules list grouped by the application's category, go to the **Apps** menu and apply grouping by category:

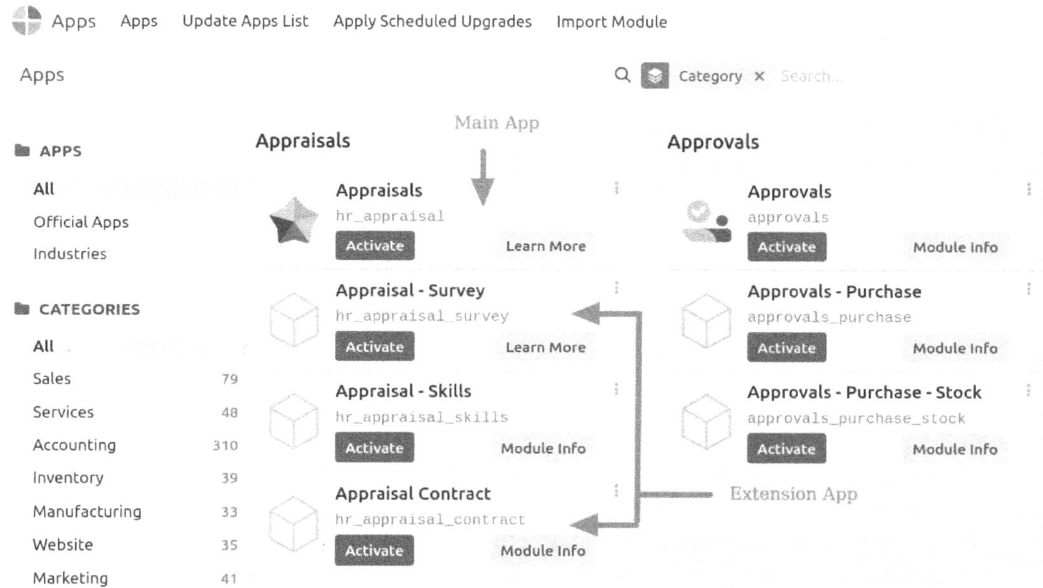

Figure 3.1 – Grouping apps by category

If you plan on developing a new application in Odoo, you should create boundaries for various features. This will be very helpful to divide your application into different add-on modules. Now that you know the purpose of the add-on module in Odoo, we can start building our own one.

Creating and installing a new add-on module

In this recipe, we will create a new module, make it available in our Odoo instance, and install it.

Getting ready

To begin, we will need an Odoo instance that's ready to use.

If you followed the *Easy installation of Odoo from the source* recipe in *Chapter 1, Installing the Odoo Development Environment*, Odoo should be available at `~/odoo-dev/odoo`. For explanation purposes, we will assume this location for Odoo, although you can use any other location of your preference.

We will also need a location to add our own Odoo modules. For the purpose of this recipe, we will use a `local-addons` directory alongside the `odoo` directory, at `~/odoo-dev/local-addons`.

You can upload your own Odoo modules on GitHub and clone them on your local system for development purposes.

How to do it...

As an example, for this chapter, we will create a small add-on module to manage a hostel.

The following steps will create and install a new add-on module:

1. Change the working directory in which we will work and create the add-ons directory where our custom module will be placed:

   ```
   $ cd ~/odoo-dev
   $ mkdir local-addons
   ```

2. Choose a technical name for the new module and create a directory with that name for the module. For our example, we will use `my_hostel`:

   ```
   $ mkdir local-addons/my_hostel
   ```

   A module's *technical name* must be a valid Python identifier. It must begin with a letter, and only contain letters, numbers, and underscore characters. It is preferable that you only use lowercase letters in the module name.

3. Make the Python module importable by adding an `__init__.py` file:

   ```
   $ touch local-addons/my_hostel/__init__.py
   ```

4. Add a minimal module manifest for Odoo to detect it as an add-on module. Inside the `my_hostel` folder, create an `__manifest__.py` file with this line:

   ```
   {'name': 'Hostel Management'}
   ```

5. Start your Odoo instance, including the module directory, in the add-on path:

   ```
   $ odoo/odoo-bin --addons-path=odoo/addons/,local-addons/
   ```

   If the `--save` option is added to the Odoo command, the add-ons path will be saved in the configuration file. The next time you start the server, if no add-on path option is provided, this will be used.

6. Make the new module available in your Odoo instance. Log in to Odoo using **admin**, enable **Developer Mode** in the **About** box, and in the **Apps** top menu, select **Update Apps List**. Now, Odoo should know about our Odoo module:

Figure 3.2 – The dialog to update the app list

7. Select the **Apps** menu at the top, and in the search bar in the top-right corner, delete the default Apps filter and search for `my_hostel`. Click on the **Activate** button, and the installation will finish.

How it works...

An Odoo module is a directory that contains code files and other assets. The directory name that's used is the module's technical name. The `name` key in the module manifest is its title.

The `__manifest__.py` file is the module manifest. This contains a Python dictionary with module metadata, including category, version, the modules it depends on, and a list of the data files that it will load. It contains important metadata about the add-on module and declares the data files that should be loaded.

In this recipe, we used a minimal manifest file, but in real modules, we will need other important keys. These are discussed in the next recipe, *Completing the add-on module manifest*.

The module directory must be Python-importable, so it also needs to have an `__init__.py` file, even if it's empty. To load a module, the Odoo server will import it. This will cause the code in the `__init__.py` file to be executed, so it works as an entry point to run the module Python code. Due to this, it will usually contain import statements to load the module Python files and submodules.

Known modules can be installed directly from the command line using the `--init` or **-i** option. For example, if you want to install the `my_hostel` app, you can use **-i** `my_hostel`. This list is initially set when you create a new database from the modules found on the add-on path provided at that time. It can be updated in an existing database with the **Update Module List** menu.

Completing the add-on module manifest

The manifest is an important piece for Odoo modules.

Getting ready

We should have a module to work with, already containing a `__manifest__.py` manifest file. You may want to follow the previous recipe to provide such a module to work with.

How to do it...

We will add a manifest file and an icon to our add-on module:

1. To create a manifest file with the most relevant keys, edit the module's `__manifest__.py` file so that it looks like this:

```
{
    'name': "Hostel Management",
    'summary': "Manage Hostel easily",
    'description': "Efficiently manage the entire residential
facility in the school.", # Supports reStructuredText(RST)
format (description is Deprecated),
    'author': "Your name",
    'website': "http://www.example.com",
    'category': 'Uncategorized',
    'version': '17.0.1.0.0',
    'depends': ['base'],
    'data': ['views/hostel.xml'],
    'assets': {
    'web.assets_backend': [
        'web/static/src/xml/**/*',
    ],
           },
     'demo': ['demo.xml'],
}
```

2. To add an icon for the module, choose a PNG image to use and copy it to `static/description/icon.png`.

How it works...

The content in the manifest file is a regular Python dictionary, with keys and values. The example manifest we used contains the most relevant keys:

- `name`: This is the title of the module.
- `summary`: This is the subtitle with a one-line description.

- `description`: This is a long description written in plaintext or **ReStructuredText (RST)** format. It is usually surrounded by triple quotes and is used in Python to delimit multiline texts. For an RST quick-start reference, visit `http://docutils.sourceforge.net/docs/user/rst/quickstart.html`.
- `author`: This is a string with the name of the authors. When there is more than one, it is common practice to use a comma to separate their names, but note that it should still be a string, not a Python list.
- `website`: This is a URL people should visit to learn more about the module or the authors.
- `category`: This is used to organize modules by areas of interest. The list of the standard category names available can be seen at `https://github.com/odoo/odoo/blob/17.0/odoo/addons/base/data/ir_module_category_data.xml`. However, it's also possible to define other new category names here.
- `version`: This is the module's version number. It can be used by the Odoo app store to detect newer versions of installed modules. If the version number does not begin with the Odoo target version (for example, `17.0`), it will be automatically added. Nevertheless, it will be more informative if you explicitly state the Odoo target version – for example, by using `17.0.1.0.0` or `17.0.1.0`, instead of `1.0.0` or `1.0`, respectively.
- `depends`: This is a list with the technical names of the modules it directly depends on. If your module does not depend on any other add-on module, then you should at least add a `base` module. Don't forget to include any module defining XML IDs, views, or models that are referenced by this module. That will ensure that they all load in the correct order, avoiding hard-to-debug errors.
- `data`: This is a list of relative paths for the data files to load during module installation or upgrade. The paths are relative to the module `root` directory. Usually, these are XML and CSV files, but it's also possible to have YAML data files. These are discussed in depth in *Chapter 6, Managing Module Data*.
- `demo`: This is the list of relative paths to the files with demonstration data to load. These will only be loaded if the database was created with the `Demo Data` flag enabled.

The image that is used as the module icon is the PNG file at `static/description/icon.png`.

Odoo is expected to have significant changes between major versions, so modules that have been built for one major version are not likely to be compatible with the next version without conversion and migration work. For this reason, it's important to be sure about a module's Odoo target version before installing it.

To ensure compatibility, we need to follow these steps:

- Firstly, check whether the installation is successful. If it is, then proceed to check whether the module's functionality works properly.
- However, if the installation is not successful, you will then need to adjust the code and functional logic based on the errors you are receiving.

There's more...

Instead of having a long description in the module manifest, it's possible to have a separate description file. Since version 8.0, it can be replaced by a `README` file, with either a `.txt`, `.rst`, or a `.md` (markdown) extension. Otherwise, include a `description/index.html` file in the module.

This HTML description will override the description that's defined in the manifest file.

There are a few more keys that are frequently used:

- `licence`: The default value is `LGPL-3`. This identifier is used for a license under the module that is made available. Other license possibilities include `AGPL-3`, `Odoo Proprietary License v1.0` (mostly used in paid apps), and `Other OSI Approved Licence`.
- `application`: If this is `True`, the module is listed as an application. Usually, this is used for the central module of a functional area.
- `auto_install`: If this is `True`, it indicates that this is a *glue* module, which is automatically installed when all of its dependencies are installed.
- `installable`: If this is `True` (the default value), it indicates that the module is available for installation.
- `external_dependencies`: Some Odoo modules internally use `Python/bin` libraries. If your modules are using such libraries, you need to put them here. This will stop users from installing the module if the listed modules are not installed on the host machine.
- `{pre_init, post_init, uninstall}_hook`: This is a Python function hook that's called during installation/uninstallation. For a more detailed example, refer to *Chapter 8, Advanced Server-Side Development Techniques*.
- `Assets`: A definition of how all static files are loaded in various asset bundles. Odoo assets are grouped by bundles. Each bundle (a list of file paths of specific types – `xml`, `js`, `css`, or `scss`) is listed in the module manifest.

There are a number of special keys that are used for app store listing:

- `price`: This key is used to set the price for your add-on module. The value of this key should be an integer value. If a price is not set, this means your app is free.
- `currency`: This is the currency for the price. Possible values are `USD` and `EUR`. The default value for this key is `EUR`.
- `live_test_url`: If you want to provide a live test URL for your app, you can use this key to show the `Live Preview` button on the app store.
- `iap`: Set your IAP developer key if the module is used to provide an IAP service.
- `images`: This gives the path of images. This image will be used as a cover image in Odoo's app store.

Organizing the add-on module file structure

An add-on module contains code files and other assets, such as XML files and images. For most of these files, we are free to choose where to place them inside the module directory.

However, Odoo uses some conventions on the module structure, so it is advisable to follow them. Proper code improves readability, eases maintenance, helps debugging, lowers complexity, and promotes reliability. These apply to every new module and all new developments.

Getting ready

We are expected to have an add-on module directory with only the `__init__.py` and `__manifest__.py` files. In this recipe, we assume this is `local-addons/my_hostel`.

How to do it...

To create a basic skeleton for the add-on module, perform the following steps:

1. Create directories for the code files:

```
$ cd local-addons/my_hostel
$ mkdir models
$ touch models/__init__.py
$ mkdir controllers
$ touch controllers/__init__.py
$ mkdir views
$ touch views/views.xml
$ mkdir security
$ mkdir wizards
$ touch wizards/__init__.py
$ mkdir reports
$ mkdir data
$ mkdir demo
$ mkdir i18n
```

2. Edit the module's top `__init__.py` file so that the code in the subdirectories is loaded:

```
from . import models
from . import controllers
from . import wizards
```

This should get us started with a structure containing the most frequently used directories, similar to this one:

```
my_hostel
├── __init__.py
├── __manifest__.py
├── controllers
│   └── __init__.py
├── data
├── demo
├── i18n
├── models
│   └── __init__.py
├── security
├── static
│   ├── description
│   └── src
│       ├─ js
│       ├─ scss
│       ├─ css
│       └ xml
├── reports
├── wizards
│   └── __init__.py
└──views
    └── __init__.py
```

How it works...

To provide some context, an Odoo add-on module can have three types of files:

- The *Python code* is loaded by the `__init__.py` files, where the `.py` files and code subdirectories are imported. Subdirectories containing Python code, in turn, need their own `__init__.py` file.
- *Data files* that are to be declared in the `data` and `demo` keys of the `__manifest__.py` module manifest in order to be loaded are usually XML and CSV files for the user interface, fixture data, and demonstration data. There may also be YAML files, which can include some procedural instructions that are run when the module is loaded – for instance, to generate or update records programmatically rather than statically in an XML file.
- *Web assets*, such as JavaScript code and libraries, CSS, SASS, and QWeb/HTML templates, are files that are used to build UI parts and manage user actions in those UI elements. These are declared through a manifest on assets key that includes new files with existing files, which adds these assets to the web client, widgets, or website pages.

The add-on files are organized into the following directories:

- `models/` contains the backend code files, thus creating the models and their business logic. One file per model is recommended with the same name as the model – for example, `hostel.py` for the `hostel.hostel` model. These are addressed in depth in *Chapter 4, Application Models*.
- `views/` contains the XML files for the user interface, with the actions, forms, lists, and so on. Like models, it is advised to have one file per model. Filenames for website templates are expected to end with the `_template` suffix. Backend views are explained in *Chapter 9, Backend Views*, and website views are addressed in *Chapter 14, CMS Website Development*.
- `data/` contains other data files with the module's initial data. Data files are explained in *Chapter 6, Managing Module Data*.
- `demo/` contains data files with demonstration data, which is useful for tests, training, or module evaluation.
- `i18n/` is where Odoo will look for the translation `.pot` and `.po` files. Refer to *Chapter 11, Internationalization*, for further details. These files don't need to be mentioned in the manifest file.
- `security/` contains the data files that define access control lists, which is usually an `ir.model.access.csv` file and, possibly, an XML file to define access *groups and record rules* for row-level security. Take a look at *Chapter 10, Security Access*, for more details on this.
- `controllers/` contains the code files for the website controllers and for modules providing that kind of feature. Web controllers are covered in *Chapter 13, Web Server Development*.
- `static/` is where all web assets are expected to be placed. Unlike other directories, this directory name is not just a convention. The files inside this directory are public and can be accessed without a user login. This directory mostly contains files such as JavaScript, style sheets, and images. They don't need to be mentioned in the module manifest but will have to be referred to in the web template. This is discussed in detail in *Chapter 14, CMS Website Development*.
- `wizards/` contains all of the files related to wizards. In Odoo, wizards are used to hold intermediate data. We learn more about wizards in *Chapter 8, Advanced Server-Side Development Techniques*.
- `reports/`: Odoo provides a feature to generate PDF documents such as sales orders and invoices. This directory holds all the files related to PDF reports. We will learn more about PDF reports in *Chapter 12, Automation, Workflows, Emails, and Printing*.

When adding new files to a module, don't forget to declare them either in the `__manifest__.py` file (for data files) or `__init__.py` file (for code files); otherwise, those files will be ignored and won't be loaded.

Adding models

Models define the data structures that will be used by our business applications. This recipe shows you how to add a basic model to a module. Models determine the logical structure of a database and how data is stored, organized, and manipulated. In other words, a model is a table of information that can be linked with other tables. A model usually represents a business concept, such as a sales order, contact, or product.

Modules contain various elements, such as models, views, data files, web controllers, and static web data.

To create a hostel module, we need to develop a model that represents the hostel.

Getting ready

We should have a module to work with. If you followed the first recipe in this chapter, *Creating and installing a new add-on module*, you will have an empty module called `my_hostel`. We will use that for our explanation.

How to do it...

To add a new `Model`, we need to add a Python file describing it and then upgrade the add-on module (or install it, if this has not already been done). The paths that are used are relative to our add-on module's location (for example, `~/odoo-dev/local-addons/my_hostel/`):

1. Add a Python file to the `models/hostel.py` module with the following code:

```
from odoo import fields, models
class Hostel(models.Model):
    _name = 'hostel.hostel'
    _description = "Information about hostel"

    name = fields.Char(string="Hostel Name", required=True)
    hostel_code = fields.Char(string="Code", required=True)
    street = fields.Char('Street')
    street2 = fields.Char('Street2')
    state_id = fields.Many2one("res.country.state",
string="State")
```

2. Add a Python initialization file with code files to be loaded by the `models/__init__.py` module with the following code:

```
from . import hostel
```

3. Edit the module's Python initialization file to have the module load the `models/` directory:

```
from . import models
```

4. Upgrade the Odoo module from the command line or the **Apps** menu in the user interface. If you look closely at the server log while upgrading the module, you should see the following line:

```
odoo.modules.registry: module my_hostel: creating or updating
database table
```

After this, the new `hostel.hostel` model should be available in our Odoo instance. There are two ways to check whether our model has been added to the database.

First, you can check it in the Odoo user interface. Activate the developer tools and open the menu at **Settings** | **Technical** | **Database Structure** | **Models**. Search for the `hostel.hostel` model here.

The second way is to check the table entry in your PostgreSQL database. You can search for the `hostel_hostel` table in the database. In the following code example, we used `test-17.0` as our database. However, you can replace your database name with the following command:

```
$ psql test-17.0
test-17.0# \d hostel_hostel;
```

How it works...

Our first step was to create a Python file where our new module was created.

The Odoo framework has its own **Object Relational Mapping** (**ORM**) framework. This `ORM` framework provides an abstraction over the PostgreSQL database. By inheriting the Odoo Python `Model` class, we can create our own model (table). When a new model is defined, it is also added to a central model registry. This makes it easier for other modules to make modifications to it later.

Models have a few generic attributes prefixed with an underscore. The most important one is `_name`, which provides a unique internal identifier that will be used throughout the Odoo instance. The ORM framework will generate the database table based on this attribute. In our recipe, we used `_name = 'hostel.hostel'`. Based on this attribute, the ORM framework will create a new table called `hostel_hostel`. Note that the ORM framework will create a table name by replacing it. with _ in the value of the _name attribute. `_description` which provides a model's informal name, we used `_name = 'hostel.hostel'` and `_description='Information about hostel'`, and `_description='Information about hostel'` only starts with an alphabetical character we can't start with a number or special symbol character.

The `model` fields are defined as class attributes. We began by defining the `name` field of the `Char` type. It is convenient for models to have this field because, by default, it is used as the record description when referenced by other models.

We also used an example of a relational field – `state_id`. This defines a many-to-one relationship between `Hostel` and `State`.

There's much more to say about models, and they will be covered in depth in *Chapter 4, Application Models*.

Next, we must make our module aware of this new Python file. This is done by the `__init__.py` files. Since we placed the code inside the `models/` subdirectory, we need the previous `__init__` file to import that directory, which should, in turn, contain another `__init__` file, importing each of the code files there (just one, in our case).

Changes to Odoo models are activated by upgrading the module. The Odoo server will handle the translation of the `model` class into database structure changes.

Although no example is provided here, business logic can also be added to these Python files, either by adding new methods to the model's class or by extending the existing methods, such as `create()` or `write()`. This is addressed in *Chapter 5, Basic Server-Side Development*.

Adding access security

When adding a new data model, you need to define who can create, read, update, and delete records. When creating a totally new application, this can involve defining new user groups. Consequently, if a user doesn't have these access rights, then Odoo will not display your menus and views. In the previous recipe, we accessed our menu by converting an `admin` user into a superuser. After completing this recipe, you will be able to access menus and views for the `Hostel` module directly as an `admin` user.

This recipe builds on the `Hostel` model from the previous recipes and defines a new security group of users to control who can access or modify the records of `Hostel`.

Getting ready

The add-on module that implements the `hostel.hostel` model, which was provided in the previous recipe, is needed because, in this recipe, we will add the security rules for it. The paths that are used are relative to our add-on module location (for example, `~/odoo-dev/local-addons/my_hostel/`).

How to do it...

The security rules we want to add to this recipe are as follows:

- Everyone will be able to read hostel records.
- A new group of users called **Hostel Manager** will have the right to create, read, update, and delete hostel records.

To implement this, you need to perform the following steps:

1. Create a file called `security/hostel_security.xml` with the following content:

```
<?xml version="1.0" encoding="utf-8"?>
<odoo>
<record id="module_category_hostel" model="ir.module.category">
    <field name="name">Hostel Management</field>
    <field name="sequence">31</field>
</record>

<record id="group_hostel_manager" model="res.groups">
    <field name="name">Hostel Manager</field>
    <field name="category_id" ref="module_category_hostel"/>
    <field name="users" eval="[(4, ref('base.user_root')),(4,
ref('base.user_admin'))]"/>
</record>

<record id="group_hostel_user" model="res.groups">
    <field name="name">Hostel User</field>
    <field name="category_id" ref="module_category_hostel"/>
</record>

</odoo>
```

2. Add a file called `security/ir.model.access.csv` with the following content:

```
id,name,model_id:id,group_id:id,perm_read,perm_write,perm_
create,perm_unlink
access_hostel_manager_id,access.hostel.manager,my_hostel.model_
hostel_hostel,my_hostel.group_hostel_manager,1,1,1,1
access_hostel_user_id,access.hostel.user,my_hostel.model_hostel_
hostel,my_hostel.group_hostel_user,1,0,0,0
```

3. Add both files to the `data` entry of `__manifest__.py`:

```
# ...
"data": [
    "security/hostel_security.xml",
    "security/ir.model.access.csv",
    "views/hostel.xml",
],
# ...
```

The newly defined security rules will be in place once you update the add-on in your instance.

How it works...

We provide two new data files that we add to the add-on module's manifest so that installing or updating the module will load them in the database:

- The `security/hostel_security.xml` file defines a new security group by creating a `res.groups` record. We also gave Hostel Manager rights to the `admin` user by using its reference ID, `base.user_admin`, so that the admin user will have rights to the `hostel.hostel` model.
- The `ir.model.access.csv` file associates permissions on models with groups. The first line has an empty `group_id:id` column, which means that the rule applies to everyone. The last line gives all privileges to members of the group we just created.

The order of the files in the data section of the manifest is important. The file for creating the security groups must be loaded before the file listing the access rights, as the access rights definition depends on the existence of the groups. Since the views can be specific to a security group, we recommend putting the group's definition file in the list to be on the safe side.

See also

This book has a chapter dedicated to security. For more information on security, refer to *Chapter 10, Security Access*.

Adding menu items and views

Once we have models for our data structure needs, we want a user interface so that our users can interact with them. Menus and views play a crucial role in structuring and enhancing the user experience. Menus, from a technical perspective, are dynamic user interface components that present a structured set of options or links, typically allowing users to access various features, functions, or content areas within an application. This recipe builds on the `Hostel` model from the previous recipe and adds a menu item to display a user interface, featuring list and form views.

Getting ready

The add-on module to implement the `hostel.hostel` model, which was provided in the previous recipe, is needed. The paths that will be used are relative to our add-on module location (for example, `~/odoo-dev/local-addons/my_hostel/`).

How to do it...

To add a view, we will add an XML file with its definition to the module. Since it is a new model, we must also add a menu option for the user to be able to access it.

For models, XML files adding views folder to create a view, action, and menu item.

Be aware that the sequence of the following steps is relevant, since some of them use references to IDs that were defined in the preceding steps:

1. Create the XML file to add the data records describing the user interface, `views/hostel.xml`:

```
<?xml version="1.0" encoding="utf-8"?>
<odoo>
<!-- Data records go here -->
</odoo>
```

2. Add the new data file to the add-on module manifest, `__manifest__.py`, by adding it to `views/hostel.xml`:

```
{
    "name": "Hostel Management",
    "summary": "Manage Hostel easily",
    "depends": ["base"],
    "data": ["views/hostel.xml"],
}
```

3. Add the action that opens the views in the `hostel.xml` file:

```
<record id="action_hostel" model="ir.actions.act_window">
        <field name="name">Hostel</field>
        <field name="type">ir.actions.act_window</field>
        <field name="res_model">hostel.hostel</field>
        <field name="view_mode">tree,form</field>
        <field name="help" type="html">
            <p class="oe_view_nocontent_create">
                Create Hostel.
            </p>
        </field>
    </record>
```

4. Add the menu items to the `hostel.xml` file, making it visible to users:

```
<menuitem id="hostel_main_menu" name="Hostel" sequence="1"/>
<menuitem id="hostel_type_menu" name="Hostel" parent="hostel_
main_menu" action="my_hostel.action_hostel" groups="my_hostel.
group_hostel_manager" sequence="1"/>
```

5. Add a custom form view to the hostel.xml file:

```
<record id="view_hostel_form_view" model="ir.ui.view">
        <field name="name">hostel.hostel.form.view</field>
        <field name="model">hostel.hostel</field>
        <field name="arch" type="xml">
            <form string="Hostel">
                <sheet>
                    <div class="oe_title">
                        <h3>
                            <table>
                                <tr>
                                    <td style="padding-
                                    right:10px;">
                                    <field name="name"
                                    required="1"
                                    placeholder="Name" />
                                  </td>
                                    <td style="padding-
                                    right:10px;">
                                    <field name="hostel_code"
                                    placeholder="Code" />
                                        </td>
                                </tr>
                            </table>
                        </h3>
                    </div>
                    <group>
                        <group>
                            <label for="street"
                            string="Address"/>
                            <div class="o_address_format">
                                <field name="street"
                                placeholder="Street..."
                                class="o_address_street"/>
                                <field name="street2"
                                placeholder="Street 2..."
                                class="o_address_street"/>
                            </div>
                        </group>
                    </group>
                </sheet>
            </form>
        </field>
    </record>
```

6. Add a custom tree (list) view to the `hostel.xml` file:

```
<record id="view_hostel_tree_view" model="ir.ui.view">
        <field    name="name">hostel.hostel.tree.view</field>
        <field name="model">hostel.hostel</field>
        <field name="arch" type="xml">
            <tree>
                <field name="name"/>
                <field name="hostel_code"/>
            </tree>
        </field>
</record>
```

7. Add custom **Search** options to the `hostel.xml` file:

```
<record id="view_hostel_search_view" model="ir.ui.view">
    <field name="name">Hostel Search</field>
    <field name="model">hostel.hostel</field>
    <field name="arch" type="xml">
        <search>
            <field name="name"/>
            <field name="hostel_code"/>
        </search>
    </field>
</record>
```

When a new model is added in Odoo, the user doesn't have any access rights by default. We must define access rights for the new model in order to get access. In our example, we haven't defined any access rights, so the user doesn't have access to our new model. Without access, our menus and views are not visible either. Luckily, there is one shortcut! By switching to superuser mode, you can see menus for our app without having access rights.

Accessing Odoo as a superuser

By converting the `admin` user into a `superuser` type, you can bypass the access rights and, therefore, access menus and views without giving default access rights. To convert the `admin` user into a superuser, activate **Developer Mode**. After doing this, from the developer tool options, click on the **Become Superuser** option.

As a developer preference, try everything without becoming a superuser; it will be very helpful to learn Odoo in depth. By becoming a Superuser, all security access and record rule checks will be bypassed.

The following screenshot has been provided as a reference:

Run JS Tests
Run JS Mobile Tests
Run Click Everywhere Test
Open View
Start Tour

Edit Action
View Fields
Manage Filters

View Access Rights
View Record Rules

Get View
Edit View: Kanban
Edit SearchView

Activate Assets Debugging
Activate Tests Assets Debugging
Regenerate Assets Bundles
Become Superuser

Figure 3.3 – The option to activate superuser mode

After becoming a superuser, your menu will have a striped background, as shown in the following screenshot:

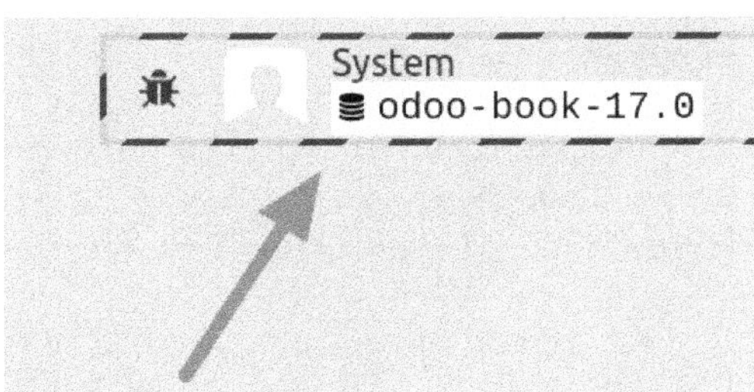

Figure 3.4 – Superuser mode activated

If you try and upgrade the module now, you should be able to see a new menu option (you might need to refresh your web browser). Clicking on the **Hostel** menu will open a list view for hostel models, as shown in the following screenshot:

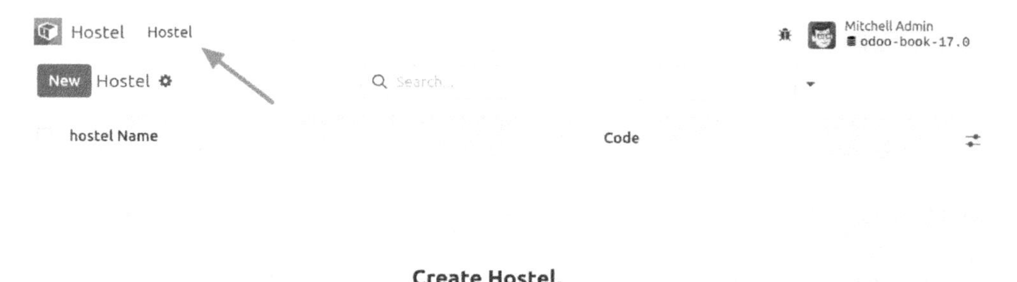

Figure 3.5 – The menu to access Hostel

How it works...

At a low level, the user interface is defined by records stored in special models. The first two steps create an empty XML file to define the records to be loaded, and then we add them to the module's list of data files to be installed.

Data files can be placed anywhere inside the module directory, but the convention is for the user interface to be defined inside a `views/` subdirectory. Usually, the name of these files is based on the name of the model. In our case, we create the user interface for the `hostel.hostel` model, so we created the `views/hostel.xml` file.

The next step is to define a window action to display the user interface in the main area of the web client. The action has a target model defined by `res_model`, and the `name` attribute is used to display the title to the user when the user opens the action. These are just the basic attributes. The window action supports additional attributes, giving much more control over how the views are rendered, such as what views are to be displayed, adding filters on the records that are available, or setting default values. These are discussed in detail in *Chapter 9*, *Backend Views*.

In general, data records are defined using a `<record>` tag, and we created a record for the `ir.actions.act_window` model in our example. This will create the window actions.

Similarly, menu items are stored in the `ir.ui.menu` model, and we can create these with the `<record>` tag. However, there is a shortcut tag called `<menuitem>` available in Odoo, so we used this in our example.

These are the menu item's main attributes:

- `name`: This is the menu item text to be displayed.
- `action`: This is the identifier of the action to be executed. We use the ID of the window action we created in the previous step.
- `sequence`: This is used to set the order in which the menu items of the same level are presented.

- `parent`: This is the identifier for the parent menu item. Our example menu item had no parent, meaning that it would be displayed at the top of the menu.
- `web_icon`: This attribute is used to show the icon for the menu. This icon is only displayed in the Odoo Enterprise edition.

At this point, we haven't defined any of the views in our module. However, if you upgrade your module at this stage, Odoo will automatically create them on the fly. Nevertheless, we will surely want to control how our views look, so in the next two steps, a form and a tree view are created.

Both views are defined with a record on the `ir.ui.view` model. The attributes we used are as follows:

- `name`: This is a title identifying the view. In the source code of Odoo, you will find the XML ID repeated here, but if you want, you can add a more human-readable title as a name.
- If the `name` field is omitted, Odoo will generate one using the model name and the type of view. This is perfectly fine for the standard view of a new model. It is recommended to have a more explicit name when you extend a view, as this will make your life easier when you look for a specific view in the user interface of Odoo.
- `model`: This is the internal identifier of the target model, as defined in its `_name` attribute.
- `arch`: This is the view architecture, where its structure is actually defined. This is where different types of views differ from each other.

Form views are defined with a top `<form>` element, and its canvas is a two-column grid. Inside the form, `<group>` elements are used to vertically compose fields. Two groups result in two columns with fields, which are added using the `<field>` element. Fields use a default widget according to their data type, but a specific widget can be used with the help of the `widget` attribute.

Tree views are simpler; they are defined with a top `<tree>` element that contains `<field>` elements for the columns to be displayed.

Finally, we added a **Search** view to expand the search option in the box at the top-right. Inside the `<search>` top-level tag, we can have the `<field>` and `<filter>` elements. Field elements are additional fields that can be searched from the input given in the search view. Filter elements are predefined filter conditions that can be activated with a click. These subjects are discussed in detail in *Chapter 9, Backend Views*.

Using the scaffold command to create a module

When creating a new Odoo module, there is some boilerplate code that needs to be set up. To help quick-start new modules, Odoo provides the `scaffold` command.

This recipe shows you how to create a new module using the `scaffold` command, which will put in place a skeleton of the file for directories to use.

Getting ready

We will create the new add-on module in a custom module directory, so we need Odoo installed and a directory for our custom modules. We will assume that Odoo is installed at `~/odoo-dev/odoo` and that our custom modules will be placed in the `~/odoo-dev/local-addons` directory.

How to do it...

We will use the `scaffold` command to create boilerplate code. Perform the following steps to create new a module using the `scaffold` command:

1. Change the working directory to where we will want our module to be. This can be whatever directory you choose, but it needs to be within an add-on path to be useful. Following the directory choices that we used in the previous recipe, this should be as follows:

   ```
   $ cd ~/odoo-dev/local-addons
   ```

2. Choose a technical name for the new module, and use the `scaffold` command to create it. For our example, we will choose `my_module`:

   ```
   $ ~/odoo-dev/odoo/odoo-bin scaffold my_module
   ```

3. Edit the `__manifest__.py` default module manifest provided and change the relevant values. You will surely want to at least change the module title in the `name` key.

This is what the generated add-on module should look like:

```
$ tree my_module
my_module/
├── __init__.py
├── __manifest__.py
├── controllers
│   ├── __init__.py
│   └── controllers.py
├── demo
│   └── demo.xml
├── models
│   ├── __init__.py
│   └── models.py
├── security
│   └── ir.model.access.csv
└── views
    ├── templates.xml
    └── views.xml
5 directories, 10 files
```

You should now edit the various generated files and adapt them to the purpose of your new module.

How it works...

The `scaffold` command creates the skeleton for a new module based on a template.

By default, the new module is created in the current working directory, but we can provide a specific directory to create the module, passing it as an additional parameter.

Consider the following example:

```
$ ~/odoo-dev/odoo/odoo-bin scaffold my_module ~/odoo-dev/local-addons
```

A `default` template is used, but a `theme` template is also available for website theme authoring. To choose a specific template, the `-t` option can be used. We are also allowed to use a path for a directory with a template.

This means that we can use our own templates with the `scaffold` command. The built-in templates can be found in the `/odoo/cli/templates` Odoo subdirectory. To use our own template, we can use something like the following command:

```
$ ~/odoo-dev/odoo/odoo-bin scaffold -t path/to/template my_module
```

By default, Odoo has two templates in the `/odoo/cli/templates` directory. One is the `default` template, and the second is the `theme` template. However, you can create your own templates or use them with `-t`, as shown in the preceding command.

4
Application Models

This chapter will guide you through some small enhancements to an existing add-on module. You already registered your add-on module in the Odoo instance in *Chapter 3, Creating Odoo Add-On Modules*. Now, you will explore the database aspects of the module in more depth. You will learn how to create a new model (database table), add new fields, and apply constraints. You will also discover how to use inheritance in Odoo to modify existing models. In this chapter, you will use the same module that you created in the previous chapter.

This chapter covers the following topics:

- Defining the model representation and order
- Adding data fields to a model
- Adding a float field with configurable precision
- Adding a monetary field to a model
- Adding relational fields to a model
- Adding a hierarchy to a model
- Adding constraint validations to a model
- Adding computed fields to a model
- Exposing related fields stored in other models
- Adding dynamic relations using reference fields
- Adding features to a model using inheritance
- Using abstract models for reusable model features
- Copying the model definition using inheritance

Technical requirements

Before proceeding with the examples in this chapter, make sure you have the module that we developed in *Chapter 3, Creating Odoo Add-On Modules*, and that it is properly installed and configured.

Defining the model representation and order

A model refers to a representation of a database table. A model defines the structure and behavior of a database table, including fields, relationships, and various methods. Models are defined in Python code using Odoo's **object-relational mapping** (**ORM**) system. ORM allows developers to interact with the database using Python classes and methods, rather than writing raw SQL queries.

Model attributes are the features of a model that will be defined when we create a new model; otherwise, we use the attributes of the model that already exists. Models use structural attributes with an underscore prefix to define their behavior.

Getting ready

The `my_hostel` instance should already contain a Python file called `models/hostel.py`, which defines a basic model. We will edit it to add new class-level attributes.

How to do it...

By utilizing these attributes effectively, developers can create well-organized, reusable, and maintainable code in Odoo, leading to a more efficient and robust application. The following are the attributes that can be used on a model:

1. `_name` : The `name` attribute is the most important one, as it determines the internal global identifier and the database table name. The model name is expressed in dot notation within the module namespace. For instance, `name="hostel.hostel"` will create the `hostel_hostel` table in the database:

   ```
   _name = 'hostel.hostel'
   ```

2. `_table`: We can define the SQL table name utilized by the model if '`_auto`' is enabled:

   ```
   _name = 'project.task.stage.personal'
   _table = 'project_task_user_rel'
   ```

3. `_description`: To assign a descriptive title to the model that reflects its purpose and functionality, insert the following code snippet:

   ```
   _description = 'Information about hostel'
   ```

> **Note**
>
> If you don't use `_description` for your model, Odoo will show a warning in the logs.

4. `_order`: The default field for ordering the search results is '`id`'. However, this can be changed so that we can use the fields of our choice, by providing an `_order` attribute with a string containing a comma-separated list of field names. A field name can be followed by the `desc` keyword to sort it in descending order. To order the records by `id` in descending order, followed by names in ascending order, use the following code syntax:

```
_order = "id desc, name"
```

> **Note**
>
> Only fields stored in the database can be used. Non-stored computed fields can't be used to sort records. The syntax for the `_order` string is similar to the `SQL ORDER BY` clauses, although it's stripped down. For instance, special clauses, such as `NULLS FIRST`, are not allowed.

5. `_rec_name`: This is used to set the field that's used as a representation or title for the records. The default field for `rec_name` is the name field. `_rec_name` is the display name of the record used by Odoo's **graphical user interface** (**GUI**) to represent that record. If you want to change `rec_name` and set `hostel_code` as a representative of the model, use the following code syntax:

```
_rec_name = 'hostel_code'
hostel_code = fields.Char(string="Code", required=True)
```

> **Note**
>
> If your model doesn't have a name field and you haven't specified `_rec_name` either, your display name will be a combination of the model name and record ID, like this – `(hostel.hostel, 1)`.

6. `_rec_names_search`: This is used to search specific records by mentioned field values. It is similar to using the `name_search` function. You can directly use this attribute instead of using the `name_search` method. To do so, use the following code syntax:

```
_rec_names_search = ['name', 'code']
```

There's more...

All models have a `display_name` field that shows the record representation in a human-readable format, which has been automatically added to all models since version 8.0. The default `_compute_display_name()` method uses the `_rec_name` attribute to determine which field contains the data for the display name. To customize the display name, you can override the `_compute_display_name()` method and provide your logic. The method should return a list of tuples, each containing the record ID and the Unicode string representation.

For example, to have the hostel name and hostel code in the representation, such as `Youth Hostel (YHG015)`, we can define the following:

Take a look at the following example. This will add a release date to the record's name:

```
@api.depends('hostel_code')
    def _compute_display_name(self):
        for record in self:
            name = record.name
            if record.hostel_code:
                name = f'{name} ({record.hostel_code})'
            record.display_name = name
```

After adding the preceding code, your `display_name` record will be updated. Suppose you have a record with the name `Bell House Hostel` and its code is `BHH101`; then, the preceding `_compute_display_name()` method will generate a name such as `Bell House Hostel (BHH101)`.

When we're done, our `hostel.py` file should appear as follows:

```
from odoo import fields, models
class Hostel(models.Model):
_name = 'hostel.hostel'
_description = "Information about hostel"
_order = "id desc, name"
_rec_name = 'hostel_code'
name = fields.Char(string="hostel Name", required=True)
hostel_code = fields.Char(string="Code", required=True)
street = fields.Char('Street')
street2 = fields.Char('Street2')
zip = fields.Char('Zip', change_default=True)
city = fields.Char('City')
state_id = fields.Many2one("res.country.state", string='State')
country_id = fields.Many2one('res.country', string='Country')

phone = fields.Char('Phone',required=True)
mobile = fields.Char('Mobile',required=True)
email = fields.Char('Email')
@api.depends('hostel_code')
```

```
    def _compute_display_name(self):
        for record in self:
            name = record.name
            if record.hostel_code:
                name = f'{name} ({record.hostel_code})'
            record.display_name = name
```

Your <form> view in the `hostel.xml` file will look as follows:

```
<form string="Hostel">
  <sheet>
    <div class="oe_title">
      <h3>
        <table>
          <tr>
            <td style="padding-right:10px;">
            <field name="name" required="1"
            placeholder="Name" /></td>
            <td style="padding-right:10px;">
            <field name="hostel_code" placeholder="Code"
            /></td>
          </tr>
        </table>
      </h3>
    </div>
    <group>
      <group>
        <label for="street" string="Address"/>
          <div class="o_address_format">
            <field name="street" placeholder="Street..."
            class="o_address_street"/>
            <field name="street2" placeholder="Street 2..."
            class="o_address_street"/>
            <field name="city" placeholder="City"
            class="o_address_city"/>
            <field name="state_id" class="o_address_state"
            placeholder="State" options='{"no_open":
            True}'/>
            <field name="zip" placeholder="ZIP"
            class="o_address_zip"/>
```

```
                <field name="country_id" placeholder="Country"
                class="o_address_country" options='{"no_open":
                True, "no_create": True}'/>
              </div>
          </group>
          <group>
            <field name="phone" widget="phone"/>
              <field name="mobile" widget="phone"/>
                <field name="email" widget="email"
                context="{'gravatar_image': True}"/>
          </group>
        </group>
      </sheet>
    </form>
```

We should then upgrade the module to activate these changes in Odoo.

To update the module, execute the following:

```
Activate developer mode ->Apps -> Update App List
```

Then, search for the `my_hostel` module and upgrade it via the dropdown, as shown in the following screenshot:

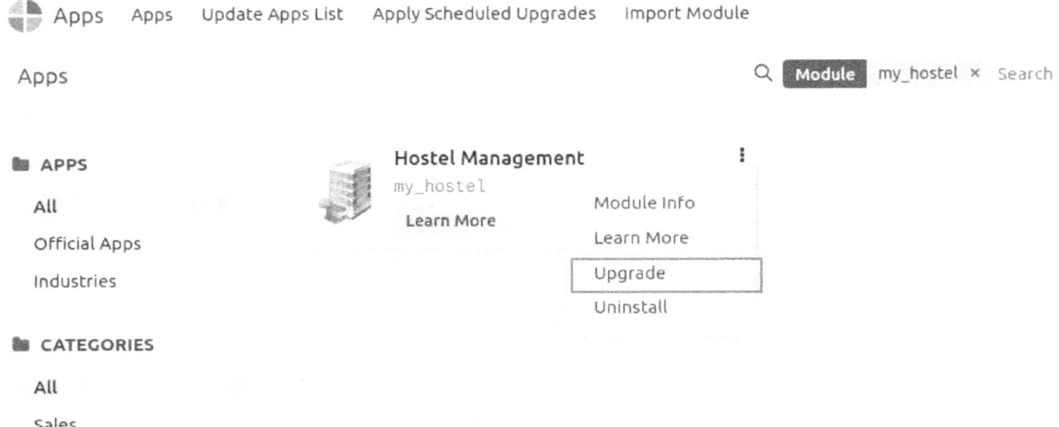

Figure 4.1 – The option to update the module

Alternatively, you can also use the `-u my_hostel` command in the command line.

Adding data fields to a model

A field represents a column in a database table and defines the structure of the data that can be stored in that column. Fields in Odoo models are used to specify the attributes and characteristics of the data that the model will store. Each field has a data type (e.g., `Char`, `Integer`, `Float`, or `Date`) and various attributes that determine how the field behaves.

In this section, you will explore the various data types that fields can support and how to add them to a model.

Getting ready

This recipe assumes that you have an instance ready with the `my_hostel` add-on module available, as described in *Chapter 3, Creating Odoo Add-On Modules.*

How to do it...

The `my_hostel` add-on module should already have `models/hostel.py`, defining a basic model. We will edit it to add new fields:

1. Use the minimal syntax to add fields to the `Hostel` model:

```
from odoo import models, fields
class Hostel(models.Model):
    # ...
    email = fields.Char('Email')
    hostel_floors = fields.Integer(string="Total Floors")
    image = fields.Binary('Hostel Image')
    active = fields.Boolean("Active", default=True,
    help="Activate/Deactivate hostel record")
    type = fields.Selection([("male", "Boys"), ("female",
"Girls"),
    ("common", "Common")], "Type", help="Type of Hostel",
    required=True, default="common")
    other_info = fields.Text("Other Information",
    help="Enter more information")
    description = fields.Html('Description')
    hostel_rating = fields.Float('Hostel Average Rating',
digits=(14, 4))
```

We have added new fields to the model. We still need to add these fields to the form view in order to reflect these changes in the user interface. Refer to the following code to add fields to the form view:

```
<field name="image" widget="image" class="oe_avatar"/>
  <group>
    <group>
    <label for="street" string="Address"/>
    <div class="o_address_format">
      <field name="street" placeholder="Street..." class="o_
address_street"/>
      <field name="street2" placeholder="Street 2..." class="o_
address_street"/>
      <field name="city" placeholder="City" class="o_address_
city"/>
      <field name="state_id" class="o_address_state"
      placeholder="State" options='{"no_open": True}'/>
      <field name="zip" placeholder="ZIP" class="o_address_
zip"/>
      <field name="country_id" placeholder="Country"
      class="o_address_country" options='{"no_open": True,
       "no_create": True}'/>
    </div>
    <field name="phone" widget="phone"/>
    <field name="mobile" widget="phone"/>
    <field name="email" widget="email" context="{'gravatar_
image': True}"/>
    </group>
    <group>
      <field name="hostel_floors"/>
      <field name="active"/>
      <field name="type"/>
      <field name="hostel_rating"/>
      <field name="other_info"/>
    </group>
  </group>
  <group>
    <field name="description"/>
  </group>
```

Upgrading the module will make these changes effective in the Odoo model.

How it works...

To add fields to models, you need to define an attribute of the corresponding type in their Python classes. The available types of non-relational fields are as follows:

- **Char**: Stores string values.
- **Text**: Stores multiline string values.
- **Selection**: Stores one value from a list of predefined values and descriptions. This has a list of values and description pairs. The value that is selected is what gets stored in the database, and it can be a string or an integer. The description is automatically translatable.

> **Note**
>
> Odoo does not display the description if the value is zero for integer keys. The `Selection` field also accepts a function reference as its `selection` attribute instead of a list. This allows you to dynamically generate lists of options. You can find an example relating to this in the *Adding dynamic relations using reference fields* recipe in this chapter, where a `selection` attribute is also used.

- **Html**: Stores rich text in the HTML format.
- **Binary**: Stores binary files, such as images or documents.
- **Boolean**: Stores `True`/`False` values.
- **Date**: Stores date values as Python date objects. Use `fields.Date.today()` to set the default value to the current date.
- **Datetime**: Stores `datetime` values as Python datetime objects in UTC time. Use `fields.Date.now()` to set the default value to the current time.
- **Integer**: Stores integer values.
- **Float**: Stores numeric values with optional precision (total digits and decimal digits).
- **Monetary**: Stores an amount in a specific currency. This will be explained further in the *Adding a monetary field to a model* recipe in this chapter.

Step 1 of this recipe shows the minimal syntax to add to each field type. The field definitions can be expanded to add other optional attributes, as shown in *step 2*. Here's an explanation of the field attributes that were used:

- `string` is the field's title and is used in UI view labels. It is optional. If not set, a label will be derived from the field name by adding a title case and replacing the underscores with spaces.

- `translate`, when set to `True`, makes the field translatable. It can hold a different value, depending on the user interface language.
- `default` is the default value. It can also be a function that is used to calculate the default value – for example, `default=_compute_default`, where `_compute_default` is a method that was defined on the model before the field definition.
- `help` is an explanation text that's displayed in the UI tooltips.
- `groups` makes the field available only to some security groups. It is a string containing a comma-separated list of XML IDs for security groups. This is addressed in more detail in *Chapter 10, Security Access*.
- `copy` flags whether the field value is copied when the record is duplicated. By default, it is `True` for non-relational and `Many2one` fields, and `False` for `One2many` and computed fields.
- `index`, when set to `True`, creates a database index for the field, which sometimes allows for faster searches. It replaces the deprecated `select=1` attribute.
- The `readonly` flag makes the field read-only by default in the user interface.
- The `required` flag makes the field mandatory by default in the user interface.
- The various whitelists that are mentioned here are defined in `odoo/fields.py`.
- The `company_dependent` flag makes the field store different values for each company. It replaces the deprecated `Property` field type.
- The value isn't stored on the model table. It is registered as `` `ir.property` ``. When the value of the `company_dependent` field is needed, an `` `ir.property` `` is searched and linked to the current company (and the current record if one property exists). If the value is changed on the record, it either modifies the existing property for the current record (if one exists) or creates a new one for the current company and `res_id`. If the value is changed on the company side, it will impact all records on which the value hasn't been changed.
- `group_operator` is an aggregate function used to display results in the group by mode.

  Possible values for this attribute include `count`, `count_distinct`, `array_agg`, `bool_and`, `bool_or`, `max`, `min`, `avg`, and `sum`. Integer, float, and monetary field types have the default `sum` value for this attribute. This field is used by the `:meth:~odoo.models.Model.read_group` method to group rows based on this field.

  The supported aggregate functions are as follows:

  - `array_agg`: Concatenates all values, including nulls, into an array
  - `count`: Counts the number of rows
  - `count_distinct`: Counts the number of distinct rows

    - `bool_and`: Returns `true` if all values are `true`, and `false` otherwise
    - `bool_or`: Returns `true` if at least one value is `true`, and `false` otherwise
    - `max`: Returns the maximum value of all values
    - `min`: Returns the minimum value of all values
    - `avg`: Returns the average (arithmetic mean) of all values
    - `sum`: Returns the sum of all values

- `Store`: This is for whether the field is stored in the database (the default is `True`, and `False` for computed fields).
- `group_expand`: This function is used to expand `read_group` results when grouping on the current field:

```
    .. code-block:: python
         @api.model
         def _read_group_selection_field(self, values, domain,
order):
             return ['choice1', 'choice2', ...] # available
selection choices.
          @api.model
         def _read_group_many2one_field(self, records, domain,
order):
             return records + self.search([custom_domain])
```

- The `sanitize` flag is employed within HTML fields to systematically remove potentially insecure tags from their content. Activation of this flag results in a comprehensive cleansing of the input. For users seeking more nuanced control over HTML sanitization, there are additional attributes available. It is important to note that these attributes are effective only when the sanitize flag is enabled.

If you need finer control in HTML sanitization, there are a few more attributes that you can use, which only work if `sanitize` is enabled:

- `sanitize_tags=True`, to remove tags that are not part of a whitelist (this is the default)
- `sanitize_attributes=True`, to remove attributes of the tags that are not part of a whitelist
- `sanitize_style=True`, to remove style properties that are not part of a whitelist
- `strip_style=True`, to remove all style elements
- `strip_class=True`, to remove the class attributes

Finally, we updated the form view according to the newly added fields in the model. We placed all fields in form view, but you can place them anywhere you want. Form views are explained in more detail in *Chapter 9*, *Backend Views*.

There's more...

The `Date` and `Datetime` field objects expose a few utility methods that can be convenient for Date and Datetime:

For `Date`, we have the following:

- `fields.Date.to_date(string_value)` parses the string into a date object.
- `fields.Date.to_string(date_value)` converts the Python Date object to a string.
- `fields.Date.today()` returns the current day in a string format. This is appropriate for use with default values.
- `fields.Date.context_today(record, timestamp)` returns the day of the timestamp (or the current day, if the timestamp is omitted) in a string format, according to the time zone of the record's (or record set's) context.

For `Datetime`, we have the following:

- `fields.Datetime.to_datetime(string_value)` parses the string into a datetime object.
- `fields.Datetime.to_string(datetime_value)` converts the datetime object to a string.
- `fields.Datetime.now()` returns the current day and time in a string format. This is appropriate to use for default values.
- `fields.Datetime.context_timestamp(record, timestamp)` converts a timestamp-naive datetime object into a time zone-aware datetime object. using the time zone in the context of a record. This is not suitable for default values but can be used for instances when you're sending data to an external system.

In addition to the basic fields, there are also few relational fields such as `Many2one`, `One2many`, and `Many2many`. These are covered in the *Adding relational fields to a model* recipe in this chapter.

You can also create fields with values that are computed automatically by using the `compute` field attribute to define the computation function. This is covered in the *Adding computed fields to a model* recipe of this chapter.

Some fields are added by default in Odoo models, so you should avoid using these names for your fields. These are as follows:

- `id` (the record's automatically generated identifier)
- `create_date` (the record creation timestamp)
- `create_uid` (the user who created the record)
- `write_date` (the last recorded timestamp edit)
- `write_uid` (the user who last edited the record)

The automatic creation of these log fields can be disabled by setting the `_log_access=False` model attribute.

Another special column that can be added to a model is `active`. It must be a `Boolean` field, allowing users to mark records as inactive. It is used to enable the `archive/unarchive` feature on the records. Its definition is as follows:

```
active = fields.Boolean('Active', default=True)
```

By default, only records with `active` set to `True` are visible. To retrieve them, we need to use a domain filter with `[('active', '=', False)]`. Alternatively, if the `'active_test': False` value is added to the environment's context, ORM will not filter out inactive records.

In some cases, you may not be able to modify the context to get both the active and the inactive records. If so, you can use the `['|', ('active', '=', True), ('active', '=', False)]` domain.

> **Tip**
>
> `[('active', 'in' (True, False))]` does not work as you might expect. Odoo explicitly looks for an `('active', '=', False)` clause in the domain. It will default to restricting the search to active records only.

Adding a float field with configurable precision

When using `float` fields, we may want to let the end user configure the decimal precision that will be used. In this recipe, we will add a `hostel_rating` field to the `hostel` model, with user-configurable decimal precision.

Getting ready

We will continue using the `my_hostel` add-on module from the previous recipe.

How to do it...

Perform the following steps to apply dynamic decimal precision to the model's `hostel_rating` field:

1. Create a data folder and add a `data.xml` file. Inside this file, add the following record for the decimal precision model. This will add a new configuration.

```
<record forcecreate="True" id="decimal_point" model="decimal.
precision">
<field name="name">Rating Value</field>
<field name="digits">3</field>
</record>
```

2. Activate **Developer Mode** from the link in the **Settings** menu (refer to the *Activating the Odoo developer tools* recipe in *Chapter 1*, *Installing the Odoo Development Environment*). This will enable the **Settings** | **Technical** menu.
3. Access the decimal precision configurations. To do this, open the **Settings** top menu and select **Technical** | **Database Structure** | **Decimal Accuracy**. We should see a list of the currently defined settings.

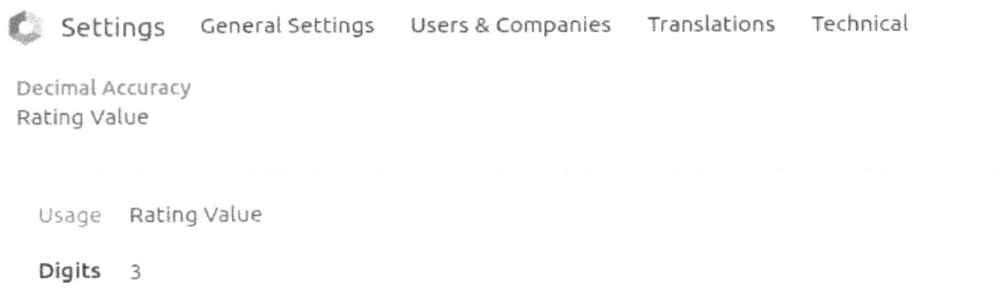

Figure 4.2 – Creating new decimal precision

4. To add the `model` field using this decimal precision setting, edit the `models/hostel.py` file by adding the following code:

```
class Hostel(models.Model):
    hostel_rating = fields.Float('Hostel Average Rating',
    # digits=(14, 4) # Method 1: Optional precision (total,
decimals),
    digits='Rating Value' # Method 2
)
```

How it works...

When you add a string value to the `digits` attribute of the field, Odoo looks up that string in the decimal accuracy model's `Usage` field and returns a tuple, with 16-digit precision and the number of decimals that were defined in the configuration. Using the field definition, instead of having it hardcoded, we allow the end user to configure it according to their needs.

Adding a monetary field to a model

To work with monetary values and currencies in a model, we can use Odoo to provide special support for working with monetary values and currencies in its models, through the use of specific field types and features. Odoo's special support for monetary values and currencies simplifies the handling of financial data, ensuring accuracy, consistency, and compliance with currency-related requirements.

Getting ready

We will use the same `my_hostel` add-on module from the previous recipe.

How to do it...

We need to add a currency field along with a monetary field to store the currency for the amounts.

We will add `models/hostel_room.py`, to add the necessary fields:

1. Create the field to store the currency for the amounts:

```
class HostelRoom(models.Model):
    _name = "hostel.room"
    ...#
    currency_id = fields.Many2one('res.currency',
string='Currency')
```

2. Add the monetary field to store the amount:

```
class HostelRoom(models.Model):
    _name = "hostel.room"
    ...#
    rent_amount = fields.Monetary('Rent Amount', help="Enter
rent amount per month") # optional attribute: currency_
field='currency_id' incase currency field have another name then
'currency_id'
```

Create a security file for the new model and a form view to show it in the UI. Upgrade the add-on module to apply the changes. The monetary field will appear like this:

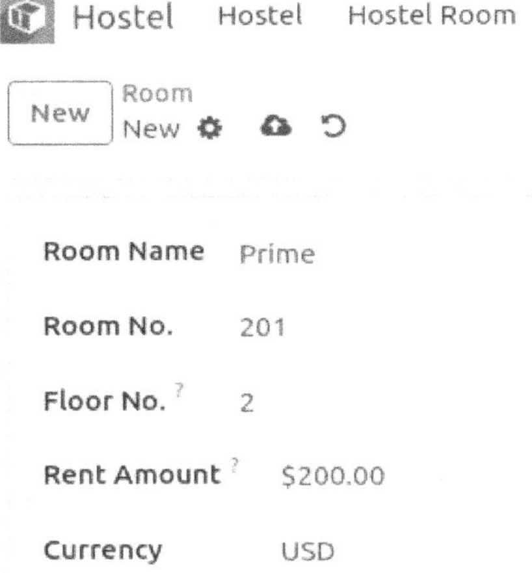

Figure 4.3 – The currency symbol in the monetary field

How it works...

Odoo can display `monetary` fields correctly in the user interface because they have a second field that indicates their currency. This field is similar to a float field.

The currency field is usually named `currency_id`, but we can use any other name as long as we specify it with the optional `currency_field` parameter.

If your currency information is stored in a field named `currency_id`, you don't need to specify the `currency_field` attribute for the monetary field.

This is helpful when you have to store amounts in different currencies in the same record. For example, if you want to have the currency of the sale order and the company, you can create two fields as `fields.Many2one(res.currency)` and use one for each amount.

The currency definition (the `decimal_precision` field of the `res.currency model`) determines the decimal precision for the amount.

Adding relational fields to a model

Relational fields are used to represent relations between Odoo models. There are three types of relations:

- `many-to-one`, or `m2o` for
- `one-to-many`, or `o2m` for short
- `many-to-many`, or `m2m` for short

To illustrate this, let's consider the hostel room model. A room belongs to a single hostel, so the relation between the hostel and the room is m2o. However, a hostel can have multiple rooms, so the opposite relationship is o2m.

We can also have a m2m relationship. For instance, a room can offer various amenities and amenities can be available in different rooms. This is a bidirectional m2m relationship.

Getting ready

We will continue using the my_hostel add-on module from the previous recipe.

How to do it...

We will edit the models/hostel_room.py file to add these fields:

1. Add the m2o field for the hostel in Hostel Room:

```
class HostelRoom(models.Model):

# ...
    hostel_id = fields.Many2one("hostel.hostel", "hostel",
help="Name of hostel")
```

2. We want to create a o2m field for a student that links to a room.
3. First, we need a new model for hostel students. We will make a hostel_student.py file and add some basic fields to the hostel student model. Then, we will add a room_id m2o field to connect the student and room models.
4. Finally, we will add an o2m field, student_ids, of the hostel.student model to the hostel.room model:

```
class HostelStudent(models.Model):
    _name = "hostel.student"
    name = fields.Char("Student Name")
    gender = fields.Selection([("male", "Male"),
    ("female", "Female"), ("other", "Other")],
    string="Gender", help="Student gender")
    active = fields.Boolean("Active", default=True,
    help="Activate/Deactivate hostel record")
    room_id = fields.Many2one("hostel.room", "Room",
    help="Select hostel room")
```

```
class HostelRoom(models.Model):
    _name = "hostel.room"
    # ...
    student_ids = fields.One2many("hostel.student", "room_id",
    string="Students", help="Enter students")
```

5. We will create a new file, `hostel_amenities.py`. Add the following code to the file:

```
class HostelAmenities(models.Model):
    _name = "hostel.amenities"
    _description = "Hostel Amenities"
    name = fields.Char("Name", help="Provided Hostel Amenity")
    active = fields.Boolean("Active",
    help="Activate/Deactivate whether the amenity should be
given or not")
```

Now, we will add an `m2m` field of amenities to the `hostel.room` model. Add the following code to `hostel_room.py`:

```
class HostelRoom(models.Model):
    _name = "hostel.room"
    # ...
    hostel_amenities_ids = fields.Many2many("hostel.amenities",
    "hostel_room_amenities_rel", "room_id", "amenitiy_id",
    string="Amenities", domain="[('active', '=', True)]",
    help="Select hostel room amenities")
```

Now, upgrade the add-on module, and the new fields should be available in the model. They won't be visible in the views until they are added to them. We will add new fields to the `hostel_room.xml` file.

We can confirm their addition by inspecting the `model` fields in **Settings** | **Technical** | **Database Structure** | **Models** in **Developer** mode.

How it works...

An m2o field stores the database ID of another record in a column of the model's table. This creates a foreign key constraint in the database, which ensures that the stored ID is a valid reference to a record in another table. By default, these relationship fields do not have a database index, but you can add one by setting the `index=True` attribute.

You can also specify what happens when the record that is referenced by an m2o field is deleted. The `ondelete` attribute controls this behavior. For instance, what should happen to students when their room record is deleted? The default option is `'set null'`, which means the field will have an empty value. Another option is `'restrict'`, which means the related record cannot be deleted. A third option is `'cascade'`, which means the linked record will be deleted as well.

You can also use `context` and `domain` for other relational fields. These attributes are mainly useful on the client side, and they provide default values for the views of the related records that are accessed through a field:

- `context` sets some variables in the client context when you click on a field to see the related record's view. For example, you can use it to set default values for new records that are created in that view.
- `domain` is a filter that limits the list of related records that you can choose from.

You can learn more about `context` and `domain` in *Chapter 9, Backend Views.*

An o2m field is the opposite of an m2o field, and it lets you access a list of related records from a model. Unlike other fields, it does not have a column in the database table. It is just a convenient way to display these related records in views. To use an o2m field, you need to have a corresponding `m2o field` in the other model. In our example, we added an o2m field to the room model. The `student_ids` o2m field has a reference to the `room_id` field of the `hostel.room` model.

A m2m field does not have a column in the model's table. Instead, it uses another table in the database to store the relationship between two models. This table has two columns for the IDs of the related records. When you link a room and its amenity with an m2m field, a new record is created in this table with the room's ID and the amenity's ID.

Odoo creates the relationship table for you. By default, the name of the relationship table is made from the names of the two models, sorted alphabetically, with a `_rel` suffix. You can change this name with the `relation` attribute.

You should use the `relation` attribute when the names of the two models are too long for the default name. PostgreSQL has a limit of 63 characters for database identifiers. So, if the names of the two models are more than 23 characters each, you should set a shorter name with the `relation` attribute. We will explain this more in the next section.

There's more...

You can also use the `auto_join` attribute for `m2o` fields. This attribute lets ORM use SQL joins on this field. This means that ORM does not check the user access control and record access rules for this field. This can help with performance issues in some cases, but it is better to avoid it.

We have seen the simplest way to define the relational fields. Now, let's look at the attributes that are specific to these fields.

These are the attributes for the `o2m` field:

- `comodel_name`: This is the name of the model that the field relates to. You need this attribute for all relational fields. You can write it without the keyword, as the first argument.
- `inverse_name`: This is only for the `o2m` fields. It is the name of the `m2o` field in the other model that links back to this model.
- `limit`: This is for the `o2m` and `m2m` fields. It sets a maximum number of records to read and display in the user interface.

These are the attributes for the `m2m` field:

- `comodel_name`: This is the name of the model that the field relates to. It is the same as for the `o2m` field.
- `relation`: This is the name of the table in the database that stores the relationship. You can use this attribute to change the default name.
- `column1`: This is the name of column 1 in the relation table that links to this model.
- `column2`: This is the name of column 2 in the relation table that links to the other model.

Odoo typically handles the creation and management of these attributes automatically. It can identify and utilize an existing relation table for an inverse `m2m` field. However, there are specific scenarios where manual intervention is required.

When dealing with multiple `m2m` fields between the same two models, it becomes necessary to assign distinct relation table names for each field.

In cases where the names of the two models exceed PostgreSQL's limit of 63 characters for database object names, you must set these attributes yourself. The default relation table name is typically <model1>_<model2>rel. However, this table includes a primary key index with a longer name (<model1><model2>rel<model1>id<model2>_id_key), which also needs to adhere to the 63-character limit. Therefore, if the combined names of the two models surpass this limit, you must opt for a shorter relation table name.

Adding a hierarchy to a model

You can use an `m2o` field to represent hierarchies, where each record has a parent record and many child records in the same model. However, Odoo also provides improved support for this type of field by using the `nested set model` (`https://en.wikipedia.org/wiki/Nested_set_model`). When activated, queries using the `child_of` operator in their domain filters will run significantly faster.

Staying with the `Hostel` example, we will build a hierarchical category tree that can be used to categorize hostels.

Getting ready

We will continue using the `my_hostel` add-on module from the previous recipe.

How to do it...

We will add a new Python file, `models/hostel_categ.py`, for the category tree, as follows:

1. To load the new Python code file, add the following line to `models/__init__.py`:

```
from . import hostel_categ
```

2. To create the `Hostel Category` model with the parent and child relationships, create the `models/hostel_categ.py` file with the following code:

```
from odoo import models, fields, api
class HostelCategory(models.Model):
    _name = "hostel.category"
    name = fields.Char('Category')
    parent_id = fields.Many2one(
        'hostel.category',
        string='Parent Category',
        ondelete='restrict',
        index=True)
    parent_path = fields.Char(index=True)
    child_ids = fields.One2many(
        'hostel.category', 'parent_id',
        string='Child Categories')
```

3. To enable the special hierarchy support, also add the following code:

```
_parent_store = True
_parent_name = "parent_id" # optional if field is 'parent_id'
parent_path = fields.Char(index=True, unaccent=False)
```

4. To add a check to prevent looping relations, add the following line to the model:

```
from odoo.exceptions import ValidationError
...
@api.constrains('parent_id')
def _check_hierarchy(self):
    if not self._check_recursion():
        raise models.ValidationError(
            'Error! You cannot create recursive categories.')
```

5. Now, we need to assign a category to a hostel. To do this, we will add a new `m2o` field to the `hostel.hostel` model:

```
category_id = fields.Many2one('hostel.category')
```

Finally, a module upgrade will make these changes effective.

To display the `hostel.category` model in the user interface, you will need to add menus, views, and security rules. For more details, refer to *Chapter 3*, *Creating Odoo Add-On Modules*. Alternatively, you can access all code at `https://github.com/PacktPublishing/Odoo-17-Development-Cookbook-Fifth-Edition/tree/main/Chapter04`.

How it works...

We want to create a new model with hierarchical relations. This means that each record can have a parent record and many child records in the same model. Here are the steps to do this:

1. We create an `m2o` field to reference the parent record. We use `index=True` to make this field indexed in the database for faster queries. We also use `ondelete='cascade'` or `ondelete='restrict'` to control what happens when the parent record is deleted.
2. We create a `o2m` field to access all the child records of a record. This field does not add anything to the database, but it is a convenient way to get the child records. We add a special support for hierarchies by using `parent_store=True` in the model attribute. This makes the queries using the `child_of` operator faster, but it also makes the write operations slower. We also add a helper field called `parent_path` to store data for hierarchical searches. If we use a different name from `parent_id` for the parent field, we need to specify it with `parent_name` in the model attribute.

3. We prevent cyclic dependencies in the hierarchy by using the `_check_recursion` method from `models.Model`. This avoids us having a record that is both an ancestor and a descendant of another record, which can cause infinite loops.
4. We add a `category_id` field with `Many2one` type to the hostel.hostel model, so that we can assign a category to each hostel. This is just to complete our example.

There's more...

You should use this technique for hierarchies that do not change much but are read and queried a lot. This is because the nested set model in the database needs to update the `parent_path` column (and the related database indexes) for all records when a category is added, deleted, or moved. This can be slow and costly, especially when there are many concurrent transactions.

If you have a hierarchy that changes a lot, you might get better performance by using the standard `parent_id` and `child_ids` relationships. This way, you can avoid table-level locks.

Adding constraints validations to a model

We want to make sure that our models do not have invalid or inconsistent data. Odoo has two kinds of constraints to do this:

- `Database-level constraints`: These are the constraints that PostgreSQL supports. The most common ones are the `UNIQUE` constraints, which prevent duplicate values. We can also use `CHECK` and `EXCLUDE` constraints for other conditions. These constraints are fast and reliable, but they are limited by what PostgreSQL can do.
- `Server-level constraints`: These are the constraints that we write in Python code. We can use these constraints when the database-level ones are not enough for our needs. These constraints are more flexible and powerful, but they are slower and more complex.

Getting ready

We will continue using the `my_hostel` add-on module from the previous recipe. We will use the hostel room model and add some constraints to it. We will use the hostel room model from *Chapter 3, Creating Odoo Add-On Modules*, and add some constraints to it.

We will use a `UNIQUE` constraint to ensure that room numbers are not repeated. We will also add a Python model constraint to check that the rent amount is positive.

How to do it...

1. SQL constraints are defined through the `_sql_constraints` model attribute. This attribute is assigned a list of triples containing strings `(name, sql_definition, message)`, where `name` is a valid SQL constraint name, `sql_definition` is a `table_constraint` expression, and `message` is the error message. We can add the following code to the `hostel.room` model:

```
_sql_constraints = [
   ("room_no_unique", "unique(room_no)", "Room number must be
unique!")]
```

2. A Python constraint is a method that checks a condition on a set of records. We use the `constrains()` decorator to mark the method as a constraint and to indicate which fields are involved in the condition. The constraint is automatically checked when any of these fields are changed. The method should raise an exception if the condition is not met:

```
from odoo.exceptions import ValidationError
...
 @api.constrains("rent_amount")
     def _check_rent_amount(self):
         """Constraint on negative rent amount"""
         if self.rent_amount < 0:
         raise ValidationError(_("Rent Amount Per Month should
not be a negative value!"))
```

   You need to upgrade the add-on module and restart the server after you make these changes to the code file.

> **Note**
>
> If you add SQL constraints to the existing model through model inheritance, make sure you don't have rows that violate the constraints. If you have such rows, then SQL constraints will not be added, and an error will be generated in the log.
>
> For more information on PostgreSQL constraints in general and table constraints in particular, take a look at `http://www.postgresql.org/docs/current/static/ddl-constraints.html`.

How it works...

We can use Python code to validate our models and prevent invalid data. To do this, we use two things:

A method that checks a condition on a set of records. We use the `constrains()` decorator to mark the method as a constraint and to indicate which fields are involved in the condition. The constraint is automatically checked when any of these fields are changed.

A `ValidationError` exception that we raise when the condition is not met. This exception shows an error message to the user and stops the operation.

Adding computed fields to a model

We may want to create a field that depends on the values of other fields in the same record or in related records. For instance, we can calculate the total amount by multiplying a unit price by a quantity. In Odoo models, we can use computed fields to do this.

To demonstrate how computed fields work, we will add one to the Hostel Room model that computes the availability of rooms based on student occupancy.

We can also make computed fields editable and searchable. We will show you how to do this in our example as well.

Getting ready

We will continue using the `my_hostel` add-on module from the previous recipe.

How to do it...

We will modify the `models/hostel_room.py` code file to include a new field and the methods that implement its logic:

1. A computed field's value usually relies on the values of other fields in the same record. ORM requires the developer to declare those dependencies on the `compute` method using the `depends()` decorator. ORM uses the given dependencies to recompute the field whenever any of its dependencies change. Start by adding the new fields to the `Hostel Rooms` model:

```
student_per_room = fields.Integer("Student Per Room",
required=True,help="Students allocated per room")'
availability = fields.Float(compute="_compute_check_
availability",string="Availability", help="Room availability in
hostel")
@api.depends("student_per_room", "student_ids")
def _compute_check_availability(self):
    """Method to check room availability"""
    for rec in self:

rec.availability = rec.student_per_room - len(rec.student_ids.
ids)
```

2. By default, computed fields are read-only because the user should not enter a value.

   However, in some cases, it might be helpful to allow the user to set a value directly. For example, in our hostel student scenario, we will add an admission date, discharge date, and duration. We would like the user to be able to enter either the duration or the discharge date and have the other value updated accordingly:

```
admission_date = fields.Date("Admission Date",
    help="Date of admission in hostel",
    default=fields.Datetime.today)
discharge_date = fields.Date("Discharge Date",
    help="Date on which student discharge")
duration = fields.Integer("Duration",  compute="_compute_check_
duration", inverse="_inverse_duration", help="Enter duration of
living")
@api.depends("admission_date", "discharge_date")
def _compute_check_duration(self):
    """Method to check duration"""
    for rec in self:
        if rec.discharge_date and rec.admission_date:
            rec.duration = (rec.discharge_date - rec.admission_
date).days
def _inverse_duration(self):
    for stu in self:
        if stu.discharge_date and stu.admission_date:
            duration = (stu.discharge_date - stu.admission_
date).days
            if duration != stu.duration:
                stu.discharge_date = (stu.admission_date +
timedelta(days=stu.duration)).strftime('%Y-%m-%d')
```

   A compute method assigns a value to the field, while an inverse method assigns values to the field's dependencies.

   Note that the inverse method is invoked when the record is saved, while the compute method is invoked whenever any of its dependencies change.

3. Computed fields are not stored in the database by default. One solution is to store the field with the `store=True` attribute:

```
availability = fields.Float(compute="_compute_check_
availability", store=True, string="Availability", help="Room
availability in hostel")
```

   As computed fields are not stored in the database by default, it is not possible to search on a computed field unless we use the `store=True` attribute or add a `search` method.

How it works...

A computed field looks like a regular field, except that it has a `compute` attribute that specifies the name of the method that computes its value.

However, computed fields are not the same as regular fields internally. Computed fields are calculated on the fly at runtime, and because of that, they are not stored in the database, so you cannot search or write on them by default. You need to do some extra work to enable writing and search support for them. Let's see how to do it.

The computation method is calculated on the fly at runtime, but ORM uses caching to avoid recalculating it unnecessarily every time its value is accessed. So, it needs to know what other fields it relies on. It uses the `@depends` decorator to determine when its cached values should be invalidated and recalculated.

Make sure that the compute method always assigns a value to the computed field. Otherwise, an error will occur. This can happen when you have conditions in your code that sometimes fail to assign a value to the computed field. This can be hard to debug.

Write support can be added by implementing the `inverse` method. This uses the value assigned to the computed field to update the source fields. Of course, this only works for simple calculations. However, there are still cases where it can be helpful. In our example, we make it possible to set the discharge date by editing the duration days, since `Duration` is a computed field.

The `inverse` attribute is optional; if you don't want to make the computed field editable, you can skip it.

It is also possible to make a non-stored computed field searchable by setting the `search` attribute to the method name (similar to `compute` and `inverse`). Like `inverse`, `search` is also optional; if you don't want to make the computed field searchable, you can skip it.

However, this method is not supposed to perform the actual search. Instead, it receives the operator and value used to search on the field as parameters and is supposed to return a domain, with the alternative search conditions to use.

The optional `store=True` flag stores the field in the database. In this case, after being computed, the field values are stored in the database, and from then on, they are retrieved in the same way as regular fields, instead of being recomputed at runtime. Thanks to the `@api.depends` decorator, ORM will know when these stored values need to be recomputed and updated. You can think of it as a persistent cache. It also has the benefit of making the field usable for search conditions, including sorting and grouping by operations. If you use `store=True` in your compute field, you no longer need to implement the `search` method because the field is stored in a database, and you can search/sort based on it.

The `compute_sudo=True` flag is for cases where the computations need to be done with higher privileges. This might be needed when the computation needs to use data that may not be accessible to the end user.

> **Note**
>
> The default value of `compute_sudo` changed in Odoo v13. Before Odoo v13, the value of `compute_sudo` was `False`, but in v13, the default value of `compute_sudo` depends on the store attribute. If the `store` attribute is `True`, then `compute_sudo` is `True`; otherwise, it is `False`. However, you can always override it by explicitly setting `compute_sudo` in your field definition.

There's more...

Odoo v13 introduced a new caching mechanism for ORM. Previously, the cache was based on the environment, but now, in Odoo v13, there is one global cache. So, if you have a computed field that relies on context values, then you may get the wrong values sometimes. To solve this problem, you need to use the `@api.depends_context` decorator. Refer to the following example:

```
@api.depends('price')
@api.depends_context('company_id')
def _compute_value(self):
    company_id = self.env.context.get('company_id')
   ...
   # other computation
```

You can see in the preceding example that our computation uses `company_id` from the context. By using `company_id` in the `depends_context` decorator, we ensure that the field value will be recomputed based on the value of `company_id` in the context.

Exposing related fields stored in other models

Odoo clients can only read data from the server for the fields that belong to the model they are querying. They cannot access data from related tables using dot notation as server-side code can.

However, we can make the data from related tables available to the clients by adding it as related fields. This is what we will do to get the hostel of the room in the student model.

Getting ready

We will continue using the `my_hostel` add-on module from the previous recipe.

How to do it...

Edit the `models/hostel_student.py` file to add the new `related` field.

Ensure that we have a field for the hostel room, and then, we add a new relation field to link the student with their hostel:

```
class HostelStudent(models.Model):
    _name = "hostel.student"
    # ...
    hostel_id = fields.Many2one("hostel.hostel", related='room_
id.hostel_id')
```

Finally, we need to upgrade the add-on module for the new field to be available in the model.

How it works...

A related field is a special type of field that references another field from a different record. To create a related field, we need to specify the `related` attribute and give it a string that shows the path of fields to follow. For example, we can create a related field that shows the hostel of the room of a student by following the `room_id.hostel_id` path.

There's more...

Related fields are, in fact, computed fields. They just provide a convenient shortcut syntax to read field values from related models. As they are computed fields, this means that the `store` attribute is also available. As a shortcut, they also have all the attributes from the referenced field, such as `name` and `translatable`, as required.

Additionally, they support a `related_sudo` flag, similar to `compute_sudo`; when set to `True`, the field chain is traversed without checking the user access rights.

Using related fields in a `create()` method can affect performance, as the computation of these fields is delayed until the end of their creation. So, if you have an `o2m` relationship, such as in the `sale.order` and `sale.order.line` models, and you have a related field on the line model referring to a field on the order model, you should explicitly read the field on the order model during record creation, instead of using the related field shortcut, especially if there are a lot of lines.

Adding dynamic relations using reference fields

With relational fields, we need to decide the relation's target model (or co-model) beforehand. However, sometimes, we may need to leave that decision to the user and first choose the model we want and then the record we want to link it to.

With Odoo, this can be achieved using reference fields.

Getting ready

We will continue using the my_hostel add-on module from the previous recipe.

How to do it...

Edit the models/hostel.py file to add the new related field:

1. First, we need to add a helper method to dynamically build a list of selectable target models:

```
from odoo import models, fields, api
class Hostel(models.Model):

_name = 'hostel.hostel'
    # ...
    @api.model
    def _referencable_models(self):
        models = self.env['ir.model'].search([
            ('field_id.name', '=', 'message_ids')])
        return [(x.model, x.name) for x in models]
```

2. Then, we need to add the reference field and use the previous function to provide a list of selectable models:

```
ref_doc_id = fields.Reference(
    selection='_referencable_models',
    string='Reference Document')
```

Since we are changing the model's structure, a module upgrade is needed to activate these changes.

How it works...

Reference fields are similar to m2o fields, except that they allow the user to select the model to link to.

The target model is selectable from a list that's provided by the selection attribute. The selection attribute must be a list of two-element tuples, where the first is the model's internal identifier and the second is a text description for it.

Here's an example:

```
[('res.users', 'User'), ('res.partner', 'Partner')]
```

However, rather than providing a fixed list, we can use the most common models. For simplicity, we used all the models that have the messaging feature. Using the _referencable_models method, we provided a model list dynamically.

Our recipe started by providing a function to browse all the model records that can be referenced, to dynamically build a list that will be provided to the `selection` attribute. Although both forms are allowed, we declared the function name inside quotes, instead of directly referencing the function without quotes. This is more flexible, and it allows for the referenced function to be defined only later in code, for example, which is something that is not possible when using a direct reference.

The function needs the `@api.model` decorator because it operates on the model level, not on the record set level.

While this feature looks nice, it comes with a significant execution overhead. Displaying the reference fields for a large number of records (for instance, in a list view) can create heavy database loads, as each value has to be looked up in a separate query. It is also unable to take advantage of database referential integrity, unlike regular relation fields.

Adding features to a model using inheritance

Odoo boasts a robust feature that significantly enhances its flexibility and functionality, which is particularly beneficial for businesses seeking tailored solutions. This feature enables the integration of module add-ons, allowing them to augment the capabilities of existing modules without the need to alter their underlying codebase. This is achieved through the addition or modification of fields and methods, as well as the extension of current methods with supplementary logic. This modular approach not only facilitates a customizable and scalable system but also ensures that upgrades and maintenance remain streamlined, preventing the complexities typically associated with custom modifications.

The official documentation describes three kinds of inheritance in Odoo:

- Class inheritance (extension)
- Prototype inheritance
- Delegation inheritance

We will see each one of these in a separate recipe. In this recipe, we will see class inheritance (extension). This is used to add new fields or methods to existing models.

We'll expand the existing partner model, `res.partner`, to include it in a computed field that calculates how many hostel rooms are assigned to each user. This will help determine which section each room is assigned to and which user occupies it.

Getting ready

We will continue using the `my_hostel` add-on module from the previous recipe.

How to do it...

We will extend the built-in partner model. If you remember, we already inherited the `res.parnter` model in the *Adding relational fields to a model* recipe in this chapter. To keep the explanation as simple as possible, we will reuse the `res.partner` model in the `models/hostel_book.py` code file:

1. First, we will ensure that the `authored_book_ids` inverse relation is in the partner model and add the computed field:

```
class ResPartner(models.Model):
   _inherit = "res.partner"
   is_hostel_rector = fields.Boolean("Hostel Rector",
help="Activate if the following person is hostel rector")
   assign_room_ids = fields.Many2many('library.
book',string='Authored Books')
   count_assign_room = fields.Integer( 'Number of Authored
Books', compute="_compute_count_room")
```

2. Next, add the method that's needed to compute the book count:

```
@api.depends('assign_room_ids')
   def _compute_count_room(self):
       for partner in self:
           partner.count_assign_room = len(partner.assign_room_
ids)
```

3. Finally, we need to upgrade the add-on module for the modifications to take effect.

How it works...

When a model class is defined with the `_inherit` attribute, it adds modifications to the inherited model, rather than replacing it.

This means that fields defined in the inheriting class are added or changed on the parent model. At the database layer, ORM adds fields to the same database table.

Fields are also incrementally modified. This means that if the field already exists in the superclass, only the attributes declared in the inherited class are modified; the other ones are kept as they are in the parent class.

Methods defined in the inheriting class replace methods in the parent class. If you don't invoke the parent method with the `super` call, the parent's version of the method will not be executed, and we will lose the features. So, whenever you add a new logic by inheriting existing methods, you should include a statement with `super` to call its version in the parent class. This is discussed in more detail in *Chapter 5, Basic Server-Side Development*.

This recipe will add new fields to the existing model. If you also want to add these new fields to existing views (the user interface), refer to the *Changing existing views – view inheritance* recipe in *Chapter 9, Backend Views*.

Copying the model definition using inheritance

We saw class inheritance (extension) in the previous recipe. Now, we will see **prototype inheritance**, which is used to copy the entire definition of the existing model. In this recipe, we will make a copy of the `hostel.room` model.

Getting ready

We will continue using the `my_hostel` add-on module from the previous recipe.

How to do it...

Prototype inheritance is executed by using the `_name` and `_inherit` class attributes at the same time. Perform the following steps to generate a copy of the `hotel.room` model:

1. Add a new file called `hostel_room_copy.py` to the `/my_hostel/models/` directory.
2. Add the following content to the `hostel_room_copy.py` file:

```
from odoo import fields, models, api, _
class HostelRoomCopy(models.Model):
   _name = "hostel.room.copy"
   _inherit="hostel.room"
   _description = "Hostel Room Information Copy"
```

3. Import a new file reference into the `/my_library/models/__init__.py` file. Following the changes, your `__init__.py` file will look like this:

```
from . import hostel_room
from . import hostel_room_copy
```

4. Finally, we need to upgrade the add-on module for the modifications to take effect.
5. To check the new model's definition, go to the **Settings** | **Technical** | **Database Structure** | **Models** menu. You will see a new entry for the `hostel.room.copy` model here.

> Tip
>
> In order to see menus and views for the new model, you need to add the XML definition of views and menus. To learn more about views and menus, refer to the *Adding menu items and views recipe* in *Chapter 3, Creating Odoo Add-On Modules*.

How it works...

By using `_name` with the `_inherit` class attribute at the same time, you can copy the definition of the model. When you use both attributes in the model, Odoo will copy the model definition of `_inherit` and create a new model with the `_name` attribute.

In our example, Odoo will copy the definition of the `Hostel.room` model and create a new model, `hostel.room.copy`. The new `hostel.room.copy` model has its own database table with its own data that is totally independent from the `hostel.room` parent model. Since it still inherits from the partner model, any subsequent modifications to it will also affect the new model.

Prototype inheritance copies all the properties of the parent class. It copies fields, attributes, and methods. If you want to modify them in the child class, you can simply do so by adding a new definition to the child class. For example, the `hostel.room` model has the `_name_get` method. If you want to use a different version of `_name_get` in the child, you need to redefine the method in the `hostel.room.copy` model.

> **Note**
>
> Prototype inheritance does not work if you use the same model name in the `_inherit` and `_name` attributes. If you do use the same model name in the `_inherit` and `_name` attributes, it will just behave like a normal extension inheritance.

There's more...

In the official documentation, this is called prototype inheritance, but in practice, it is rarely used. The reason for this is that delegation inheritance usually answers to that need in a more efficient way, without the need to **duplicate data structures**. For more information on this, you can refer to the next recipe, Using delegation inheritance to copy features to another model.

Using delegation inheritance to copy features to another model

The third type of inheritance is Delegation inheritance. Instead of `_inherit`, it uses the `_inherits` class attribute. There are cases where, rather than modifying an existing model, we want to create a new model based on an existing one to use the features it already has. We can copy a model's definitions with prototype inheritance, but this will generate duplicate data structures. If you want to copy a model's definitions without duplicating data structures, then the answer lies in Odoo's delegation inheritance, which uses the `_inherits` model attribute (note the additional `s`).

Traditional inheritance is quite different from the similarly named concept in object-oriented programming. Delegation inheritance, in turn, is similar, in that a new model can be created to include the features from a parent model. It also supports polymorphic inheritance, where we inherit from two or more other models.

We operate a hostel that accommodates both rooms and students. To better manage our accommodations, it's essential to integrate student-related information into our system. Specifically, for each student, we require comprehensive identification and address details, similar to those captured in the partner model. Additionally, it's crucial to maintain records related to room allocation, including the start and end dates of each student's stay and their card number.

Directly adding these fields to the existing partner model isn't an ideal approach, as it would unnecessarily clutter the model with student-specific data that is irrelevant for non-student partners. A more effective solution would be to enhance the partner model by creating a new model that inherits from it and introduces the additional fields required to manage student information. This approach ensures a cleaner, more organized, and functionally efficient system to cater to our hostel's unique needs.

Getting ready

We will continue using the `my_hostel` add-on module from the previous recipe.

How to do it...

The new library member model should be in its own Python code file, but to keep the explanation as simple as possible, we will reuse the `models/hostel_student.py` file:

1. Add the new model, inheriting from `res.partner`:

```
class HostelStudent(models.Model):
    _name = "hostel.student"
    _inherits = {'res.partner': 'partner_id'}
    _description = "Hostel Student Information"
    .........
    partner_id = fields.Many2one('res.partner',
ondelete='cascade')
```

2. Next, we will add the fields that are specific to each student:

```
gender = fields.Selection([("male", "Male"),
        ("female", "Female"), ("other", "Other")],
        string="Gender", help="Student gender")
    room_id = fields.Many2one("hostel.room", "Room",
        help="Select hostel room")
```

Now, we should upgrade the add-on module to activate the changes.

How it works...

The `_inherits` model attribute sets the parent models that we want to inherit from. In this case, we just have one – `res.partner`. Its value is a key-value dictionary, where the keys are the inherited models and the values are the field names that were used to link to them. These are `m2o` fields that we must also define in the model. In our example, `partner_id` is the field that will be used to link with the `Partner` parent model.

To better understand how this works, let's look at what happens at a database level when we create a new member:

- A new record is created in the `res_partner` table.
- A new record is created in the `hostel_student` table.
- The `partner_id` field of the `hostel_student` table is set to the ID of the `res_partner` record that is created for it

The member record is automatically linked to a new partner record. It's just an `m2o` relationship, but the delegation mechanism adds some magic so that the partner's fields are seen as if they belong to the member record, and a new partner record is also automatically created with the new member.

You may be interested to know that this automatically created partner record has nothing special about it. It's a regular partner, and if you browse the partner model, you will be able to find that record (without the additional member data, of course). All members are partners, but only some partners are also members. So, what happens if you delete a partner record that is also a member? You decide by choosing the `ondelete` value for the relation field. For `partner_id`, we used `cascade`. This means that deleting the partner will also delete the corresponding member. We could have used the more conservative setting, `restrict`, to prohibit deleting the partner while it has a linked member. In this case, only deleting the member will work.

It's important to note that delegation inheritance only works for fields, not for methods. So, if the partner model has a `do_something()` method, the members model will not automatically inherit it.

There's more...

There is a shortcut for this inheritance delegation. Instead of creating an `_inherits` dictionary, you can use the `delegate=True` attribute in the `m2o` field definition. This will work exactly like the `_inherits` option. The main advantage is that this is simpler. In the given example, we performed the same inheritance delegation as in the previous one, but in this case, instead of creating an `_inherits` dictionary, we used the `delegate=True` option in the `partner_id` field:

```
class HostelStudent(models.Model):
    _name = "hostel.student"
    _description = "Hostel Student Information"

    partner_id = fields.Many2one('res.partner', ondelete='cascade',
delegate=True)
```

A noteworthy case of delegation inheritance is the users model, `res.users`. It inherits from partners (`res.partner`). This means that some of the fields that you can see on the user are actually stored in the partner model (notably, the `name` field). When a new user is created, we also get a new, automatically created partner.

We should also mention that traditional inheritance with `_inherit` can also copy features into a new model, although in a less efficient way. This was discussed in the *Adding features to a model using inheritance* recipe.

Using abstract models for reusable model features

Sometimes, there is a particular feature that we want to be able to add to several different models. Repeating the same code in different files is a bad programming practice; it would be better to implement it once and reuse it.

Abstract models allow us to create a generic model that implements some features that can then be inherited by regular models, in order to make that feature available.

As an example, we will implement a simple archive feature. This adds the `active` field to the model (if it doesn't exist already) and makes an archive method available to toggle the `active` flag. This works because `active` is a magic field. If it is present in a model by default, the records with `active=False` will be filtered out from the queries. We will then add it to the `hostel room` model.

Getting ready

We will continue using the `my_hostel` add-on module from the previous recipe.

How to do it...

The archive feature certainly deserves its own add-on module, or at least its own Python code file. However, to keep the explanation as simple as possible, we will cram it into the `models/hostel_room.py` file:

1. Add the abstract model for the archive feature. It must be defined in the library book model, where it will be used:

```
class BaseArchive(models.AbstractModel):
    _name = 'base.archive'
    active = fields.Boolean(default=True)

    def do_archive(self):
        for record in self:
            record.active = not record.active
```

2. Now, we will edit the Hostel Room model to inherit the archive model:

```
class HostelRoom(models.Model):
    _name = "hostel.room"
    _inherit = ['base.archive']
```

An upgrade of the add-on module is required in order for the changes to be activated.

How it works...

An abstract model is created by a class based on `models.AbstractModel`, instead of the usual `models.Model`. It has all the attributes and capabilities of regular models; the difference is that ORM will not create an actual representation for it in the database. This means that it can't have any data stored in it. It only serves as a template for a reusable feature that will be added to regular models.

Our archive abstract model is quite simple. It just adds the `active` field and a method to toggle the value of the `active` flag, which we expect to be used later, via a button on the user interface. When a model class is defined with the `_inherit` attribute, it inherits the attribute methods of those classes, and the attribute methods that are defined in the current class add modifications to those inherited features.

The mechanism at play here is the same as that of a regular model extension (as per the *Adding features to a model using inheritance* recipe). You may have noticed that `_inherit` uses a list of model identifiers instead of a string with one model identifier. In fact, `_inherit` can have both forms. Using the list form allows us to inherit from multiple (usually `Abstract`) classes. In this case, we inherit just one, so a text string would be fine. A list was used instead, for illustration purposes.

There's more...

A noteworthy built-in abstract model is `mail.thread`, which is provided by the `mail (Discuss)` add-on module. On models, it enables the discussion features that power the message wall that's seen at the bottom of many forms.

Other than `AbstractModel`, a third model type is available – `models.TransientModel`. This has a database representation like `models.Model`, but the records that are created there are supposed to be temporary and regularly purged by a server-scheduled job. Other than that, transient models work just like regular models.

`models.TransientModel` is useful for more complex user interactions, known as **wizards**. A wizard is used to request inputs from the user. In *Chapter 8, Advanced Server-Side Development Techniques*, we will explore how to use these for advanced user interaction.

5
Basic Server-Side Development

We learned how to declare or extend business models in custom modules in *Chapter 4, Application Models*. Writing methods for calculated fields and ways to restrict the field values are both addressed in that chapter's tutorials. This chapter focuses on the fundamentals of server-side programming in Odoo method declarations, record set manipulation, and extending inherited methods. You may use this to create or alter business logins in the Odoo module.

In this chapter, we will cover the following tutorials:

- Specifying model methods and implementing API decorators
- Notifying errors to the user
- Getting a blank recordset for a different model
- Creating new records
- Updating values of recordset records
- Searching for records
- Combining recordsets
- Filtering recordsets
- Traversing recordset relations
- Sorting recordsets
- Extending a model's established business logic
- Extending `write()` and `create()`
- Customizing how records are searched
- Fetching data in groups using `read_group()`

Technical requirements

The online platform for Odoo is one of the prerequisites for this chapter.

You can obtain all the code used in this chapter from the following GitHub repository: `https://github.com/PacktPublishing/Odoo-17-Development-Cookbook-Fifth-Edition/tree/main/Chapter05`

Specifying model methods and using API decorators

A class in Odoo models consists of both business logic methods and field declarations. We learned how to add fields to a model in *Chapter 4, Application Models*. We will now see how to include business logic and methods in a model.

In this tutorial, we'll learn how to create a function that may be used by our application's user interface buttons or another piece of code. This method will operate on `HostelRoom` and take the necessary steps to modify the state of a number of rooms.

Getting ready

This tutorial assumes that you have an instance ready, with the `my_hostel` add-on module available, as described in *Chapter 3, Creating Odoo Add-On Modules*. You will need to add a `state` field to the `HostelRoom` model, which is defined as follows:

```
from odoo import api, fields, models
class HostelRoom(models.Model):
    # [...]
    state = fields.Selection([
        ('draft', 'Unavailable'),
        ('available', 'Available'),
        ('closed', 'Closed')],
        'State', default="draft")
```

Refer to the *Adding models* tutorial in *Chapter 3, Creating Odoo Add-On Modules*, for more information.

How to do it...

To define a method for hostel rooms to change the state of a selection of rooms, you need to add the following code to the model definition:

1. Add a helper method to check whether a state transition is allowed:

```
@api.model
def is_allowed_transition(self, old_state, new_state):
    allowed = [('draft', 'available'),
               ('available', 'closed'),
```

```
                ('closed', 'draft')]
        return (old_state, new_state) in allowed
```

2. Add a method to change the state of a room to a new state that is passed as an argument:

```
    def change_state(self, new_state):
        for room in self:
            if room.is_allowed_transition(room.state,\
            new_state):
                room.state = new_state
            else:
                continue
```

3. Add a method to change the room state by calling the `change_state` method:

```
    def make_available(self):
        self.change_state('available')
    def make_closed(self):
        self.change_state('closed')
```

4. Add a button and status bar in the `<form>` view. This will help us trigger these methods from the user interface:

```
<form>
...
    <button name="make_available" string="Make Available"
type="object"/>
    <button name="make_closed" string="Make Borrowed"
type="object"/>
    <field name="state" widget="statusbar"/>
...
</form>
```

To access these updates, you must update the module or install it.

How it works...

Several methods are defined in the tutorial's code. They are typical Python methods with `self` as their first argument and the option of receiving additional arguments. The `odoo.api` module's **decorators** are used to adorn some methods.

> TIP
>
> In Odoo 9.0, the API decorators were first added to support both the old and new frameworks. The previous API is no longer supported as of Odoo 10.0, however, some decorators, such @ `api.model`, are still in use.

When writing a new method, if you don't use a decorator, then the method is executed on a recordset. In such methods, `self` is a recordset that can refer to an arbitrary number of database records (this includes empty recordsets), and the code will often loop over the records in `self` to do something on each individual record.

The `@api.model` decorator is similar, but it's used on methods for which only the model is important, not the contents of the recordset, which is not acted upon by the method. The concept is similar to Python's `@classmethod` decorator.

In *Step 1*, we created the `is_allowed_transition()` method. The purpose of this method is to verify whether a transition from one state to another is valid. The tuples in the `allowed` list are the available transitions. For example, we don't want to allow a transition from `closed` to `available`, which is why we haven't put `('closed', 'available')`.

In *Step 2*, we created the `change_state()` method. The purpose of this method is to change the status of the room. When this method is called, it changes the status of the room to the state given by the `new_state` parameter. It only changes the room status if the transition is allowed. We used a `for` loop here because `self` can contain multiple recordsets.

In *Step 3*, we created the methods that change the state of the room by calling the `change_state()` method. In our case, this method will be triggered by the buttons that were added to the user interface.

In *Step 4*, we added `<button>` in the `<form>` view. Upon clicking this button, the Odoo web client will invoke the Python function mentioned in the `name` attribute. Refer to the *Adding buttons to forms* tutorial in *Chapter 9*, *Backend Views*, to learn how to call such a method from the user interface. We have also added the `state` field with the `statusbar` widget to display the status of the room in the `<form>` view.

When the user clicks on the button from the user interface, one of the methods from *Step 3* will be called. Here, `self` will be the recordset that contains the record of the `hostel.room` model. After that, we call the `change_state()` method and pass the appropriate parameter based on the button that was clicked.

When `change_state()` is called, `self` is the same recordset of the `hostel.room` model. The body of the `change_state()` method loops over `self` to process each room in the recordset. Looping on `self` looks strange at first, but you will get used to this pattern very quickly.

Inside the loop, `change_state()` calls `is_allowed_transition()`. The call is made using the `room` local variable, but it can be made on any recordset for the `hostel.room` model, including, for example, `self`, since `is_allowed_transition()` is decorated with `@api.model`. If the transition is allowed, `change_state()` assigns the new state to the room by assigning a value to the attribute of the recordset. This is only valid on recordsets with a length of `1`, which is guaranteed to be the case when iterating over `self`.

Reporting errors to the user

Sometimes, it's required to stop processing during method execution because the user's activity is invalid or an error condition has been satisfied. By displaying an informative error message, this tutorial demonstrates how to handle these situations.

The `UserError` exception is commonly utilized to inform users about errors or exceptional situations. It is typically employed when the user's input fails to meet the expected criteria or when a particular operation cannot be executed due to specific conditions.

Getting ready

This tutorial requires that you set up an instance with the `my_hostel` add-on module installed, as per the instructions from before.

How to do it...

We will make a change to the `change_state` method from the previous tutorial and display a helpful message when the user is trying to change the state that is not allowed by the `is_allowed_transition` method. Perform the following steps to get started:

1. Add the following import at the beginning of the Python file:

```
from odoo.exceptions import UserError
from odoo.tools.translate import _
```

2. Modify the `change_state` method and raise a `UserError` exception from the `else` part:

```
def change_state(self, new_state):
    for room in self:
        if room.is_allowed_transition(room.state, new_state):
            room.state = new_state
        else:
            msg = _('Moving from %s to %s is not
allowed') % (room.state, new_state)
            raise UserError(msg)
```

How it works...

When an exception is raised in Python, it propagates up the call stack until it is processed. In Odoo, the **remote procedure call** (**RPC**) layer that answers the calls made by the web client catches all exceptions and, depending on the exception class, triggers different possible behaviors on the web client.

Any exception not defined in `odoo.exceptions` will be handled as an internal server error (**HTTP status 500**) with the stack trace. `UserError` will display an error message in the user interface. The code of the tutorial raises `UserError` to ensure that the message is displayed in a user-friendly way. In all cases, the current database transaction is rolled back.

We are using a function with a strange name, `_()`, which is defined in `odoo.tools.translate`. This function is used to mark a string as translatable and to retrieve the translated string at runtime, given the language of the end user that's found in the execution context. More information on this is available in *Chapter 11, Internationalisation*.

> **Important note**
>
> When using the `_()` function, ensure that you pass only strings with the interpolation placeholder, not the whole interpolated string. For example, `_('Warning: could not find %s') % value` is correct, but `_('Warning: could not find %s' % value)` is incorrect because the first one will not find the string with the substituted value in the translation database.

There's more...

Sometimes, you are working on error-prone code, meaning that the operation you are performing may generate an error. Odoo will catch this error and display a traceback to the user. If you don't want to show a full error log to the user, you can catch the error and raise a custom exception with a meaningful message. In the example provided, we are generating `UserError` from the `try...catch` block so that instead of showing a full error log, Odoo will now show a warning with a meaningful message:

```
def post_to_webservice(self, data):
    try:
        req = requests.post('http://my-test-service.com', data=data,
timeout=10)
        content = req.json()
    except IOError:
        error_msg = _("Something went wrong during data submission")
        raise UserError(error_msg)
    return content
```

There are a few more exception classes defined in `odoo.exceptions`, all deriving from the base legacy `except_orm` exception class. Most of them are only used internally, apart from the following:

- `ValidationError`: This exception is raised when a Python constraint on a field is not respected. In *Chapter 4, Application Models*, refer to the *Adding constraint validations to a model* tutorial for more information.
- `AccessError`: This error is usually generated automatically when the user tries to access something that is not allowed. You can raise the error manually if you want to show the access error from your code.
- `RedirectWarning`: With this error, you can show a redirection button with the error message. You need to pass two parameters to this exception: the first parameter is the action ID, and the second parameter is the error message.

- `Warning`: In Odoo 8.0, `odoo.exceptions.Warning` played the same role as `UserError` in 9.0 and later. It is now deprecated because the name was deceptive (it is an error, not a warning) and it collided with the Python built-in `Warning` class. It is kept for backward compatibility only, and you should use `UserError` in your code.

Obtaining an empty recordset for a different model

The current model's methods are accessible through `self` while creating Odoo code. It is not feasible to start working on a different model by simply instantiating its class; you must first obtain a recordset for that model.

This tutorial shows you how to get an empty recordset for any model that's registered in Odoo inside a model method.

Getting ready

This tutorial will reuse the setup of the library example in the `my_hostel` add-on module.

We will write a small method in the `hostel.room` model and search for all `hostel.room.members`. To do this, we need to get an empty recordset for `hostel.room.members`. Make sure you have added the `hostel.room.members` model and access rights for that model.

How to do it...

To get a recordset for `hostel.room.members` in a method of `hostel.room`, you need to perform the following steps:

```
def log_all_room_members(self):
    hostel_room_obj = self.env['hostel.room.member']
    all_members = hostel_room_obj.search([])
    print("ALL MEMBERS:", all_members)
    return True
```

Figure 5.1 – log_all_room_members

1. In the `HostelRoom` class, write a method called `log_all_room_members`:

```
class HostelRoom(models.Model):
    # ...
    def log_all_room_members(self):
        # This is an empty recordset of model hostel.room.member
        hostel_room_obj = self.env['hostel.room.member']

        all_members = hostel_room_obj.search([])
        print("ALL MEMBERS:", all_members)
        return True
```

2. Add a button to the <form> view to invoke our method:

```
<button name="log_all_room_members"  string="Log Members"
type="object"/>
```

Update the module to apply the changes. After that, you will see the **Log Members** button in the room's <form> view. You may view the member's recordset in the server log by clicking that button.

How it works...

At startup, Odoo loads all the modules and combines the various classes that derive from `Model`, and also defines or extends the given model. These classes are stored in the **Odoo registry**, indexed by name. The `env` attribute of any recordset, available as `self.env`, is an instance of the `Environment` class defined in the `odoo.api` module.

The `Environment` class plays a central role in Odoo development:

- It provides shortcut access to the registry by emulating a Python dictionary. If you know the name of the model you're looking for, `self.env[model_name]` will get you an empty recordset for that model. Moreover, the recordset will share the environment of `self`.
- It has a `cr` attribute, which is a database cursor you may use to pass raw SQL queries. Refer to the *Executing raw SQL queries* tutorial in *Chapter 8, Advanced Server-Side Development Techniques*, for more information on this.
- It has a `user` attribute, which is a reference to the current user performing the call. Take a look at *Chapter 8, Advanced Server-Side Development Techniques*, and the *Changing the user performing an action* tutorial for more on this.
- It has a `context` attribute, which is a dictionary that contains the context of the call. This includes information about the language of the user, the time zone, and the current selection of records. Refer to the *Calling a method with a modified context* tutorial in *Chapter 8, Advanced Server-Side Development Techniques*, for more on this.

The call to `search()` is explained in the *Searching for records* tutorial later.

See also

Sometimes, you want to use a modified version of the environment. One such example is that you want an environment with a different user and language. In *Chapter 8, Advanced Server-Side Development Techniques*, you will learn how to modify the environment at runtime.

Creating new records

Creating new records is a regular requirement when putting business logic processes into practice. How you can build records for the `hostel.room.category` model is included in this tutorial. We'll add a function to the `hostel.room.category` model to generate dummy categories for the purposes of our example. We will add the <form> view to activate this approach.

Getting ready

You need to understand the structure of the models for which you want to create a record, especially their names and types, as well as any constraints that exist on these fields (for example, whether some of them are mandatory).

For this tutorial, we will reuse the `my_hostel` module from *Chapter 4, Application Models*. Take a look at the following example to quickly recall the `hostel.room.category` model:

```
class RoomCategory(models.Model):
    _name = 'hostel.room.category'
    _description = 'Hostel Room Category'
    name = fields.Char('Category')
    description = fields.Text('Description')
    parent_id = fields.Many2one(
        'hostel.room.category',
        string='Parent Category',
        ondelete='restrict',
        index=True
    )
    child_ids = fields.One2many(
        'hostel.room.category', 'parent_id',
        string='Child Categories')
```

Make sure you have added menus, views, and access rights for the `hostel.room.category` model.

How to do it...

To create a category with some child categories, you need to perform the following steps:

```
def create_categories(self):
    categ1 = {
        'name': 'Child category 1',
        'description': 'Description for child 1'
    }
    categ2 = {
        'name': 'Child category 2',
        'description': 'Description for child 2'
    }
    parent_category_val = {
        'name': 'Parent category',
        'description': 'Description for parent category',
        'child_ids': [
            (0, 0, categ1),
            (0, 0, categ2),
        ]
    }
    self.env['hostel.room.category'].create(parent_category_val)
    return True
```

Figure 5.2 – Create a category

1. Create a method in the `hostel.room.category` model with the name `create_categories`:

```
def create_categories(self):
    ......
```

2. Inside the body of this method, prepare a dictionary of values for the fields of the first child category:

```
categ1 = {
    'name': 'Child category 1',
    'description': 'Description for child 1'
}
```

3. Prepare a dictionary of values for the fields of the second category:

```
categ2 = {
    'name': 'Child category 2',
    'description': 'Description for child 2'
}
```

4. Prepare a dictionary of values for the fields of the parent category:

```
parent_category_val = {
    'name': 'Parent category',
    'description': 'Description for parent category',
    'child_ids': [
        (0, 0, categ1),
        (0, 0, categ2),
    ]
}
```

5. Call the `create()` method to create the new records:

```
record = self.env['hostel.room.category'].create(parent_
category_val)
```

6. Add a button in the `<form>` view to trigger the `create_categories` method from the user interface:

```
<button name="create_categories" string="Create Categories"
type="object"/>
```

How it works...

To add a new record for a model, we can call the `create(values)` method on any recordset related to the model. This method returns a new recordset with a length of `1` and contains the new record, with the field values specified in the `values` dictionary.

The keys in the dictionary identify the fields by name, while the accompanying values reflect the field's value. Depending on the field type, you need to pass different Python types for the values:

- A `Text` field value is given with Python strings.
- The `Float` and `Integer` field values are given using Python floats or integers.
- A `boolean` field value is given preferably using Python Booleans or integers.
- A `Date` field value is given with the Python `datetime.date` object.
- A `Datetime` field value is given with the Python `datetime.datetime` object.
- A `Binary` field value is passed as a Base64-encoded string. The `base64` module from the Python standard library provides methods such as `encodebytes(bytestring)` to encode a string in Base64.
- A `Many2one` field value is given with an integer, which has to be the database ID of the related record.
- `One2many` and `Many2many` fields use a special syntax. The value is a list that contains tuples of three elements, as follows:

Tuple	Effect
`(0, 0, dict_val)`	Creates a new record that will be related to the main record.
`(6, 0, id_list)`	Creates a relation between the record being created and existing records, whose IDs are in the Python list called `id_list`. Caution: When used on a `One2many` field, this will remove the records from any previous relation.

Table 5.1 – Relational field write

In this tutorial, we create the dictionaries for two categories in the hostel room we want to create, and then we use these dictionaries in the `child_ids` entry of the dictionary for the hostel room categories being created by using the `(0, 0, dict_val)` syntax we explained earlier.

When `create()` is called in *Step 5*, three records are created:

- One for the parent room category, which is returned by `create`
- Two records for the child room category, which are available in `record.child_ids`

There's more...

If the model defined some default values for some fields, nothing special needs to be done. `create()` will take care of computing the default values for the fields that aren't present in the supplied dictionary.

The `create()` method also supports the creation of records in a batch. To create multiple records in a batch, you need to pass a list of multiple values to the `create()` method, as shown in the following example:

```
categ1 = {
    'name': 'Category 1',
    'description': 'Description for Category 1'
}
categ2 = {
    'name': 'Category 2',
    'description': 'Description for Category 2'
}
multiple_records = self.env['hostel.room.category'].create([categ1,
categ2])
```

This code will return the recordset of created categories of the hostel room category.

Updating values of recordset records

Business logic often requires us to update records by changing the values of some of their fields. This tutorial shows you how to modify the `room_no` field of the partner as we go.

Getting ready

This tutorial will use the same simplified `hostel.room` definition of the *Creating new records* tutorial. You may refer to this simplified definition to find out about the fields.

We have the `room_no` field in the `hostel.room` model. For illustration purposes, we will write in this field with the click of a button.

How to do it...

1. To update a room's `room_no` field, you can write a new method called `update_room_no()`, which is defined as follows:

```
def update_room_no(self):
    self.ensure_one()
    self.room_no = "RM002"
```

2. Then, you can add a button to the room's `<form>` view in `xml`, as follows:

```
<button name="update_room_no" string="Update Room No"
type="object"/>
```

3. Restart the server and update the `my_hostel` module to see the changes. Upon clicking the **Update Room No** button, `room_no` will be changed.

How it works...

The method starts by checking whether the room recordset that's passed as `self` contains exactly one record by calling `ensure_one()`. If this is not the case, this procedure will generate an exception, and processing will stop. This is necessary because we don't want to change the room number of multiple records. If you want to update multiple values, you can remove `ensure_one()` and update the attribute using a loop on the recordset.

Finally, the method modifies the values of the attributes of the room record. It updates the `room_no` field with the defined room number. Just by modifying the field attributes of the recordset, you can perform write operations.

There's more...

There are three options available if you want to add new values to the fields of records:

- Option one is the one that was explained in this tutorial. It works in all contexts by assigning values directly to the attribute representing the field of the record. It isn't possible to assign a value to all recordset elements in one go, so, you need to iterate on the recordset, unless you are certain that you are only handling a single record.
- Option two is to use the `update()` method by passing dictionary mapping field names to the values you want to set. This also only works for recordsets with a length of `1`. It can save some typing when you need to update the values of several fields at once on the same record. Here's *Step 2* of the tutorial, rewritten to use this option:

```
def change_room_no(self):
    self.ensure_one()
    self.update({
        'room_no': "RM002",
        'another_field': 'value'
        ...
    })
```

- Option three is to call the `write()` method, passing a dictionary that maps the field names to the values you want to set. This method works for recordsets of arbitrary size and will update all records with the specified values in one single database operation when the two previous options perform one database call per record and per field. However, it has some limitations: it does not work if the records are not yet present in the database (refer to the *Writing on change methods* tutorial in *Chapter 8, Advanced Server-Side Development Techniques*, for more information on this). Also, it requires a special format when writing relational fields, similar to the one used by the `create()` method. Check the following table for the format that's used to generate different values for the relational fields:

Tuple	Effect
`(0, 0, dict_val)`	This creates a new record that will be related to the main record.
`(1, id, dict_val)`	This updates the related record with the specified ID with the values supplied.
`(2, id)`	This removes the record with the specified ID from the related records and deletes it from the database.
`(3, id)`	This removes the record with the specified ID from the related records. The record is not deleted from the database.
`(4, id)`	This adds an existing record with the supplied ID to the list of related records.
`(5, )`	This removes all the related records, equivalent to calling `(3, id)` for each related `id`.
`(6, 0, id_list)`	This creates a relation between the record being updated and the existing record, whose IDs are in the Python list called `id_list`.

Table 5.2 – Relational field update

Important note

The `1`, `2`, `3`, and `5` operation types cannot be used with the `create()` method.

Searching for records

Searching for records is also a common operation in business logic methods. There are many cases where we need to search the data based on different criteria. Finding the room by name and category is demonstrated in this tutorial.

Getting ready

This tutorial will use the same `hostel.room` definition as the *Creating new records* tutorial did previously. We will write the code in a method called `find_room(self)`.

How to do it...

To find the rooms, you need to perform the following steps:

1. Add the `find_room` method to the `hostel.room` model:

```
def find_room(self):
    ...
```

2. Write the search domain for your criteria:

```
domain = [
    '|',
            '&', ('name', 'ilike', 'Room Name'),
                 ('category_id.name', 'ilike', 'Category Name'),
            '&', ('name', 'ilike', 'Second Room Name 2'),
                 ('category_id.name', 'ilike', 'SecondCategory Name
2')
]
```

3. Call the `search()` method with the domain, which will return the recordset:

```
rooms = self.search(domain)
```

The `rooms` variable will have a recordset of searched rooms. You can print or log that variable to see the result in the server log.

How it works...

Step 1 defines the method name prefixed with the `def` keyword.

Step 2 creates a search domain in a local variable. Often, you'll see this creation inline in the call to search, but with complex domains, it is good practice to define it separately.

For a full explanation of the search domain syntax, refer to the *Defining filters on record lists – domain* tutorial in *Chapter 9, Backend Views.*

Step 3 calls the `search()` method with the domain. The method returns a recordset that contains all the records that match the domain, which can then be processed further. In this tutorial, we call the method with just the domain, but the following keyword arguments are also supported:

- `offset=N`: This is used to skip the first `N` records that match the query. This can be used along with `limit` to implement pagination or to reduce memory consumption when processing a very large number of records. It defaults to `0`.
- `limit=N`: This indicates that, at most, `N` records should be returned. By default, there is no limit.
- `order=sort_specification`: This is used to force the order in the recordset returned. By default, the order is given by the `_order` attribute of the model class.
- `count=boolean`: If `True`, this returns the number of records instead of the recordset. It defaults to `False`.

> **Important note**
>
> We recommend using the `search_count(domain)` method rather than `search(domain, count=True)`, as the name of the method conveys the behavior in a much clearer way. Both will give the same result.

Sometimes, you need to search from another model so that searching for `self` will return a recordset of the current model. To search from another model, we need to get an empty recordset for the model. For example, let's say we want to search for some contacts. To do that, we will need to use the `search()` method on the `res.partner` model. Refer to the following code. Here, we get the empty recordset of `res.partner` to search the contacts:

```
def find_partner(self):
    PartnerObj = self.env['res.partner']
    domain = [
        '&', ('name', 'ilike', 'SerpentCS'),
             ('company_id.name', '=', 'SCS')
    ]
    partner = PartnerObj.search(domain)
```

In the preceding code, we have two conditions in the domain. You can omit the `'&'` from the domain when you have two conditions to compare, because when you do not specify the domain; then, Odoo will take `'&'` as a default.

There's more...

We said previously that the `search()` method returned all the records matching the domain. This is not actually completely true. The security rules ensure that the user only gets those records to which they have `read` access rights. Additionally, if the model has a `boolean` field called `active` and no term of the search domain specifies a condition on that field, then an implicit condition is added by search to only return `active=True` records. So, if you expect a search to return something, but you only get empty recordsets, ensure that you check the value of the `active` field (if present) to check for record rules.

Refer to the *Calling a method with a different context* tutorial in *Chapter 8, Advanced Server-Side Development Techniques*, for a way to not have the implicit `active=True` condition added. Take a look at the *Limiting record access using record rules* tutorial in *Chapter 10, Security Access*, for more information about record-level access rules.

If, for some reason, you find yourself writing raw SQL queries to find record IDs, ensure that you use `self.env['record.model'].search([('id', 'in', tuple(ids))]).ids` after retrieving the IDs to ensure that security rules are applied. This is especially important in **multi-company** Odoo instances where the record rules are used to ensure proper discrimination between companies.

Combining recordsets

Sometimes, you will find that you have obtained recordsets that are not exactly what you need. This tutorial shows various ways of combining them.

Getting ready

To use this tutorial, you need to have two or more recordsets for the same model.

How to do it...

Follow these steps to perform common operations on recordsets:

1. To merge two recordsets into one while preserving their order, use the following operation:

```
result = recordset1 + recordset2
```

2. To merge two recordsets into one while ensuring that there are no duplicates in the result, use the following operation:

```
result = recordset1 | recordset2
```

3. To find the records that are common to two recordsets, use the following operation:

```
result = recordset1 & recordset2
```

How it works...

The class for recordsets implements various Python operator redefinitions, which are used here. Here's a summary table of the most useful Python operators that can be used on recordsets:

Operator	Action performed
`R1 + R2`	This returns a new recordset containing the records from `R1`, followed by the records from `R2`. This can generate duplicate records in the recordset.
`R1 - R2`	This returns a new recordset consisting of the records from `R1` that are not in `R2`. The order is preserved.
`R1 & R2`	This returns a new recordset with all the records that belong to both `R1` and `R2` (intersection of recordsets). The order is *not* preserved here, but there are no duplicates.
`R1 | R2`	This returns a new recordset with the records belonging to either `R1` or `R2` (union of recordsets). The order is *not* preserved, but there are no duplicates.
`R1 == R2`	`True` if both recordsets contain the same records.
`R1 <= R2` `R1 < R2`	`True` if all records in `R1` are a subset of `R2`. Both syntaxes are equivalent.
`R1 >= R2`	`True` if all records in `R1` are a superset of `R2`. Both syntaxes are equivalent.
`R1 != R2`	`True` if `R1` and `R2` do not contain the same records.
`R1 in R2`	`True` if `R1` (must be one record) is part of `R2`.
`R1 not in R2`	`True` if `R1` (must be one record) is not part of `R2`.

Table 5.3 – Operators used with the domain

There are also in-place operators, +=, -=, &=, and |=, which modify the left-hand side operand instead of creating a new recordset. These are very useful when updating a record's `One2many` or `Many2many` fields. Refer to the *Updating values of recordset records* tutorial for an example of this.

Filtering recordsets

Sometimes, you already have a recordset, but you just need to work on a subset of those records. Of course, you may iterate over the recordset, checking for the condition each time and taking action in accordance with the outcome of the check. The construction of a new recordset comprising only the interesting records and the use of a single operation on that recordset can be simpler and, in certain situations, more efficient.

This tutorial shows you how to use the `filter()` method to extract a subset of recordsets based on a condition.

Getting ready

We will reuse the simplified `hostel.room` model that was shown in the *Creating new records* tutorial. This tutorial defines a method to extract rooms that have multiple members from a supplied recordset.

How to do it...

To extract records that have multiple members from a recordset, you need to perform the following steps:

1. Define the method to filter the recordset:

```
def filter_members(room):
 all_rooms = self.search([])
 filtered_rooms = self.rooms_with_multiple_members(all_
rooms)
```

2. Define the method to accept the original recordset:

```
@api.model
def room_with_multiple_members(self, all_rooms):
```

3. Define an inner `predicate` function:

```
def predicate(room):
    if len(room.member_ids) > 1:
        return True
    return False
```

4. Call `filter()`, as follows:

```
return all_room.filter (predicate)
```

The outcome of this procedure can be printed or logged so that a server log can include it. For further information, see the tutorial's sample code.

How it works...

The recordset that is created by the `filter()` method's implementation is empty. This blank recordset receives all the records that the predicate function evaluates to `True`. Finally, the fresh recordset is given back. Records in the original recordset are still in the same sequence.

A named internal function was used in the last tutorial. You will frequently see an anonymous Lambda function being utilized for such straightforward predicates:

```
@api.model
def room_with_multiple_rooms(self, all_rooms):
    return all_rooms.filter(lambda b: len(b.member_ids) > 1)
```

Actually, you need to filter a recordset based on the fact that the value of a field is *truthy* in the Python sense (non-empty strings, non-zero numbers, non-empty containers, and so on). So, if you want to filter records that have a category set, you can pass the field name to filter like this: `all_rooms.filter('category_id')`.

There's more...

Remember that `filter()` uses memory to work. Use a search domain or even switch to SQL to improve the speed of a method on the critical route at the cost of readability.

Traversing recordset relations

When working with a recordset with a length of `1`, various fields are available as record attributes. Relational attributes (`One2many`, `Many2one`, and `Many2many`) are also available with values that are recordsets, too. As an example, let's say we want to access the name of the category from the recordset of the `hostel.room` model. You can access the category name by traversing through the `Many2one` field's `category_id` as follows: `room.category_id.name`. However, when working with recordsets with more than one record, the attributes cannot be used.

This tutorial demonstrates how to navigate recordset relations using the `mapped()` function. We'll create a function to extract the members' names from the list of rooms that are supplied as input.

Getting ready

We will reuse the `hostel.room` model that was shown in the *Creating new records* tutorial in this chapter.

How to do it...

You must do the following actions in order to retrieve the names of members from the room recordset:

1. Define a method called `get_members_names()`:

   ```
   @api.model
   def get_members_names(self, rooms):
   ```

2. Call `mapped()` to get the name of the contacts of the member:

   ```
   return rooms.mapped('member_ids.name')
   ```

How it works...

Simply defining the method is *Step 1*. The fields of the recordset are traversed in *Step 2* by calling the `mapped(path)` function; `path` is a string that comprises field names separated by dots. The next element in the route is applied to the new recordset created by `mapped()` for each field in the path. This new recordset comprises all the records connected by that field to every element in the current recordset. A recordset is returned by `mapped()` if the final field in the route is a relational field; otherwise, a Python list is returned.

The `mapped()` method has two useful properties:

- When the route is a single scalar field name, the returned list is in the same chronological order as the recordset that was processed
- If a relational field is present in the route, the result's order is not retained, but duplicates are eliminated

> **Important note**
> This second property is very useful when you want to perform an operation on all the records that are pointed to by a `Many2many` field for all the records in `self`, but you need to ensure that the action is performed only once (even if two records of `self` share the same target record).

There's more...

When using `mapped()`, keep in mind that it operates in memory inside the Odoo server by repeatedly traversing relations and therefore making SQL queries, which may not be efficient. However, the code is terse and expressive. If you are trying to optimize a method on the critical path of the performance of your instance, you may want to rewrite the call to `mapped()` and express it as `search()` with the appropriate domain, or even move to SQL (at the cost of readability).

The `mapped()` method can also be called with a function as an argument. In this case, it returns a list containing the result of the function that's applied to each record of `self`, or the union of the recordsets that's returned by the function, if the function returns a recordset.

See also

For more information, refer to the following:

- The *Searching for records* tutorial in this chapter
- The *Executing raw SQL queries* tutorial in *Chapter 8, Advanced Server-Side Development Techniques*

Sorting recordsets

When you fetch a recordset with the `search()` method, you can pass an optional argument order to get a recordset that's in a particular order. This is useful if you already have a recordset from a previous bit of code and you want to sort it. It may also be useful if you use a set operation to combine two recordsets, for example, which would cause the order to be lost.

This tutorial shows you how to use the `sorted()` method to sort an existing recordset. We will sort rooms by rating.

Getting ready

We will reuse the `hostel.room` model that was shown in the *Creating new records* tutorial in this chapter.

How to do it...

You need to perform the following steps to get the sorted recordset of rooms based on `rating`:

1. Define a method called `sort_rooms_by_rating()`:

```
@api.model
def sort_rooms_by_rating(self, rooms):
```

2. Use the `sorted()` method, as in the given example, to sort room records based on the `room_rating` field:

```
return rooms.sorted(key='room_rating')
```

How it works...

Simply defining the method is *Step 1*. In *Step 2*, we use the recordset of the rooms' `sorted()` function. The field that is supplied as the key parameter will have its data fetched internally by the `sorted()` function. Then, it returns a `sorted` recordset using Python's native sorted method.

It also has one optional argument, `reverse=True`, which returns a recordset in reverse order. `reverse` is used as follows:

```
rooms.sorted(key='room_rating', reverse=True)
```

There's more...

The `sorted()` method will sort the records in a recordset. Called without arguments, the `_order` attribute of the model will be used. Otherwise, a function can be passed to compute a comparison key in the same way as the Python built-in sorted (sequence, key) function.

> **Important note**
>
> When the default `_order` parameter of the model is used, the sorting is delegated to the database, and a new `SELECT` function is performed to get the order. Otherwise, the sorting is performed by Odoo. Depending on what is being manipulated, and depending on the size of the recordsets, there might be some important performance differences.

Extending the business logic defined in a model

Dividing application functionalities into various modules is a popular practice in Odoo. You may easily accomplish this by installing or uninstalling the application, which will enable or disable functionalities. Furthermore, you must modify the behavior of some methods that were predefined in the original app when you add new features to it. An old model could occasionally benefit from the addition of fresh fields. This is one of the most useful functions of the underlying framework and the process is quite simple in Odoo.

In this tutorial, we will see how you can extend the business logic of one method from the method in another module. Additionally, we will use the new module to add new fields to an existing module.

Getting ready

For this tutorial, we will continue to use the `my_hostel` module from the last tutorial. Make sure that you have the `hostel.room.category` model in the `my_hostel` module.

For this tutorial, we will create a new module called `my_hostel_terminate`, which depends on the `my_ hostel` module. In this module, we will manage termination dates from the hostel. We will also automatically calculate the withdrawal date based on the category.

In the *How to add features to a model using inheritance* tutorial in *Chapter 4, Application Models*, we saw how to add a field to the existing model. In this module, extend the `hostel.room` model as follows:

```
class HostelRoom(models.Model):
    _inherit = 'hostel.room'
    date_terminate = fields.Date('Date of Termination')
```

Then, extend the hostel.room.category model, as follows:

```
class RoomCategory(models.Model):
    _inherit = 'hostel.room.category'
    max_allow_days = fields.Integer(
        'Maximum allows days',
        help="For how many days room can be borrowed",
        default=365)
```

To add this field in views, you need to follow the *Changing existing views – view inheritance* tutorial from *Chapter 9, Backend Views*. You can find a full example of the code at https://github.com/PacktPublishing/Odoo-17-Development-Cookbook-Fifth-Edition.

How to do it...

To extend the business logic in the hostel.room model, you need to perform the following steps:

1. From my_hostel_terminate, we want to set date_terminate in the rooms record when we change the room status to Closed. For this, we will override the make_closed method from the my_ hostel_terminate module:

```
def make_closed(self):
    day_to_allocate = self.category_id.max_allow_days or 10
    self.date_return = fields.Date.today() + timedelta(days=day_
to_allocate)
    return super(HostelRoom, self).make_closed()
```

2. We also want to reset date_terminate when the room is returned and available to borrow, so we will override the make_available method to reset the date:

```
def make_available(self):
    self.date_terminate = False
    return super(HostelRoom, self).make_available()
```

How it works...

Steps 1 and *2*, in the preceding section, carry out the extension of the business logic. We define a model that extends hostel.room and redefines the make_closed() and make_available() methods. In the last line of both methods, the result that was implemented by the parent class is returned:

```
return super(HostelRoom, self).make_closed()
```

In the case of Odoo models, the parent class is not what you'd expect by looking at the Python class definition. The framework has dynamically generated a class hierarchy for our recordset, and the parent class is the definition of the model from the modules that we depend on. So, the call to super()

brings back the implementation of `hostel.room` from `my_hostel`. In this implementation, `make_closed()` changes the state of the room to `Closed`. So, calling `super()` will invoke the parent method and it will set the room state to `Closed`.

There's more...

In this tutorial, we choose to extend the default implementation of the methods. In the `make_closed()` and `make_available()` methods, we modified the returned result *before* the `super()` call. Note that, when you call `super()`, it will execute the default implementation. It is also possible to perform some actions *after* the `super()` call. Of course, we can also do both at the same time.

To alter a method's behavior in the midst, though, is more challenging. To do this, we must restructure the code in order to extract an extension point to a different function, which we can then override in the extension module.

You might be inspired to rewrite a function from scratch. Always proceed with extreme caution. The extension mechanism and maybe the add-ons that extend the method are broken if you do not use the `super()` implementation of your method, which means that the extension methods will never be invoked. Avoid doing this unless you are working in a controlled environment where you are certain which add-ons are installed and you have verified that you are not breaking them. Additionally, if necessary, make sure to clearly document everything you do.

What can you do before and after calling the original implementation of the method? There are lots of things, including (but not limited to) the following:

- Change the arguments that are sent to the initial implementation (in the past)
- Alter the context that was previously provided to the original implementation
- Change the outcome that the initial implementation returned (after)
- Call another method (before and after)
- Create records (before and after)
- Raise a `UserError` error to cancel the execution in forbidden cases (before and after)
- Split `self` into smaller recordsets and call the original implementation on each of the subsets in a different way (before)

Extending write() and create()

Extending the business logic defined in a model tutorial from this chapter showed us how to extend methods that are defined on a model class. If you think about it, methods that are defined on the parent class of the model are also part of the model. This means that all the base methods that are defined on `models.Model` (actually, on `models.BaseModel`, which is the parent class of `models.Model`) are also available and can be extended.

This tutorial shows you how to extend `create()` and `write()` to control access to some fields of the records.

Getting ready

We will extend the library example from the `my_hostel` add-on module in *Chapter 3, Creating Odoo Add-On Modules.*

Add a `remarks` field to the `hostel.room` model. We only want members of the `Hostel Managers` group to be able to write to that field:

```
from odoo import models, api, exceptions
class HostelRoom(models.Model):
    _name = 'hostel.room'
    remarks = fields.Text('Remarks')
```

Add the `remarks` field to the `<form>` view of the `view/hostel_room.xml` file to access this field from the user interface:

```
<field name="remarks"/>
```

Modify the `security/ir.model.access.csv` file to give write access to library users:

```
id,name,model_id:id,group_id:id,perm_read,perm_write,perm_create,perm_
unlink
access_hostel,hostel.room.user,model_hostel_room,base.group_
user,1,1,0,0
```

How to do it...

To prevent users who are not members of the manager group from modifying the value of `remarks`, you need to perform the following steps:

1. Extend the `create()` method, as follows:

```
    @api.model
    def create(self, values):
        if not self.user_has_groups('my_hostel.group_hostel_
manager'):
            if values.get('remarks'):
                raise UserError(
                    'You are not allowed to modify '
                    'remarks'
                )
        return super(HostelRoom, self).create(values)
```

2. Extend the `write()` method, as follows:

```
        def write(self, values):
            if not self.user_has_groups('my_hostel.group_hostel_
    manager'):
                if values.get('remarks'):
                    raise UserError(
                        'You are not allowed to modify '
                        'manager_remarks'
                    )
            return super(HostelRoom, self).write(values)
```

Install the module to see the code in action. Now, only a manager type of user can modify the `remarks` field. To test this implementation, you can log in as a demo user or revoke manager access from the current user.

How it works...

Step 1 in the preceding section redefines the `create()` method. Before calling the base implementation of `create()`, our method uses the `user_has_groups()` method to check whether the user belongs to the `my_hostel.group_hostel_manager` group (this is the XML ID of the group). If this is not the case and a value is passed for `remarks`, a `UserError` exception is raised, preventing the creation of the record. This check is performed before the base implementation is called.

Step 2 does the same thing for the `write()` method. Prior to writing, we check the group and the presence of the field in the values so we can write and raise a `UserError` exception if there is a problem.

> **Important note**
>
> Having the field set to read only in the web client does not prevent RPC calls from writing it. This is why we extend `create()` and `write()`.

In this tutorial, you have seen how you can override the `create()` and `write()` methods. However, note that this is not limited to the `create()` and `write()` methods. You can override any model method. For example, let's say you want to do something when the record is deleted. To do so, you need to override the `unlink()` method (the `unlink()` method will be called when the record is deleted). Here is the small code snippet to override the `unlink()` method:

```
def unlink(self):
    # your logic
    return super(HostelRoom, self).unlink()
```

> **Warning**
>
> When overriding a method in Odoo, never forget to call the `super()` method, otherwise, you will encounter an issue. This is because when you don't use the `super()` method, the code in the original method is never executed. If, in our previous code snippet, we didn't call `super(...).unlink()`, records would not be deleted.

There's more...

When extending `write()`, note that, before calling the `super()` implementation of `write()`, `self` is still unmodified. You can use this to compare the current values of the fields to the ones in the `values` dictionary.

In this tutorial, we chose to raise an exception, but we could have also chosen to remove the offending field from the `values` dictionary and silently skipped updating that field in the record:

```
def write(self, values):
    if not self.user_has_groups('my_hostel.group_hostel_manager'):
        if values.get('remarks'):
            del values['remarks']
    return super(HostelRoom, self).write(values)
```

After calling `super().write()`, if you want to perform additional actions, you have to be wary of anything that can cause another call to `write()`, or you will create an infinite recursion loop. The workaround is to put a marker in the context that will be checked to break the recursion:

```
class MyModel(models.Model):
    def write(self, values):
        sup = super(MyModel, self).write(values)
        if self.env.context.get('MyModelLoopBreaker'):
            return
        self = self.with_context(MyModelLoopBreaker=True)
        self.compute_things() # can cause calls to writes
        return sup
```

In the preceding example, we have added the `MyModelLoopBreaker` key before calling the `compute_things()` method. So, if the `write()` method is called again, it doesn't go in an infinite loop.

Customizing how records are searched

The *Defining the model representation and order* tutorial in *Chapter 3, Creating Odoo Add-On Modules*, introduced the `name_get()` method, which is used to compute a representation of the record in various places, including in the widget that's used to display `Many2one` relations in the web client.

This tutorial will show you how to search for a room in the `Many2one` widget by room number and name by redefining `name_search`.

Getting ready

For this tutorial, we will use the following model definition:

```
class HostelRoom(models.Model):
    def name_get(self):
        result = []
        for room in self:
            member = room.member_ids.mapped('name')
            name = '%s (%s)' % (room.name, ', '.join(member))
            result.append((room.id, name))
            return result
```

When using this model, a room in a `Many2one` widget is displayed as **Room Title** (**Member1, Member2,..**). Users expect to be able to type in a member's name and find the list filtered according to this name, but this will not work since the default implementation of `name_search` only uses the attribute referred to by the `_rec_name` attribute of the model class, which, in our case, is `'name'`. We also want to allow filtering by room number.

How to do it...

You need to perform the following steps in order to execute this tutorial:

1. To be able to search for `hostel.room` either by the room's name, one of the members, or the room number, you need to define the `_name_search()` method in the `HostelRoom` class, as follows:

```
@api.model
def _name_search(self, name='', args=None, operator='ilike',
                 limit=100, name_get_uid=None):
    args = [] if args is None else args.copy()
    if not(name == '' and operator == 'ilike'):
        args += ['|', '|',
                 ('name', operator, name),
                 ('isbn', operator, name),
                 ('author_ids.name', operator, name)
                 ]
    return super(HostelRoom, self)._name_search(
        name=name, args=args, operator=operator,
        limit=limit, name_get_uid=name_get_uid)
```

2. Add the `previous_room_id` `Many2one` field in the `hostel.room` model to test the `_name_search` implementation:

```
previous_room = fields.Many2one('hostel.room', string='Previous Room')
```

3. Add the following field to the user interface:

```
<field name="previous_room_id" />
```

4. Restart and update the module to reflect these changes.

You can invoke the `_name_search` method by searching in the `previous_room_id` `Many2one` field.

How it works...

The default implementation of `name_search()` actually only calls the `_name_search()` method, which does the real job. This `_name_search()` method has an additional argument, `name_get_uid`, which is used in some corner cases such as if you want to compute the results using `sudo()` or with a different user.

We pass most of the arguments that we receive unchanged to the `super()` implementation of the method:

- `name` is a string that contains the value the user has typed so far.
- `args` is either `None` or a search domain that's used as a prefilter for the possible records. (It can come from the domain parameter of the `Many2one` relation, for instance.)
- `operator` is a string containing the match operator. Generally, you will have `'ilike'` or `'='`.
- `limit` is the maximum number of rows to retrieve.
- `name_get_uid` can be used to specify a different user when calling `name_get()` to compute the strings to display in the widget.

Our implementation of the method does the following:

1. It generates a new empty list if `args` is `None`, and makes a copy of `args` otherwise. We make a copy to avoid our modifications to the list having side effects on the caller.
2. Then, we check that `name` is not an empty string or that `operator` is not `'ilike'`. This is to avoid generating a dumb domain, such as `[('name', ilike, '')]`, that doesn't filter anything. In this case, we jump straight to the `super()` call implementation.
3. If we have `name`, or if `operator` is not `'ilike'`, then we add some filtering criteria to `args`. In our case, we add clauses that will search for the supplied name in the title of the rooms, in their room number, or the members' names.

4. Finally, we call the `super()` implementation with the modified domain in `args` and force `name` to be `''` and `operator` to be `ilike`. We do this to force the default implementation of `_name_search()` to not alter the domain it receives, so, the one we specified will be used.

There's more...

We mentioned in the introduction that this method is used in the `Many2one` widget. For completeness, it is also used in the following parts of Odoo:

- When using the `in` operator on the `One2many` and `Many2many` fields in the domain
- To search for records in the `many2many_tags` widget
- To search for records in the CSV file import

See also

The *Defining the model representation and order* tutorial in *Chapter 3, Creating Odoo Add-On Modules*, demonstrates how to define the `name_get()` method, which is used to create a text representation of a record.

The *Defining filters on record lists – domain* tutorial in *Chapter 9, Backend Views*, provides more information about search domain syntax.

Fetching data in groups using read_group()

In the previous tutorials, we saw how we can search and fetch data from the database. However, sometimes, you want results by aggregating records, such as the average cost of last month's sales order. Usually, we use `group by` and the `aggregate` function in SQL queries for such a result. Luckily, in Odoo, we have the `read_group()` method. In this tutorial, you will learn how to use the `read_group()` method to get the aggregate result.

Getting ready

In this tutorial, we will use the `my_hostel` add-on module from *Chapter 3, Creating Odoo Add-On Modules*.

Modify the `hostel.room` model, as shown in the following model definition:

```
class HostelRoom(models.Model):
    _name = 'hostel.room'
    name = fields.Char('Name', required=True)
    cost_price = fields.Float('Room Cost')
    category_id = fields.Many2one('hostel.room.category')
```

Add the hostel.room.category model. For simplicity, we will just add it to the same hostel_room.py file:

```
class HostelCategory(models.Model):
    _name = 'hostel.room.category'
    name = fields.Char('Category')
    description = fields.Text('Description')
```

We will be using the hostel.room model and getting an average cost price per category.

How to do it...

To extract grouped results, we will add the _get_average_cost method to the hostel.room model, which will use the read_group() method to fetch the data in a group:

```
@api.model
def _get_average_cost(self):
    grouped_result = self.read_group(
        [('cost_price', "!=", False)], # Domain
        ['category_id', 'cost_price:avg'], # Fields to access
        ['category_id'] # group_by
        )
    return grouped_result
```

To test this implementation, you need to add a button to the user interface that triggers this method. Then, you can print the result in the server log.

How it works...

The read_group() method internally uses the SQL groupby and aggregate functions to fetch the data. The most common arguments that are passed to the read_group() method are as follows:

- domain: This is used to filter records for grouping. For more information on domain, refer to the *Searching views* tutorial in *Chapter 9*, *Backend Views*.
- fields: This passes the names of fields you want to fetch with the grouped data. Possible values for this argument are as follows:
    - field name: You can pass the field name into the fields argument, but if you are using this option, then you must pass this field name to the groupby parameter too, otherwise it will generate an error.

    - `field_name:agg`: You can pass the field name with the `aggregate` function. For example, in `cost_price:avg`, `avg` is an SQL aggregate function. A list of PostgreSQL aggregate functions can be found at `https://www.postgresql.org/docs/current/static/functions-aggregate.html`.
    - `name:agg(field_name)`: This is the same as the previous one, but, with this syntax, you can provide column aliases, such as `average_price:avg(cost_price)`.

- `groupby`: This argument accepts a list of field descriptions. Records will be grouped based on these fields. For the `date` and `datetime` column, you can pass `groupby_function` to apply date groupings based on different time durations. You can do grouping based on months for date type fields.
- `read_group()` also supports some optional arguments, as follows:
    - `offset`: This indicates an optional number of records to skip.
    - `limit`: This indicates an optional maximum number of records to return.
    - `orderby`: If this option is passed, the result will be sorted based on the given fields.
    - `lazy`: This accepts Boolean values and, by default, is `True`. If `True` is passed, the results are only grouped by the first `groupby`, and the remaining `groupby` arguments are put in the `__context` key. If `False`, all `groupby` functions are done in one call.

Performance tip

`read_group()` is a lot faster than reading and processing values from a recordset. So, for KPIs or graphs, you should always use `read_group()`.

6

Managing Module Data

In Odoo, managing module data involves various tasks such as creating, updating, and deleting records in a database upon installation, upgrade, or removal of a module. This is typically done through XML files called data files.

We'll study how add-on modules might offer data during installation in this chapter. This helps us when we provide metadata, such as view descriptions, menus, or actions, or when we provide default settings. Another important usage is providing demonstration data, which is loaded when a database is created with the **Load demonstration data** checkbox checked.

In this chapter, we will cover the following topics:

- Using external IDs and namespaces
- Loading data using XML files
- Using the `noupdate` and `forcecreate` flags
- Loading data using CSV files
- Add-on updates and data migration
- Deleting records from XML files
- Invoking functions from XML files

Technical requirements

The technical requirements for this chapter include the online Odoo platform.

All the code that's used in this chapter can be downloaded from the following GitHub repository: `https://github.com/PacktPublishing/Odoo-17-Development-Cookbook-Fifth-Edition/tree/main/Chapter06`.

In order to avoid repeating a lot of code, we'll make use of the models that were defined in *Chapter 4, Application Models*. To follow these examples, make sure you grab the code from the `my_hostel` module from `Chapter05/my_hostel`.

Using external IDs and namespaces

Records in Odoo are identified using external IDs or XML IDs. We have utilized XML IDs so far in this book in areas such as views, menus, and actions, but we still don't know what an XML ID is. This recipe will give you more clarity about it.

How to do it...

We will write in the already-existing records to demonstrate how to use cross-module references:

1. Update the manifest file of the `my_hostel` module by registering a data file like this:

```
'data': [
'data/data.xml',
],
```

2. Create a new room in the `hostel.room` model:

```
<record id="hostel_room" model="hostel.room">
<field name="name"> Hostel Room 01 </field>
</record>
```

3. Change the name of the main company:

```
<record id="base.main_company" model="res.company">
<field name="name">Packt Publishing</field>
</record>
```

Install the module to apply the changes. After installation, a new record for the `Hostel Room 01` room will be created, and the company will be renamed `Packt Publishing`.

How it works...

An XML ID is a string that refers to a record in the database. The IDs themselves are records of their `ir.model.data` model. This model includes information such as the module name that declares the XML ID, the ID string, the referred model, and the referred ID.

Every time we use an XML ID on a `<record>` tag, Odoo checks whether the string is namespaced (that is, whether it contains exactly one dot), and if not, it adds the current module name as a namespace. Then, it looks up whether there is already a record in `ir.model.data` with the specified name. If so, an `UPDATE` statement for the listed fields is executed; if not, a `CREATE` statement is executed. This is how you can provide partial data when a record already exists, as we did earlier.

In the first example of this tutorial, the record has the ID `hostel_room_1`. As it is not namespaced, the final external ID will have a module name like this – `my_hostel.hostel_room_1`. Then, Odoo will try to find a record for `my_hostel.hostel_room_1`. As Odoo doesn't have a record for that external ID yet, it will generate a new record in the `hostel.room` model.

In the second example, we have used the external ID of the main company, which is `base.main_company`. As its namespace suggests, it is loaded from the base module. As the external ID is already present, instead of creating a record, Odoo will perform the write `(UPDATE)` operation so that the company name will change to `Packt Publishing`.

> **Important note**
>
> A widespread application for partial data, apart from changing records defined by other modules, is using a shortcut element to create a record conveniently and writing a field on that record, which is not supported by the shortcut element – `<act_window id="my_action" name="My action" model="res.partner" /><record id="my_action" model="ir.actions.act_window"> <field name="auto_search" eval="False" /></record>`.

In Odoo, the `ref` function is used to establish relationships between different records within the system. It allows you to create references from one record to another, typically using a *many2one* relationship.

The `ref` function, as used in the *Loading data using XML files* tutorial of this chapter, also adds the current module as a namespace if appropriate but raises an error if the resulting XML ID does not exist already. This also applies to the `id` attribute if it is not namespaced already.

If you want to see the list of all external identifiers, start developer mode and open the menu at **Settings** | **Technical** | **Sequence & Identifiers** | **External Identifiers**.

There's more...

You will probably need to access records with an XML ID from your Python code sooner or later. Use the `self.env.ref()` function in these cases. This returns a browse record (`recordset`) of the referenced record. Note that, here, you always have to pass the full XML ID. Here's an example of a full XML ID – `<module_name>.<record_id>`.

Sooner or later, you'll probably need to use Python code to retrieve records that have an XML ID. In these circumstances, use the `self.env.ref()` method. This gives you access to the linked record's browsing record (`recordset`). Keep in mind that you must always pass the complete XML ID here.

You can see the XML ID of any record from the user interface. For that, you need to activate developer mode in Odoo; refer to *Chapter 1, Installing the Odoo Development Environment*, to do so. After activating developer mode, open the **Form** view of the record for which you want to find out the XML ID. You will see a bug icon in the top bar. From that menu, click on the **View Metadata** option. See the following screenshot for reference:

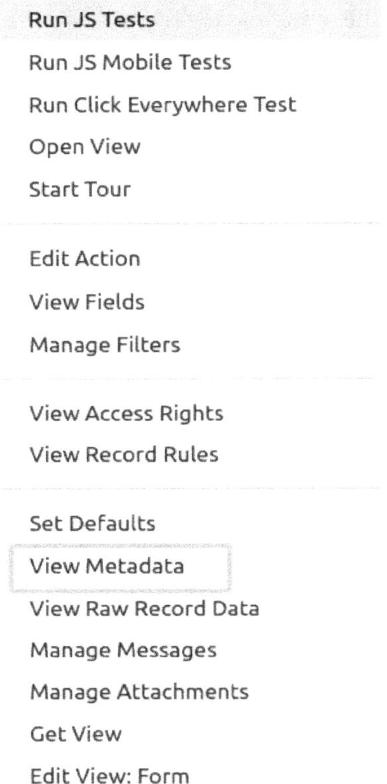

Figure 6.1 – The menu to open a record's metadata

See also

Consult the *Using the noupdate and forcecreate flags* tutorial of this chapter to find out why the company's name is only changed during the installation of the module.

Loading data using XML files

In the previous tutorial, we created the new room record with the `hostel_room_1` external identifier. In this tutorial, we will add a different type of data from the XML file. We'll add a room and an author as demonstration data. We'll also add a well-known publisher as normal data to our module.

How to do it...

Follow the following steps to create two data XML files and link them in your__manifest__.py file:

1. Add a file called data/demo.xml to your manifest, in the demo section:

```
'demo': [
'data/demo.xml',
],
```

2. Add the following content to this file:

```
<odoo>
<record id="member_hda" model="res.partner">
<field name="name">Husen Daudi</field>
</record>
<record id="member_jvo" model="res.partner">
<field name="name">Jay Vora</field>
</record>
<record id="hostel_room_1" model="hostel.room">
<field name="name">Hostel Room 01</field>
<field name="room_no">HR001</field>
<field name="author_ids"
eval="[(6, 0, [ref('author_hda'), ref('author_jvo')])]"
/>
</record>
</odoo>
```

3. Add a file called data/data.xml to your manifest, in the data section:

```
'data': [
'data/data.xml',
...
],
```

4. Add the following XML content to the data/data.xml file:

```
<odoo>
<record id="res_partner_packt" model="res.partner">
<field name="name">Packt Publishing</field>
<field name="city">Birmingham</field>
<field name="country_id" ref="base.uk" />
</record>
</odoo>
```

When you update your module now, you'll see the publisher we created, and if your database has demo data enabled, as pointed out in *Chapter 3*, *Creating Add-On Odoo Modules*, you'll also find this room and its members.

How it works...

The data XML files uses the `<record>` tag to create a row in the database table. The `<record>` tag has two mandatory attributes, `id` and `model`. For the `id` attribute, consult the *Using external IDs and namespaces* tutorial; the `model` attribute refers to a model's `_name` property. Then, we use the `<field>` element to fill the columns in the database, as defined by the model you named. The model also decides which fields it is mandatory to fill and also defines the default values. In this case, you don't need to give those fields a value explicitly.

There are two ways to register data XML files in a module manifest. The first is with the `data` key and the second is with the `demo` key. The XML files in the `data` key are loaded every time you install or update the module, while XML files with `demo` keys are loaded only if you enabled demo data for your database.

In *step 1*, we registered a `data XML` file in the manifest with the `demo` key. Because we are using the `demo` key, the `XML` file will be loaded only if you have enabled demo data for the database.

In *step 2*, the `<field>` element can contain a value as simple text in the case of scalar values. If you need to pass the content of a file (to set an image, for example), use the `file` attribute on the `<field>` element and pass the file's name relative to the add-ons path.

To set up references, there are two possibilities. The simplest is using the `ref` attribute, which works for *many2one* fields and just contains the XML ID of the record to be referenced. For *one2many* and *many2many* fields, we need to use the `eval` attribute. Use the `eval` attribute in XML to evaluate expressions dynamically. This is a general-purpose attribute that can be used to evaluate Python code to use as the field's value. Normally, content within `<field>` tags is treated as strings – for example, `<field name="value">4.5</field>`. This will evaluate to the string `4.5` and not `float`. If you want to evaluate the value to a float, a Boolean, or another type, except `string`, you need to use the eval attribute, such as `<field name="value" eval="4.5" /> <field name="value" eval="False" />`.

Here's another example – think of `strftime('%Y-01-01')` as a way to populate a `date` field. `X2many` fields expect to be populated by a list of three tuples, where the first value of the tuple determines the operation to be carried out. Within an `eval` attribute, we have access to a function called `ref`, which returns the database ID of an XML ID, given as a string. This allows us to refer to a record without knowing its concrete ID, which is probably different in different databases, as shown here:

- `(2, id, False)`: This deletes the linked record with `id` from the database. The third element of the tuple is ignored.

- `(3, id, False)`: This detaches the record with `id` from the `one2many` field. Note that this operation does not delete the record – it just leaves the existing record as it is. The last element of the tuple is also ignored.
- `(4, id, False)`: This adds a link to the existing record `id`, and the last element of the tuple is ignored. This should be what you use most of the time, usually accompanied by the `ref` function to get the database ID of a record known by its XML ID.
- `(5, False, False)`: This cuts all links but keeps the linked records intact.
- `(6, False, [id, ...])`: This clears out currently referenced records to replace them with the ones mentioned in the list of IDs. The second element of the tuple is ignored.

Important note

Note that order matters in data files and that records within data files can only refer to records defined in data files earlier in the list. This is why you should always check whether your module installs in an empty database because, during development, you often add records all over the place, and the records defined afterward are already in the database from an earlier update.

Demo data is always loaded after the files from the `data` key, which is why the reference in this example works.

There's more...

While you can do basically anything with the `record` element, there are shortcut elements that make it more convenient for a developer to create certain kinds of records. These include menu items, templates, and act windows. Refer to *Chapter 9*, *Backend Views*, and *Chapter 14*, *CMS Website Development*, for more information about these.

A `field` element can also contain the `function` element, which calls a function defined on a model to provide a field's value. Refer to the *Invoking functions from XML files* tutorial for an application in which we simply call a function to directly write to the database, circumventing the loading mechanism.

The preceding list misses entries for `0` and `1` because they are not very useful when loading the data. They are entered, as follows, for the sake of completeness:

- `(0, False, {'key': value})`: This creates a new record of the referenced model, with its fields filled from the dictionary at position three. The second element of the tuple is ignored. As these records don't have an XML ID and are evaluated every time the module is updated, leading to double entries, it's better to avoid this. Instead, create the record in its own record element, and link it as explained in the *How it works...* section of this tutorial.
- `(1, id, {'key': value})`: This can be used to write on an existing linked record. For the same reasons that we mentioned earlier, you should avoid this syntax in your XML files.

These syntaxes are the same as the ones we explained in the *Creating new records* and *Updating values of records* tutorials in *Chapter 5, Basic Server-Side Development*.

Using the noupdate and forcecreate flags

Most add-on modules have different types of data. Some data simply needs to exist for the module to work properly, other data shouldn't be changed by the user, and most data can be changed as the user wants and is only provided as a convenience. This tutorial will detail how to address the different types. First, we'll write a field in an already-existing record, and then we'll create a record that is supposed to be recreated during a module update.

How to do it...

We can enforce different behaviors from Odoo when loading data by setting certain attributes on the enclosing `<odoo>` element, or the `<record>` element itself:

1. Add a publisher that will be created at installation time but not updated on subsequent updates. However, if the user deletes it, it will be recreated:

```
<odoo noupdate="1">
    <record id="res_partner_packt" model="res.partner">
        <field name="name">Packt Publishing</field>
        <field name="city">Birmingham</field>
        <field name="country_id" ref="base.uk"/>
    </record>
</odoo>
```

2. Add a `room` category that is not changed during add-on updates and is not recreated if the user deletes it:

```
<odoo noupdate="1">
    <record id="room_category_all" model="hostel.room.category"
            forcecreate="false">
        <field name="name">All rooms</field>
    </record>
</odoo>
```

How it works...

The `<odoo>` element can have a `noupdate` attribute, which is propagated to the `ir.model.data` records that are created when reading the enclosed data records for the first time, thus ending up as a column in this table.

When Odoo installs an add-on (called `init` mode), all records are written, whether `noupdate` is `true` or `false`. When you update an add-on (called `update` mode), the existing XML IDs are checked to see whether they have the `noupdate` flag set, and if so, elements that try to write to this XML ID are ignored. This is not the case if the record in question was deleted by the user, which is why you can force `notrecreate` `noupdate` records in `update` mode by setting the `forcecreate` flag on the record to `false`.

> **Important note**
>
> In legacy add-ons (prior to and including version 8.0), you'll often find an `<openerp>` element enclosing a `<data>` element, which contains `<record>` and other elements. This is still possible but deprecated. Now, `<odoo>`, `<openerp>`, and `<data>` have exactly the same semantics; they are meant as a bracket to enclose XML data.

There's more...

If you want to load records even with the `noupdate` flag, you can run the Odoo server with the `--init=your_addon` or `-i your_addon` parameter. This will force Odoo to reload your records. However, this will also cause deleted records to be recreated. Note that this can cause double records and related installation errors if a module circumvents the XML ID mechanism – for example, by creating records in Python code called by the `<function>` tag.

With this code, you can circumvent any `noupdate` flag, but first, make sure that this is really what you want. Another option to solve the scenario presented here is to write a migration script, as outlined in the *Add-on updates and data migration* tutorial.

See also

Odoo also uses XML IDs to keep track of which data is to be deleted after an add-on update. If a record has an XML ID from the module's namespace before the update but the XML ID is not reinstated during the update, the record and its XML ID will be deleted from the database because they're considered obsolete. For a more in-depth discussion of this mechanism, refer to the *Add-on updates and data migration* tutorial.

Loading data using CSV files

While you can do everything you need to with XML files, this format is not the most convenient when you need to provide larger amounts of data, especially given that many people are more comfortable preprocessing data in Calc or other spreadsheet software. Another advantage of the CSV format is that it is what you get when you use the standard `export` function. In this tutorial, we'll take a look at importing table-like data.

How to do it...

Traditionally, **access-control lists** (**ACLs**) in Odoo are used to manage record and operation access rights. ACLs specify who can execute particular actions (such as read, write, create, and delete) on specified entries using predefined rules. ACLs are commonly defined in Odoo modules via XML files. For more details on ACLs, check out the ACLs tutorial in *Chapter 10, Security Access*

Add `security/ir.model.access.csv` to your data files:

```
'data': [
    ...
    'security/ir.model.access.csv',
],
```

1. Add access security to the module in the `ir.model.data` CSV file:

```
id,name,model_id:id,group_id:id,perm_read,perm_write,perm_
create,perm_unlink
access_hostel_manager,hostel.room.manager,model_hostel_
room,group_hostel_manager,1,1,1,1
```

We now have an ACL that permits hostel managers to read book records, and it also allows them to edit, create, or delete them.

How it works...

You simply drop all your data files into your manifest's *data* list. Odoo will use the file extension to decide which type of file it is. A specialty of CSV files is that their filenames must match the name of the model to be imported – in our case, `ir.model.access`. The first line needs to be a header with column names that match the model's field names exactly.

For scalar values, you can use a quoted (if necessary, because the string contains quotes or commas itself) or an unquoted string.

When writing *many2one* fields with a CSV file, Odoo first tries to interpret the column value as an XML ID. If there's no dot, Odoo adds the current module name as a namespace and looks up the result in `ir.model.data`. If this fails, the model's `name_search` function is called with the column's value as a parameter, and the first returned result wins. If this also fails, the line is considered invalid and Odoo raises an error.

Add-on updates and data migration

The data model you choose when writing an add-on module might turn out to have some weaknesses, so you may need to adjust it during the life cycle of your add-on module. In order to allow that without a lot of hacks, Odoo supports versioning in add-on modules and running migrations if necessary.

How to do it...

We assume that in an earlier version of our module, the `allocation_date` field was a character field, where people wrote whatever they saw fit as the date. We now realize that we need this field for comparisons and aggregations, which is why we want to change its type to `Date`.

Odoo does a great job at type conversions, but in this case, we're on our own, which is why we need to provide instructions as to how to transform a database with the previous version of our module installed on it, where the current version can run. Let's try this with the following steps:

1. Bump the version in your `__manifest__.py` file:

```
'version': '17.0.2.0.1',
```

2. Provide the pre-migration code in `migrations/17.0.1.0.1/pre-migrate.py`:

```
def migrate(cr, version):
    cr.execute('ALTER TABLE hostel_room RENAME COLUMN
allocation_date TO allocation_date_char')
```

3. Provide the post-migration code in `migrations/17.0.1.0.1/post-migrate.py`:

```
from odoo import fields
from datetime import date
def migrate(cr, version):
    cr.execute('SELECT id, allocation_date_char FROM
    hostel_room')
    for record_id, old_date in cr.fetchall():
        # check if the field happens to be set in Odoo's
        internal
        # format
        new_date = None
        try:
            new_date = fields.Date.to_date(old_date)
        except ValueError:
            if len(old_date) == 4 and old_date.isdigit():
                # probably a year
                new_date = date(int(old_date), 1, 1)
        if new_date:
            cr.execute('UPDATE hostel_room SET allocation_
date=%s WHERE id=2',
                       (new_date,))
```

Without this code, Odoo would have renamed the old `allocation_date` column `allocation_date_moved` and created a new one, as there's no automatic conversion from character fields to date fields. From the point of view of the user, the data in `allocation_date` is simply gone.

How it works...

The first crucial point is that you increase the version number of your add-on, as migrations run only between different versions. During every update, Odoo writes the version number from the manifest at the time of the update into the `ir_module_module` table. The version number is prefixed with Odoo's major and minor versions if the version number has three or fewer components. In the preceding example, we explicitly named Odoo's major and minor version, which is good practice, but a value of `1.0.1` would have had the same effect because, internally, Odoo prefixes short version numbers for add-ons with its own major and minor version numbers. Generally, using the long notation is a good idea because you can see at a glance which version of Odoo an add-on is meant for.

The two migration files are just code files that don't need to be registered anywhere. When updating an add-on, Odoo compares the add-on's version, as noted in `ir_module_module`, with the version in the add-on's manifest. If the manifest's version is higher (after adding Odoo's major and minor version), this add-on's `migrations` folder will be searched to see whether it contains folders with the version(s) in between, up to and including the version that is currently updated.

Then, within the folders found, Odoo searches for Python files whose names start with `pre-`, loads them, and expects them to define a function called `migrate`, which has two parameters. This function is called with a database cursor as the first argument and the currently installed version as the second argument. This happens before Odoo even looks at the rest of the code that the add-on defines, so you can assume that nothing changes in your database layout compared to the previous version.

After all the `pre-migrate` functions run successfully, Odoo loads the models and the data declared in the add-on, which can cause changes in the database layout. Given that we renamed `date_release in pre-migrate.py`, Odoo will just create a new column with that name but with the correct data type.

After that, with the same search algorithm, the `post-migrate` files will be searched and executed if found. In our case, we need to look at every value to see whether we can make something usable out of it; otherwise, we keep the data as `NULL`. Don't write scripts that iterate over a whole table if not absolutely necessary; in this case, we would have written a very big, unreadable SQL switch.

> **Important tip**
>
> If you simply want to rename a column, you don't need a migration script. In this case, you can set the `oldname` parameter of the field in question to the field's original column name; Odoo then takes care of the renaming itself.

There's more...

In both the pre- and post-migration steps, you only have access to a cursor, which is not very convenient if you're used to Odoo environments. It can lead to unexpected results to use models at this stage because, in the pre-migration step, the add-on's models are not yet loaded, and also, in the post-migration step, the models defined by add-ons that depend on the current add-on are not yet loaded either. However, if this is not a problem for you, either because you want to use a model that your add-on doesn't touch or a model for which you know that this issue is not a problem, you can create the environment you're used to by writing the following:

```
from odoo import api, SUPERUSER_ID
def migrate(cr, version):
    env = api.Environment(cr, SUPERUSER_ID, {})
    # env holds all currently loaded models
```

See also

When writing migration scripts, you'll often be confronted with repetitive tasks, such as checking whether a column or table exists, renaming things, or mapping some old values to new values. It's frustrating and error-prone to reinvent the wheel here; consider using `https://github.com/OCA/openupgradelib` if you can afford the extra dependency.

Deleting records from XML files

We learned how to generate or change records from the XML file in the previous tutorials. You may occasionally wish to remove the records that have already been created from the dependent module. The `<delete>` tag can be used.

Getting ready

In this tutorial, we will add some categories from the XML file and then delete them. In real situations, you will create this record from another module. But for simplicity, we will just add some categories to the same XML file, as follows:

```
<record id="room_category_to_remove" model="hostel.room.category">
    <field name="name">Single sharing</field>
</record>
<record id="room_category_not_remove" model="hostel.room.category">
    <field name="name">Double Sharing</field>
</record>
```

How to do it...

There are two ways to remove records from the XML file:

- Using the XML ID of previously created records:

  ```
  <delete model="hostel.room.category" id="room_category_to_
  remove"/>
  ```

- With the search domain:

```
<delete model="hostel.room.category" search="[('name', 'ilike',
'Single Room Category')]"/>
```

How it works...

You will need to use the `<delete>` tag. To remove a record from a model, you need to provide the name of the model in the `model` attribute. This is a mandatory attribute.

The XML IDs of the records that had been generated from the data files of another module must be supplied in the first method. Odoo will look for the record while installing the module. If the specified XML ID matches a record, the record will be deleted; otherwise, an error will be raised. Only records that were generated from XML files (or records with XML IDs) are able to be deleted.

In the second method, you need to pass the domain in the `domain` attribute. During the installation of the module, Odoo will search the records by this domain. If records are found, it deletes them. This option will not raise an error if no records match the given domain. Use this option with extreme caution because it might delete your user's data, since the search option deletes all the records that match the domain.

> **Warning**
> `<delete>` is rarely used in Odoo, as it is dangerous. If you are not careful with this, you might break the system. Avoid it if possible.

Invoking functions from XML files

You can create all types of records from XML files, but sometimes, it is difficult to generate data that includes some business logic. You might want to modify records when a user installs a dependent module in production. In this case, you can invoke the `model` method through the `<function>` tag.

How to do it...

For this tutorial, we will use the code from the previous tutorial. As an example, we will increase the existing room price by $10 USD. Note that you might use another currency based on company configurations.

Follow these steps to invoke the Python method from the XML file:

1. Add the `_update_room_price()` method to the `hostel.room` model:

```
@api.model
def _update_room_price(self):
    all_rooms = self.search([])
    for room in all_rooms:
        room.cost_price += 10
```

2. Add `<function>` to the data XML file:

```
<function model="hostel.room" name="_update_room_price"/>
```

How it works...

In *step 1*, we added the `_update_room_price()` method, which searches for all books and increases the price by $10 USD. We started the method name with _, as this is considered private by ORM and cannot be invoked through RPC.

In *step 2*, we used the `<function>` tag with two attributes:

- `model`: The model name with which the method is declared
- `name`: The name of the method you want to invoke

When you install this module, `_update_room_price()` will be called and the price of books will increase by $10 USD.

> **Important note**
> Always use this function with the `noupdate` options. Otherwise, it will be invoked every time you update your module.

There's more...

With `<function>`, it is also possible to send parameters to the functions. Let's say you only want to increase the price of rooms in a particular category and you want to send that amount as a parameter.

To do that, you need to create a method that accepts the category as a parameter, as follows:

```
@api.model
def update_room_price(self, category, amount_to_increase):
    category_rooms = self.search([('category_id', '=', category.
id)])
    for room in category_rooms:
        room.cost_price += amount_to_increase
```

To pass the category and amount as a parameter, you need to use the `eval` attribute, as follows:

```
<function model="hostel.room"
    name="update_room_price"
    eval="(ref('category_xml_id'), 20)"/>
```

When you install the module, it will increase the price of the rooms of the given category by $20 USD.

7
Debugging Modules

In *Chapter 5, Basic Server-Side Development*, we saw how to write model methods to implement the logic of our module. However, we may get stuck when we encounter errors or logical issues. In order to resolve these errors, we need to perform a detailed inspection and this may take time. Luckily, Odoo provides you with some debugging tools that can help you find the root cause of various issues. In this chapter, we will look at various debugging tools and techniques in detail.

In this chapter, we will cover the following recipes:

- The auto-reload and `--dev` options
- Producing server logs to help debug methods
- Using the Odoo shell to interactively call methods
- Using the Python debugger to trace method execution
- Understanding the debug mode options

The auto-reload and --dev options

In the previous chapters, we saw how to add a model, fields, and views. Whenever we make changes to Python files, we need to restart the server to apply those changes. If we make changes in XML files, we need to restart the server and update the module to reflect those changes in the user interface. If you are developing a large application, this can be time-consuming and frustrating. Odoo provides a command-line option, `--dev`, to overcome these issues. The `--dev` option has several possible values, and, in this recipe, we will see each of them.

Getting ready

Install `inotify or watchdog` in your developer environment with the following command in the shell. Without `inotify or watchdog`, the auto-reload feature will not work:

```
$ pip3 install inotify
$ pip3 install watchdog
```

How to do it...

To enable the `dev` option, you need to use `--dev=value` from the command line. Possible values for this option are `all`, `reload`, `pudb|wdb|ipdb|pdb`, `qweb`, `werkzeug`, and `xml`. Take a look at the following recipe for more information.

How it works...

Check the following list for all `--dev` options and their purposes:

- `reload`: Whenever you make changes in Python, you need to restart the server to reflect those changes in Odoo. The `--dev=reload` option will reload the Odoo server automatically when you make changes in any Python file. This feature will not work if you have not installed the Python `inotify` package. When you run an Odoo server with this option, you will see a log like this: `AutoReload watcher running with inotify`.
- `qweb`: You can create dynamic website pages in Odoo using QWeb templates. In *Chapter 14, CMS Website Development*, we will see how to develop a web page with the QWeb template. You can debug issues in the QWeb template with the `t-debug` attribute. The `t-debug` options will only work if you enable the `dev` mode with `--dev=qweb`.
- `werkzeug`: Odoo uses `werkzeug` to handle HTTP requests. Internally, Odoo will catch and suppress all exceptions generated by `werkzeug`. If you use `--dev=werkzeug`, werkzeug's interactive debugger will be displayed on the web page when an exception is generated.
- `xml`: Whenever you make changes in the view structure, you need to reload the server and update the module to apply those changes. With the `--dev=xml` option, you just need to reload Odoo from the browser. There is no need to restart the server or update the module.
- `pudb|wdb|ipdb|pdb`: You can use the **Python debugger** (**PDB**) to get more information about the errors. When you use the `--dev=pdb` option, it will activate the PDB whenever an exception is generated in Odoo. Odoo supports four Python debuggers: `pudb`, `wdb`, `ipdb`, and `pdb`.
- `all`: If you use `--dev=all`, all of the preceding options will be enabled.

```
$ odoo/odoo-bin -c ~/odoo-dev/my-instance.cfg --dev=all
```

If you want to enable only a few options, you can use comma-separated values, as follows:

```
$ odoo/odoo-bin -c ~/odoo-dev/my-instance.cfg --dev=reload,qweb
```

> **Important note**
>
> If you have made changes to the database structure, such as if you have added new fields, the `--dev=reload` option will not reflect these in the database schema. You need to update the module manually; it only works for Python business logic. If you add a new view or menu, the `--dev=xml` option will not reflect this in the user interface. You need to update the module manually. This is very helpful when you are designing the structure of the view or the website page. If users have made changes in the view from the GUI, then `--dev=xml` will not load the XML from the file. Odoo will use the view structure, which the user changes.

Producing server logs to help debug methods

Server logs are useful when trying to figure out what has been happening at runtime before a crash. They can also be added to provide additional information when debugging is an issue. This recipe shows you how to add logging to an existing method.

Getting ready

We will add some logging statements to the following method, which saves the stock levels of products to a file (you will also need to add the dependencies of the `product` and `stock` modules to the manifest):

```
from os.path import join as opj
from odoo import models, api, exceptions
EXPORTS_DIR = '/srv/exports'
class ProductProduct(models.Model):
    _inherit = 'product.product'
    @api.model
    def export_stock_level(self, stock_location):
        products = self.with_context(
            location=stock_location.id
        ).search([])
        products = products.filtered('qty_available')
        fname = opj(EXPORTS_DIR, 'stock_level.txt')
        try:
            with open(fname, 'w') as fobj:
                for prod in products:
                    fobj.write('%s\t%f\n' % (prod.name,
                                              prod.qty_available))
```

```
        except IOError:
            raise exceptions.UserError('unable to save file')
```

How to do it...

In order to get some logs when this method is being executed, perform the following steps:

1. At the beginning of the code, import the `logging` module:

```
import logging
```

2. Before the definition of the model class, get a logger for the module:

```
_logger = logging.getLogger(__name__)
```

3. Modify the code of the `export_stock_level()` method, as follows:

```
@api.model
def export_stock_level(self, stock_location):
        _logger.info('export stock level for %s', stock_
location.name)
        products = self.with_context(
            location=stock_location.id).search([])
        products = products.filtered('qty_available')
        _logger.debug('%d products in the location',
len(products))
        fname = join(EXPORTS_DIR, 'stock_level.txt')
        try:
            with open(fname, 'w') as fobj:
                for prod in products:
                    fobj.write('%s\t%f\n' % (
                        prod.name, prod.qty_available))
        except IOError:
            _logger.exception(
                'Error while writing to %s in %s',
                'stock_level.txt', EXPORTS_DIR)
            raise exceptions.UserError('unable to save file')
```

How it works...

Step 1 imports the **logging** module from the Python standard library. Odoo uses this module to manage its logs.

Step 2 sets up a logger for the Python module. We use the common idiom `__name__` in Odoo as an automatic variable for the name of the logger and to call the logger by `_logger`.

> **Important note**
>
> The `__name__` variable is set automatically by the Python interpreter at module-import time, and its value is the full name of the module. Since Odoo does a little trick with the imports, the add-on modules are seen by Python as belonging to the `odoo.addons` Python package. So, if the code of the recipe is in `my_hostel/models/hostel.py`, `__name__` will be `odoo.addons.my_hostel.models.hostel`.

By doing this, we get two benefits:

- The global logging configuration set on the **odoo** logger is applied to our logger because of the hierarchical structure of loggers in the **logging** module
- The logs will be prefixed with the full module path, which is a great help when trying to find where a given log line is produced

Step 3 uses the logger to produce log messages. The available methods for this are (by increasing log level) **debug**, **info**, **warning**, **error**, and **critical**. All these methods accept a message in which you can have `%` substitutions and additional arguments to be inserted into the message. You do not need to handle the `%` substitution yourself; the logging module is smart enough to perform this operation if the log has to be produced. If you are running with a log level of **INFO**, then **DEBUG** logs will avoid substitutions that will consume CPU resources in the long run.

Another useful method shown in this recipe is `_logger.exception()`, which can be used in an exception handler. The message will be logged with a level of `ERROR`, and the stack trace is also printed in the application log.

There's more...

You can control the **logging level** of the application from the command line or from the configuration file. There are two main ways of doing this:

The first way is to use the `--log-handler` option. Its basic syntax is like this: `--log-handler=prefix:level`. In this case, the prefix is a piece of the path of the logger name, and the level is **DEBUG**, **INFO**, **WARNING**, **ERROR**, or **CRITICAL**. If you omit the prefix, you set the default level for all loggers. For instance, to set the logging level of `my_hostel` loggers to **DEBUG** and keep the default log level for the other add-ons, you can start Odoo as follows:

```
$ python odoo.py --log-handler=odoo.addons.my_hostel:DEBUG
```

It is possible to specify `--log-handler` multiple times on the command line. You can also configure the **log handler** in the configuration file of your Odoo instance. In that case, you can use a comma-separated list of **prefix:level** pairs. For example, the following line is the same configuration for a minimal logging output as before. We maintain the most important messages and the error messages by default, except for messages produced by **werkzeug**, for which we only want critical messages, and `odoo.service.server`, for which we keep info-level messages, including server startup notifications:

```
log_handler = :ERROR,werkzeug:CRITICAL,odoo.service.server:INFO
```

The second way is to use the `--log-level` option. To control the log level globally, you can use `--log-level` as the command-line option. Possible values for this option are `critical`, `error`, `warn`, `debug`, `debug_rpc`, `debug_rpc_answer`, `debug_sql`, and `test`.

There are some shortcuts for setting logging levels. Here is a list of them:

- `--log-request` is a shortcut for `--log-handler=odoo.http.rpc.request:DEBUG`
- `--log-response` is a shortcut for `--log-handler=odoo.http.rpc.response:DEBUG`
- `--log-web` is a shortcut for `--log-handler=odoo.http:DEBUG`
- `--log-sql` is a shortcut for `--log-handler=odoo.sql_db:DEBUG`

Using the Odoo shell to interactively call methods

The Odoo web interface is meant for end users, although the developer mode unlocks a number of powerful features. However, testing and debugging through the web interface is not the easiest way to do things, as you need to manually prepare the data, navigate in the menus to perform actions, and so on. The Odoo shell is a **command-line interface**, which you can use to issue calls. This recipe shows how to start the Odoo shell and perform actions such as calling a method inside the shell.

Getting ready

We will reuse the same code as in the previous recipe to produce server logs to help debug methods. This allows the `product.product` model to add a new method. We will assume that you have an instance with the add-on installed and available. In this recipe, we expect that you have an Odoo configuration file for this instance called `project.conf`.

How to do it...

In order to call the `export_stock_level()` method from the Odoo shell, perform the following steps:

1. Start the Odoo shell and specify your project configuration file:

```
$ ./odoo-bin shell -c project.conf --log-level=error
```

2. Check for error messages and read the information text that's displayed before the usual Python command-line prompt:

```
env: <odoo.api.Environment object at 0x7f48cc0868c0>
odoo: <module 'odoo' from '/home/serpentcs/workspace/17.0/
odoo/__init__.py'>
openerp: <module 'odoo' from '/home/serpentcs/workspace/17.0/
odoo/__init__.py'>
self: res.users(1,)
Python 3.10.13 (main, Aug 25 2023, 13:20:03) [GCC 9.4.0]
```

```
Type 'copyright', 'credits' or 'license' for more information
IPython 7.13.0 -- An enhanced Interactive Python. Type '?' for
help.
```

3. Get a record set for `product.product`:

```
>>> product = env['product.product']
```

4. Get the main stock location record:

```
>>> location_stock = env.ref('stock.stock_location_stock')
```

5. Call the `export_stock_level()` method:

```
>>> product.export_stock_level(location_stock)
```

6. Commit the transaction before exiting:

```
>>> env.cr.commit()
```

7. Exit the shell by pressing *Ctrl* + *D*.

How it works...

Step 1 uses `odoo-bin shell` to start the Odoo shell. All the usual command-line arguments are available. We use `-c` to specify a project configuration file and `--log-level` to reduce the verbosity of the logs. When debugging, you may want to have a logging level of `DEBUG` for some specific add-ons.

Before providing you with a Python command-line prompt, `odoo-bin shell` starts an Odoo instance that does not listen on the network and initializes some global variables, which are mentioned in the output:

- `env` is an environment that's connected to the database and specified on the command line or in the configuration file.
- `odoo` is the `odoo` package that's imported for you. You get access to all the Python modules within that package to do what you want.
- `openerp` is an alias for the `odoo` package for backward compatibility.
- `self` is a record set of `res.users` that contains a single record for the Odoo superuser (administrator), which is linked to the `env` environment.

Steps 3 and *4* use `env` to get an empty record set and find a record according to the XML ID. *Step 5* calls the method on the `product.product` record set. These operations are identical to what you would use inside a method, with a minor difference being that we use `env` and not `self.env` (although we can have both, as they are identical). Take a look at *Chapter 5, Basic Server-Side Development*, for more information on what is available.

Step 6 commits the database transaction. This is not strictly necessary here because we did not modify any record in the database, but if we had done so and wanted these changes to persist, this is necessary; when you use Odoo through the web interface, each RPC call runs in its own database transaction, and Odoo manages these for you. When running in shell mode, this no longer happens and you have to call `env.cr.commit()` or `env.cr.rollback()` yourself. Otherwise, when you exit the shell, any transaction in progress is automatically rolled back. When testing, this is fine, but if you use the shell, for example, to script the configuration of an instance, don't forget to commit your work!

There's more...

In shell mode, by default, Odoo opens Python's REPL shell interface. You can use the REPL of your choice using the `--shell-interface` option. The supported REPLs are `ipython`, `ptpython`, `bpython`, and `python`:

```
$ ./odoo-bin shell -c project.conf  --shell-interface=ptpython
```

Using the Python debugger to trace method execution

Sometimes, application logs are not enough to figure out what is going wrong. Fortunately, we also have the Python debugger. This recipe shows us how to insert a breakpoint in a method and trace the execution by hand.

Getting ready

We will reuse the `export_stock_level()` method that was shown in the *Using the Odoo shell to interactively call methods* recipe of this chapter. Ensure that you have a copy to hand.

How to do it...

To trace the execution of `export_stock_level()` with pdb, perform the following steps:

1. Edit the code of the method, and insert the line highlighted here:

```
def export_stock_level(self, stock_location):
    import pdb; pdb.set_trace()
    products = self.with_context( location=stock_location.id
).search([])
    fname = join(EXPORTS_DIR, 'stock_level.txt')
    try:
        with open(fname, 'w') as fobj:
            for prod in products.filtered('qty_available'):
                fobj.write('%s\t%f\n' % (prod.name, prod.qty_
available))
    except IOError:
         raise exceptions.UserError('unable to save file')
```

2. Run the method. We will use the Odoo shell, as explained in the *Using the Odoo shell to interactively call methods* recipe:

```
$ ./odoo-bin shell -c project.cfg --log-level=error
    [...]
    >>> product = env['product.product']
    >>> location_stock = env.ref('stock.stock_location_stock')
    >>> product.export_stock_level(location_stock)
    > /home/cookbook/stock_level/models.py(18)export_stock_
level()
    -> products = self.with_context(
    (Pdb)
```

3. At the `(Pdb)` prompt, issue the `args` command (the shortcut for which is `a`) to get the values of the arguments that were passed to the method:

```
(Pdb) a
self = product.product()
stock_location = stock.location(14,)
```

4. Enter the `list` command to check where in the code you are standing:

```
(Pdb) list
13        @api.model
14        def export_stock_level(self, stock_location):
15        _logger.info('export stock level for %s',
16                     stock_location.name)
17        import pdb; pdb.set_trace()
18 ->     products = self.with_context(
19        location=stock_location.id).search([])
20        products = products.filtered('qty_available')
21        _logger.debug('%d products in the location',
22                      len(products))
23        fname = join(EXPORTS_DIR, 'stock_level.txt')
(Pdb)
```

5. Enter the `next` command three times to walk through the first lines of the method. You may also use `n`, which is a shortcut:

```
(Pdb) next
> /home/cookbook/stock_level/models.py(19)export_stock_level()
-> location=stock_location.id).search([])
(Pdb) n
> /home/cookbook/stock_level/models.py(20)export_stock_level()
-> products = products.filtered('qty_available')
(Pdb) n
> /home/cookbook/stock_level/models.py(21)export_stock_level()
-> _logger.debug('%d products in the location',
```

```
(Pdb) n
> /home/cookbook/stock_level/models.py(22)export_stock_level()
-> len(products))
(Pdb) n
> /home/cookbook/stock_level/models.py(23)export_stock_level()
-> fname = join(EXPORTS_DIR, 'stock_level.txt')
(Pdb) n
> /home/cookbook/stock_level/models.py(24)export_stock_level()
-> try:
```

6. Use the `p` command to display the values of the `products` and `fname` variables:

```
(Pdb) p products
product.product(32, 14, 17, 19, 21, 22, 23, 29, 34, 33, 26, 27,
42)
(Pdb) p fname
'/srv/exports/stock_level.txt'
```

7. Change the value of `fname` to point to the `/tmp` directory:

```
(Pdb) !fname = '/tmp/stock_level.txt'
```

8. Use the `return` (shortcut: `r`) command to execute the current function:

```
(Pdb) return
--Return--
> /home/cookbook/stock_level/models.py(26)export_stock_level()-
>None
-> for product in products:
```

9. Use the `cont` (shortcut: `c`) command to resume the execution of the program:

```
(Pdb) c
>>>
```

How it works...

In *step 1*, we hardcoded a breakpoint in the source code of the method by calling the `set_trace()` method of the `pdb` module from the Python standard library. When this method is executed, the normal flow of the program stops, and you get a `(Pdb)` prompt into which you can enter `pdb` commands.

Step 2 calls the `stock_level_export()` method using shell mode. It is also possible to restart the server normally and use the web interface to generate a call to the method you need to trace by clicking on the appropriate elements of the user interface.

When you need to manually step through some code using the Python debugger, here are a few tips that will make your life easier:

- Reduce the logging level to avoid having too many log lines, which pollutes the output of the debugger. Starting at the `ERROR` level is generally fine. You may want to enable some specific loggers with a higher verbosity, which you can do using the `--log-handler` command-line option (refer to the *Producing server logs to help debug methods* recipe).
- Run the server with `--workers=0` to avoid any multiprocessing issues that can cause the same breakpoint to be reached twice in two different processes.
- Run the server with `--max-cron-threads=0` to disable the processing of the `ir.cron` periodic tasks, which may otherwise trigger while you are stepping through the method, which produces unwanted logs and side effects.

Steps 3 to *8* use several `pdb` commands to step through the execution of the method. Here's a summary of the main commands of `pdb`. Most of these are also available using the first letter as a shortcut. We indicate this in the following list by having the optional letters between parentheses:

- `h(elp)`: This displays help with the `pdb` commands.
- `a(rgs)`: This shows the value of the arguments of the current function/methods.
- `l(ist)`: This displays the source code being executed in chunks of 11 lines, initially centered on the current line. Successive calls will move further in the source code file. Optionally, you can pass two integers at the start and end, which specify the region to display.
- `p`: This prints a variable.
- `pp`: This pretty-prints a variable (useful with lists and dictionaries).
- `w(here)`: This shows the call stack, with the current line at the bottom and the Python interpreter at the top.
- `u(p)`: This moves up one level in the call stack.
- `d(own)`: This moves down one level in the call stack.
- `n(ext)`: This executes the current line of code and then stops.
- `s(tep)`: This is to step inside the execution of a method call.
- `r(eturn)`: This resumes the execution of the current method until it returns.
- `c(ont(inue))`: This resumes the execution of the program until the next breakpoint is hit.
- `b(reak) <args>`: This creates a new breakpoint and displays its identifier; `args` can be one of the following:
  - `<empty>`: This lists all breakpoints.
  - `line_number`: This breaks at the specified line in the current file.

    - `filename:line_number`: This breaks at the specified line of the specified file (which is searched for in the directories of `sys.path`).
    - `function_name`: This breaks at the first line of the specified function.
    - `tbreak <args>`: This is similar to break, but the breakpoint will be canceled after it has been reached, so successive execution of the line won't trigger it twice.
    - `disable bp_id`: This disables a breakpoint by ID.
    - `enable bl_id`: This enables a disabled breakpoint by ID.
- `j(ump) lineno`: The next line to execute will be the one specified. This can be used to rerun or skip some lines.
- `(!) statement`: This executes a Python statement. The `!` character can be omitted if the command does not look like a `pdb` command. For instance, you need it if you want to set the value of a variable named `a`, because `a` is the shortcut for the `args` command.

There's more...

In the recipe, we inserted a `pdb.set_trace()` statement to break into `pdb` for debugging. We can also start `pdb` directly from within the Odoo shell, which is very useful when you cannot easily modify the code of the project using `pdb.runcall()`. This function takes a method as the first argument and the arguments to pass to the function as the next argument. So, inside the Odoo shell, you do the following:

```
>>> import pdb
>>> product = env['product.product']
>>> location_stock = env.ref('stock.stock_location_stock')
>>> pdb.runcall(product.export_stock_level, location_stock)
> /home/cookbook/stock_level/models.py(16)export_stock_level()
-> products = self.with_context((Pdb)
```

In this recipe, we focused on the Python debugger from the Python standard library, `pdb`. It is very useful to know about this tool because it is guaranteed to be available on any Python distribution. There are other Python debuggers available, such as `ipdb` (`https://pypi.python.org/pypi/ipdb`) and `pudb` (`https://pypi.python.org/pypi/pudb`), which can be used as drop-in replacements for `pdb`. They share the same API, and most of the commands that you saw in this recipe were unchanged. Also, of course, if you develop for Odoo using a Python IDE, you will have access to a debugger that was integrated with it.

See also

If you want to learn more about the `pdb` debugger, refer to the full documentation of `pdb` at `https://docs.python.org/3.9/library/pdb.html`.

Understanding the debug mode options

In *Chapter 1*, *Installing the Odoo Development Environment*, we saw how to enable debug/developer options in Odoo. These options are very helpful in debugging and reveal some further technical information. In this recipe, we will look at these options in detail.

How to do it...

Check the *Activating the Odoo developer tools* recipe of *Chapter 1*, *Installing the Odoo Development Environment*, and activate developer mode. After activating developer mode, you will see a drop-down menu with a bug icon in the top bar, as shown here:

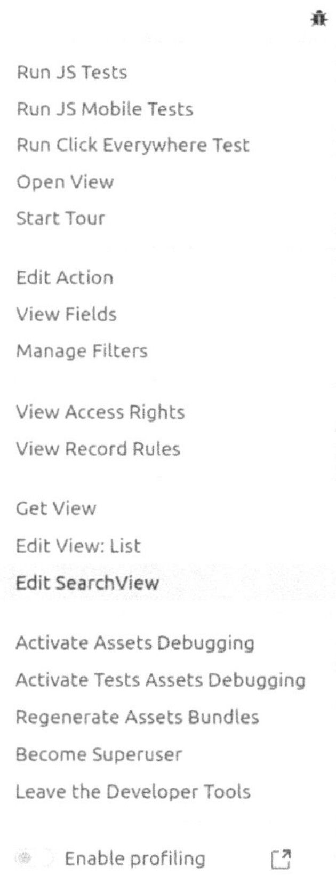

Figure 7.1 – Available options after activating debug mode

In this menu, you will see various options. Give them a go to see them in action. The next section will explain these options in more detail.

How it works...

Let's learn more about the options in the following points:

- **Run JS Tests**: This option will redirect you to the JavaScript QUnit test case page, as shown in the following screenshot. It will start running all test cases one by one. Here, you can see the progress and the status of the test cases. In *Chapter 18*, *Automated Test Cases*, we will see how can we create our own QUnit JavaScript test cases:

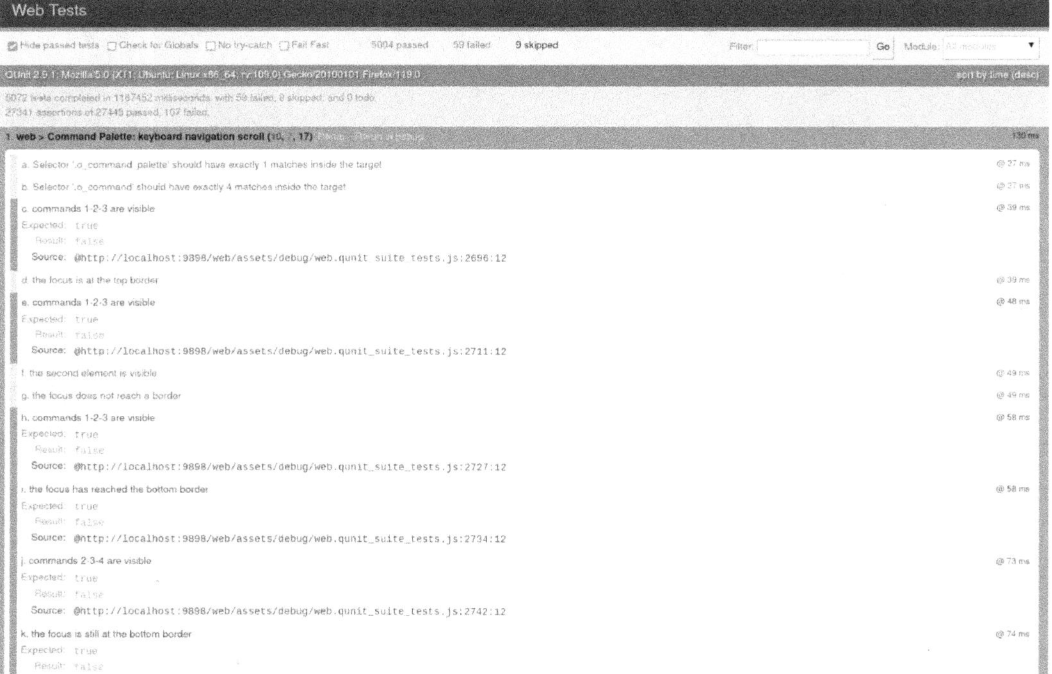

Figure 7.2 – QUnit test case result screen

- **Run JS Mobile Tests**: Similar to the preceding option, but this one runs a QUnit test case for a mobile environment.
- **Run Click Anywhere Tests**: This option will start clicking on all menus one by one. It will click in all the views and search filters. If something is broken or there is any regression, it will show the tracebacks. To stop this test, you will need to reload the page.
- **Open View**: This option will open a list of all available views. By selecting any of them, you can open that view without defining any menus or actions.
- **Disable Tours**: Odoo uses tours to improve the onboarding of new users. If you want to disable it, you can do it by using this option.

- **Start Tour**: Odoo also uses tours for automated testing. We will create a custom onboarding tour in *Chapter 15*, *Web Client Development*. This option will open a dialog box with a list of all tours, as shown in the following screenshot. By clicking on the play button next to a tour, Odoo will automatically perform all the steps of the tour:

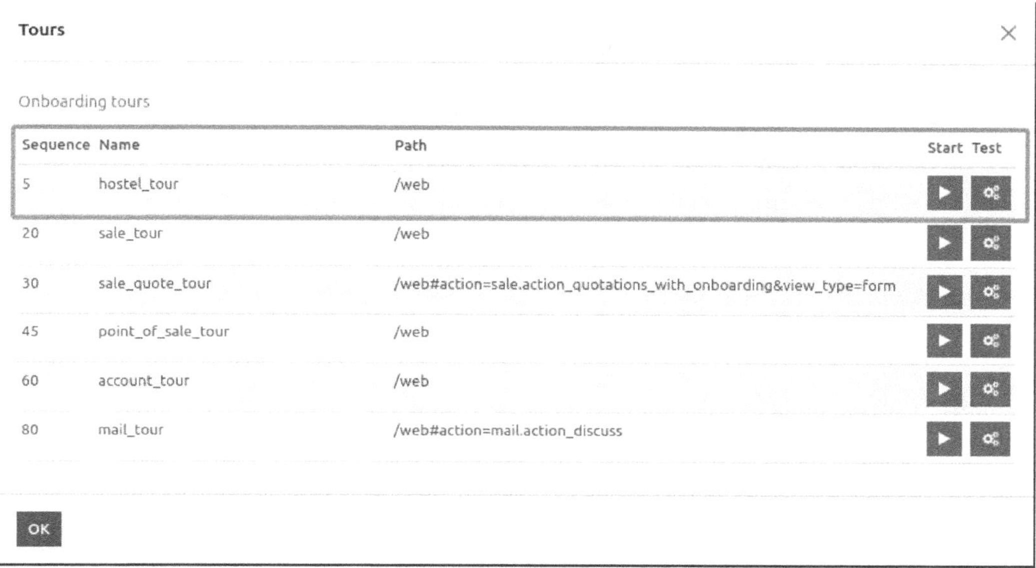

Figure 7.3 – Dialog to manually launch tours

- **Edit Action**: In the *Adding menu items and views* recipe of *Chapter 3*, *Creating Odoo Add-On Modules*, we added a menu item and an action to open views in Odoo. Details of these actions are also stored in the database as a record. This option will open the record details of the action we open to display the current view.
- **View Fields**: This option is used when you want to see the details of fields from the user interface. It will show a list of fields for the current model. For example, if you open a tree or form view for a `hostel.hostel` model, this option will show a list of fields for the `hostel.hostel` model.
- **Manage Filters**: In Odoo, users can create custom filters from the search view. This option will open a list of custom filters for the current model. Here, you can modify the custom filters.
- **Technical Translations**: This option will open a list of translated terms for the current model. You can modify the technical translation terms for your model from here. You can refer to *Chapter 11*, *Internationalization*, to learn more about translations.
- **View Access Rights**: This option will show a list of security access rights for the current model.
- **View Record Rules**: This option will show a list of security record rules for the current model.

- **Fields View Get**: You can extend and modify an existing view from other add-on modules. In some applications, these views are inherited by several add-on modules. Because of this, it is very difficult to get a clear idea of the whole view definition. With this option, you will get the final view definition after applying all view inheritances. Internally, it uses the `fields_view_get()` method.
- **Edit View: <view type>**: This option will open the dialog with the `ir.ui.view` record of the current view. This option is dynamic and it will show an option based on the view that is currently open. This means that if you open **Kanban View**, you will get an **Edit View: Kanban** option, and if you open **Form View**, you will get an **Edit View: Form** option.

> **Important tip**
>
> You can modify the view definition from the **Edit View** option. This updated definition will be applicable on the current database and these changes will be removed when you update the module. It's therefore better to modify views from modules.

- **Edit ControlPanelView**: This option is the same as the preceding one, but it will open the `ir.ui.view` record of the current model's search view.
- **Activate Assets Debugging**: Odoo provides two types of developer mode: *Developer mode* and *Developer mode with assets*. With this option, you can switch from *Developer* mode to *Developer mode with assets* mode. Check the *Activating the Odoo developer tools* recipe in *Chapter 1, Installing the Odoo Development Environment,* for more details.
- **Activate Test Assets Debugging**: As we know, Odoo uses tours for testing. Enabling this mode will load test assets in Odoo. This option will show some more tours in the **Start tour** dialog.
- **Regenerate Assets Bundles**: Odoo manages all CSS and JavaScript through asset bundles. This option deletes the old JavaScript and CSS assets and generates new ones. This option is helpful when you are getting issues because of asset caching. We will learn more about asset bundles in *Chapter 14, CMS Website Development.*
- **Become Super User**: This is a new option added from version 12. By activating this option, you switch to a super user. You can access the records even if you don't have access rights. This option is not available for all users; it is only available for users who have **Administration: settings** access rights. After activating this mode, you will see a striped top menu, as shown here:

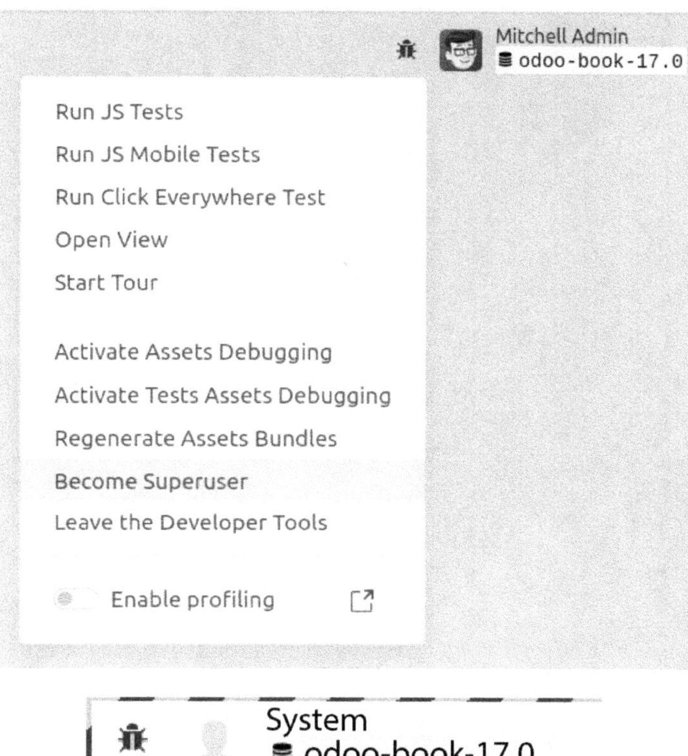

Figure 7.4 – Menu after activating a super user

- **Leave Developer Tools**: This option allows you to leave developer mode.

We have seen all of the options that are available under the debug menu. These options can be used in several ways, such as debugging, testing, and fixing issues. They can also be used to explore the source code for views.

8

Advanced Server-Side Development Techniques

In *Chapter 5, Basic Server-Side Development*, you learned how to write methods for a model class, how to extend methods from inherited models, and how to work with record sets. This chapter will deal with more advanced topics, such as working with the environment of a record set, calling a method upon a button click, and working with `onchange` methods. The recipes in this chapter will help you manage more complex business problems. You will learn how to create an understanding by incorporating visual elements and clarifying the process of creating interactive features within Odoo's application development process.

In this chapter, we will look at the following recipes:

- Changing the user that performs an action
- Calling a method with a modified context
- Executing raw SQL queries
- Writing a wizard to guide the user
- Defining `onchange` methods
- Calling `onchange` methods on the server side
- Defining `onchange` with the `compute` method
- Defining a model based on a SQL view
- Adding custom Settings options
- Implementing `init` hooks

Technical requirements

For this chapter, you'll require the Odoo online platform.

All the code used in this chapter can be downloaded from this book's GitHub repository at `https://github.com/PacktPublishing/Odoo-17-Development-Cookbook-Fifth-Edition/tree/main/Chapter08`.

Changing the user that performs an action

When writing business logic code, you may have to perform some actions with a different security context. A typical case is performing an action with `superuser` rights, bypassing security checks. Such a requirement arises when business requirements necessitate operating on records for which users do not have security access rights.

This recipe will show you how to allow normal users to create the `room` record by using `sudo()`. Put simply, we will allow users to create `room` by themselves, even if they do not have the right to create a assign the `room` record.

Getting ready

For easier understanding, we will add a new model to manage the hostel room. We will add a new model called `hostel.student`. You can refer to the following definition to add this model:

```
class HostelStudent(models.Model):
    _name = "hostel.student"
    _description = "Hostel Student Information"

    name = fields.Char("Student Name")
    gender = fields.Selection([("male", "Male"),
        ("female", "Female"), ("other", "Other")],
        string="Gender", help="Student gender")
    active = fields.Boolean("Active", default=True,
        help="Activate/Deactivate hostel record")
    hostel_id = fields.Many2one("hostel.hostel", "hostel", help="Name
of hostel")
    room_id = fields.Many2one("hostel.room", "Room",
        help="Select hostel room")
    status = fields.Selection([("draft", "Draft"),
        ("reservation", "Reservation"), ("pending", "Pending"),
        ("paid", "Done"),("discharge", "Discharge"), ("cancel",
"Cancel")],
        string="Status", copy=False, default="draft",
        help="State of the student hostel")
    admission_date = fields.Date("Admission Date",
        help="Date of admission in hostel",
```

```
        default=fields.Datetime.today)
    discharge_date = fields.Date("Discharge Date",
        help="Date on which student discharge")
    duration = fields.Integer("Duration", compute="_compute_check_
duration", inverse="_inverse_duration",
                                   help="Enter duration of living")
```

You will need to add a form view, an action, and a menu item to see this new model from the user interface. You will also need to add security rules for the hostel so that they can issue the hostel student. Please refer to *Chapter 3*, *Creating Odoo Add-On Modules*, if you don't know how to add these things.

Alternatively, you can use the ready-made initial module from our GitHub code examples to save time. This module is available in the `Chapter08/00_initial_module` folder. The GitHub code examples are available at `https://github.com/PacktPublishing/Odoo-17-Development-Cookbook-Fifth-Edition/tree/main/Chapter08/00_initial_module`.

How to do it...

If you have tested the module, you will find that only users who have `hostel.room` access rights can mark a room as a manager. Non-hostel users cannot create a room by themselves; they need to ask a manager user:

1. This user has **Hostel Manager** access rights, which means they can create **Hostel Room** records:

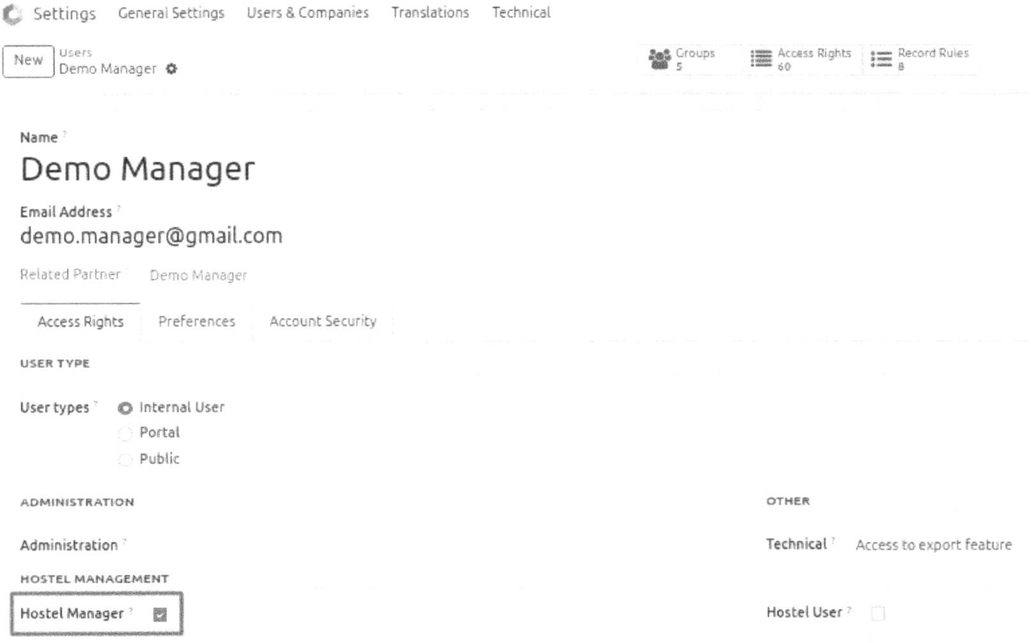

Figure 8.1 – This user has Hostel Manager access rights

As shown in the following screenshot, **Hostel Manager** can also create room records:

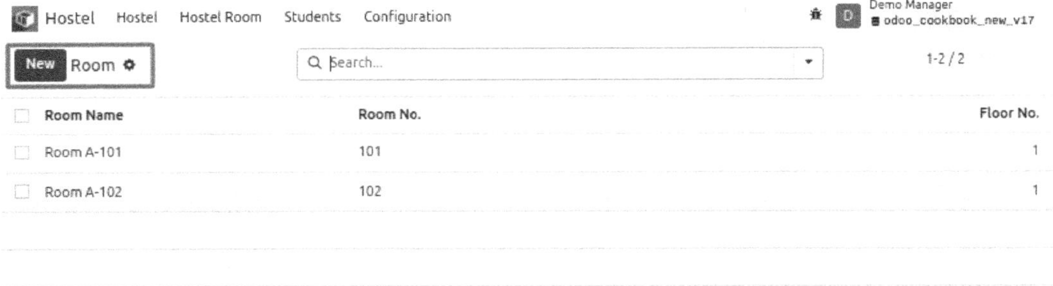

Figure 8.2 – Hostel Manager can create room records

2. This user has **Hostel User** access rights:

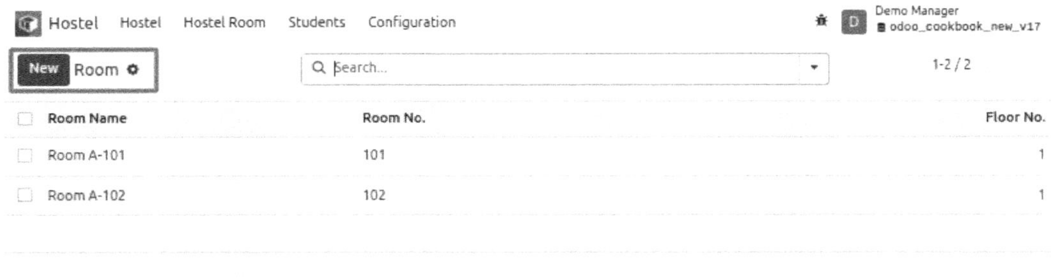

Figure 8.3 – This user has Hostel User access rights

They can only see **Hostel Room** records:

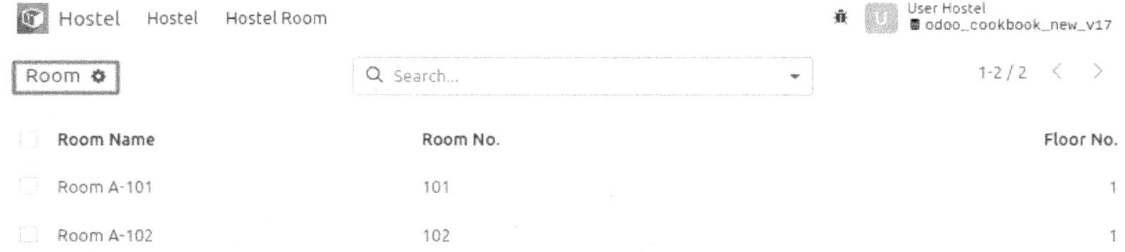

Figure 8.4 – Hostel User can see only Hostel Room records

Suppose that we want to add a new feature so that non-hostel users can create a room by themselves, for themselves. We will do this without giving them the access rights for the `hostel.room` model.

So, let's learn how to let normal hostel users student.

3. Add the `action_assign_room()` method to the `hostel.room` model:

```
class HostelStudent(models.Model):
   _name = "hostel.student"
    ...
    def action_assign_room(self):
```

4. In the method, ensure that we are acting on a single record:

```
self.ensure_one()
```

5. Raise a warning if a student is not paid (make sure you have imported `UserError` at the top):

```
if self.status != "paid":
           raise UserError(_("You can't assign a room if it's
not paid."))
```

6. Get the empty recordset of `hostel.room` as a superuser:

```
room_as_superuser = self.env['hostel.room'].sudo()
```

7. Create a new `room` record with the appropriate values:

```
room_rec = room_as_superuser.create({
           "name": "Room A-103",
           "room_no": "A-103",
           "floor_no": 1,
           "room_category_id": self.env.ref("my_hostel.single_
room_categ").id,
           "hostel_id": self.hostel_id.id,
       })
```

8. To trigger this method from the user interface, add the button to the student form view:

```
<button name="action_assign_room"
                          string="Assign Room"
                          type="object"
                          class="btn-primary"
                      />
```

9. Restart the server and update `my_hostel` to apply the given changes. After the update, you will see an **Assign Room** button on the student form view, as shown here:

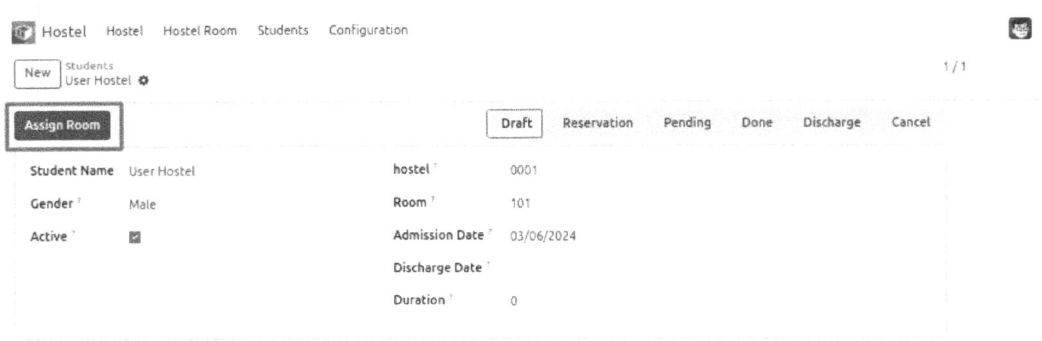

Figure 8.5 – The Assign Room button on the student form view

When you click on that, a new room record will be created. This will also work for non-hostel users. You can test this by accessing Odoo as a demo user.

How it works...

In the first three steps, we added a new method called `action_assign_room()`. This method will be called when the user clicks on the **Assign Room** button on the student form view.

In *Step 4*, we used `sudo()`. This method returns a new recordset with a modified `environment` in which the user has `superuser` rights. When `recordset` is called with `sudo()`, the environment will modify the `environment` attribute to `su`, which indicates the `superuser` state of the environment. You can access its status via `recordset.env.su`. All method calls through this `sudo` recordset are made with superuser privileges. To get a better idea of this, remove `.sudo()` from the method and then click on the **Assign Room** button. It will raise `Access Error` and the user will no longer have access to the model. Simply using `sudo()` will bypass all security rules.

If you need a specific user, you can pass a recordset containing either that user or the database ID of the user, as follows:

```
public_user = self.env.ref('base.public_user')
hostel_room = self.env['hostel.room'].with_user(public_user)
hostel_room.search([('name', 'ilike', 'Room 101')])
```

This code snippet allows you to search for rooms that are visible using the `public` user.

There's more...

Using `sudo()`, you can bypass the access rights and security record rules. Sometimes, you can access multiple records that are meant to be isolated, such as records from different companies in multi-company environments. The `sudo()` recordset bypasses all the security rules of Odoo.

If you are not careful, records that are searched for in this environment may be linked to any company present in the database, which means that you may be leaking information to a user; worse, you may be silently corrupting the database by linking records that belong to different companies.

When using `sudo()`, exercise caution to avoid unintended consequences, such as inadvertently linking records from different companies. Ensure proper data segregation and consider the potential impact on data integrity and security rules before bypassing access rights.

> **Important tip**
>
> When using `sudo()`, always double-check to ensure that your calls to `search()` do not rely on the standard record rules to filter the results.
>
> Without using `sudo()`, `search()` calls would respect standard record rules, potentially restricting access to records based on user permissions. This could lead to incomplete or inaccurate search results, affecting data visibility and application functionality.

See also

Check out these references for more information:

- If you want to learn more about environments, refer to the *Obtaining an empty recordset for a model* recipe in *Chapter 5, Basic Server-Side Development*
- For more information about access control lists and record rules, check out *Chapter 10, Security Access*

Calling a method with a modified context

`context` is part of the environment of a recordset. It is used to pass extra information, such as the time zone and the language of the user, from the user interface. You can also use the context to pass the parameters specified in the actions. Several methods in the standard Odoo add-ons use the context to adapt their business logic based on these context values. It is sometimes necessary to modify the context on a `recordset` value to get the desired results from a method call or the desired value for a computed field.

This recipe will show you how to change the behavior of a method based on values in the environmental context.

Getting ready

For this recipe, we will use the `my_hostel` module from the previous recipe. On the form view of the `hostel.room` model, we will add a button to remove room members. If a regular resident of a hostel removes other occupants from their assigned room without permission or authorization, it could create disruptions and issues within the accommodation. Note that we already have the same button in the form view of the room, but here, we will explore context usage in Odoo, gaining insights into how it influences system operations and outcomes.

How to do it...

To add a button, you need to perform the following steps:

1. Add a **Remove Room Members** button to the form view of `hostel.room`:

```
<button name="action_remove_room_members"
                            string="Remove Room Members"
                            type="object"
                            class="btn-primary"
                        />
```

2. Add the `action_remove_room_members()` method to the `hostel.room` model:

```
def action_remove_room_members(self):
    ...
```

3. Add the following code to the method to change the context of the environment and call the method to remove the room members:

```
student.with_context(is_hostel_room=True).action_remove_room()
```

4. Update the `action_remove_room()` method of the `hostel.student` model so that a different behavior is exhibited:

```
def action_remove_room(self):
        if self.env.context.get("is_hostel_room"):
            self.room_id = False
```

How it works...

In Odoo, to modify behavior influenced by context, we did the following:

1. Identified the target behavior.
2. Defined the contextual parameters.
3. Adapted the relevant code sections.
4. Tested the changes thoroughly.
5. Ensured compatibility across modules.

In *Step 1*, we removed the room members.

In *Step 2*, we added a new button, **Remove Room Members**. The user will use this button to **Remove the Members**.

In *Steps 3* and *4*, we added a method that will be called when the user clicks on the **Remove Room Members** button.

In *Step 5*, we called `student.with_context()` with some keyword arguments. This returns a new version of the `room_id` recordset with an updated context. We are adding one key to the context here, `is_hostel_room=True`, but you can add multiple keys if you want. We used `sudo()` here.

In *Step 6*, we checked whether the context had a positive value for the `is_hostel_room` key.

Now, when the hostel room removes room members in the student form view, the `room` recordset is `False`.

This is just a simple example of a modified context, but you can use any method, such as `create()`, `write()`, `unlink()`, and so on. You can also use any custom method based on your requirements.

There's more...

It is also possible to pass a dictionary to `with_context()`. In this case, the dictionary is used as the new context, which overwrites the current one. So, *Step 5* can also be written as follows:

```
new_context = self.env.context.copy()
new_context.update({'is_hostel_room': True})
student.with_context(new_context)
```

See also

Refer to the following recipes to learn more about contexts in Odoo:

- The *Obtaining an empty recordset for a model* recipe in *Chapter 5, Basic Server-Side Development*, explains what the environment is
- The *Passing parameters to forms and actions – context* recipe in *Chapter 9, Backend Views*, explains how to modify the context in action definitions
- The *Search for records* recipe in *Chapter 5, Basic Server-Side Development*, explains active records

Executing raw SQL queries

Most of the time, you can perform the operations you want by using Odoo's ORM – for example, you can use the `search()` method to fetch records. However, sometimes, you need more; either you cannot express what you want using the domain syntax (for which some operations are tricky, if not downright impossible) or your query requires several calls to `search()`, which ends up being inefficient.

This recipe shows you how to use raw SQL queries to get the name and amount a user keeps in a particular room.

Getting ready

For this recipe, we will use the `my_hostel` module from the previous recipe. For simplicity, we will just print the results in a log, but in real scenarios, you will need to use the query result in your business logic. In *Chapter 9*, *Backend Views*, we will display the result of this query in the user interface.

How to do it...

To get information about the name and amount a user keeps in a particular room, you need to perform the following steps:

1. Add the `action_category_with_amount()` method to `hostel.room`:

```
def action_category_with_amount(self):
    ...
```

2. In the method, write the following SQL query:

```
"""
            SELECT
                hrc.name,
                hrc.amount
            FROM
                hostel_room AS hostel_room
            JOIN
                hostel_room_category as hrc ON hrc.id = hostel_
room.room_category_id
            WHERE hostel_room.room_category_id = %(cate_id)s;""",
            {'cate_id': self.room_category_id.id}
```

3. Execute the query:

```
self.env.cr.execute("""
            SELECT
                hrc.name,
                hrc.amount
            FROM
                hostel_room AS hostel_room
            JOIN
                hostel_room_category as hrc ON hrc.id = hostel_
room.room_category_id
            WHERE hostel_room.room_category_id = %(cate_id)s;""",
            {'cate_id': self.room_category_id.id})
```

4. Fetch the result and log it (make sure you have imported `logger`):

```
result = self.env.cr.fetchall()
        _logger.warning("Hostel Room With Amount: %s", result)
```

5. Add a button in the form view of the `hostel.room` mode to trigger our method:

```
<button name="action_category_with_amount"
                              string="Log Category With Amount"
                              type="object"
                              class="btn-primary"/>
```

Don't forget to import `logger` in this file. Then, restart and update the `my_hostel` module.

How it works...

In *Step 1*, we added the `action_category_with_amount()` method, which will be called when the user clicks on the **Log Category With Amount.** button.

In *Step 2*, we declared a SQL **SELECT** query. This will return the category that states the amount in a hostel room. If you run this query in the PostgreSQL CLI, you will get a result based on your room data. Here is the sample date based on my database:

```
+---------------------------------------+-------+
| name                                  | amount|
|---------------------------------------+-------|
| Single Room                           | 3000  |
+---------------------------------------+-------+
```

In *Step 4*, we called the `execute()` method on the database cursor stored in `self.env.cr`. This sends the query to PostgreSQL and executes it.

In *Step 5*, we used the `fetchall()` method of the cursor to retrieve a list of rows selected by the query. This method returns a list of rows. In my case, this is `[('Single Room', 3000)]`. From the form of the query we execute, we know that each row will have exactly two values, the first being `name` and the other being the amount a user holds in a particular room. Then, we simply log it.

In *step 6*, we added an Add button to handle user actions.

> **Important note**
>
> If you are executing an `UPDATE` query, you need to manually invalidate the cache since Odoo ORM's cache is unaware of the changes you made with the `UPDATE` query. To invalidate the cache, you can use `self.invalidate_cache()`.

There's more...

The object in `self.env.cr` is a thin wrapper around a `psycopg2` cursor. The following methods are the ones that you will want to use most of the time:

- `execute(query, params)`: This executes the SQL query with the parameters marked as `%s` in the query substituted with the values in `params`, which is a tuple

> **Warning**
> Never do the substitution yourself; always use formatting options such as `%s`. If you use a technique such as string concatenation, it can make the code vulnerable to SQL injection.

- `fetchone()`: This returns one row from the database, wrapped in a tuple (even if only one column has been selected by the query)
- `fetchall()`: This returns all the rows from the database as a list of tuples
- `dictfetchall()`: This returns all the rows from the database as a list of dictionaries mapping column names to values

Be very careful when dealing with raw SQL queries:

- You are bypassing all the security of the application. Ensure that you call `search([('id', 'in', tuple(ids)])` with any list of IDs you are retrieving to filter out records to which the user has no access.
- Any modifications you are making are bypassing the constraints set by the add-on modules, except the `NOT NULL`, `UNIQUE`, and `FOREIGN KEY` constraints, which are enforced at the database level. This is also the case for any computed field recomputation triggers, so you may end up corrupting the database.
- Avoid the `INSERT/UPDATE` query – inserting or updating records via queries will not run any business logic written by overriding the `create()` and `write()` methods. It will not update stored compute fields and the ORM constraints will be bypassed too.

See also

For access rights management, refer to *Chapter 10, Security Access*.

Writing a wizard to guide the user

In the *Using abstract models for reusable model features* recipe in *Chapter 4, Application Models*, the `models.TransientModel` base class was introduced. This class has a lot in common with normal models, except that the records of transient models are periodically cleaned up in the database, hence the name transient. These are used to create wizards or dialogue boxes, which are filled in the user interface by the users and are generally used to perform actions on the persistent records of the database.

Getting ready

For this recipe, we will use the `my_hostel` module from the previous recipes. This recipe will add a new wizard. With this wizard, the user will be assigned the room.

How to do it...

Follow these steps to add a new wizard for updating the assign room and hostel records:

1. Add a new transient model to the module with the following definition:

```
class AssignRoomStudentWizard(models.TransientModel):
   _name = 'assign.room.student.wizard'

   room_id = fields.Many2one("hostel.room", "Room",
required=True)
```

2. Add the `callback` method that performs the action on the transient model. Add the following code to the `AssignRoomStudentWizard` class:

```
def add_room_in_student(self):
        hostel_room_student = self.env['hostel.student'].browse(
            self.env.context.get('active_id'))
        if hostel_room_student:
            hostel_room_student.update({
                'hostel_id': self.room_id.hostel_id.id,
                'room_id': self.room_id.id,
                'admission_date': datetime.today(),
            })
```

3. Create a form view for the model. Add the following view definition to the module views:

```
<record id='assign_room_student_wizard_form' model='ir.ui.view'>
   <field name='name'>assign room student wizard form view</
field>
   <field name='model'>assign.room.student.wizard</field>
   <field name='arch' type='xml'>
       <form string="Assign Room">
           <sheet>
               <group>
                   <field name='room_id'/>
               </group>
           </sheet>
           <footer>
               <button string='Update' name='add_room_in_
student' class='btn-primary' type='object'/>
               <button string='Cancel' class='btn-default'
special='cancel'/>
```

```
            </footer>
        </form>
    </field>
</record>
```

4. Create an action and a menu entry to display the wizard. Add the following declarations to the module menu file:

```
<record model="ir.actions.act_window" id="action_assign_room_
student_wizard">
    <field name="name">Assign Room</field>
    <field name="res_model">assign.room.student.wizard</field>
    <field name="view_mode">form</field>
    <field name="target">new</field>
</record>
```

5. Add access rights for `assign.room.student.wizard` in the `ir.model.access.csv` file:

```
access_assign_room_student_wizard_manager,access.assign.room.
student.wizard.manager,model_assign_room_student_wizard,my_
hostel.group_hostel_manager,1,1,1,1
```

6. Update the `my_hostel` module to apply the changes.

How it works...

In *Step 1*, we defined a new model. It is no different from other models, apart from the base class, which is `TransientModel` instead of `Model`. Both `TransientModel` and `Model` share a common base class, called `BaseModel`, and if you check the source code of Odoo, you will see that 99% of the work is in `BaseModel` and that both `Model` and `TransientModel` are almost empty.

The only things that change for the `TransientModel` records are as follows:

- Records are periodically removed from the database so that the tables for transient models do not grow over time
- You are not allowed to define `one2many` fields on a `TransientModel` instance that refers to a normal model as this will add a column on the persistent model that links to transient data

Use *many2many* relations in this case. You can, of course, use *one2many* fields if the related model in *one2many* is also `TransientModel`.

We define one field in the model for storing the room. We can add other scalar fields so that we can record a scheduled return date, for instance.

In *Step 2*, we added the code to the wizard class that will be called when the button defined in *Step 3* is clicked on. This code reads the values from the wizard and updates the `hostel.student` record.

In *Step 3*, we defined a view for our wizard. Refer to the *Document-style forms* recipe in *Chapter 9, Backend Views*, for details. The important point here is the button in the footer; the `type` attribute is set to `'object'`, which means that when the user clicks on the button, the method with the name specified by the name attribute of the button will be called.

In *Step 4*, we ensured that we had an entry point for our wizard in the menu of the application. We use `target='new'` in the action so that the form view is displayed as a dialogue box over the current form. Refer to the *Adding a menu item and window action* recipe in *Chapter 9, Backend Views*, for details:

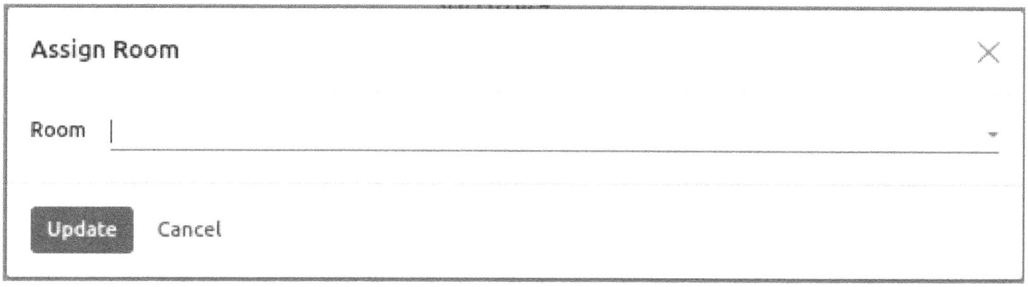

Figure 8.6 – Wizard for assigning a room to a student

In *Step 5*, we added access rights for the `assign.room.student.wizard` model. With this, the manager user will get full rights to the `assign.room.student.wizard` model.

> **Note**
>
> Before Odoo v14, `TransientModel` didn't require any access rules. Anyone can create a record, and they can only access records created by themselves. With Odoo v14, access rights are compulsory for `TransientModel`.

There's more...

Here are some tips to enhance your wizards.

Using the context to compute default values

The wizard we are presenting requires the user to fill in the name of the member in the form. There is a feature of the web client that we can use to save some typing. When an action is executed, `context` is updated with some values that can be used by wizards:

- `active_model`: This is the name of the model related to the action. This is generally the model being displayed onscreen.
- `active_id`: This indicates that a single record is active and provides the ID of that record.

- `active_ids`: If several records are selected, this will be a list containing the IDs. This happens when several items are selected in a tree view when the action is triggered. In a form view, you get `[active_id]`.
- `active_domain`: This is an additional domain on which the wizard will operate.

These values can be used to compute the default values of the model or even directly in the method called by the button. To improve on the example in this recipe, if we had a button displayed on the form view of a `hostel.room` model to launch the wizard, the context of the creation of the wizard would contain `{'active_model': 'hostel.room', 'active_id': <hostel_room_id>}`. In that case, you could define the `room_id` field to get a default value computed by the following method:

```
def _default_member(self):
    if self.context.get('active_model') == 'hostel.room':
        return self.context.get('active_id', False)
```

Wizards and code reuse

In *Step 2*, we can add `self.ensure_one()` at the beginning of the method, as follows:

```
def add_room_in_student(self):
        hostel_room_student = self.env['hostel.student'].browse(
            self.env.context.get('active_id'))
        if hostel_room_student:
            hostel_room_student.update({
                'hostel_id': self.room_id.hostel_id.id,
                'room_id': self.room_id.id,
                'admission_date': datetime.today(),
            })
```

We recommend using v17 in this recipe. It will allow us to reuse the wizard from other parts of the code by creating records for the wizard and putting them in a single recordset (refer to the *Combining recordsets* recipe in *Chapter 5, Basic Server-Side Development*, to see how to do this) before calling `add_room_in_student()` on the recordset. Here, the code is trivial, and you don't need to jump through all those hoops to record that some rooms have been assigned by different students. However, in an Odoo instance, some operations are much more complex, and it is always nice to have a wizard available that does the right thing. When using these wizards, ensure that you check the source code for any possible use of the `active_model/active_id/active_ids` keys from the context. If this is the case, you need to pass a custom context (refer to the *Calling a method with a modified context* recipe).

Redirecting the user

The method in *Step 2* doesn't return anything. This will cause the wizard dialogue to be closed after the action is performed. Another possibility is to have the method return a dictionary with the fields of `ir.action`. In this case, the web client will process the action as if a menu entry had been clicked on by the user. The `get_formview_action()` method defined in the `BaseModel` class can be used to achieve this. For instance, if we wanted to display the form view of the hostel room, we could have written the following:

```
def add_room_in_student(self):
        hostel_room_student = self.env['hostel.student'].browse(
            self.env.context.get('active_id'))
        if hostel_room_student:
            hostel_room_student.update({
                'hostel_id': self.room_id.hostel_id.id,
                'room_id': self.room_id.id,
                'admission_date': datetime.today(),
            })
        rooms = self.mapped('room_id')
        action = rooms.get_formview_action()
        if len(rooms.ids) > 1:
            action['domain'] = [('id', 'in', tuple(rooms.ids))]
            action['view_mode'] = 'tree,form'
        return action
```

This builds a list of rooms that have rooms from this wizard (in practice, there will only be one such room when the wizard is called from the user interface) and creates a dynamic action, which displays the room with the specified IDs.

The *redirecting the user* technique can be used to create a wizard that must perform several steps one after the other. Each step in the wizard can use the values of the previous steps by providing a **Next** button. This will call a method defined on the wizard that updates some fields on the wizard, returns an action that will redisplay the same updated wizard, and gets ready for the next step.

See also

Please refer to the following recipes for more details:

- Refer to the *Document-style forms* recipe in *Chapter 9, Backend Views*, for more details on defining a view for a wizard
- To understand more about views and calling server-side methods, refer to the *Adding a menu item and window action* recipe in *Chapter 9, Backend Views*
- For more details on creating records for the wizard and putting them in a single recordset, refer to the *Combining recordsets* recipe in *Chapter 5, Basic Server-Side Development*

Defining onchange methods

When writing business logic, it is often the case that some fields are interrelated. We looked at how to specify constraints between fields in the *Adding constraint validations to a model* recipe in *Chapter 4, Application Models*. This recipe illustrates a slightly different concept. Here, `onchange` methods are called when a field is modified in the user interface to update the values of other fields of the record in the web client, usually in a form view.

We will illustrate this by providing a wizard similar to the one defined in the *Writing a wizard to guide the user* recipe, but that can be used to record duration returns. When a date is set in the form view, the duration is updated for the student. While we are demonstrating `onchange` methods on `Model`, these features are also available on normal `Transient` models.

Getting ready

For this recipe, we will use the `my_hostel` module from the *Writing a form to guide the user* recipe of this chapter. We will create a hostel student and add an `onchange` method that will auto-fill the duration when a user selects a discharge date or admission date field.

You will also want to prepare your work by defining the following model for the form view:

```
class HostelStudent(models.Model):
    _name = "hostel.student"
    _description = "Hostel Student Information"

    admission_date = fields.Date("Admission Date",
        help="Date of admission in hostel",
        default=fields.Datetime.today)
    discharge_date = fields.Date("Discharge Date",
        help="Date on which student discharge")
    duration = fields.Integer("Duration",          inverse="_inverse_
duration",help="Enter duration of living")
```

Finally, you will need to define a view. These steps will be left as an exercise for you to carry out.

How to do it...

To automatically populate the duration to return when the user is changed, you need to add an `onchange` method in the `HostelStudent` step, with the following definition:

```
@api.onchange('admission_date', 'discharge_date')
def onchange_duration(self):
    if self.discharge_date and self.admission_date:
        self.duration = (self.discharge_date.year - \
                        self.admission_date.year) * 12 + \
```

```
                (self.discharge_date.month - \
                self.admission_date.month)
```

How it works...

An onchange method uses the @api.onchange decorator, which is passed the names of the fields that change and will thus trigger the call to the method. In our case, we say that whenever admission_date or discharge_date is modified in the user interface, the method must be called.

In the body of the method, we calculated the duration, and we used an attribute assignment to update the duration attribute of the from view.

There's more...

As we have seen in this recipe, the basic use of onchange methods is to compute new values for fields when some other fields are changed in the user interface.

Inside the body of the method, you get access to the fields that are displayed in the current view of the record, but not necessarily all the fields of the model. This is because onchange methods can be called while the record is being created in the user interface before it is stored in the database! Inside an onchange method, self is in a special state, denoted by the fact that self.id is not an integer, but an instance of odoo.models.NewId. Therefore, you must not make any changes to the database in an onchange method since the user may end up canceling the creation of the record, which will not roll back any changes made by the onchange method during the process of editing.

Calling onchange methods on the server side

The onchange method has a limitation: it will not be invoked when you are performing operations on the server side. onchange is only invoked automatically when the dependent operations are performed through the Odoo user interface. Yet, in several cases, these onchange methods must be called because they update important fields in the created or updated record. Of course, you can do the required computation yourself, but this is not always possible since the onchange method can be added or modified by a third-party add-on module that's been installed on the instance that you don't know about.

This recipe explains how to call the onchange methods on a record by manually invoking the onchange method before creating a record.

Getting ready

In the *Changing the user that performs an action* recipe, we added a **Return Room** button so that users can update the room and hostel by themselves. We now want to do the same for returning the room and hostel; we will just use the **Assign Room** return wizard.

How to do it...

In this recipe, we will manually update a record of the `hostel.room` model. To do this, you need to perform the following steps:

1. Import `Form` from the `tests` utility in the `hostel.student.py` file:

```
from odoo.tests.common import Form
```

2. Create the `return_room` method in the `hostel.room` model:

```
def return_room(self):
    self.ensure_one()
```

3. Get an empty recordset for `assign.room.student.wizard`:

```
wizard = self.env['assign.room.student.wizard']
```

4. Create a wizard `Form` block, like this:

```
with Form(wizard) as return_form:
```

5. Trigger `onchange` by assigning a room and then return the updated value of `room_id`:

```
return_form.room_id = self.env.ref('my_hostel.101_room')
        record = return_form.
save()                            record.with_context(active_
id=self.id).add_room_in_student()
```

How it works...

For an explanation of *Steps 1* to *3*, refer to the *Creating new records* recipe in *Chapter 5, Basic Server-Side Development*.

In *Step 4*, we created a virtual form to handle onchange specifications, such as the GUI.

Step 5 contains the full logic to return the room and hostel. In the first line, we assigned `room_id` in the wizard. Then, we called the `save()` method of the form, which returned a wizard record. After that, we called the `add_room_in_student()` method to execute the logic to return the updated room and hostel.

The `onchange` method is mostly invoked from the user interface. But in this recipe, we have learned how to use/trigger the business logic of the `onchange` method on the server side. This way, we can create records without bypassing any business logic.

See also

If you want to learn more about creating and updating records, refer to the *Creating new records* and *Updating the values of recordset records* recipes in *Chapter 5, Basic Server-Side Development*.

Defining onchange with the compute method

In the last two recipes, we saw how to define and call the onchange method. We also saw its limitation, which is that it can only be invoked automatically from the user interface. As a solution to this problem, Odoo v13 introduced a new way to define onchange behavior. In this recipe, we will learn how to use the compute method to produce behavior similar to that of the onchange method.

Getting ready

For this recipe, we will use the my_hostel module from the previous recipe. We will replace the onchange method of hostel.student with the compute method.

How to do it...

Follow these steps to modify the onchange method with the compute method:

1. Replace api.onchange in the onchange_duration() method with depends, like this:

```
@api.depends('admission_date', 'discharge_date')
 def onchange_duration(self):
        ...
```

2. Add the compute parameter in the definition of the field, like this:

```
duration = fields.Integer("Duration", compute="onchange_
duration", inverse="_inverse_duration",
                              help="Enter duration of living")
```

Upgrade the my_hostel module to apply the code, then test the return duration form to see the change.

How it works...

Functionally, our computed onchange works like the normal onchange method. The only difference is that now, onchange will be trigged upon backend changes too.

In *Step 1*, we replaced @api.onchange with @api.depends. This is required to recompute the method when the field value changes.

In *Step 2*, we registered the compute method with the field. As you may have noticed, we used readonly=False with the compute field definition. By default, compute methods are read-only, but by setting readonly=False, we are making sure that the field is editable and can be stored.

See also

To learn more about computed fields, refer to the *Adding computed fields to a model* recipe in *Chapter 4, Application Models*.

Defining a model based on a SQL view

When working on the design of an *add-on* module, we model the data in classes that are then mapped to database tables by Odoo's ORM. We apply some well-known design principles, such as *separation of concerns* and *data normalization*. However, at later stages of the module design, it can be useful to aggregate data from several models in a single table and to maybe perform some operations on them on the way, especially for reporting or producing dashboards. To make this easier, and to make use of the full power of the underlying `PostgreSQL` database engine in Odoo, it is possible to define a read-only model backed by a PostgreSQL view, rather than a table.

In this recipe, we will reuse the room model from the *Writing a wizard to guide the user* recipe in this chapter, and we will create a new model to make it easier to gather availability about rooms and authors.

Getting ready

For this recipe, we will use the `my_hostel` module from the previous recipe. We will create a new model called `hostel.room.availability` to hold the availability data.

How to do it...

To create a new model backed by a PostgreSQL view, follow these steps:

1. Create a new model with the `_auto` class attribute set to `False`:

```
class HostelRoomAvailability(models.Model):
    _name = 'hostel.room.availability'
    _auto = False
```

2. Declare the fields you want to see in the model, setting them as `readonly`:

```
room_id = fields.Many2one('hostel.room', 'Room', readonly=True)
student_per_room = fields.Integer(string="Student Per
Rooom",                         readonly=True)
availability = fields.
Integer(string="Availability",readonly=True)
amount = fields.Integer(string="Amount", readonly=True)
```

3. Define the `init()` method to create the view:

```
def init(self):
        tools.drop_view_if_exists(self.env.cr, self._table)
        query = """
        CREATE OR REPLACE VIEW hostel_room_availability AS (
        SELECT
                min(h_room.id) as id,
                h_room.id as room_id,
```

```
            h_room.student_per_room as student_per_room,
            h_room.availability as availability,
            h_room.rent_amount as amount
        FROM
            hostel_room AS h_room
        GROUP BY h_room.id
    );
    """
    self.env.cr.execute(query)
```

4. You can now define views for the new model. A pivot view is especially useful to explore data (refer to *Chapter 9*, *Backend Views*).
5. Don't forget to define some access rules for the new model (take a look at *Chapter 10*, *Security Access*).

How it works...

Normally, Odoo will create a new table for the model you are defining by using the field definitions for the columns. This is because, in the `BaseModel` class, the `_auto` attribute defaults to `True`. In *Step 1*, by setting this class attribute to `False`, we tell Odoo that we will manage this by ourselves.

In *Step 2*, we defined some fields that will be used by Odoo to generate a table. We took care to flag them as `readonly=True` so that the views do not enable modifications that you will not be able to save since PostgreSQL views are read-only.

In *Step 3*, we defined the `init()` method. This method normally does nothing; it is called after `_auto_init()` (which is responsible for table creation when `_auto = True` but does nothing otherwise), and we use it to create a new SQL view (or to update the existing view in the case of a module upgrade). The view creation query must create a view with column names that match the field names of the model.

Important tip

It is a common mistake to forget to rename the columns in the view definition query. This will cause an error message when Odoo cannot find the column.

Note that we also need to provide an `integer column` value called `ID` that contains unique values.

There's more...

It is also possible to have some computed and related fields on such models. The only restriction is that the fields cannot be stored (and therefore, you cannot use them to group records or search).

If you need to group by base user, you need to store the field by adding it to the view definition, rather than using a related field.

See also

To learn more, take a look at the following recipes:

- To learn more about UI views for user actions, refer to *Chapter 9, Backend Views*
- For a better understanding of access control and record rules, take a look at *Chapter 10, Security Access*

Adding custom Settings options

In Odoo, you can provide optional features through the **Settings** options. The user can enable or disable this option at any time. We will illustrate how to create **Settings** options in this recipe.

Getting ready

In previous recipes, we added buttons so that hostel users can click and return rooms. This is not the case for every hostel; however, we will create a **Settings** option to enable and disable this feature. We will do this by hiding these buttons. In this recipe, we will use the same `my_hostel` module from the previous recipes.

How to do it...

To create custom **Settings** options, follow these steps:

1. Add a new field by inheriting the `res.config.settings` model:

```
class ResConfigSettings(models.TransientModel):
   _inherit = 'res.config.settings'

   group_hostel_user = fields.Boolean(string="Hostel User",
implied_group='my_hostel.group_hostel_user')
```

2. Add this field to the existing **Settings** view with `xpath` (for more details, refer to *Chapter 9, Backend Views*):

```
<record id="res_config_settings_view_form" model="ir.ui.view">
       <field name="name">res.config.settings.view.form.inherit.
hostel</field>
       <field name="model">res.config.settings</field>
       <field name="priority" eval="5"/>
```

```
        <field name="inherit_id" ref="base.res_config_settings_
view_form"/>
        <field name="arch" type="xml">
            <xpath expr="//div[hasclass('settings')]"
position="inside">
                <div class="app_settings_block" data-
string="Hostel" string="Hostel" data-key="my_hostel" groups="my_
hostel.group_hostel_manager">
                    <h2>Hostel</h2>
                    <div class="row mt16 o_settings_container">
                        <div class="col-12 col-lg-6 o_setting_
box" id="hostel">
                            <div class="o_setting_left_pane">
                                <field name="group_hostel_user"/>
                            </div>
                            <div class="o_setting_right_pane">
                                <label for="group_hostel_user"/>
                                <div class="text-muted">
                                    Allow users to hostel user
                                </div>
                            </div>
                        </div>
                    </div>
                </div>
            </xpath>
        </field>
    </record>
```

3. Add some actions and a menu for **Settings**:

```
<record id="hostel_config_settings_action" model="ir.actions.
act_window">
        <field name="name">Settings</field>
        <field name="type">ir.actions.act_window</field>
        <field name="res_model">res.config.settings</field>
        <field name="view_id" ref="res_config_settings_view_
form"/>
        <field name="view_mode">form</field>
        <field name="target">inline</field>
        <field name="context">{'module' : 'my_hostel'}</field>
    </record>

    <menuitem name="Settings" id="hostel_setting_menu"
parent="hostel_main_menu" action="hostel_config_settings_action"
sequence="50"/>
```

4. Restart the server and update the `my_hostel` module to apply the changes, as shown here:

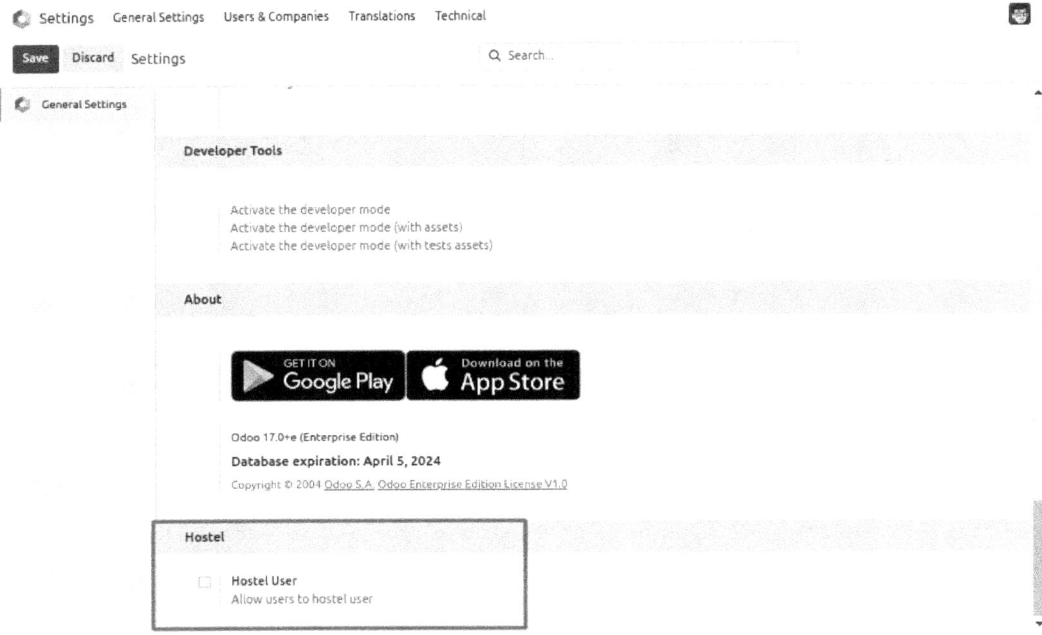

Figure 8.7 – The hostel user access right settings option to enable and disable this feature

How it works...

In Odoo, all settings options are added in the `res.config.settings` model. `res.config.settings` is a transient model. In *Step 1*, we created a new security group. We will use this group to create the **Hide** and **Show** buttons.

In *step 2*, we added a new `Boolean` field in the `res.config.settings` model by inheriting it. We added an `implied_group` attribute with a value of `my_hostel.group_hostel_user`. This group will be assigned to all `odoo` users when the admin enables or disables options with the `Boolean` field.

Settings uses a form view to display settings options on a user interface. All of these options are added in a single form view with an external ID of `base.res_config_settings_view_form`.

In *Step 3*, we added our option to the user interface by inheriting this setting from the view. We used `xpath` to add our `setting` option. We will cover this in more detail in *Chapter 9*, *Backend Views*. In the form definition, you will find that the attribute data-key value of this option will be your module name. This is only needed when you are adding a whole new tab in **Settings**. Otherwise, you can just add your option to the **Settings** tab of the existing module with `xpath`.

In *Step 4*, we added an action and a menu to access the configuration options from the user interface. You will need to pass the `{'module': 'my_hostel'}` context from the action to open the **Settings** tab of the `my_hostel` module by default when the menu is clicked.

In *Step 5*, we added `my_hostel.group_hostel_user` groups to the buttons. Because of this group, the **Hostel User** and **Return** buttons will be hidden or shown, based on the **Settings** options.

After this, you will see a separate **Settings** tab for the hostel, and, in the tab, you will see a `Boolean` field to enable or disable the **Hostel User** option. When you enable or disable this option, in the background, Odoo will apply or remove `implied_group` to or from all `odoo` users. Because we added the groups to buttons, the buttons will be displayed if the user has groups and will be hidden if the user doesn't have groups. We will look at security groups in detail in *Chapter 10, Security Access*.

There's more...

There are a few other ways to manage the **Settings** options. One of them is to separate features in the new module and `install` or `uninstall` them through various options. To do this, you will need to add a `Boolean` field with the name of the module prefixed with `module_`. If, for example, we create a new module called `my_hostel_extras`, you will need to add a `Boolean` field, as follows:

```
module_my_hostel_extras = fields.Boolean(
    string='Hostel Extra Features')
```

When you enable or disable this option, `odoo` will `install` or `uninstall` the `my_hostel_extras` module.

Another way to manage settings is to use system parameters. Such data is stored in the `ir.config_parameter` model. Here's how you can create system-wide global parameters:

```
digest_emails = fields.Boolean(
        string="Digest Emails",
        config_parameter='digest.default_digest_emails')
```

The `config_parameter` attribute in the fields will ensure the user data is stored in **System Parameters**, under the **Settings** | **Technical** | **Parameters** | **System Parameters** menu. The data will be stored with the `digest.default_digest_emails` key.

Settings options are used to make your application generic. These options give freedom to users and allow them to enable or disable features on the fly. When you convert a feature into options, you can serve more customers with one module and your customers can enable the feature whenever they like.

Implementing init hooks

In *Chapter 6, Managing Module Data*, you learned how to add, update, and delete records from XML or CSV files. Sometimes, however, the business case is complex, and it can't be solved using data files. In such cases, you can use the `init` hook from the manifest file to perform the operations you want.

Complex business cases may require dynamic initialization of data beyond standard XML or CSV files. Examples include integrating with external systems, performing complex calculations, or configuring records based on runtime conditions, facilitated by the `init` hook in the manifest file.

Getting ready

We will use the same `my_hostel` module from the previous recipe. For simplicity, in this recipe, we will just create some room records through `post_init_hook`.

How to do it...

To add `post_init_hook`, follow these steps:

1. Register the hook in the `__manifest__.py` file with the `post_init_hook` key:

```
...
'post_init_hook': 'add_room_hook',
...
```

2. Add the `add_room_hook()` method to the `__init__.py` file:

```
from odoo import api, SUPERUSER_ID


def add_room_hook(cr, registry):
    env = api.Environment(cr, SUPERUSER_ID, {})
    room_data1 = {'name': 'Room 1', 'room_no': '01'}
    room_data2 = {'name': 'Room 2', 'room_no': '02'}
    env['hostel.room'].create([room_data1, room_data2])
```

How it works...

In *Step 1*, we registered `post_init_hook` in the manifest file with the `add_room_hook` value. This means that after the module is installed, Odoo will look for the `add_room_hook` method in `__init__.py`. The `post_init_hook` value receives the environment as an argument, showcasing an instance of the `add_room_hook` function that executes after module installation.

In *Step 2*, we declared the `add_room_hook()` method, which will be called after the module is installed. We created two records from this method. In real-life scenarios, you can write complex business logic here.

In this example, we looked at `post_init_hook`, but Odoo supports two more hooks:

- `pre_init_hook`: This hook will be invoked when you start installing a module. It is the opposite of `post_init_hook`; it will be invoked after installing the current module:

  A. Register the hook in the `__manifest__.py` file with the `pre_init_hook` key:

  ```
  ...
  'pre_init_hook': 'pre_init_hook_hostel',
  ...
  ```

  B. Add the `pre_init_hook_hostel()` method to the `__init__.py` file:

  ```
  def pre_init_hook_hostel(env):
      env['ir.model.data'].search([
          ('model', 'like', 'hostel.hostel'),
      ]).unlink()
  ```

- `uninstall_hook`: This hook will be invoked when you uninstall the module. This is mostly used when your module needs a garbage-collection mechanism:

  C. Register the hook in the `__manifest__.py` file with the `uninstall_hook` key:

  ```
  ...
  'uninstall_hook': 'uninstall_hook_user',
  ...
  ```

  D. Add `uninstall_hook_user()` method to the `__init__.py` file:

  ```
  def uninstall_hook_user(env):
      hostel = env['res.users'].search([])
      hostel.write({'active': False})
  ```

Hooks are functions that run before, after, or in place of existing code. Hooks – functions that are displayed as strings – are contained in the `__init__ .py` file of an Odoo module.

9

Backend Views

In all previous chapters, you have seen the server and database side of Odoo. In this chapter, you will see the UI side of Odoo. You will learn how to create different types of views. Aside from the views, this chapter also covers other components, such as action buttons, menus, and widgets, which will help you make your application more user-friendly. After completing this chapter, you will be able to design the UI of an Odoo backend. Note that this chapter does not cover the website part of Odoo; we have a separate chapter (*14*) for that.

In this chapter, we will cover the following recipes:

- Adding a menu item and window actions
- Having an action open a specific view
- Adding content and widgets to a form view
- Adding buttons to forms
- Passing parameters to forms and actions – Context
- Defining filters on record lists – Domain
- Defining list views
- Defining search views
- Adding a search filter side panel
- Changing existing views – View inheritance
- Defining document-style forms
- Dynamic form elements using attributes
- Defining embedded views
- Displaying attachments on the side of the form view
- Defining kanban views

- Showing kanban cards in columns according to their state
- Defining calendar views
- Defining graph view and pivot view
- Defining the cohort view
- Defining the gantt view
- Defining the activity view
- Defining the map view

Technical requirements

Throughout this chapter, we will assume that you have a database with the base add-on installed and an empty Odoo add-on module where you can add XML code from the recipes to a data file referenced in the add-on's manifest. Refer to *Chapter 3*, *Creating Odoo Add-On Modules*, for more information on how to activate changes in your add-on.

The technical requirements for this chapter include an online Odoo platform.

All of the code used in this chapter can be downloaded from the GitHub repository at `https://github.com/PacktPublishing/Odoo-17-Development-Cookbook-Fifth-Edition/tree/main/Chapter09`.

Adding a menu item and window actions

The most obvious way to make a new feature available to users is by adding a menu item. When you click on a **Menu** item, something happens. This recipe walks you through how to define that something.

We will create a top-level menu and its sub-menu, which will open a list of all hostel rooms.

This can also be done using the **web user interface** through the **Settings** menu, but we prefer to use XML data files since this is what we'll have to use when creating our add-on modules.

Getting ready

In this recipe, we will need a module with a dependency on the `base` module, as the `my_hostel` module adds new models to the `hostel.room`. So, if you are using an existing module, please add the `base` dependency in the manifest. Alternatively, you can grab the initial module from `https://github.com/PacktPublishing/Odoo-17-Development-Cookbook-Fifth-Edition/tree/main/Chapter09/00_initial_module`.

How to do it...

In an XML data file of our add-on module, perform the following steps:

1. Define an action to be executed:

```
<record id="action_hostel_room" model="ir.actions.act_window">
    <field name="name">All Hostel Room</field>
    <field name="res_model">hostel.room</field>
    <field name="view_mode">tree,form</field>
</record>
```

2. Create the top-level menu, which will be as follows:

```
    <menuitem id="menu_custom_hostel_room"
    name="Hostel Room" web_icon="my_hostel,static/description/
icon.png"/>
```

3. Refer to our action in the menu:

```
<menuitem id="menu_all_hostel_room"
    parent="menu_custom_hostel_room" action="action_hostel_
room"       sequence="10" groups="" />
```

If we now upgrade the module, we will see a top-level menu with the label **Hostel Room** that opens a sub-menu called **All Hostel Room**. Clicking on that menu item will open a list of all hostel rooms.

How it works...

The first XML element, `record model="ir.actions.act_window"`, declares a window action to display a list view with all the hostel rooms. We used the most important attributes:

- `name`: To be used as the title for the views opened by the action.
- `res_model`: This is the model to be used. We are using `hostel.room`, where Odoo stores all `hostel room`.
- `view_mode`: This lists the view types to make available. It is a comma-separated values file of the views type. The default value is `tree, form`, which makes a list and form view available. If you just want to show the calendar and form views, then the value of `view_mode` should be `calendar, form`. Other possible view choices are `kanban`, `graph`, `pivot`, `calendar`, and `cohort`. You will learn more about these views in forthcoming recipes.
- `domain`: This is optional and allows you to set a filter on the records to be made available in the views. We will see all of these views in more detail in the *Defining filters on record lists – Domain* recipe of this chapter.

- `context`: This can set values made available to the opened views, affecting their behavior. In our example, on new records, we want the room rank's default value to be `1`. This will be covered in more depth in the *Passing parameters to forms and actions – Context* recipe of this chapter.
- `limit`: This sets the default amount of records that can be seen on list views. In our example, we have given a limit of `20`, but if you don't give a `limit` value, Odoo will use the default value of `80`.

Next, we create the menu item hierarchy from the top-level menu to the clickable end menu item. The most important attributes of the `menuitem` element are as follows:

- `name`: This is used as the text that the menu items display. If your menu item links to an action, you can leave this out because the action's name will be used in that case.
- `parent` (`parent_id` if using the `record` element): This is the XML ID that references the parent menu item. Items with no parents are top-level menus.
- `action`: This is the XML ID that references the action to be called.
- `sequence`: This is used to order the sibling menu items.
- `groups` (`groups_id` with the `record` tag): This is an optional list of user groups that can access the menu item. If empty, it will be available to all users.
- `web_icon`: This option only works on the top-level menu. It will display an icon of your application in the Enterprise edition.

Window actions automatically determine the view to be used by looking up views for the target model with the intended type (`form`, `tree`, and so on) and picking the one with the lowest sequence number. `ir.actions.act_window` and `menuitem` are convenient shortcut XML tags that hide what you're actually doing. If you don't want to use the shortcut XML tags, then you can create a record of the `ir.actions.act_window` and `ir.ui.menu` models via the `<record>` tag. For example, if you want to load `act_window` with `<record>`, you can do so as follows:

```
<record id="action_hostel_room" model="ir.actions.act_window">
    <field name="name">All Hostel Room</field>
    <field name="res_model">hostel.room</field>
    <field name="view_mode">tree,form</field>
    <field name="context">{'default_room_rating': 1.0}</field>
    <field name="domain">[('state', '=', 'draft')]</field>
</record>
```

In the same way, you can create a `menuitem` instance through `<record>`.

> **Important note**
>
> Be aware that names used with the `menuitem` shortcut may not map to the field names that are used when using a `record` element; `parent` should be `parent_id` and `groups` should be `groups_id`.

To build the menu, the web client reads all the records from `ir.ui.menu` and infers their hierarchy from the `parent_id` field. The menus are also filtered based on user permissions to models and groups assigned to menus and actions. When a user clicks on a menu item, its `action` is executed.

There's more...

Window actions also support a `target` attribute to specify how the view is to be presented. The possible choices are as follows:

- **current**: This is the default and opens the view in the web client's main content area.
- **new**: This opens the view in a popup.
- **inline**: This is like **current,** but it opens a form in edit mode and disables the **Action** menu.
- **Fullscreen**: The action will cover the whole browser window, so this will overlay the menus, too. Sometimes, this is called **tablet mode**.
- **main**: This is like **current**, but it also clears out the breadcrumbs.

There are also some additional attributes available for window actions that are not supported by the `ir.actions.act_window` shortcut tag. To use them, we must use the `record` element with the following fields:

- `res_id`: If opening a form, you can use it to open a specific record by setting its ID here. This can be useful for multi-step wizards or in cases when you have to view or edit a specific record frequently.
- `search_view_id`: This specifies a specific search view to use for tree and graph views.

Keep in mind that the menu in the top left (or the apps icon in the Enterprise version) and the menu in the bar at the top are both made up of menu items. The only difference is that the items in the menu in the top left don't have any parent menus, while the ones on the top bar have the respective menu items from the top bar as a parent. In the left bar, the hierarchical structure is more obvious.

Additionally, bear in mind that for design reasons, the first-level menus will open the dropdown menu if your second-level menu has child menus. In any case, Odoo will open the first menu item's action based on the sequence of child menu items.

Refer to the following to learn more about menus and views:

- The `ir.actions.act_window` action type is the most common action type, but a menu can refer to any type of action. Technically, it is the same if you link to a client action, a server action, or any other model defined in the `ir.actions.*` namespace. It just differs in what the backend makes of the action.
- If you need just a tiny bit more flexibility in the concrete action to be called, look into server actions that return a window action. If you need complete flexibility, take a look at the client actions (`ir.actions.client`), which allow you to have a completely custom user interface. However, only do this as a last resort as you lose a lot of Odoo's convenient helpers when using them.

See also

- For a detailed explanation of filters in all of the views, have a look at the *Defining filters on record lists – Domain* recipe in this chapter.

Having an action open a specific view

Window actions automatically determine the view to be used if none is given, but sometimes, we want an action to open a specific view.

We will create a basic form view for the `hostel.room` model, and then we will create a new window action specifically to open that form view.

How to do it...

1. Define the `hostel room` minimal tree and form view:

```
<record id="hostel_room_view_tree" model="ir.ui.view">
    <field name="name">Hostel Room List</field>
    <field name="model">hostel.room</field>
    <field name="arch" type="xml">
        <tree>
        <field name="name"/>
            <field name="room_no"/>
            <field name="state"/>
        </tree>
    </field>
</record>

<record id="hostel_room_view_form" model="ir.ui.view">
```

```
    <field name="name">Hostel Room Form</field>
    <field name="model">hostel.room</field>
    <field name="arch" type="xml">
        <form>
            <header>
                <button name="make_available" string="Make
Available" type="object"/>
                <button name="make_closed"  string="Make Closed"
type="object"/>
                <field name="state" widget="statusbar"/>
            </header>
            <group>
                <group>
                    <field name="name"/>
                    <field name="room_no"/>
                </group>
                <group>
                    <field name="description"/>
                </group>
            </group>
        </form>
    </field>
</record>
```

2. Update the action from the *Adding a menu item and window action* recipe to use a new form view:

```
<record id="action_hostel_room_tree" model="ir.actions.act_
window.view">
    <field name="act_window_id" ref="action_hostel_room" />
    <field name="view_id" ref="hostel_room_view_tree" />
    <field name="view_mode">tree</field>
    <field name="sequence" eval="1"/>
</record>

<record id="action_hostel_room_form" model="ir.actions.act_
window.view">
    <field name="act_window_id" ref="action_hostel_room" />
    <field name="view_id" ref="hostel_room_view_form" />
    <field name="view_mode">form</field>
    <field name="sequence" eval="2"/>
</record>
```

Now, if you open your menu and click on a **partner** in the list, you should see the very minimal form and tree that we just defined.

How it works...

This time, we used the generic XML code for any type of record, that is, the `record` element with the required `id` and `model` attributes. The `id` attribute on the `record` element is an arbitrary string that must be unique for your add-on. The `model` attribute refers to the name of the model you want to create. Given that we want to create a view, we need to create a record of the `ir.ui.view` model. Within this element, you set fields as defined in the model you chose through the `model` attribute. For `ir.ui.view`, the crucial fields are `model` and `arch`. The `model` field contains the model for which you want to define a view, while the `arch` field contains the definition of the view itself. We'll come to its contents in a short while.

The `name` field, while not strictly necessary, is helpful when debugging problems with views. So, set it to a string that tells you what this view is intended to do. This field's content is not shown to the user, so you can fill in any technical hints that you deem sensible. If you set nothing here, you'll get a default name that contains the model name and view type.

ir.actions.act_window.view

The second record we defined works in tandem with `act_window`, which we defined earlier in the *Adding a menu item and window action* recipe. We already know that by setting the `view_id` field there, we can select which view is used for the first view mode. However, given that we set the `view_mode` field to the `tree, form` view, `view_id` would have to pick a tree view, but we want to set the form view, which comes second here.

If you find yourself in a situation like this, use the `ir.actions.act_window.view` model, which gives you fine-grained control over which views to load for which view type. The first two fields defined here are examples of the generic way to refer to other objects; you keep the element's body empty but add an attribute called `ref`, which contains the XML ID of the object you want to reference. So, what happens here is we refer to our action from the previous recipe in the `act_window_id` field and refer to the view we just created in the `view_id` field. Then, though not strictly necessary, we add a sequence number to position this view assignment relative to the other view assignments for the same action. This is only relevant if you assign views for different view modes by creating multiple `ir.actions.act_window.view` records.

> **Important note**
>
> Once you define the `ir.actions.act_window.view` records, they take precedence over what you filled in the action's `view_mode` field. So, with the preceding records, you won't see a list at all, but only a form. You should add another `ir.actions.act_window.view` record that points to a list view for the `hostel.room` model.

There's more...

As we saw in the *Adding a menu item and window action* recipe, we can replace `act_window` with `<record>`. If you want to use a custom view, you can follow the given syntax:

```
<record id="action_hostel_room" model="ir.actions.act_window">
    <field name="name">All Hostel Room</field>
    <field name="res_model">hostel.room</field>
    <field name="view_id" ref="hostel_room_view_tree"/>
    <field name="view_mode">tree,form</field>
</record>
```

This example is just an alternative to `act_window`. In the code base of Odoo, you will find both types of action.

Adding content and widgets to a form view

The preceding recipe showed how to pick a specific view for an action. Now, we'll demonstrate how to make the form view more useful. In this recipe, we will use the **form view** that we defined earlier in the *Having an action open a specific view* recipe. In the form view, we will add the widgets and content.

How to do it...

1. Define the basic structure of the form view:

```
<record id="hostel_room_view_form" model="ir.ui.view">
    <field name="name">Hostel Room Form</field>
    <field name="model">hostel.room</field>
    <field name="arch" type="xml">
    <form>
        <!--form content goes here -->
    </form>
</record>
```

2. To add a head bar, which is usually used for action buttons and stage pipelines, add this inside the form:

```
<header>
    <button name="make_available" string="Make Available"
type="object"/>
    <button name="make_closed"  string="Make Closed"
type="object"/>
    <field name="state" widget="statusbar"/>
</header>
```

3. Add fields to the form, using `group` tags to organize them visually:

```
<group string="Content" name="my_content">
    <field name="name"/>
    <field name="room_no"/>
</group>
<group>
    <field name="description"/>
</group>
<notebook>
    <page string="Other Information" name="other_information">
        <field name="other_info" widget="html"/>
    </page>
</notebook>
```

Now, the form should display a top bar with a button and two vertically aligned fields, as shown in the following screenshot:

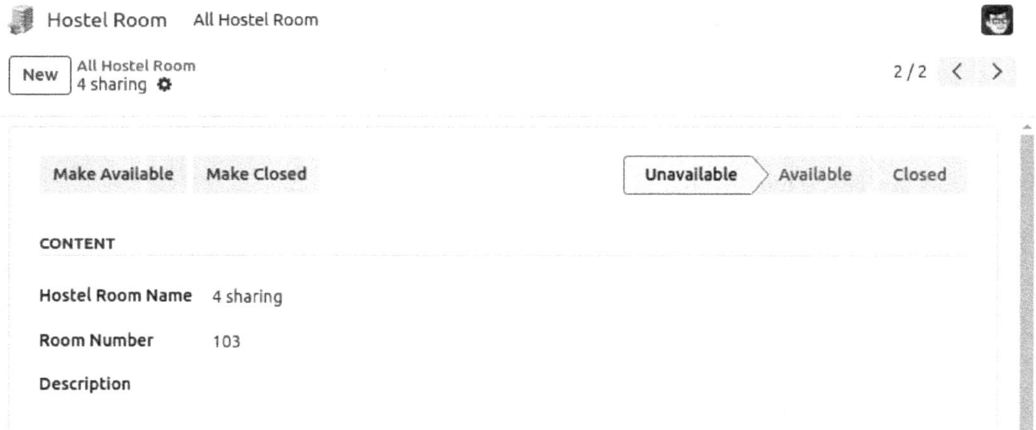

Figure 9.1 – Screenshot of the form view

How it works...

We'll look at the `arch` field of the `ir.ui.view` model first. First, note that views are defined in XML, so you need to pass the `type="xml"` attribute for the `arch` field; otherwise, the parser will be confused. It is also mandatory that your view definition contains well-formed XML; otherwise, you'll get an error such as "Element odoo has extra content" when you upgrade/install the module.

We'll now walk through the tags that we used previously and summarize the others that are available.

form

When you define a form view, it is mandatory that the first element within the `arch` field is a `form` element. This is used internally to derive the record's `type` field.

In addition to the following elements, you can use arbitrary HTML within the form tag. The algorithm has it that every element unknown to Odoo is considered plain HTML and is simply passed through to the browser. Be careful with that, as the HTML you fill in can interact with the HTML code the Odoo elements generate, which might distort the rendering.

header

This element is a container for elements that should be shown in a form's header, which is rendered as a white bar. Usually, as in this example, you place action buttons here. Alternatively, if your model has a `state` field, you could opt for a **status bar**.

button

The `button` element is used to allow the user to trigger an action. Refer to the *Adding buttons to forms* recipe for details.

<group>

The `<group>` element is Odoo's main element and is used for organizing content. Fields placed within a `<group>` element are rendered with their title, and all fields within the same group are aligned so that there's also a visual indicator that they belong together. You can also nest `<group>` elements; this causes Odoo to render the contained fields in adjacent columns.

In general, you should use the `<group>` mechanism to display all of your fields in the form view and only revert to the other elements, such as `<notebook>`, `<label>`, `<newline>`, and more, when necessary.

If you assign the `string` attribute to a group, its content will be rendered as a heading for the group.

You should develop the habit of assigning a name to every logical group of fields, too. This name is not visible to the user but is very helpful when we override views in the following recipes. Keep the name unique within the form definition to avoid confusion about which group you refer to. Don't use the `string` attribute for this because the value of the string will eventually change because of translations.

field

In order to actually show and manipulate data, your form view should contain some `field` elements. Here is an example:

```
<field name="other_info" widget="html"/>
```

These have one mandatory attribute, called `name`, which refers to the field's name in the model. Earlier, we offered the user the ability to edit the partner's categories. If we only want to disable the editing feature on a field, we can set the `readonly` attribute to `1` or `True`. This attribute may actually contain a small subset of Python code, so `readonly="2>1"` will make the field read-only too. This also applies to the `invisible` attribute, for which you used to obtain a value that is read from the database but is not shown to the user. Later, we'll take a look at which situations this can be used in.

You must have noticed the `widget` attribute in the `categories` field. This defines how the data in the field are supposed to be presented to the user. Every type of field has a standard widget, so you don't have to explicitly choose a widget. However, several types provide multiple ways of representation, so you might opt for something other than the default. As a complete list of available widgets would exceed the scope of this recipe, consult **Odoo's source code** to try them out. Take a look at *Chapter 14, CMS Website Development*, for details on how to make your own.

<notebook> and <page>

If your model has too many fields, then you can use the `<notebook>` and `<page>` tags to create tabs. Each `<page>` in the `<notebook>` tag will create a new tab, and the content inside the page will be the tab content. The following example will create two tabs with three fields in each tab:

```
<notebook>
    <page string="Tab 1">
        <field name="field1"/>
        <field name="field2"/>
        <field name="field3"/>
    </page>
    <page string="Tab 2">
        <field name="field4"/>
        <field name="field5"/>
        <field name="field6"/>
    </page>
</notebook>
```

The `string` attribute in the `<page>` tag will be the name of the tab. You can only use `<page>` tags in the `<notebook>` tag, but in the `<page>` tag, you can use any other elements.

General attributes

On most elements (this includes `group`, `field`, and `button`), you can set the `attributes` and `groups` attributes. Here is a small example:

```
<field name="other_info"
    readonly="state == 'available'"
    groups="base.group_no_one"/>
```

While `attributes` are discussed in the *Dynamic form elements using attributes* recipe, the `groups` attribute gives you the possibility to show some elements only to members of certain groups. Simply put, the group's full XML ID (separated by commas for multiple groups) is the attribute, and the element will be hidden for everyone who is not a member of at least one of the groups mentioned.

Other tags

There are situations in which you might want to deviate from the strict layout groups prescribed. For example, if you want the `name` field of a record to be rendered as a heading, the field's label will interfere with the appearance. In this case, don't put your field into a `group` element but, instead, put it into a plain HTML `h1` element. Then, before the `h1` element, put a `label` element with the `for` attribute set to your `field name`:

```
<label for="name" />
<h1><field name="name" /></h1>
```

This will be rendered with the field's content as a big heading, but the field's name will be written in a smaller type above the big heading. This is basically what the standard partner form does.

If you need a line break within a group, use the `newline` element. It's always empty:

```
<newline />
```

Another useful element is `footer`. When you open a form as a popup, this is a good place to put the action buttons. It will be rendered as a separate bar too, analogous to the `header` element.

The form view also has special widgets, such as `web_ribbon`. You can use it with the `<widget>` tag as follows:

```
<widget name="web_ribbon" title="Archived" bg_color="bg-danger"
invisible="active"/>
```

You can use `attributes` to hide and show the ribbon based on a condition. Don't worry if you are not aware of `attributes`. It will be covered in the *Dynamic form elements using attributes* recipe of this chapter.

> **Important note**
>
> Don't address **XML nodes** using their `string` attribute (or any other translated attribute, for that matter), as your view overrides will break for other languages because views are translated before inheritance is applied.

There's more...

Since form views are basically HTML with some extensions, Odoo also makes extensive use of CSS classes. Two very useful ones are `oe_read_only` and `oe_edit_only`. Elements with these classes will be visible only in **read-only mode** or **edit mode**, respectively. For example, to have the label visible only in edit mode, use the following:

```
<label f"r="n"me" cla"s="oe_edit_o"ly" />
```

Another very useful class is `oe_inline`, which you can use on fields to make them render as an inline element to avoid causing unwanted line breaks. Use this class when you embed a field into text or other markup tags.

Furthermore, the `form` element can have the `create`, `edit`, and `delete` attributes. If you set one of these to `false`, the corresponding action 'on't be available for this form. Without this being explicitly set, the availability of the action is inferred from the u'er's permissions. Note that this is purely for straightening up the UI; 'on't use this for security.

See also

The widgets and views already offer a lot of functionality, but sooner or later, you will have requirements that cannot be fulfilled with the existing widgets and views. Refer to the following recipes to create your own views and widgets:

- Refer to the *Adding buttons to forms* recipe in this chapter for more details about using the `button` element to trigger an action.
- To define your own widgets, refer to the *Creating custom widgets* recipe of *Chapter 15, Web Client Development*.
- Refer to the *Creating a new view* recipe of *Chapter 15, Web Client Development*, to create your own view.

Adding buttons to forms

Buttons are used in the form view to handle user actions. We added a button in the form view in the previous recipe, but there are quite a few different types of buttons that we can use. This recipe will add another button that will help the user to open another view. It will also put the following code in the recipe's `header` element.

How to do it...

Add a button that refers to an action:

```
<button type="action" name="%(my_hostel.hostel_room_category_action)d"
string="Open Hotel Room Category" />
```

How it works...

The button's `type` attribute determines the semantics of the other fields, so we'll first take a look at the possible values:

- `action`: This makes the button call an action, as defined in the `ir.actions.*` namespace. The `name` attribute needs to contain the action's database ID, which you can conveniently have Odoo look up with a Python-format string that contains the XML ID of the action in question.
- `object`: This calls a method from the current model. The `name` attribute contains the function's name.
- `string`: The `string` attribute is used to assign the text the user sees.

There's more...

Use the `btn-primary` CSS classes to render highlighted button and `btn-default` to render a normal button. This is commonly used for cancel buttons in wizards or to offer secondary actions in a visually unobtrusive way. Setting the `oe_link` class causes the button to look like a link. You can also use other bootstrap button classes to get different button colors.

A call with a button of the **object** type can return a dictionary that describes an action, which will then be executed on the client side. This way, you can implement multiscreen wizards or just open another record.

> **Important note**
>
> Note that clicking on a button always causes the client to issue a `write` or `create` call before running the method.

You can also add content within the `button` tag by replacing the `string` attribute. This is commonly used in button boxes, as described in the *Document style forms* recipe.

Passing parameters to forms and actions – context

Internally, every method in Odoo has access to a dictionary called **context**, which is propagated from every action to the methods involved in delivering that action. The UI also has access to it, and it can be modified in various ways by setting values in the context. In this recipe, we'll explore some of the applications of this mechanism by toying with the language, default values, and implicit filters.

Getting ready

While not strictly necessary, this recipe will be more fun if you install the French language if you haven't got this already. Consult *Chapter 11, Internationalization*, for how to do this. If you have a French database, change `fr_FR` to some other language, e.g., `en_US` will do for English. Additionally, click on the **Active** button (changing to **Archive** when you hover over it) for one of the hostel rooms in order to archive it and verify that this partner doesn't show up in the list anymore.

How to do it...

1. Create a new action, very similar to the one from the *Adding a menu item and window action* recipe:

```
<record id="action_hostel_room" model="ir.actions.act_window">
    <field name="name">All Hostel Room</field>
    <field name="res_model">hostel.room</field>
    <field name="view_id" ref="hostel_room_view_tree"/>
    <field name="view_mode">tree,form</field>
    <field name="context">{'lang': 'fr_FR','default_lang': 'fr_
FR', 'active_test': False, 'default_room_rating': 1.0}</field>
</record>
```

2. Add a menu that calls this action. This is left as an exercise for the reader.

When you open this menu, the views will show up in French, and if you create a new partner, they will have French as their pre-selected language. A less obvious difference is that you will also see **deactivated (archived) partner records**.

How it works...

The context dictionary is populated from several sources. First, some values from the current user's record (`lang` and `tz` for the user's language and the user's time zone, respectively) are read. Then, we have some add-ons that add keys for their own purposes. Furthermore, the UI adds keys about which model and which record we're busy using at the moment (`active_id`, `active_ids`, `active_model`). Moreover, as seen in the *Having an action open a specific view* recipe, we can add our own keys in actions. These are merged together and passed to the underlying server functions and the client-side UI.

So, by setting the `lang` context key, we force the display language to be **French**. You will note that this doesn't change the whole UI language; this is because only the list view that we open lies within the scope of this context. The rest of the UI was loaded already with another context that contained the user's original language. However, if you open a record in this list view, it will be presented in French, too, and if you open a linked record on the form or press a button that executes an action, the language will be propagated, too.

By setting `default_lang`, we set a default value for every record created within the scope of this context. The general pattern is `default_$fieldname: my_default_value`, which enables you to set default values for newly created partners in this case. Given that our menu is about hostel rooms, we have added `default_room_rating: 1` as the value for the `Hostel Average Rating` field by default. However, this is a model-wide default for `hostel.room`, so this wouldn't have changed anything. For scalar fields, the syntax for this is the same as what you would write in Python code: `string` fields go in quotes, `number` fields stay as they are, and `Boolean` fields are either `True` or `False`. For relational fields, the syntax is slightly more complicated; refer to *Chapter 6, Managing Module Data*, to learn how to write them.

> **Important note**
> Note that the default values set in the context override the default values set in the model definition, so you can have different default values in different situations.

The last key is `active_test`, which has very special semantics. For every model that has a field called `active`, Odoo automatically filters out records where this field is `False`. This is why the partner from where you unchecked this field disappeared from the list. By setting this key, we can suppress this behavior.

> **Important note**
> This is useful for the UI in its own right but even more useful in your Python code when you need to ensure that an operation is applied to all the records, not just the active ones.

There's more...

When defining a context, you have access to some variables, with the most important one being `uid`, which evaluates the current user's ID. You'll need this to set default filters (refer to the next recipe, *Defining filters on record lists – Domain*). Furthermore, you have access to the `context_today` function and the `current_date` variable, where the first is a `date` object that represents the current date, as seen from the user's time zone, and the latter is the current date, as seen in UTC, formatted as `YYYY-MM-DD`. To set a default value for a `date` field to the current date, use `current_date`, and for default filters, use `context_today()`.

Furthermore, you can do some date calculations with a subset of Python's `datetime`, `time`, and `relativedelta` classes.

> **Important note**
>
> Most of the domains are evaluated on the client side. The server-side domain evaluation is restricted for security reasons. When client-side evaluation was introduced, the best option in order to not break the whole system was to implement a part of Python in JavaScript. There is a small JavaScript Python interpreter built into Odoo that works well for simple expressions, and that is usually enough.
>
> Beware of the use of the `context` variable in the `<record id="action_name" model="ir.actions.act_window.view">` shortcut. These are evaluated at installation time, which is nearly never what you want. If you need variables in your context, use the `<record />` syntax.

We can also add different contexts for the buttons. It works the same way as how we added context keys to our action. This causes the function or action that the button calls to be run in the context given.

Most form element attributes that are evaluated as Python also have access to the context dictionary. The `invisible` and `readonly` attributes are examples of these. So, in cases where you want an element to show up in a form sometimes but not at other times, set the `invisible` attribute to `context.get('my_key')`. For actions that lead to a case in which the field is supposed to be invisible, set the context key to `my_key: True`. This strategy enables you to adapt your form without having to rewrite it for different occasions.

You can also set a context for relational fields, which influences how the field is loaded. By setting the `form_view_ref` or `tree_view_ref` keys to the full XML ID of a view, you can select a specific view for this field. This is necessary when you have multiple views of the same type for the same object. Without this key, you get the view with the lowest sequence number, which might not always be desirable.

See also

- The context is also used to set a default search filter. You can learn more about the default search filter in the *Defining search views* recipe of this chapter.
- For more details on setting default recipes, refer to the next recipe, *Defining filters on record lists – Domain*.
- To learn how to install the French language, consult *Chapter 11*, *Internationalization*.
- You can refer to *Chapter 6*, *Managing Module Data* to learn how to write the syntax for relational fields.

Defining filters on record lists – domain

We've already seen an example of a domain in the first recipe of this chapter, which was `[('state', '=', 'draft')]`. Often, you need to display a subset of all available records from an action or allow only a subset of possible records to be the target of a many2one relation. The way to describe these filters in Odoo is by using domains. This recipe illustrates how to use a domain to display a selection of partners.

How to do it...

To display a subset of partners from your action, you need to perform the following steps:

1. Create an action for when "state" is set to "draft:"

```
<record id="action_hostel_room" model="ir.actions.act_window">
    <field name="name">All Hostel Room</field>
    <field name="res_model">hostel.room</field>
    <field name="view_id" ref="hostel_room_view_tree"/>
    <field name="view_mode">tree,form</field>
    <field name="context">{'lang': 'fr_FR','default_lang': 'fr_
FR', 'active_test': False, 'default_room_rating': 1.0}</field>
    <field name="domain">[('state', '=', 'draft'), ('room_
rating', '&gt;', '0.0')]</field>
</record>
```

2. Add menus that call these actions. This is left as an exercise for the reader.

How it works...

The simplest form of a domain is a list of three tuples that contain a field name (of the model in question) as `string` in the first element, an operator as `string` in the second element, and the value that the field is to be checked against as the third element. This is what we did before, and this is interpreted as, "*All those conditions have to apply to the records we're interested in.*" This is actually a shortcut because the domains know the two prefix operators—`&` and `|`—where `&` is the default. So, in normalized form, the first domain will be written as follows:

```
['&',('state', '=', 'draft'), ('room_rating', '&gt;', '0.0')]
```

While these can be a bit hard to read for bigger expressions, the advantage of prefix operators is that their scope is rigidly defined, which saves you from having to worry about operator precedence and brackets. It's always two expressions: the first `&` applies to `'&',('state', '=', 'draft'),` with `('room_rating', '>', '0.0')` as the first operand and `('room_rating', '>', '0.0')` as the second. Then, we have the first operand and `('room_rating', '>', '0.0')` as the second.

In the second step, we have to write out the full form because we need the `|` operator.

For example, say we have a complex domain such as this: `['|', ('user_id', '=', uid), '&', ('lang', '!=', 'fr_FR'), '|', ('phone', '=', False), ('email', '=', False)]`. See the following figure to learn about how this domain is evaluated:

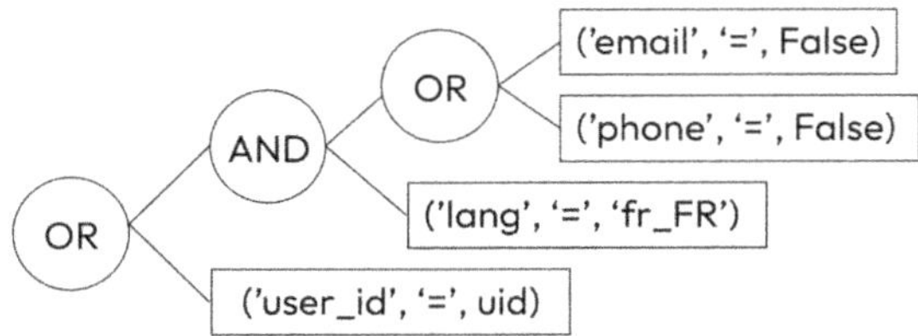

Figure 9.2 – The evaluation of a domain

There is also a `!` operator for negation, but given logical equivalences and negated comparison operators such as `!=` and `not in`, it is not really necessary.

> **Important note**
>
> Note that this is a unary prefix operator, so it only applies to the following expression in the domain and not to everything that follows.

Note that the right operand doesn't need to be a fixed value when you write a domain for a window action or other client-side domains. You can use the same minimal Python as is used in the *Passing parameters to forms and actions – Context* recipe, so you can write filters such as **changed last week** or **my partners**.

There's more...

The preceding domains work only for the fields of the model itself, while we often need to filter based on the properties of linked records. To do this, you can use the notation that's also used in `@api.depends` definitions or related fields: create a dotted path from the current model to the model you want to filter for. To search partners that have a salesperson who is a member of a group starting with the letter G, you would use the `[('user_id.groups_id.name', '=like', 'G%')]` domain. The path can be long, so you only have to be sure that there are relation fields between the current model and the model you want to filter for.

Operators

The following table lists the available operators and their semantics:

Operator (equivalent)	Semantics
`=`, `!=`, `<>`	The first one is for an exact match, the second one is for not equal, and the last one is the deprecated notation of not equal.
`in`, `not in`	This checks whether the value is one of the values named in a list in the right operand. It is given as a Python list: `[('uid', 'in', [1, 2, 3])]` or `[('uid', 'not in', [1, 2, 3])]`.
`<`, `<=`	Greater than, greater than or equal to.
`>`, `>=`	Less than, less than or equal to.
`like`, `not like`	Checks whether the right operand is contained (as a substring) in the value.
`ilike`, `not ilike`	The same as the preceding one, but case insensitive.
`=like`, `=ilike`	You can search for patterns here: `%` matches any string and `_` matches one character. This is the equivalent of PostgreSQL's `like`.
`child_of`	For models with a `parent_id` field, this searches for children of the right operand. The right operand is included in the results.
`=?`	Evaluates to `true` if the right operand is `false`; otherwise, it behaves like `=`. This is useful when you generate domains programmatically and want to filter by a value if one is set but ignore the value otherwise.

Table 9.1 – Operators and their semantics

Note that some of the operators work only with certain fields and values. For example, the domain `[('category_id', 'in', 1)]` is invalid and will generate an error, while the domain `[('category_id', 'in', [1])]` is valid.

Pitfalls of searching using domains

This all works fine for traditional fields, but a notorious problem is searching for the value of a non-stored function field. People often omit the search function. This is simple enough to fix by providing the search function in your own code, as described in *Chapter 4, Application Models.*

Another issue that might baffle developers is Odoo's behavior when searching through `one2many` or `many2many` fields with a negative operator. Imagine that you have a partner with the `A` tag, and you search for `[('category_id.name', '!=', 'B')]`. Your partner shows up in the result, and this is what you expected, but if you add the `B` tag to this partner, it still shows up in your results because, for the search algorithm, it is enough that there is one linked record (`A` in this case) that does not fulfill the criterion. Now, if you remove the `A` tag so that `B` is the only tag, the partner will be filtered out. If you also remove the `B` tag so that the partner has no tags, it is still filtered out because the conditions of the linked records presuppose the existence of this record. In other situations, though, this is the behavior you want, so it is not really an option to change the standard behavior. If you need a different behavior here, provide a search function that interprets the negation the way you need.

> **Important note**
>
> People often forget that they are writing XML files when it comes to domains. You need to escape the less-than operator. Searching for records that have been created before the current day will have to be written as `[('create_date', '<', current_date)]` in XML.

Domains are used widely in Odoo. You will find them everywhere in Odoo; they are used for searching, filtering, security rules, search views, user actions, and more.

If you ever need to manipulate a domain that you didn't create programmatically, use the utility functions provided in `odoo.osv.expression`. The `is_leaf`, `normalize_domain`, `AND`, and `OR` functions will allow you to combine domains exactly the way that Odoo does. Don't do this yourself because there are many corner cases that you have to take into account, and it is likely that you'll overlook one.

See also

- For the standard application of domains, see the *Defining search views* recipe.

Defining list views

After having spent quite some time on the form view, we'll now take a quick look at how to define list views. Internally, these are called tree views in some places and list views in others, but given that there is another construction within the Odoo view framework called **tree**, we'll stick to **list** here.

How to do it...

1. Define your list view:

```
<record id="hostel_room_view_tree" model="ir.ui.view">
    <field name="name">Hostel Room List</field>
    <field name="model">hostel.room</field>
    <field name="arch" type="xml">
        <tree>
            <field name="name"/>
            <field name="room_no"/>
```

```
                <field name="state"/>
            </tree>
        </field>
    </record>
```

2. Register a tree view in the action that we created in the *Adding a menu item and window action* recipe of this chapter:

```
<record id="action_hostel_room" model="ir.actions.act_window">
    <field name="name">All Hostel Room</field>
    <field name="res_model">hostel.room</field>
    <field name="view_id" ref="hostel_room_view_tree"/>
    <field name="view_mode">tree,form</field>
    <field name="context">{'tree_view_ref': 'my_hostel.hostel_
room_view_tree', 'lang': 'fr_FR','default_lang': 'fr_FR',
'active_test': False, 'default_room_rating': 1.0}</field>
    <field name="domain">[('state', '=', 'draft')]</field>
</record>
```

3. Add menus that call these actions. This is left as an exercise for the reader.

Install/Upgrade the module. After that, you will see our tree view for the hostel room, and if you check it, it will show different row styles based on our conditions.

How it works...

You already know most of what happens here. We define a view, using the `tree` type this time, and attach it to our action with an `ir.actions.act_window.view` element. So, the only thing left to discuss is the `tree` element and its semantics. With a list, you don't have many design choices, so the only valid children of this element are the `field` and `button` elements. You can also use some widgets in the list view; in our example, we have used the `many2one_avatar_user` widget. The tree view has the support of a special widget called `handle`. This is specific to list views. It is meant for integer fields and renders a drag handle that the user can use to drag a row to a different position in the list, thereby updating the field's value. This is useful for sequence or priority fields.

By using the `optional` attribute, you can show fields optionally. Adding the `optional` attribute to a field will allow the user to hide and show the column at any time from the UI. In our example, we have used it for the `country` and `state` fields.

What is new here are the `decoration` attributes in the `tree` element. This contains rules as to which font and/or color is chosen for the row, given in the form of `decoration-$name="Python code"`. We made these invisible because we only need the data and don't want to bother our users with the two extra columns. The possible classes are `decoration-bf` (bold) and `decoration-it` (italic), and the semantic bootstrap classes are `decoration-danger`, `decoration-info`, `decoration-muted`, `decoration-primary`, `decoration-success`, and `decoration-warning`.

There's more...

For numeric fields, you can add a `sum` attribute that causes this column to be summed up with the text you set in the attribute as a tooltip. Less common are the `avg`, `min`, and `max` attributes, which display the average, minimum, and maximum, respectively. Note that these four only work on the records that are currently visible, so you might want to adjust the action's `limit` (covered earlier in the *Adding a menu item and window action* recipe) in order for the user to see all the records immediately.

A very interesting attribute for the `tree` element is `editable`. If you set this to top or bottom, the list behaves entirely differently. Without it, clicking on a row opens a form view for the row. With it, clicking on a row makes it editable inline, with the visible fields rendered as form fields. This is particularly useful in embedded list views, which are discussed later in the *Defining embedded views* recipe of this chapter. The choice of top or bottom relates to whether new lines will be added to the top or bottom of the list.

By default, records are ordered according to the `_order` property of the displayed model. The user can change the ordering by clicking on a column header, but you can also set a different initial order by setting the `default_order` property in the `tree` element. The syntax is the same as in `_order`.

> **Important note**
> Ordering is often a source of frustration for new developers. As Odoo lets PostgreSQL do the work here, you can only order according to the fields that PostgreSQL knows about and only the fields that live in the same database table. So, if you want to order according to a function or a related field, ensure that you set `store=True`. If you need to order according to a field inherited from another model, declare a stored related field.

The `create`, `edit`, and `delete` attributes of the `tree` element work the same as for the `form` element we described earlier in the *Adding content and widgets to a form view* recipe of this chapter. They also determine the available controls if the `editable` attribute is set.

Defining search views

When opening your list view, you'll notice the search field in the upper right. If you type something there, you will receive suggestions about what to search for, and there is also a set of predefined filters to choose from. This recipe will walk you through how to define these suggestions and options.

How to do it...

1. Define your search view:

```
<record id="hostel_room_view_search" model="ir.ui.view">
    <field name="model">hostel.room</field>
    <field name="arch" type="xml">
```

```
        <search>
            <field name="name"/>
            <field name="room_no"/>
            <group expand="0" string="Group By">
                <filter string="State" name="state"
context="{'group_by':'state'}"/>
           </group>
        </search>
    </field>
</record>
```

2. Tell your action to use it:

```
<record id="action_hostel_room" model="ir.actions.act_window">
    <field name="name">All Hostel Room</field>
    <field name="res_model">hostel.room</field>
    <field name="search_view_id" ref="hostel_room_view_search"
/>
    <field name="view_mode">tree,form</field>
    <field name="context">{'tree_view_ref': 'my_hostel.hostel_
room_view_tree', 'lang': 'fr_FR','default_lang': 'fr_FR',
'active_test': False, 'default_room_rating': 1.0}</field>
    <field name="domain">[('state', '=', 'draft')]</field>
</record>
```

When you type something into the search bar now, you'll be offered the ability to search for this term in the `name`, `room no`, and `state` fields. If your term happens to be a substring of a bank account number in your system, you'll even be offered the option to search exactly for this bank account.

How it works...

In the case of `name`, we simply listed the field as the one to be offered to the user to search for. We left the semantics as the default, which is a substring search for character fields.

For categories, we do something more interesting. By default, your search term is applied to a `many2many` field trigger called `name_search`, which would be a substring search in the category names in this case. However, depending on your category structure, it can be very convenient to search for partners who have the category you're interested in or a child of it. Think of a main category, **newsletter subscribers**, with the subcategories **weekly newsletter**, **monthly newsletter**, and a couple of other newsletter types. Searching for **newsletter subscribers** with the preceding search view definition will give you everyone who is subscribed to any of those newsletters in one go, which is a lot more convenient than searching for every single type and combining the results.

The `filter_domain` attribute can contain an arbitrary domain, so you're not restricted to searching for the same field you named in the `name` attribute nor to using only one term. The `self` variable is what the user filled in and is also the only variable that you can use here.

Here's a more elaborate example from the default search view for hostel room:

```
<field name="name"
    filter_domain="[
        '|',
        ('display_name', 'ilike', self),
        ('room_no', '=', self)]"/>
```

This means that the user doesn't have to think about what to search for. All they need to do is type in some letters, press *Enter*, and, with a bit of luck, one of the fields mentioned contains the string we're looking for.

For the `child_ids` field, we used another trick. The type of field not only decides the default way of searching for the user's input but also defines the way in which Odoo presents the suggestions. Additionally, given that `many2one` fields are the only ones that offer auto-completion, we force Odoo to do that, even though `child_ids` is a `one2many` field, by setting the `widget` attribute. Without this, we will have to search in this field without suggestions for completion. The same applies to `many2many` fields.

> **Important note**
>
> Note that every field with a `many2one` widget set will trigger a search on its model for every one of the user's keystrokes; don't use too many of them.

You should also put the most-used fields on the top because the first field is what is searched for if the user just types something and presses *Enter*. The search bar can also be used with the keyboard; select a suggestion by pressing the down arrow and open the completion suggestion of `many2one` by pressing the right arrow. If you educate your users on this and pay attention to the sensible ordering of fields in the search view, this will be much more efficient than typing something first, grabbing the mouse, and selecting an option.

The `filter` element creates a button that adds the content of the filter's `domain` attribute to the search domain. You should add a logical internal `name` and a `string` attribute to describe the filter to your users.

The `<group>` tag is used to provide a grouping option under the **Group by** button. In this recipe, we have added an option to group records based on the `country_id` field.

There's more...

You can group filters using the `group` tag, which causes them to be rendered slightly closer together than the other filters, but this has semantic implications, too. If you put multiple filters in the same group and activate more than one of them, their domains will be combined with the | operator, while the filters and fields not in the same group are combined with the `&` operator. Sometimes, you might want disjunction for your filters, which is where they filter for mutually exclusive sets, in which case, selecting both of them will always lead to an empty result set. Within the same group, you can achieve the same effect with the `separator` element.

> **Important note**
>
> Note that if the user fills in multiple queries for the same field, they will be combined with `|`, too, so you don't need to worry about that.

Apart from the `field` attribute, the `filter` element can have a `context` attribute, whose content will be merged with the current context and eventually other context attributes in the search view. This is essential for views that support grouping (refer to the *Defining kanban view* and *Defining graph view* recipes) because the resulting context determines the field(s) to be grouped using the `group_by` key. We'll look into the details of grouping in the appropriate recipes, but the context has other uses, too. For example, you can write a function field that returns different values depending on the context, and then you can change the values by activating a filter.

The search view itself also responds to context keys. In a very similar way to default values when creating records, you can pass default values for a search view through the context. If we had set a context of `{'search_default_room_rating': 1}` in our previous action, the `room_rating` filter would have been pre-selected in the search view. This works only if the filter has a name, though, which is why you should always set it. To set defaults for fields in the search view, use `search_default_$fieldname`.

Furthermore, the `field` and `filter` elements can have a `groups` property with the same semantics as in the form views in order to make the element only visible to certain groups.

See also

- For further details about manipulating the context, see the *Passing parameters to forms and actions – Context* recipe.
- Users who speak languages with heavy use of diacritical marks will probably want to have Odoo search for `e`, `è`, `é`, and `ê` when filling in the `e` character. This is a configuration of the **PostgreSQL server** called **unaccent**, which Odoo has special support for, but this is outside the scope of this book. Refer to `https://www.postgresql.org/docs/10/unaccent.html` for more information about unaccent.

Adding a search filter side panel

Odoo provides one more way to display search filters, which is a **search filter side panel**. This panel shows a list of filters on the side of the view. A search panel is very useful when search filters are used frequently by the end user.

Getting ready

The search panel is part of the search view. So, for this recipe, we will continue using the `my_module` add-on from the previous recipe. We will add our search panel to the previously designed search view.

How to do it...

Add `<searchpanel>` in the search view, as shown here:

```
<record id="hostel_room_view_search" model="ir.ui.view">
    <field name="model">hostel.room</field>
    <field name="arch" type="xml">
        <search>
            <field name="name"/>
            <field name="room_no"/>
            <field name="state"/>
            <searchpanel>
                <field name="state" expand="1" select="multi"
icon="fa-check-square-o" enable_counters="1"/>
            </searchpanel>
        </search>
    </field>
</record>
```

Update the module to apply the modification. After the update, you will see the search panel on the left side of the view.

How it works...

To add the search panel, you will need to use the `<searchpanel>` tag in the search view. To add your filter, you will need to add a field in the search panel.

In our example, first, we added a `state` field. You also need to add an `icon` attribute to the field. This icon will be displayed before the title of the filter. Once you add the field to the search panel, it will display the title with an icon, and, below that, a list of all the users. Upon clicking on a user, the records in the list view will be filtered, and you will only see the contacts of the selected user. In this filter, only one item can be active, meaning once you click on another user's filter, the previous user's filter will be removed. If you want to activate multi-user filters, you can use the `select="multi"` attribute. If you use that attribute, you will find the checkbox for each filter option, and you will be able to activate multiple filters at a time. We have used the `select="multi"` attribute on the `state` filter. This will allow us to select and filter by multiple categories at once.

> **Important note**
> Be careful when you are using the side panel filter on `many2one` or `many2many`. If the relation model has too many records, only the **top 200 records** will be displayed to avoid performance issues.

There's more...

If you want to display search panel items in groups, you can use the `groupby` attribute on a field. For example, if you want to group a category based on its parent hierarchy, you can add the `groupby` attribute with the `parent_id` field, as can be seen here:

```
<field name="state"
    icon="fa-check-square-o"
    select="multi"
    groupby="parent_id"/>
```

This will show the category filters grouped according to the parent category of the record.

Changing existing views – view inheritance

So far, we have ignored the existing views and declared completely new ones. While this is didactically sensible, you'll rarely be in situations where you'll want to define a new view for an existing model. Instead, you'll want to slightly modify the existing views, be it to simply have them show a field that you added to the model in your add-on or to customize them according to your needs or your customers' needs.

In this recipe, we'll change the default partner form to show the record's last modification date and make the `mobile` field searchable by modifying the search view. Then, we'll change the position of one column in the partners' list view.

How to do it...

1. Inject the field into the default form view:

```
<record id="hostel_room_view_form_inherit" model="ir.ui.view">
    <field name="name">Hostel Room Form Inherit</field>
    <field name="model">hostel.room</field>
    <field name="inherit_id" ref="my_hostel.hostel_room_view_
form" />
```

```
    <field name="arch" type="xml">
        <xpath expr="//group[@name='my_content']/group"
position="inside">
            <field name="room_no"/>
        </xpath>
        <xpath expr="//group[@name='my_content']/group"
position="after">
            <group>
                <field name="description"/>
                <field name="room_rating"/>
            </group>
        </xpath>
    </field>
</record>
```

2. Add the field to the default search view:

```
<record id="hostel_room_view_search_inherit" model="ir.ui.view">
    <field name="name">Hostel Room Search inherit</field>
    <field name="model">hostel.room</field>
    <field name="inherit_id" ref="my_hostel.hostel_room_view_
search" />
    <field name="arch" type="xml">
        <xpath expr="." position="inside">
            <field name="room_rating"></field>
        </xpath>
    </field>
</record>
```

3. Add the field to the default list view:

```
<record id="hostel_room_view_tree_inherit" model="ir.ui.view">
    <field name="name">Hostel Room List Inherit</field>
    <field name="model">hostel.room</field>
    <field name="inherit_id" ref="my_hostel.hostel_room_view_
tree" />
    <field name="arch" type="xml">
        <xpath expr="//field[@name='name']" position="after">
            <field name="room_no"/>
        </xpath>
    </field>
</record>
```

After updating your module, you should see the **Last updated on** field beneath the website field on the partner form. When you type something into the search box, it should suggest that you search for the partners on the mobile field, and in the partner's list view, you will see that the order of the phone number and email has changed.

How it works...

In *step 1*, we added a basic structure for form inheritance. The crucial field here is, as you've probably guessed, `inherit_id`. You need to pass the XML ID of the view you want to modify (inherit from) to it. The `arch` field contains instructions on how to modify the existing XML nodes within the view you're inheriting from. You should actually think of the whole process as simple XML processing because all the semantic parts only come a lot later.

The most canonical instruction within the `arch` field of an inherited view is the `field` element, which has the required attributes: `name` and `position`. As you can only have every field appear once in a form, the name already uniquely identifies a field. With the `position` attribute, we can place whatever we put within the field element, either `before`, `inside`, or `after` regarding the field we named. The default is `inside`, but for readability, you should always name the position you require. Remember that we're not talking semantics here; this is about the position in the XML tree relative to the field we have named. How this will be rendered afterward is a completely different matter.

Step 2 demonstrates a different approach. The `xpath` element selects the first element that matches the XPath expression named in the `expr` attribute. Here, the `position` attribute tells the processor where to put the contents of the `xpath` element.

> **Important note**
>
> If you want to create an XPath expression based on a CSS class, Odoo provides a special function called `hasclass`. For example, if you want to select a `<div>` element with the `test_class` CSS class, then the expression will be `expr="//div[hasclass('test_class')]"`.

Step 3 shows how you can change the position of an element. This option was introduced in **version 12**, and it is rarely used. In our example, we moved the `phone` field so that it came after the `email` field using the `position=move` option.

XPath might look somewhat scary, but it is a very efficient means of selecting the node you need to work on. Take the time to look through some simple expressions; it's worth it. You'll likely stumble upon the term **context node**, to which some expressions are relative. In Odoo's view inheritance system, this is always the root element of the view you're inheriting from.

For all the other elements found in the `arch` field of an inheriting view, the processor looks for the first element with the same node name and matching attributes (with the attribute position excluded, as this is part of the instruction). Use this only in cases where it is very unlikely that this combination is not unique, such as a group element combined with a `name` attribute.

> **Important note**
>
> Note that you can have as many instruction elements within the `arch` field as you need. We only used one per inherited view because there's nothing else we want to change currently.

There's more...

The `position` attribute has two other possible values: `replace` and `attributes`. Using `replace` causes the selected element to be replaced with the content of the instruction element. Consequently, if you don't have any content, the selected element can simply be removed. The preceding list or form view would cause the `state` field to be removed:

```
<field name="state" position="replace" />
```

> **Important note**
>
> Removing fields can cause other inheriting views to break and several other undesirable side effects, so avoid that if possible. If you really need to remove fields, do so in a view that comes late in the order of evaluation (refer to the next section, *Order of evaluation in view inheritance*, for more information).

`attributes` has very different semantics from the preceding examples. The processor expects the element to contain the `attribute` elements with a `name` attribute. These elements will then be used to set attributes for the selected element. If you want to heed the earlier warning, you should set the `invisible` attribute to `1` for the `state` field:

```
<field name="state" position="attributes">
    <attribute name="invisible">1</attribute>
</field>
```

An `attribute` node can have `add` and `remove` attributes, which, in turn, should contain the value to be removed from or added to the space-separated list. This is very useful for the `class` attribute, where you'd add a class (instead of overwriting the whole attribute) by using the following:

```
<field name="description" position="attributes">
    <attribute name="class" add="oe_inline" separator=" "/>
</field>
```

This code adds the `oe_inline` class to the `description` field. If the field already has a class attribute present, Odoo will join the value with the value of the `separator` attribute.

Order of evaluation in view inheritance

As we currently have only one parent view and one inheriting view, we don't run into any problems with conflicting view overrides. When you have installed a couple of modules, you'll find a lot of overrides for the partner form. This is fine as long as they change different things in a view, but there are occasions where it is important to understand how overriding works in order to avoid conflicts.

The direct descendants of a view are evaluated in ascending order of their `priority` field, so views with a lower priority are applied first. Every step of inheritance is applied to the result of the first, so if a view with priority `3` changes a field and another one with priority `5` removes it, this is fine. This does not work, however, if the priorities are reversed.

You can also inherit an inheriting view itself from a view. In this case, the second-level inheriting view is applied to the result of the view it inherits from. So, if you have four views, `A`, `B`, `C`, and `D`, where `A` is a standalone form, `B` and `C` inherit from `A`, and `D` inherits from `B`, the order of evaluation is `A`, `B`, `D`, and `C`. Use this to enforce an order without having to rely on priorities; this is safer in general. If an inheriting view adds a field and you need to apply changes to this field, inherit from the inheriting view and not from the standalone one.

> **Important note**
>
> This kind of inheritance always works on the complete XML tree from the original view, with modifications from the previous inheriting views applied.

The following points provide information on some advanced tricks that are used to tweak the behavior of view inheritance:

- For inheriting views, a very useful and not very well-known field is `groups_id`. This field causes inheritance to take place only if the user requesting the parent view is a member of one of the groups mentioned there. This can save you a lot of work when adapting the user interface for different levels of access because, with inheritance, you can have more complex operations than just showing or not showing the elements based on group membership, as is possible with the `groups` attribute for form elements.
- You can, for example, remove elements if the user is a member of a group (which is the inverse of what the `groups` attribute does). You can also carry out some elaborate tricks, such as adding attributes based on group membership. Think about simple things, such as making a field read-only for certain groups, or more interesting concepts, such as using different widgets for different groups.
- What was described in this recipe relates to the `mode` field of the original view being set to primary, while the inheriting views have the mode extension, which is the default. We will investigate the case in which the mode of an inheriting view is set to `primary` later, where the rules are slightly different.

Defining document-style forms

In this recipe, we'll review some design guidelines in order to present a uniform user experience.

How to do it...

1. Start your form with a `header` element:

```
<header>
    <button name="make_available" string="Make Available"
type="object"/>
    <button name="make_closed"  string="Make Closed"
type="object"/>
    <button type="action" name="%(my_hostel.hostel_room_
category_action)d"        string="Open Hotel Room Category" />
    <field name="state" widget="statusbar"/>
</header>
```

2. Add a `sheet` element for content:

```
<sheet>
```

3. Put in the `stat` button, which will be used to show total `Hostel Room` and will redirect to hostel room:

```
<div class="oe_button_box" name="button_box">
    <button type="object" class="oe_stat_button" icon="fa-
pencil-square-o" name="action_open_related_hostel_room">
        <div class="o_form_field o_stat_info">
            <span class="o_stat_value">
                <field name="related_hostel_room"/>
            </span>
            <span class="o_stat_text">Hostel Room</span>
        </div>
    </button>
</div>
```

4. Add some prominent field(s):

```
<div class="oe_title">
    <h1>
        <field name="name"/>
    </h1>
</div>
```

5. Add your content; you can use a notebook if there are a lot of fields:

```
<group>
    <field name="child_ids"/>
    <field name="hoste_room_ids" widget="many2many_tags"/>
</group>
```

6. After the sheet, add the `chatter` widget (if applicable):

```
</sheet>
    <div class="oe_chatter">
         <field name="message_follower_ids" widget="mail_
followers"/>
         <field name="message_ids" widget="mail_thread"/>
         <field name="activity_ids" widget="mail_activity"/>
    </div>
```

Let's have a look at how this recipe works.

How it works...

The header should contain buttons that execute actions on the object that the user currently sees. Use the `btn-primary` class to make buttons visually stand out (in purple at the time of writing), which is a good way to guide the user regarding which is the most logical action to execute at the moment. Try to have all the highlighted buttons to the left of the non-highlighted buttons and hide the buttons that are not relevant in the current state (if applicable). If the model has a state, show it in the header using the `statusbar` widget. This will be rendered as right-aligned in the header.

The `sheet` element is rendered as a stylized sheet, and the most important fields should be the first thing the user sees when looking at it. Use the `oe_title` classes to have them rendered in a prominent place (floating left with slightly adjusted font sizes at the time of writing).

If there are other records of interest concerning the record the user currently sees (such as the partner's invoices on a partner form), put them in an element with the `oe_right` and `oe_button_box` classes; this aligns the buttons in it to the right. On the buttons themselves, use the `oe_stat_button` class to enforce a uniform rendering of the buttons. It's also customary to assign an icon class from the **Font Awesome** icons for the `icon` attribute. You can learn more about Font Awesome at `https://fontawesome.com/v4.7.0/icons/`.

You can use the `oe_chatter` class and **Chatter widgets** to get the default chatter at the bottom of the form view. For this, you need to use the `mail.thread` mixin. We will see this in detail in *Chapter 23, Managing Emails in Odoo.*

> **Important note**
>
> Even if you do not like this layout, stick to the element and class names described here and adjust what you need with CSS and possibly JavaScript. This will make the user interface more compatible with existing add-ons and allow you to integrate better with core add-ons.

See also

- To find out more about Font Awesome, go to `https://fontawesome.com/v4.7.0/icons/`.
- For more details on the `mail.thread` mixin, refer to *Chapter 23, Managing Emails in Odoo*.

Dynamic form elements using attributes

So far, we have only looked into changing forms depending on the user's groups (the `groups` attribute for elements and the `groups_id` field for inherited views) and nothing more. This recipe will show you how to modify the form view based on the value of the fields in it.

How to do it...

1. Define an attribute called `attributes` on a form element:

```
<field name="child_ids"
    invisible="not parent_id"
    required="parent_id"/>
```

2. Ensure that all the fields you refer to are available in your form:

```
<field name="parent_id"/>
```

This will make the `child_ids` field invisible if the `parent_id` is not `hostel room category`, and it will be required if it's a hostel room category.

How it works...

attributes contains a dictionary with `invisible`, `required`, and `readonly` keys (all of which are optional). The values are domains that may refer to the fields that exist on the form (and really only those, so there are no dotted paths), and the whole dictionary is evaluated according to the rules for client-side Python, as described earlier in the *Passing parameters to forms and actions – Context* recipe of this chapter. So, for example, you can access the context in the right-hand operand.

There's more...

While this mechanism is quite straightforward for scalar fields, it's less obvious how to handle the `one2many` and `many2many` fields. In fact, in standard Odoo, you can't do much with those fields within **attributes**. However, if you only need to check whether such a field is empty, use `[[6, False, []]]` as your right-hand operand.

Defining embedded views

When you show a `one2many` or a `many2many` field on a form, you don't have much control over how it is rendered if you haven't used one of the specialized widgets. Additionally, in the case of the `many2one` fields, it is sometimes desirable to be able to influence the way the linked record is opened. In this recipe, we'll look at how to define private views for those fields.

How to do it...

1. Define your field as usual, but don't close the tag:

```
<field name="hostel_room_ids">
```

2. Write the view definition(s) into the tag:

```
<tree>
    <field name="name"/>
    <field name="room_no"/>
</tree>
<form>
    <sheet>
        <group>
            <field name="name"/>
            <field name="room_no"/>
        </group>
    </sheet>
</form>
```

3. Close the tag:

```
</field>
```

How it works...

When Odoo loads a form view, it first checks whether the `relational` type fields have embedded views in the field, as outlined previously. Those embedded views can have the exact same elements as the views we defined before. Only if Odoo doesn't find an embedded view of some type does it use the model's default view of this type.

There's more...

While embedded views might seem like a great feature, they complicate view inheritance a lot. For example, as soon as embedded views are involved, the field names are not guaranteed to be unique, and you'll usually have to use some elaborate XPaths to select elements within an embedded view.

So, in general, you should better define standalone views and use the `form_view_ref` and `tree_view_ref` keys, as described earlier in the *Having an action open a specific view* recipe of this chapter.

Displaying attachments on the side of the form view

In some applications, such as invoicing, you need to fill in data based on a document. To ease the data-filling process, a new feature was added to Odoo version 12 to display the document on the side of the form view.

In this recipe, we will learn how to display the form view and the document side by side:

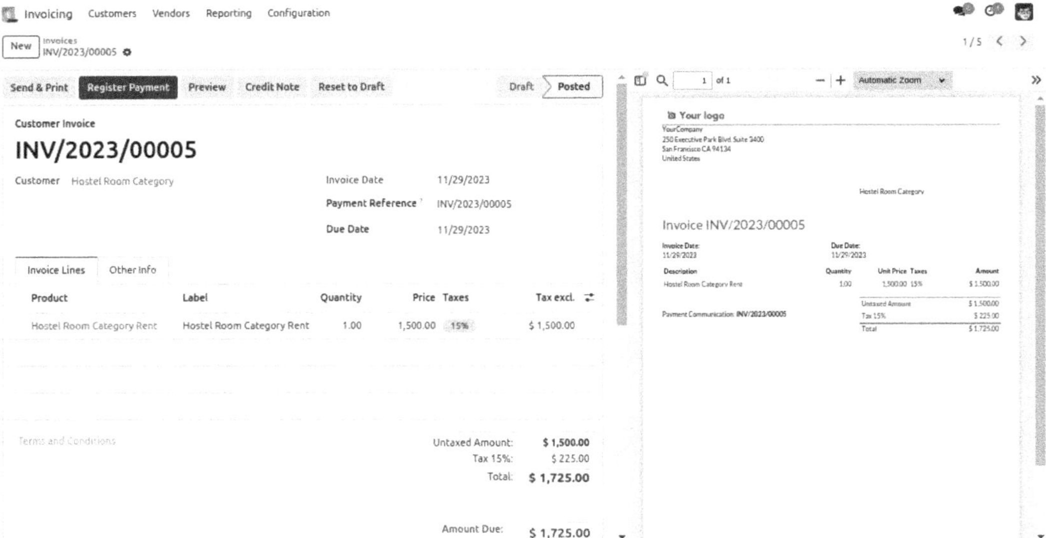

Figure 9.3 – Cascading attachments and the form view

> **Important note**
>
> This feature is only meant for large displays (`>1534px`), so if you have a small viewport, this feature will be hidden. Internally, this feature uses some responsive utilities, so this feature only works in the **Enterprise** edition. However, you can still use this code in your module. Odoo will automatically handle this, so if the module is installed in the Enterprise edition, it will show the document, while in the Community edition, it will hide everything without any side effects.

How to do it...

We will enable this feature to modify a form view for the `hostel.room.category` model, as follows:

```
<record id="hostel_room_category_view_form" model="ir.ui.view">
    <field name="name">Hostel Room Categories Form</field>
    <field name="model">hostel.room.category</field>
    <field name="arch" type="xml">
        <form>
            <sheet>
                <div class="oe_button_box" name="button_box">
                    <button type="object" class="oe_stat_button"
icon="fa-pencil-square-o" name="action_open_related_hostel_room">
                        <div class="o_form_field o_stat_info">
                            <span class="o_stat_value">
                                <field name="related_hostel_room"/>
                            </span>
                            <span class="o_stat_text">Hostel Room</
span>
                        </div>
                    </button>
                </div>
                <div class="oe_title">
                   <h1>
                        <field name="name"/>
                    </h1>
                </div>
                <group>
                    <group>
                        <field name="description"/>
                    </group>
                    <group>
                        <field name="parent_id"/>
                    </group>
                </group>
                <group>
```

```
                    <field name="child_ids"
                           invisible="not parent_id"
                           required="parent_id"/>
                    <field name="hoste_room_ids">
                        <tree>
                            <field name="name"/>
                            <field name="room_no"/>
                        </tree>
                        <form>
                            <sheet>
                                <group>
                                    <field name="name" />
                                    <field name="room_no"/>
                                </group>
                            </sheet>
                        </form>
                    </field>
                </group>
            </sheet>
            <div class="o_attachment_preview" options="{'types':
['image', 'pdf'], 'order': 'desc'}"/>
            <div class="oe_chatter">
                <field name="message_follower_ids" widget="mail_
followers"/>
                <field name="message_ids" widget="mail_thread"/>
                <field name="activity_ids" widget="mail_activity"/>
            </div>
        </form>
    </field>
</record>
```

Update the module to apply the changes. You need to upload a PDF or image via the record chatter. When you upload it, Odoo will display the attachment on the side.

How it works...

This feature only works if your model has inherited the `mail.thread` model. To show the document on the side of any form view, you will need to add an empty `<div>` with the `o_attachment_preview` class before the chatter elements. That's it; the documents attached in the chatter will be displayed on the side of the form view.

By default, the `pdf` and `image` documents will be displayed in ascending order by date. You can change this behavior by providing extra options, which include the following:

- `type`: You need to pass the list of document types you want to allow. Only two values are possible: `pdf` and `image`. For example, if you want to display only `pdf`-type images, you can pass `{'type': ['pdf']}`.
- `order`: The possible values are `asc` and `desc`. These allow you to show documents in ascending order or descending order of the document creation date.

There's more...

In most cases, you want to display documents on the side of the initial state of any record. If you want to hide the attachment preview based on domain, you can use `attributes` on the `<div>` tag to hide the preview.

Take a look at the following example: it will hide the PDF preview if the value of the `state` field is not `draft`:

```
<div class="o_attachment_preview"
    invisible="state != 'draft'"/>
```

This is how you can hide attachments when they are not needed. Usually, this feature is used to fill data from PDFs and is only activated in draft mode.

Defining kanban views

So far, we have presented you with a list of records that can be opened to show a form. While those lists are efficient when presenting a lot of information, they tend to be slightly boring, given the lack of design possibilities. In this recipe, we'll take a look at **kanban views**, which allow us to present lists of records in a more appealing way.

How to do it...

1. Define a view of the `kanban` type:

```
<record id="hostel_room_category_view_kanban" model="ir.
ui.view">
    <field name="name">Hostel Room Categories kanban</field>
    <field name="model">hostel.room.category</field>
    <field name="arch" type="xml">
        <kanban class="o_kanban_mobile" sample="1">
```

2. List the fields you'll use in your view:

```
        <field name="name"/>
        <field name="description"/>
        <field name="parent_id"/>
```

3. Implement a design:

```
    <templates>
        <t t-name="kanban-box">
            <div t-attf-class="oe_kanban_global_click">
                 <div class="row mb4">
                     <div class="col-6 o_kanban_record_headings">
                         <strong>
                             <span>
                                 <field name="name"/>
                             </span>
                         </strong>
                      </div>
                      <div class="col-6 text-end">
                          <strong><i role="img"
    title="description"/>
                              <t t-esc="record.description.
    value"/></strong>
                         </div>
                    </div>
                    <div class="row">
                         <div class="col-12">
                             <span><field name="parent_id"/></span>
                         </div>
                    </div>
               </div>
         </t>
    </templates>
```

4. Close all the tags:

```
            </kanban>
        </field>
    </record>
```

5. Add this view to one of your actions. This is left as an exercise for the reader. You will find a full working example in the GitHub example files: `https://github.com/PacktPublishing/Odoo-13-Development-Cookbook-Fourth-Edition/tree/master/Chapter09/15_kanban_view/my_module`.

How it works...

We need to give a list of fields to load in *step 2* in order to be able to access them later. The content of the `templates` element must be a single `t` element with the `t-name` attribute set to `kanban-box`.

What you write inside this element will be repeated for each record, with special semantics for `t` elements and `t-*` attributes. For details about this, refer to the *Using client-side QWeb templates* recipe from *Chapter 15, Web Client Development* because, technically speaking, kanban views are just an application of QWeb templates.

There are a few modifications that are particular to kanban views. You have access to the `read_only_mode`, `record`, and `widget` variables during evaluation. Fields can be accessed using `record.fieldname`, which is an object with the `value` and `raw_value` properties, where `value` is the field's value that has been formatted in a way that is presentable to the user, and `raw_value` is the field's value, as it comes from the database.

> **Important note**
>
> `many2many` fields make an exception here. You'll only get an ID list through the `record` variable. For a user-readable representation, you must use the `field` element.

Note the `type` attribute of the link at the top of the template. This attribute makes Odoo generate a link that opens the record in view mode (**open**) or edit mode (**edit**), or it deletes the record (**delete**). The `type` attribute can also be `object` or `action`, which will render the links that call a function from the model or an action. In both cases, you need to supplement the attributes for buttons in form views, as outlined in the *Adding buttons to forms* recipe of this chapter. Instead of the `a` element, you can also use the `button` element; the `type` attribute has the same semantics here.

There's more...

There are a few more helper functions worth mentioning. If you need to generate a pseudo-random color for an element, use the `kanban_color(some_variable)` function, which will return a CSS class that sets the `background` and `color` properties. This is usually used in the `t-att-class` elements.

If you want to display an image stored in a binary field, use `kanban_image(modelname, fieldname, record.id.raw_value)`, which returns a data URI if you included the field in your fields list; the field is set, is a placeholder if the field is not set, or is a URL that makes Odoo stream the field's contents if you didn't include the field in your fields list. Do not include the field in the fields list if you need to display a lot of records simultaneously or if you expect very big images. Usually, you'd use this in a `t-att-src` attribute of an `img` element.

> **Important note**
>
> Doing design in kanban views can be a bit trying. What often works better is generating HTML using a function field of the HTML type and generating this HTML from a Qweb view. In this way, you're still doing QWeb but doing so on the server side, which is a lot more convenient when you need to work on a lot of data.

See also

- To know more about template elements, refer to the *Using client-side QWeb templates* recipe from *Chapter 15, Web Client Development.*

Showing kanban cards in columns according to their state

This recipe shows you how to set up a kanban view where the user can drag and drop a record from one column to the other, thereby pushing the record in question into another state.

Getting ready

From now on, we'll make use of the hostel module here, as this defines models that lend themselves better to date- and state-based views than those defined in the base module. So, before proceeding, add `base` to the dependencies list of your add-on.

How to do it...

1. Define a kanban view for the hostel room category:

```
<record id="hostel_room_category_view_kanban" model="ir.
ui.view">
    <field name="name">Hostel Room Categories kanban</field>
    <field name="model">hostel.room.category</field>
    <field name="arch" type="xml">
        <kanban class="o_kanban_mobile" sample="1" default_
group_by="parent_id">
            <field name="name"/>
            <field name="description"/>
            <templates>
                <t t-name="kanban-box">
                    <div t-attf-class="oe_kanban_global_click">
                        <div class="row mb4">
                            <div class="col-6 o_kanban_record_
headings">
                                <strong>
```

```
                                        <span>
                                            <field name="name"/>
                                        </span>
                                    </strong>
                                </div>
                                <div class="col-6 text-end">
                                    <strong><i role="img"
title="description"/> <t t-esc="record.description.value"/></
strong>
                                </div>
                            </div>
                            <div class="row">
                                <div class="col-12">
                                    <span><field name="parent_id"/></
span>
                                </div>
                            </div>
                        </div>
                    </t>
                </templates>
            </kanban>
        </field>
    </record>
```

2. Add a menu and an action using this view. This is left as an exercise for the reader.

How it works...

Kanban views support grouping, which allows you to display records that have a group field in common in the same column. This is commonly used for a `parent hotel room category` or `parent_id` field because it allows the user to change this field's value for a record by simply dragging it into another column. Set the `default_group_by` attribute on the `kanban` element to the name of the field you want to group by in order to make use of this functionality.

To control the behavior of kanban grouping, there are a few options available in Odoo:

- `group_create`: This option is used to hide or show the **Add a new column** option in grouped kanban. The default value is `true`.
- `group_delete`: This option enables or disables the **Column delete** option in the kanban group context menu. The default value is `true`.
- `group_edit`: This option enables or disables the **Column edit** option in the kanban group context menu. The default value is `true`.

- `archivable`: This option enables or disables the option to archive and restore the records from the kanban group context menu. This only works if the `active` Boolean field is present in your model.
- `quick_create`: With this option, you can create records directly from the kanban view.
- `quick_create_view`: By default, the `quick_create` option displays only the name field in kanban. However, with the `quick_create_view` option, you can give the reference to the minimal form view so as to display it in kanban.
- `on_create`: If you don't want to use `quick_create` when creating a new record and you don't want to redirect the user to the form view either, you can give the reference of the wizard so that it will open the wizard when clicking the **Create** button.

There's more...

If not defined in the dedicated attribute, any search filter can add grouping by setting a context key named `group_by` to the field name(s) to group by.

Defining calendar views

This recipe walks you through how to display and edit information about dates and duration in your records in a visual way.

How to do it...

Follow these steps to add a `calendar` view for the `hostel.room.category` model:

1. Define a `calendar` view:

```
<record id="hostel_room_category_view_calendar" model="ir.
ui.view">
    <field name="name">Hostel Room Categories Calendar</field>
    <field name="model">hostel.room.category</field>
    <field name="arch" type="xml">
        <calendar date_start="date_assign" date_stop="date_end"
color="parent_id">
            <field name="name" />
            <field name="parent_id" />
        </calendar>
    </field>
</record>
```

2. Add menus and actions using this view. This is left as an exercise for the reader.

How it works...

The `calendar` view needs to pass the field names in the `date_start` and `date_stop` attributes to indicate which fields to look at when building the visual representation. Only use fields with the `Datetime` or `Date` type; other types of fields will not work and will instead generate an error. While `date_start` is required, you can leave out `date_stop` and set the `date_delay` attribute instead, which is expected to be a `Float` field that represents the duration in hours.

The `calendar` view allows you to give records that have the same value in a field the same (arbitrarily assigned) color. To use this functionality, set the `color` attribute to the name of the field you need. In our example, we can see at a glance which hostel room category belongs to the same hostel room category because we assigned `parent_id` as the field to determine the color groups.

The fields you name in the `calendar` element's body are shown within the block that represents the time interval covered, separated by commas.

There's more...

The `calendar` view has some other helpful attributes. If you want to open calendar entries in a popup instead of the standard form view, set `event_open_popup` to `1`. By default, you create a new entry by just filling in some text, which internally calls the model's `name_create` function to actually create the record. If you want to disable this behavior, set `quick_add` to `0`.

If your model covers a whole day, set `all_day` to a field's name that is `true` if the record covers the whole day and `false` otherwise.

Defining graph view and pivot view

In this recipe, we'll take a look at Odoo's business intelligence views. These are read-only views that are meant to present data.

Getting ready

We're still making use of the `hostel` module here. You can configure a graph and pivot views to get different statistics. For our example, we will focus on the assigned user. We will generate a graph and pivot view to see the users of the hostel room category. By the way, the end user can generate statistics of their choice by modifying the view options.

How to do it...

1. Define a graph view using bars:

```
<record id="hostel_room_category_view_graph" model="ir.ui.view">
    <field name="name">Hostel Room Categories Graph</field>
```

```
        <field name="model">hostel.room.category</field>
        <field name="arch" type="xml">
            <graph type="bar">
                <field name="parent_id"/>
                <field name="child_ids"/>
            </graph>
        </field>
    </record>
```

2. Define a pivot view:

```
<record id="hostel_room_category_view_pivot" model="ir.ui.view">
    <field name="name">Hostel Room Categories Pivot</field>
    <field name="model">hostel.room.category</field>
    <field name="arch" type="xml">
        <pivot>
            <field name="parent_id" type="row"/>
            <field name="name" type="col"/>
        </pivot>
    </field>
</record>
```

3. Add menus and actions using this view. This is left as an exercise for the reader.

If everything went well, you should see graphs that show how many parent hostel room categories are assigned to which hostel room categories and the state of those hostel room categories.

How it works...

The graph view is declared with a root element, `graph`. The `type` attribute on a `graph` element determines the initial mode of a graph view. The possible values are `bar`, `line`, and `chart`, but `bar` is the default. The graph view is highly interactive, so the user can switch between the different modes and also add and remove fields. If you use `type="bar",`, you can also use `stacked="1"` to show a stacked bar chart during grouping.

The `field` elements tell Odoo what to display on which axis. For all graph modes, you need at least one field with the `row` type and one with the `measure` type to see anything useful. Fields of the `row` type determine the grouping, while those of the `measure` type stand for the value(s) to be shown. Line graphs only support one field of each type, while charts and bars handle two group fields with one measure nicely.

Pivot views have their own root element, `pivot`. The pivot view supports an arbitrary amount of group and measure fields. Nothing will break if you switch to a mode that doesn't support the number of groups and measures you defined; some fields will just be ignored, and the result might not be as interesting as it could be.

There's more...

For all graph types, `Datetime` fields are tricky to group because you'll rarely encounter the same field value here. So, if you have a `Datetime` field of the `row` type, also specify the `interval` attribute with one of the following values: `day`, `week`, `month`, `quarter`, or `year`. This will cause the grouping to take place in the given interval.

> **Important note**
>
> Grouping, like sorting, relies heavily on PostgreSQL. So, the rule applies here also that a field must live in the database and in the current table in order to be usable.
>
> It is a common practice to define database views that collect all the data you need and define a model on top of this view in order to have all the necessary fields available.
>
> Depending on the complexity of your view and the grouping, building the graph can be quite an expensive exercise. Consider setting the `auto_search` attribute to `False` in these cases so that the user can first adjust all the parameters and only then trigger a search.

The pivot table also supports grouping in columns. Use the `col` type for the fields you want to have there.

Defining the cohort view

For the cohort analysis of records, the new cohort view was added in Odoo version 12. The cohort view is used to find out the life cycle of a record over a particular time span. With the cohort view, you can see the churn and retention rate of any object for a particular time.

Getting ready

The `cohort` view is part of the **Odoo Enterprise edition**, so you cannot use it with only the Community edition. If you are using the Enterprise edition, you need to add `web_cohort` in the manifest file of your module. For our example, we will create a view to see the cohort analysis for hostel room category.

How to do it...

Follow these steps to add the `cohort` view for the `hostel.room.category` model:

1. Define a `cohort` view:

```
<record id="hostel_room_category_view_cohort" model="ir.
ui.view">
    <field name="name">Hostel Room Categories Cohort</field>
    <field name="model">hostel.room.category</field>
    <field name="arch" type="xml">
```

```
        <cohort date_start="date_assign" date_stop="date_end"
interval="month" string="Categories Cohort" />
    </field>
</record>
```

2. Add menus and actions using this view. This is left as an exercise for the reader.

How it works...

To create a cohort view, you need to provide `date_start` and `date_stop`. These will be used in the view to determine the time span of any record. For example, if you are managing a subscription of a service, the start date of the subscription will be `date_start` and the date when the subscription is going to expire will be `date_stop`.

By default, the `cohort` view will be displayed in the `retention` mode at intervals of a month. You can use the given options to obtain different behaviors in the `cohort` view:

- `mode`: You can use cohort with two modes: `retention (default)` or `churn`. The `retention` mode starts at 100% and decreases with time, while the `churn` mode starts at 0% and increases with time.
- `timeline`: This option accepts two values: `forward (default)` or `backward`. In most cases, you need to use the forward timeline. However, if `date_start` is in the future, you will need to use the backward timeline. An example of when we would use the backward timeline would be for the registration of an event attendee where the event date is in the future, and the registration date is in the past.
- `interval`: By default, the cohort is grouped by month, but you can change this in the interval options. Other than months, the cohort also supports day, week, and year intervals.
- `measure`: Just like graph and pivot, measure is used to display the aggregated value of a given field. If no option is given, the cohort will display the count of records.

Defining the gantt view

Odoo version 13 added a new `gantt` view with new options. The `gantt` view is useful for seeing overall progress and scheduling business processes. In this recipe, we will create a new `gantt` view and look at its options.

Getting ready

The `gantt` view is part of the Odoo Enterprise edition, so you can't use it with the Community edition. If you are using the Enterprise edition, you need to add the `web_gantt` dependency in the manifest file of your module.

In our example, we will continue using the `my_hostel` module from the previous recipe. We will create a new `gantt` view for the hostel room category.

How to do it...

1. Define a `gantt` view for the hostel room category model as follows:

```
<record id="hostel_room_category_view_gantt" model="ir.ui.view">
    <field name="name">Hostel Room Categories Gantt</field>
    <field name="model">hostel.room.category</field>
    <field name="arch" type="xml">
        <gantt date_start="date_assign" date_stop="date_end"
string="Hostel Room Category" default_group_by="parent_id"
color="parent_id">
            <field name="name"/>
            <field name="parent_id"/>
            <templates>
                <div t-name="gantt-popover" >
                    <ul class="pl-1 mb-0 list-unstyled">
                        <li>
                            <strong>Name: </strong>
                            <t t-esc="name"/>
                        </li>
                        <li>
                            <strong>Parent Category: </strong>
                            <t t-esc="parent_id[1]"/>
                        </li>
                    </ul>
                </div>
            </templates>
        </gantt>
    </field>
</record>
```

2. Add menus and actions using this view. This is left as an exercise for the reader.

Install and update the module to apply the changes; after the update, you will see the `gantt` view on the hostel room category.

How it works...

With the `gantt` view, you can display an overall schedule on one screen. In our example, we have created a `gantt` view for the hostel room category grouped by parent. Typically, you need two attributes to create a `gantt` view, `start_date`, and `stop_date`, but there are some other attributes that extend the functionality of the `gantt` view. Let's see all the options:

- `start_date`: Defines the starting time of the `gantt` item. It must be a date or date-time field.
- `default_group_by`: Use this attribute if you want to group the `gantt` items based on field.
- `color`: This attribute is used to decide the color of a `gantt` item.
- `progress`: This attribute is used to indicate the progress of a `gantt` item.
- `decoration-*`: Decoration attributes are used to decide the color of a gantt item based on conditions. It can be used like this: `decoration-danger="state == 'lost'"`. Its other values are `decoration-success`, `decoration-info`, `decoration-warning`, and `decoration-secondary`.
- `scales`: Use the `scales` attribute if you want to enable the `gantt` view only for a few scales. For example, if you only want day and week scales, you can use `scales="day,week"`.
- By default, `gantt` view items are resizable and draggable, but if you want to disable that, you can use the `edit="0"` attribute.

There's more...

When you hover over a `gantt` view item, you will see the name and date for the item. If you want to customize that popup, you can define a **QWeb template** in the `gantt` view definition like this:

```
<gantt date_start="date_assign" date_stop="date_end" string="Hostel
Room Category" default_group_by="parent_id" color="parent_id">
    <field name="name"/>
    <field name="parent_id"/>
    <templates>
        <div t-name="gantt-popover" >
             <ul class="pl-1 mb-0 list-unstyled">
                  <li>
                       <strong>Name: </strong>
                           <t t-esc="name"/>
                        </li>
                        <li>
                            <strong>Parent Category: </strong>
                            <t t-esc="parent_id[1]"/>
                        </li>
                   </ul>
              </div>
          </templates>
      </gantt>
```

Note that you will need to add the fields that you want to use in the template via the `<field>` tag.

Defining the activity view

Activities are an important part of Odoo apps. They are used to schedule to-do actions for different business objects. The `activity` view helps you to see the statuses and schedules of all activities on the model.

Getting ready

In our example, we will continue using the `my_hostel` module from the previous recipe. We will create a new `activity` view for the hostel room category.

How to do it...

1. Define a `activity` view for the `hostel room category` model as follows:

```
<record id="hostel_room_category_view_activity" model="ir.
ui.view">
    <field name="name">Hostel Room Categories Activity</field>
    <field name="model">hostel.room.category</field>
    <field name="arch" type="xml">
        <activity string="Hostel Room Category">
            <templates>
                <div t-name="activity-box">
                    <div>
                        <field name="name" display="full"/>
                        <field name="parent_id" muted="1"
display="full"/>
                    </div>
                </div>
            </templates>
        </activity>
    </field>
</record>
```

2. Add menus and actions using this view. This is left as an exercise for the reader.

How it works...

The `activity` view is simplistic; most of the things are managed automatically. You just have the option to customize the first column. To display your data in the first column, you need to create a QWeb template with the name `activity-box`, and that's it; Odoo will manage the rest.

The `activity` view will display your template in the first column, and other columns will show the scheduled activities grouped by activity type.

Defining the map view

Odoo version 13 adds a new view called a `map` view. As its name suggests, it is used to show a map with a marker. They are very useful for on-site services.

Getting ready

In our example, we will continue using the `my_hostel` module from the previous recipe. We will create the new `map` view for the hostel room category. The `map` view is part of the **Odoo Enterprise edition**, so you can't use it with the Community edition. If you are using the Enterprise edition, you need to add the `web_map` dependency in the manifest file of your module.

Odoo uses the API from `https://www.mapbox.com/` to display maps in the view. In order to see the map in Odoo, you will need to generate the access token from the **mapbox**. Make sure you have generated an access token and set it in the Odoo configuration.

How to do it...

1. Define a `map` view for the hostel room category model as follows:

```
<record id="hostel_room_category_view_map" model="ir.ui.view">
    <field name="name">Hostel Room Categories Map</field>
    <field name="model">hostel.room.category</field>
    <field name="arch" type="xml">
        <map>
            <field name="name" string="Title "/>
            <field name="parent_id" string="Hostel Room Category
"/>
        </map>
    </field>
ss</record>
```

2. Add menus and actions using this view. This is left as an exercise for the reader.

How it works...

Creating a map view is pretty simple; you just need a `many2one` field that refers to the `hostel.room.category` model. The `hostel.room.category` model has `address` fields, which are used by the map view to display the marker for the address. You will need to use the `res_partner` attribute to map the address for the `map` view. In our case, we have used the `parent_id` field as the hostel room category parent record set in the `parent_id` field.

10
Security Access

Odoo is typically used by multi-user organizations. Each user has a distinct position in every organization, and they have varying access based on their function. The HR manager, for example, does not have access to the company's accounting information. You may determine which information a user can access in Odoo using access rights and record rules. We will learn how to set access rights rules and record rules in this chapter.

Such compartmentalization of access and security requires that we provide access to roles based on their permission levels. We will learn about this in this chapter.

In this chapter, we will cover the following recipes:

- Creating security groups and assigning them to users
- Adding security access to models
- Limiting access to fields in models
- Using record rules to restrict record access
- Activating features with security groups
- Accessing recordsets as a superuser
- Hiding view elements and menus based on groups

To concisely get the point across, the recipes in this chapter make small additions to an example existing module.

Technical requirements

The technical requirements for this chapter include using the module that we created by following the tutorials in *Chapter 3*, *Creating Odoo Add-On Modules*. To follow the examples here, you should have that module created and ready to use.

All the code that will be used in this chapter can be downloaded from this book's GitHub repository at `https://github.com/PacktPublishing/Odoo-17-Development-Cookbook-Fifth-Edition/tree/main/Chapter10`.

Creating security groups and assigning them to users

Security access in Odoo is configured through security groups: permissions are given to groups and then groups are assigned to users. Each functional area has base security groups provided by a central application.

When add-on modules augment an existing application, they should add rights to the respective groups, as explained in the *Adding security access to models* recipe.

When add-on modules introduce a new functional area that is not yet covered by an existing core application, the associated security groups should be added. We should usually have at least user and management roles.

Take the Hostel example we introduced in *Chapter 3*, *Creating Odoo Add-On Modules*, as an example – it doesn't fit neatly into any of the Odoo core apps, so we will add security groups for it.

Getting ready

This tutorial assumes that you have an Odoo instance ready with `my_hostel` available, as described in *Chapter 3*, *Creating Odoo Add-On Modules*.

How to do it...

To add new access security groups to a module, perform the following steps:

1. Ensure that the `__manifest__.py` add-on module manifest has the `category` key defined:

```
'category': Hostel,
```

2. Add the new `security/groups.xml` file to the manifest `data` key:

```
'data': [
    'security/groups.xml',
],
```

3. Add the new XML file for the data records to the `security/groups.xml` file, starting with an empty structure:

```
<?xml version="1.0" encoding="utf-8"?>
<odoo>
    <!--  Add step 4 goes here -->
</odoo>
```

4. Add the `<record>` tags for the two new groups inside the data XML element:

```
<record id="group_hostel_user" model="res.groups">
     <field name="name">User</field>
     <field name="category_id"
            ref="base.module_category_hostel"/>
     <field name="implied_ids"
            eval="[(4, ref('base.group_user'))]"/>
</record>
<record id="group_hostel_manager" model="res.groups">
    <field name="name">Manager</field>
    <field name="category_id"
           ref="base.module_category_hostel"/>
    <field name="implied_ids"
           eval="[(4, ref('group_hostel_user'))]"/>
    <field name="users" eval="[(4, ref('base.user_admin'))]"/>
</record>
```

If we upgrade the add-on module, these two records will be loaded. To see these groups in the UI, you need to activate developer mode. You'll then be able to see them through the **Settings** | **Users** | **Groups** menu option, like so:

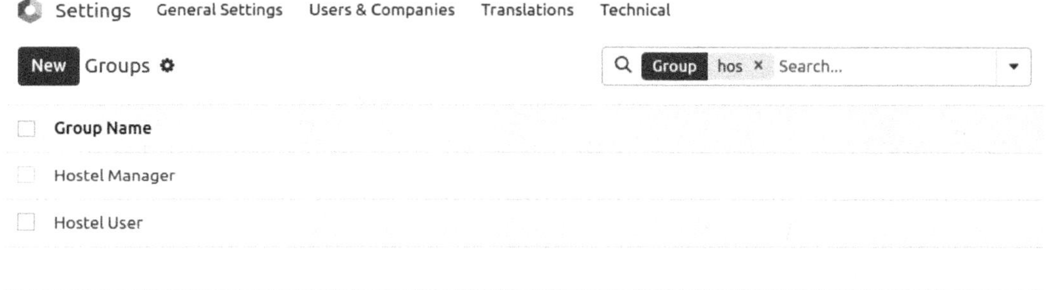

Figure 10.1 – Newly added security groups

> **Important information**
>
> When you add a new model, the admin user does not have access to that model. This implies that the admin user cannot see the menus and views that have been added for that model. To show it, you must first add access rules to that model, something we will accomplish in the Adding security access to models recipe. Note that you may access newly added models as a superuser; for more information, see the *Accessing Odoo as a superuser* recipe in *Chapter 3, Creating Odoo Add-On Modules*.

How it works...

Add-on modules are organized into functional areas, or major *applications*, such as *accounting and finance*, *sales*, or *human resources*. These are defined by the `category` key in the manifest file.

If a category name does not exist yet, it will be automatically created. For convenience, a `base.module_category_<category_name_in_manifest>` XML ID will also be generated for the new category name in lowercase letters, replacing the spaces with underscores. This is useful for relating security groups to application categories.

In our example, we used the **Hostel** category name in the manifest, and it generated a `base.module_category_hostel` XML identifier.

By convention, data files that contain security-related elements should be placed inside a `security` subdirectory.

A manifest file must also be used to register security files. The sequence in which files are specified in the module manifest's **data** key is critical since you cannot utilize a reference to a security group in other views or **ACL** files until the group has been created. It's advisable to put the security data file first, followed by the ACL files and other user interface data files.

In our example, we created groups with the `<record>` tag, which will create a record of the `res.groups` model. The most important columns of the *res.group* model are as follows:

- `name`: This is the group's display name.
- `category_id`: This is a reference to the application category and is used to organize the groups in the user's form.
- `implied_ids`: These are other groups to inherit permissions from.
- `users`: This is the list of users that belong to this group. In new add-on modules, we usually want the admin user to belong to the application's manager group.

The first security group uses `implied_ids` as the `base.group_user` group. This is the `Employee` user group and is the basic security group all the backend users are expected to share.

The second security group sets a value on the `users` field to assign it to the administrator user, which has the `base.user_admin` XML ID.

Users that belong to a security group will automatically belong to its implied groups. For example, if you assign a *Hostel Manager* group to any user, that user will also be included in the *User* group. This is because the *Hostel Manager* group has the *User* group in its `implied_ids` column.

Furthermore, security groups' access rights are cumulative. A user has permission if any of the groups to which they belong (directly or indirectly) grant it to them.

Some security groups are displayed as a selection box in the user form rather than distinct tick boxes. This occurs when the groups involved are in the same application category and are linearly interconnected by `implied_ids`. Group A, for example, implies Group B, while Group B implies Group C. If a group is not associated with any other groups via `implied_ids`, a checkbox will appear instead of a selection box.

> **Note**
>
> Note that the relationships that were defined in the preceding fields also have reverse relationships that can be edited in the related models, such as security groups and users.

Setting values on reference fields, such as `category_id` and `implied_ids`, can be done using the related records' XML IDs and some special syntax. This syntax was explained in detail in *Chapter 6, Managing Module Data*.

There's more...

The special `base.group_no_one` security group called **Extra Rights** is also noteworthy. In previous Odoo versions, it was used for advanced features hidden by default and was only made visible when the `Technical Features` flag was activated. From version 9.0, this has changed, and the features are visible so long as `Developer Mode` is active.

Security groups only provide cumulative access rights. There is no method to deny a group's access. This implies that a manually established group used to customize rights should inherit from the nearest group with fewer permissions than intended (if any), and then add all remaining permissions required.

Groups also have these additional fields available:

- **Menus (the `menu_access` field)**: These are the menu items the group has access to
- **Views (the `view_access` field)**: These are the UI views the group has access to
- **Access rights (the `model_access` field)**: This is the access it has to models, as detailed in the *Adding security access to models* recipe

- **Rules (the `rule_groups` field)**: These are the record-level access rules that apply to the group, as detailed in the *Limiting record access using record rules* recipe
- **Notes (the `comment` field)**: This is a description or commented piece of text for the group

With that, we've learned how to build security groups and assign them via the GUI. We will utilize these groups to establish an access control list and record rules in the next few recipes.

See also

To learn how to access newly added models through the *superuser*, please refer to the *Accessing Odoo as a superuser* recipe in *Chapter 3, Creating Odoo Add-On Modules*.

Adding security access to models

It's common for add-on modules to add new models. For example, in *Chapter 3, Creating Odoo Add-On Modules*, we added a new Hostel model. It is easy to miss out on creating security access for new models during development, and you might find it hard to see menus and views that have been created. This is because, from *Odoo version 12*, admin users don't get default access rights to new models. To see the views and menus for the new model, you need to add security **ACLs**.

However, models with no ACLs will trigger a warning log message upon loading, informing the user about the missing ACL definitions:

```
WARNING The model hostel.hostel has no access rules, consider adding
one example, access_hostel_hostel, access_hostel_hostel, model_hostel_
hostel, base.group_user,1,0,0,0
```

You may also access freshly uploaded models as a superuser, which circumvents all security requirements. For further information, see the *Accessing Odoo as a superuser* recipe in *Chapter 3, Creating Odoo Add-On Modules*. Administrators have access to the superuser functionality. So, for new models to be useable by non-admin users, we must establish their access control lists so that Odoo understands how to access them and what activities each user group is permitted to conduct.

Getting ready

We will continue using the `my_hostel` module from the previous tutorial and add the missing ACLs to it.

How to do it...

`my_hostel` should already contain the `models/hostel.py` Python file that creates the `hostel.hostel` model. We will now add a data file that describes this model's security access control by performing the following steps:

1. Edit the `__manifest__.py` file to declare a new data file:

```
data: [
    # ...Security Groups
    'security/ir.model.access.csv',
    # ...Other data files
]
```

2. Add a new `security/ir.model.access.csv` file to the module with the following lines:

```
id,name,model_id:id,group_id:id,perm_read,perm_write,perm_
create,perm_unlink
acl_hostel,hostel_hostel_default,model_hostel_hostel,base_group_
user,1,0,0,0
acl_hostel_manager,hostel_manager,model_hostel_hostel,group_
hostel_manager,1,1,1,1
```

We should then upgrade the module so that these ACL records are added to our Odoo database. More importantly, if we sign into a demonstration database using the `demo` user, we should be able to access the **My Hostel** menu option without receiving any security errors.

How it works...

Security ACLs are stored in the core `ir.model.access` model. We just need to add the records that describe the intended access rights for each user group.

Any sort of data file will do, although the most popular is a CSV file. The file can be placed anywhere inside the add-on module directory; however, it is common to keep all **security**-related files under a security subfolder.

This new data file is added to the manifest in the first phase of our tutorial. The next step is to include the files that explain the security access control rules. The CSV file must be named after the model into which the entries will be imported, so the name we've chosen isn't simply a convention; it's required. For further information, see *Chapter 6, Managing Module Data*.

If the module also creates new security groups, its data file should be declared in the manifest before the ACLs' data files, since you may want to use them for the ACLs. They must already be created when the ACL file is processed.

The columns in the CSV file are as follows:

- `id`: This is the internal XML ID identification for this rule. Any unique name inside the module is acceptable, but the best practice is to use `access_<model>_<group>`.
- `name`: This is a title for the access rule. It is a common practice to use an `access.<model>.<group>` name.
- `model_id:id`: This is the XML ID for the model. Odoo automatically assigns this kind of ID to models with a `model_<name>` format, using the model's `_name` with underscores instead of dots. If the model was created in a different add-on module, a fully qualified XML ID that includes the module name is needed.
- `group_id:id`: This is the XML ID for the user group. If left empty, it applies to all users. The base module provides some basic groups, such as `base.group_user` for all employees and `base.group_system` for the administration user. Other apps can add their own user groups.
- `perm_read`: Members of the preceding group can read the model's records. It accepts two values: `0` or `1`. Use `0` to restrict read access on the model and `1` to provide read access.
- `perm_write`: Members of the preceding group can update the model's records. It accepts two values: `0` or `1`. Use `0` to restrict write access on the model and `1` to provide write access.
- `perm_create`: Members of the preceding group can add new records of this model. It accepts two values: `0` or `1`. Use `0` to restrict create access on the model and `1` to provide create access.
- `perm_unlink`: Members of the preceding group can delete records of this model. It accepts two values: `0` or `1`. Use `0` to restrict unlink access on the model and `1` to provide unlink access.

The CSV file we used adds read-only access to the **Employees | Employee** standard security group and full write access to the **Administration | Settings** group.

The **Employee** user group, `base.group_user`, is particularly important because the user groups that are added by the Odoo standard apps inherit from it. This means that if we need a new model to be accessible by all the backend users, regardless of the specific apps they work with, we should add that permission to the **Employee** group.

The **Employee** user group, `base.group_user`, is particularly essential since it is inherited by the user groups introduced by the Odoo standard applications. This implies that if we want a new model to be accessible to all backend users, independent of the applications they use, we need to add it to the **Employee** group.

The resulting ACLs can be viewed from the GUI in debug mode by navigating to **Settings | Technical | Security | Access Controls List**, as shown in the following screenshot:

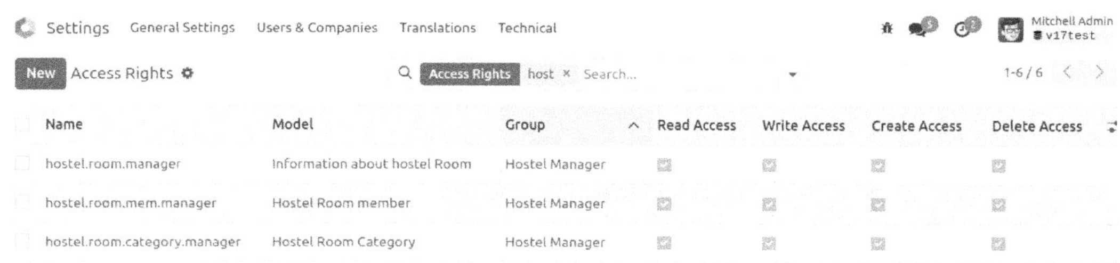

Figure 10.2 – ACL list view

Some people find it easier to use this user interface to define ACLs and then use the **Export** feature to produce a CSV file.

There's more...

It seems reasonable to provide this access to the Hostel user and the Hostel Manager groups specified in the *Creating security groups and assigning them to users* recipe. If you went through that lesson, it's a nice exercise to go through this one while changing the group identities to Hostel ones.

It is crucial to remember that access lists given by add-on modules should not be directly customized because they will be reloaded on the next module update, erasing any modification done from the GUI.

There are two methods for customizing ACLs. One option is to build new security groups that inherit from the modules and add extra rights to them, but this only enables us to add permissions and not remove them. A more adaptable way would be to uncheck the **Active** checkbox on certain ACL lines to disable them. Because the active field is not shown by default, we must alter the tree view so that it includes the `<field name="active" />` column. We may also create new ACL lines to add or alter permissions. The deactivated ACLs will not be revived after a module update, and the newly inserted ACL lines will not be impacted.

It's also worth mentioning that ACLs only apply to conventional models and aren't required for abstract or transient models. If these are defined, they will be ignored, and a warning message will be logged in the server log.

See also

You can also access newly added models through a superuser since this bypasses all security rules. To learn more about this, please refer to the *Accessing Odoo as a superuser* recipe in *Chapter 3, Creating Odoo Add-On Modules*.

Limiting access to fields in models

In other circumstances, we may want additional fine-grained access control, as well as the ability to restrict access to individual fields in a model.

Using the `groups` property, it is possible to restrict access to a field to certain security **groups**. This recipe will demonstrate how to add a field with restricted access to the Hostels model.

Getting ready

We will continue using the `my_hostel` module from the previous tutorial.

How to do it...

To add a field with access that's limited to specific security groups, perform the following steps:

1. Edit the model file to add the field:

```
is_public = fields.Boolean(groups='my_hostel.group_hostel_
manager')
notes = fields.Text(groups='my_hostel.group_hostel_manager')
```

2. Edit the view in the XML file to add the field:

```
<field name="is_public" />
<field name="notes" />
```

That's it. Now, upgrade the add-on module for the changes in the model to take place. If you sign in with a user with no system configuration access, such as demo in a database with demonstration data, the Hostel form won't display the field.

How it works...

Fields containing the `groups` property are processed differently to determine whether the user belongs to any of the security groups specified in the attribute. If a user is not a member of a certain group, Odoo will remove the field from the UI and limit ORM operations on that field.

Note that this security is not superficial. The field is not only hidden in the UI but is also made unavailable to the user in the other ORM operations, such as `read` and `write`. This is also true for *XML-RPC* or *JSON-RPC* calls.

Be careful when using these fields in business logic or on-change UI events (`@api.onchange` methods); they can raise errors for users with no access to the field. One workaround for this is to use privilege elevation, such as the `sudo()` model method or the `compute_sudo` field attribute for computed fields.

The `groups` value is a string that contains a comma-separated list of valid XML IDs for security groups. The simplest way to find the XML ID for a particular group is to activate developer mode and navigate to the group's form, at **Settings** | **Users** | **Groups**, and then access the **View Metadata** option from the debug menu, as shown in the following screenshot:

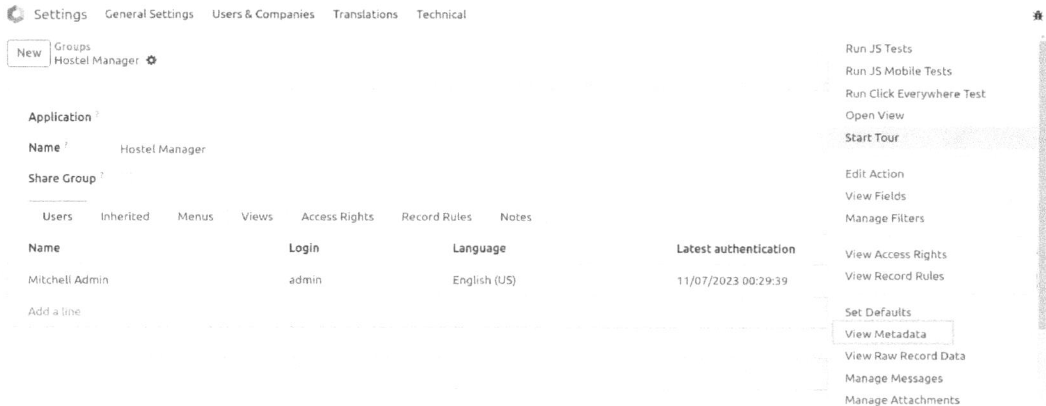

Figure 10.3 – Menu for viewing a group's XML ID

You may also view a security group's XML ID through code by utilizing the `<record>` tag that formed the group. However, looking at the information, as shown in the preceding screenshot, is the most simple approach to figuring out a group's XML ID.

There's more...

In some circumstances, we want a field to be available or unavailable based on specific requirements, such as the values in a field, such as `stage_id` or `state`. Typically, this is handled at the view level by utilizing characteristics such as states or attributes to dynamically display or hide the field based on particular criteria. For a more complete explanation, see *Chapter 9*, *Backend Views*.

Note that these techniques work at the user interface level only and don't provide actual access security. To do this, you should add checks to the business logic layer. Either add model methods decorated with `@constrains`, implementing the specific validations intended, or extend the `create`, `write`, or `unlink` methods to add validation logic. You can get further insights into how to do this by going back to *Chapter 5*, *Basic Server-Side Development*.

See also

Please see *Chapter 9*, *Backend Views*, for further information on how to hide and reveal a field using criteria.

For further insights into the business logic layer, please refer to *Chapter 5*, *Basic Server-Side Development*.

Limiting record access using record rules

A basic requirement for every application is the ability to restrict which records are exposed to each user on a certain model.

This is accomplished through the use of **record rules**. A record rule is a domain filter expression specified on a model that is subsequently applied to each data query performed by the impacted users.

As an example, we will add a record rule to the *Hostel* model so that users in the `Employee` group will only have access to the public hostel.

Getting ready

We will continue using the `my_hostel` module from the previous recipe.

How to do it...

Record rules can be added using a data XML file. To do this, perform the following steps:

1. Ensure that the `security/security_rules.xml` file is referenced by the manifest `data` key:

    ```
    'data': [
        'security/security_rules.xml',
        # ...
    ],
    ```

2. We should have a `security/security_rules.xml` data file with a `<odoo>` section that creates the security group:

```
<odoo noupdate="1">
    <record model="ir.rule" id="hostel_user_rule">
        <field name="name">Hostel: see only own hostel</field>
        <field name="model_id" ref="model_hostel_hostel"/>
        <field name="groups" eval="[(4, ref('my_hostel.group_hostel_user'))]"/>
        <field name="domain_force">
            [('is_public', '=', True)]
        </field>
    </record>
    <record model="ir.rule" id="hostel_all_rule">
        <field name="name">Hostel: see all hostels</field>
        <field name="model_id" ref="model_hostel_hostel"/>
        <field name="groups" eval="[(4, ref('my_hostel.group_hostel_manager'))]"/>
        <field name="domain_force">[(1, '=', 1)]</field>
    </record>
</odoo>
```

Figure 10.4 – Record rule for the hostel user

Upgrading the add-on module will load the record rules inside the Odoo instance. If you are using demo data, you can test it through the default `demo` user to give hostel user rights to the demo user. If you are not using demo data, you can create a new user with hostel user rights.

How it works...

Record rules are just data records that are placed in the `ir.rule` core model. While the file containing them can be located anywhere in the module, the `security` subfolder is the preferred location. A single XML file including both security groups and record rules is usual.

In contrast to groups, record rules in standard modules are imported into an `odoo` section with the `noupdate="1"` property. Because certain records will not be reloaded after a module update, manually customizing them is safe and will survive further upgrades.

To be consistent with the standard modules, our record rules should also be contained within an `<odoo noupdate="1">` section.

Record rules can be seen from the GUI via the **Settings**| **Technical** | **Security** | **Record Rules** menu option, as shown in the following screenshot:

Figure 10.5 – ACLs for the Hostel model

The following are the most important record rule fields that were used in this example:

- **Name** (`name`): A descriptive title for the rule.
- **Object** (`model_id`): A reference to the model to which the rule applies.
- **Groups** (`groups`): The security groups affected by the rule. If no security group is mentioned, the rule is deemed global and is enforced differently (keep reading to understand more about these groups).
- **Domain** (`domain`): A domain expression that is used to filter the records. The rule is only going to apply to these filtered records.

The first record rule we created was for the `Hostel User` security group. It uses the `[('is_public', '=', True)]` domain expression to select only the hostel that is available publicly. Thus, users with the `Hostel User` security group will only be able to see public hostels.

> **Note**
>
> *The domain expressions used in the record rules are executed on the server using ORM objects. As a result, dot notation may be used on the fields on the left (the first tuple member). The* `[('country_id.code', '=', 'IN')]` *domain expression, for example, will only return entries containing the country of India.*

Since record rules are mostly based on the current user, you can use the `user` recordset on the right-hand side (the third tuple element) of the domain. So, if you want to show the records for the company of the current user, you can use the `[('conpany_id', '=', user.company_id.id)]` domain. Alternatively, if you want to show the records that are created by the current user, you can use the `[('user_id', '=', user.id)]` domain.

We want the `Hostel Manager` security group to have access to all hostels, regardless of whether they are public or private. Because it is a descendant of the `Hostel User` group, it will be able to see only public hostels until we intervene.

The non-global record rules are joined using the `OR` logical operator; each rule adds access and never removes this access. For the `Hostel Manager` security group to have access to all the hostels, we must add a record rule to it so that it can add access for all hostels, as follows:

```
[('is_public', 'in', [True, False])]
```

We chose to do this differently here and use the `[(1, '=', 1)]` special rule instead to unconditionally give access to all hostel records. While this may seem redundant, remember that if we don't do this, the `Hostel User` rule can be customized in a way that will keep some hostel out of reach from the Settings user. The domain is special because the first element of a domain tuple must be a field name; this exact case is one of two cases where that is not true. The special domain of `[(1, '=', 0)]` is never true, but also not very useful in the case of record rules. This is because this type of rule is used to restrict access to all the records. The same thing is also possible with access lists.

> **Important information**
>
> Record rules are ignored if you've activated `SUPERUSER` mode. When testing your record rules, ensure that you use another user for that.

There's more...

When a record rule is not assigned to a security group, it is labeled as global and treated differently from the other rules.

Global record rules have a greater impact than group-level record rules and establish access limits that cannot be overridden. They are connected technically via an `AND` operator. They are used in standard modules to create multi-company security access so that each user may only see data from their own business.

In summary, standard non-global record rules are combined with an `OR` operator, and a record is accessible if any of the rules grant that access. When using an `AND` operator, global record rules add limits to the access provided by conventional record rules. Regular record rules cannot override restrictions imposed by global record rules.

Activating features with security groups

Some functions can be restricted by security groups so that they can only be accessible to people who belong to these groups. Security groups can inherit other groups, granting them permissions as well.

These two features are utilized to provide a feature toggling functionality in Odoo. Security groups may also be used to activate or disable functionality for certain users or the whole Odoo instance.

This recipe demonstrates how to add options to configuration settings and demonstrates the two approaches for enabling extra features: making them visible through security groups or adding them by installing an additional module.

For the first case, we will make the hostel start dates an optional additional feature while for the second, as an example, we will provide an option for installing the *Notes* module.

Getting ready

This tutorial uses the `my_hostel` module, which was described in *Chapter 3, Creating Odoo Add-On Modules*. We will need security groups to work with, so you also need to have followed the *Adding security access to models* recipe in this chapter.

In this recipe, some identifiers need to refer to the add-on module's technical name. We will assume that this is `my_hostel`. If you are using a different name, replace `my_hostel` with the actual technical name of your add-on module.

How to do it...

To add the configuration options, follow these steps:

1. To add the necessary dependency and the new XML data files, edit the `__manifest__.py` file like this and check that it depends on `base_setup`:

```
{   'name': 'Cookbook code',
    'category': 'Hostel',
    'depends': ['base_setup'],
    'data': [
        'security/ir.model.access.csv',
        'security/groups.xml',
        'views/hostel_hostel.xml',
        'views/res_config_settings.xml',
    ],
}
```

2. To add the new security group that's used for feature activation, edit the `security/groups.xml` file and add the following record to it:

```
    <record id="group_start_date" model="res.groups">
        <field name="name">Hostel: Start date feature</field>
        <field name="category_id" ref="base.module_category_
hidden" />
    </record>
```

3. To make the hostel start date visible only when this option is enabled, edit the field definition in the `models/hostel.py` file:

```
class HostelHostel(models.Model):
    # ...
    date_start = fields.Date(
        'Start Date',
        groups='my_hostel.group_start_date',     )
```

4. Edit the `models/__init__.py` file to add a new Python file for the configuration settings model:

```
from . import hostel
from . import res_config_settings
```

5. To extend the core configuration wizard by adding new options to it, add the `models/res_config_settings.py` file with the following code:

```
from odoo import models, fields
class ConfigSettings(models.TransientModel):
    _inherit = 'res.config.settings'
    group_start_date = fields.Boolean(
            "Manage Hostel Start dates",
            group='base.group_user',
            implied_group='my_hostel.group_start_dates',
    )
    module_note = fields.Boolean("Install Notes app")
```

6. To make the options available in the UI, add `views/res_config_settings.xml`, which extends the settings form view:

```
<?xml version="1.0" encoding="utf-8"?>
<odoo>
    <record id="view_general_config_hostel" model="ir.ui.view">
        <field name="name">Configuration: add Hostel options</
field>
        <field name="model">res.config.settings</field>
        <field name="inherit_id" ref="base_setup.res_config_
settings_view_form" />
        <field name="arch" type="xml">
            <div id="business_documents" position="before">
             <h2>Hostel</h2>
                <div class="row mt16 o_settings_container">
                    <!-- Add Step 7 and 8 goes here -->
                </div>
            </div>
```

```
        </field>
    </record>
</odoo>
```

7. In the settings form view, add the option to add a start date feature:

```
<!-- Start Dates option -->
<div class="col-12 col-lg-6 o_setting_box">
    <div class="o_setting_left_pane">
        <field name="group_start_date" class="oe_inline"/>
    </div>
    <div class="o_setting_right_pane">
        <label for="group_start_date"/>
        <div class="text-muted">
            Enable Start date feature on hostels
        </div>
    </div>
</div>
```

8. In the settings form view, add the option to install the Note module:

```
<!-- Note module option -->
<div class="col-12 col-lg-6 o_setting_box">
    <div class="o_setting_left_pane">
        <field name="module_note" class="oe_inline"/>
    </div>
    <div class="o_setting_right_pane">
        <label for="module_note"/>
        <div class="text-muted">
            Install note module
        </div>
    </div>
</div>
```

After upgrading the add-on module, the two new configuration options should be available at **Settings** | **General Settings**. The screen should look like this:

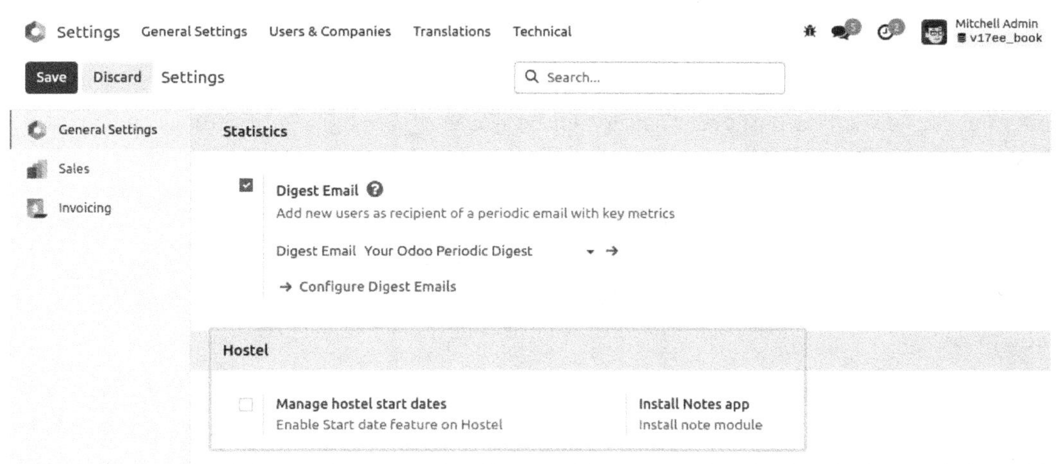

Figure 10.6 – Hostel config in General Settings

As shown in the preceding screenshot, you will have new settings in the **Hostel** section. The first option, **Manage hostel start date**, will enable the start date feature for the hostel record. The second option, **Install Notes app**, will install Odoo's Notes app.

How it works...

The core `base` module provides the `res.config.settings` model, which provides the business logic behind the activation option. The `base_setup` add-on module uses the `res.config.settings` model to provide several basic configuration options that can be made available inside a new database. It also makes the **Settings** | **General Settings** menu available.

The `base_setup` module adapts `res.config.settings` to a central management dashboard, so we need to extend it to add configuration settings.

If we decide to create a specific settings form for the Hostel app, we can still inherit from the `res.config.settings` model with a different `_name` and then, for the new model, provide the menu option and form view for just those settings. We already saw this method in the *Adding your own settings options* recipe of *Chapter 8, Advanced Server-Side Development Techniques*.

We activated these functionalities in two ways: by activating a security group and making the functionality visible to the user, and by installing an add-on module that offers this feature. The fundamental `res.config.settings` model provides the logic necessary to handle both of these scenarios.

The first step in this tutorial adds the `base_setup` add-on module to the dependencies since it provides extensions to the `res.config.settings` model we want to use. It also adds an additional XML data file that we will need to add the new options to the **General Settings** form.

In the second step, we created a new security group, **Hostel: start date feature**. The feature that needs to be activated should only be visible to that group, so it will be hidden until the group is enabled.

In our example, we want the hostel start date to only be available when the corresponding configuration option is enabled. To achieve this, we can use the `groups` attribute on the field so that it is made available only for this security group. We did this at the model level so that it is automatically applied to all the UI views where the field is used.

Finally, we extended the `res.config.settings` model to add the new options. Each option is a Boolean field, and its name must begin either with `group_` or `module_`, according to what we want it to do.

The `group_` option field should have an `implied_group` attribute and should be a string that contains a comma-separated list of XML IDs for the security groups to activate when it is enabled. The XML IDs must be complete, with the module name, dot, and identifier name provided; for example, `module_name.identifier`.

We can also provide a `group` attribute to specify which security groups the feature will be enabled for. It will be enabled for all the Employee-based groups if no groups are defined. Thus, the related groups won't apply to portal security groups, since these don't inherit from the Employee base security group like the other regular security groups do.

The mechanism behind the activation is quite simple: it adds the security group in the `group` attribute to `implied_group`, thus making the related feature visible to the corresponding users.

The `module_` option field does not require any additional attributes. The remaining part of the field name identifies the module to be installed when this option has been activated. In our example, `module_note` will install the Note module.

> **Important information**
> Unchecking the box will uninstall the module without warning, which can cause data loss (models or fields and module data will be removed as a consequence). To avoid unchecking the box by accident, the `secure_uninstall` community module (from `https://github.com/OCA/server-tools`) prompts the user for a password before they uninstall the add-on module.

There's more...

Configuration settings can also have fields named with the `default_` prefix. When one of these has a value, the ORM will set it as a global default. The `settings` field should have a `default_model` attribute to identify the model that's been affected, and the field name after the `default_` prefix identifies the `model` field that will have the default value set.

Additionally, fields with none of the three prefixes mentioned can be used for other settings, but you will need to implement the logic to populate their values, using the `get_default_` name prefixed methods, and have them act when their values are edited using the `set_` name prefixed methods.

For those who would like to dive deeper into the details of the configuration settings, take a look at Odoo's source code in `./odoo/addons/base/models/res_config.py`, which is extensively commented on.

Accessing recordsets as a superuser

We looked at security strategies including access rules, security groups, and record rules in prior recipes. You can avoid unauthorized access by using these approaches. However, in some complicated business scenarios, you may need to view or edit records, even if the user does not have access to them. For example, suppose the public user does not have access to the leads records, but the user may produce leads records in the backend by submitting the website form.

You may access recordsets as a superuser by using `sudo()`. We covered `sudo()` in *Chapter 8, Advanced Server-Side Development Techniques*, in the *Changing the user that performs an action* recipe. We'll see here that even if you've set ACL rules or assigned a security group to the field, you may still acquire access using `sudo()`.

How to do it...

We will use the same `my_hostel` module from the previous tutorial. We already have an ALC rule that gives read-only access to normal users. We will add a new field with security groups so that only the Manager user has access to it. After that, we will modify the field value for the normal user. Follow these steps to achieve this:

1. Add the new field to the `hostel.hostel` model:

```
details_added = fields.Text(
      string="Details",
      groups='my_hostel.group_hostel_manager')
```

2. Add the field to the form view:

```
<field name="details_added"/>
```

3. Add the `add_details()` method to the `hostel.hostel` model:

```
    def add_details(self):
        self.ensure_one()
        message = "Details are(added by: %s)" % self.env.user.
name
        self.sudo().write({
```

```
            'details_added': message
        })
```

4. Add the button to the form view so that we can trigger our method from the user interface. This should be placed inside the `<header>` tag:

```
<button name="add_details"
    string="Add Details"
    type="object"/>
```

Restart the server and update the module to apply these changes.

How it works...

In *Steps 1* and *2*, we added a new field called `details_added` to the model and form view. Note that we put the `my_hostel.group_hostel_manager` group on the field in Python, so this field can only be accessed by the Manager user.

In the next step, we added the `add_details()` method. We updated the value of the `details_added()` field inside this method's body. Note that we used `sudo()` before calling the write method.

Finally, we added a button in the form view to trigger the method from the user interface.

To test this implementation, you need to log in with the non-manager user. If you have loaded the database with demonstration data, you can log in with the demo user and then click on the **Add Details** report button in the form view of the hostel. Upon clicking that button, the `add_details()` method will be called, and this will write the message into the `details_added` field, even if the user doesn't have proper rights. You can check the value of the field through the admin user because this field will be hidden from the demo user.

Upon clicking the **Add Details** button, we will get the recordset of the current hostel in the `add_details()` method as an argument, `self`. Before we wrote the values into the hostel recordset, we used `self.sudo()`. This returns the same recordset but with superuser rights. This recordset will have the `su=True` environment attribute, and it will bypass all access rules and record rules. Because of that, the non-manager user will be able to write in the hostel record.

There's more...

You need to be extra careful when you use `sudo()` because it bypasses all access rights. If you want to access the recordset as another user, you can pass the ID of that user inside `sudo` – for example, `self.sudo(uid)`. This will return a recordset containing the environment of that user. This way, it will not bypass all the access rules and record rules, but you can perform all the actions that are allowed for that user.

Hiding view elements and menus based on groups

We covered how to hide fields from some users using group parameters in the Python field declaration in previous recipes. Another method for hiding fields in the user interface is to add security groups to the XML elements in the view specification. You may also conceal security groups from a certain user by using menus.

Getting ready

For this recipe, we will reuse the `my_hostel` add-on module from the previous recipe. In the previous recipe, we added a button to the `<header>` tag. We will hide that whole header from a few users by adding a group attribute to it.

Add the model, the views, and the menus for the `hostel.room.category` model. We will hide the category menus from a user. Please refer to *Chapter 4*, *Application Models*, to learn how to add model views and menus.

How to do it...

Follow these steps to hide elements based on security groups:

1. Add a `groups` attribute to the `<header>` tag to hide it from other users:

```
...
<header groups="my_hostel.group_hostel_manager">
...
```

2. Add the `groups` attribute to the `<menuitem>` hostel category so that it's only displayed for librarian users:

```
<menuitem name="Hostel Room Categories"
    id="hostel_room_category_menu"
    parent="hostel_base_menu"
    action="hostel_room_category_action"
    groups="my_hostel.group_hostel_manager"/>
```

Restart the server and update the module to apply these changes.

How it works...

In *Step 1*, we added `groups="my_hostel.group_hostel_manager"` to the `<header>` tag. This means that the whole header part will only be visible to hostel users and hostel managers. Normal backend users who don't have `group_hostel_manager` will not see the header part.

In *Step 2*, we added the `groups="my_hostel.group_hostel_manager"` attribute to `menuitem`. This means that this menu is only visible to hostel users.

You can use the `groups` attribute almost everywhere, including `<field>`, `<notebook>`, `<group>`, and `<menuitems>`, or on any tag from the view architecture. Odoo will hide those elements if the user doesn't have that group. You can use the same group attributes in web pages and *QWeb reports*, which will be covered in *Chapter 12, Automation, Workflows, Emails, and Printing*, and *Chapter 14, CMS Website Development*.

As we saw in the *Accessing recordsets as a superuser* recipe of this chapter, we can hide fields from some users using the `groups` argument in the Python field definition. Note that there is a big difference between using security groups on fields and using Python security groups in views. Security groups in Python provide real security; unauthorized users can't even access the fields through ORM or RPC calls. However, the groups in view are just for improving usability. Fields that are hidden through groups in the XML file can still be accessed through RPC or ORM.

See also

Please refer to *Chapter 4, Application Models*, to learn how to add model views and menus.

11
Internationalization

Odoo supports a variety of languages and enables users to use the language(s) with which they are most comfortable. The Odoo i18n features that are already built-in help with this. With string translations, Odoo also supports date and time formatting.

In this chapter, you will discover how to upload translation files to your modules and enable various languages. Due to the diversity of countries and the prevalence of local languages, users often find it easier to connect with a system when it's presented in their native tongue. To accommodate this, Odoo offers a feature that enables software text to be translated into the user's preferred language. This functionality enhances the user experience by ensuring that the interface is accessible and comprehensible to individuals, regardless of their linguistic background, thereby promoting wider adoption and usability of the software across various regions and demographics. Utilizing these new functionalities will enhance the Odoo user experience.

The following recipes will be covered in this chapter:

- Setting up a language installation and user preference settings
- Setting up options relating to language
- Text translation using a web client user interface
- Exporting translation into a file
- Using `gettext` tools to make translations easier
- Importing translation files into Odoo
- Altering a website's custom language URL code

Many of these recipes can be completed either from the web client user interface or from the command line. Wherever possible, we will see how to use both of these options. Odoo uses Transifex(Odoo) and Weblate (OCA) translation platforms.

Setting up a language installation and user preference settings

Odoo can be localized to accommodate various languages and locale settings, including date and number formats.

The only language that is initially installed is the standard English language. We need to install various localities and languages so that people may utilize them. This recipe describes how user preferences are implemented, as well as how they may be established.

How to do it...

Activate developer mode and follow these steps to install a new language in an Odoo instance:

1. Go to **Settings** | **General Settings** | **Language**. Here, you will see the **Add Language** link, as shown in the following screenshot. Click on that link; a dialogue box will open where you can load languages:

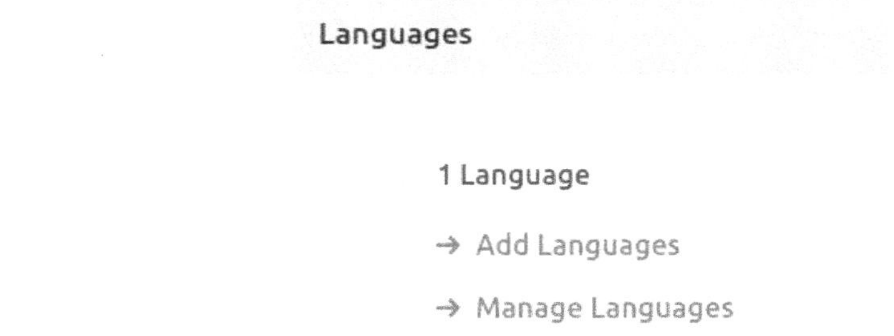

Figure 11.1 – Language options in the general settings

2. Select the language you want to load:

Figure 11.2 – Dialogue to load a language

3. Clicking on **Add** will load the selected language, and the confirmation dialogue box will open, as follows:

Odoo

French (BE) / Français (BE) has been successfully installed. Users can choose their favorite language in their preferences.

Close | Switch to French (BE) / Français (BE) & Close

Figure 11.3 – Dialogue that shows a language has been loaded

4. New languages can also be installed from the command line. The equivalent command for the preceding steps is as follows:

```
$ ./odoo-bin -d mydb --load-language=es_ES
```

5. To set the language that's used by a user, go to **Settings** | **Users & Companies** | **Users**, and, in the **Preferences** tab of the **User** form, set the **Language** field value:

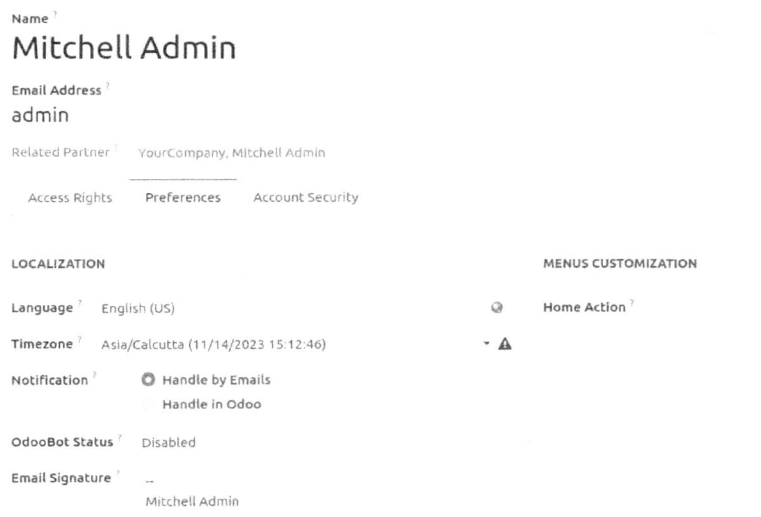

Figure 11.4 – User's form to set the language

Through the **Preferences** menu item, users may easily change these variables on their own. They may access this by clicking on their username in the top-right corner of the web client window:

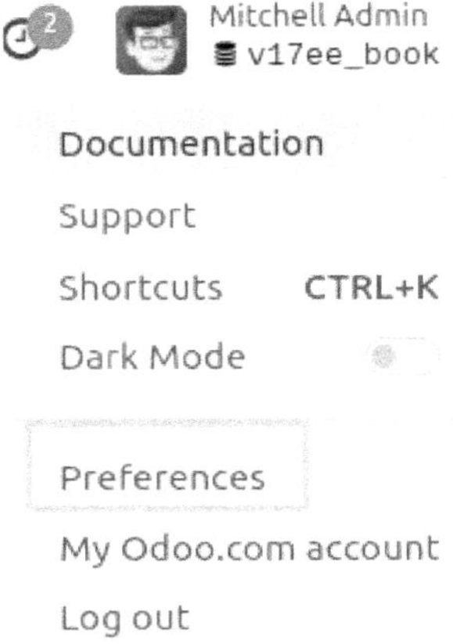

Figure 11.5– Preferences option to set the language

How it works...

Users can have their own language and time zone preferences. The language settings are used to translate user interface text into the chosen language and apply local conventions for float and monetary fields.

Before a language is made available for the user to select, it must be installed with the **Add language** option. The list of available languages can be seen by going to the **Settings** | **Translations** | **Languages** menu option in developer mode. The ones with the active flag set are installed.

Each Odoo add-on module is responsible for providing translation resources, which should be placed inside an `i18n` subdirectory. Each language's data should be in a `.po` file. In our example, for the Spanish language, the translation data is loaded from the `es_ES.po` data file.

Odoo also supports the notion of a **base language**. For example, if we have an `es.po` file for Spanish and an `es_MX.po` file for Mexican Spanish, then `es.po` is detected as the base language for `es_MX.po`. When the Mexican Spanish language is installed, both data files are loaded; first the one for the base language and then the specific language. Therefore, in our case, the Mexican Spanish translation file simply has to contain the strings that are unique to the language variety.

The `i18n` subdirectory is also expected to have a `<module_name>.pot` file, providing a template for translations and containing all the translatable strings. The *Exporting translation strings to a file* recipe of this chapter explains how to export the translatable strings to generate this file.

In previous versions of Odoo, when an additional language is installed, the corresponding resources are loaded from all installed add-on modules and stored in the **Translated Terms** model. Its data can be viewed (and edited) within the **Settings** | **Translations** | **Application Terms** | **Translated Terms** menu option (note that this menu is only visible in developer mode).

From Odoo version 17 onwards, you won't be able to find this menu as translated terms are now stored as native terms. Any field that is translatable now stores JSON data representing all translations of all translated language values. For example, translations of product names are now stored directly in the `name` field. The process of translations has not changed – you just can't see the **Settings** | **Translations** | **Application Terms** | **Translated Terms** menu item with all translated terms shown in a list.

Translation files for the installed languages are also loaded when a new add-on module is installed or an existing add-on module is upgraded.

There's more...

By selecting the refresh symbol again next to the languages, translation files may be refreshed without you having to upgrade the add-on modules. You can do this if your translation files have been changed and you don't want to deal with updating the modules (and all of their dependencies).

If the **Overwrite Existing Terms** checkbox is left empty, only the newly translated strings are loaded. Thus, the changed translated strings won't be loaded. Check the box if you want the already-existing translations to also be loaded and overwrite the currently loaded translations. Note that this can potentially be problematic if someone changes the translations manually through the interface.

The **Overwrite Existing Terms** checkbox exists because we can edit specific translations by going to the **Settings** | **Translations** | **Application Terms** | **Translated Terms** menu item, or by using the **Technical Translation** shortcut option in the **Debug** menu. Translations that are added or modified in this way won't be overwritten unless the language is reloaded with the **Overwrite Existing Terms** checkbox enabled.

It can be useful to know that add-on modules can also have an `i18n_extra` subdirectory with extra translations. First, the `.po` files in the `i18n` subdirectory are downloaded. Then, Odoo ORM downloads files for the base language and, after that, for the language variant. Following this, the `.po` files in the `i18n_extra` subdirectory are downloaded, first for the base language and then for the language variant. The final string translation that is loaded is the one that ultimately takes precedence.

Setting up options relating to language

The locale settings should be the right ones, so long as the user is using the correct language, because they come with suitable defaults.

You may still want to change a language's settings, though. For instance, you could opt to use the user interface's default English language setting while changing the American date and number formats to better suit your needs.

Additionally, locale settings such as date and number formats are provided by languages and their variants (such as `es_MX` for Mexican Spanish).

Getting ready

We will require developer mode to be turned on. If it hasn't previously been enabled, do it in the manner described in *Chapter 1*, *Installing the Odoo Development Environment*, in the *Activating the Odoo developer tools* recipe, where you installed the Odoo development environment.

How to do it...

Follow these steps to change a language's locale settings:

1. Select the **Settings** | **Translations** | **Languages** menu option to view the installed languages and their options. A form with the necessary options will open when you click on one of the installed languages:

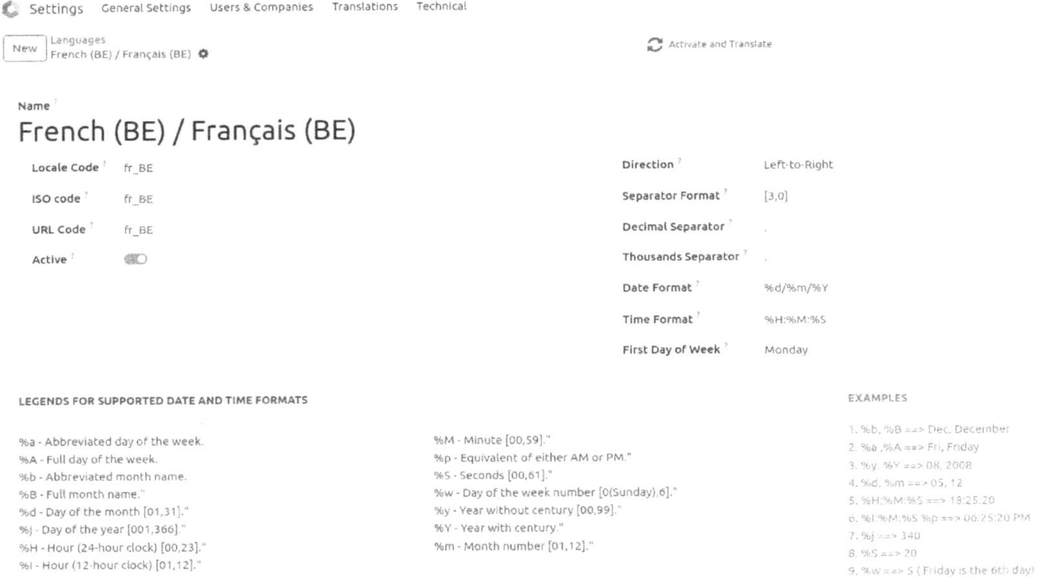

Figure 11.6 – Form to configure language settings

2. Edit the language settings. To change the date to the ISO format, change **Date Format** to `%Y-%m-%d`. To change the number format to use a comma as a decimal separator, modify the **Decimal Separator** and **Thousands Separator** fields accordingly.

How it works...

The user language is selected in the user preferences and placed in the **lang** context key when logging in and initiating a new Odoo user session. By translating the source texts into the user language and formatting the dates and numbers as per the language's current locale settings, the output is prepared accordingly.

There's more...

Server-side processes can modify the context in which actions are run. For example, to get a record where the dates are formatted according to the American English format, independent of the current user's language preference, you can do the following:

```
en_records = self.with_context(lang='en_US').search([])
```

For more details, refer to the *Calling a method with a modified context* recipe in *Chapter 8, Advanced Server-Side Development Techniques*.

Text translation using a web client user interface

The simplest way to translate is to use the translation feature provided by the web client. These translation strings are stored in the database and can later be exported to a `.po` file, either to be included in an add-on module or just to be imported back manually.

Text fields can have translatable content, meaning that their value will depend on the current user's language. We will also see how to set the language-dependent values on these fields.

Getting ready

We will need to have developer mode activated. If it's not, activate it, as shown in the *Activating the Odoo developer tools* recipe in *Chapter 1, Installing the Odoo Development Environment*.

How to do it...

We will demonstrate how to translate terms through the web client using the **User Groups** feature as an example:

1. Navigate to the screen you want to translate. As an example, we will open the **Groups** view via the **Settings** | **Users & Companies** | **Groups** menu item:

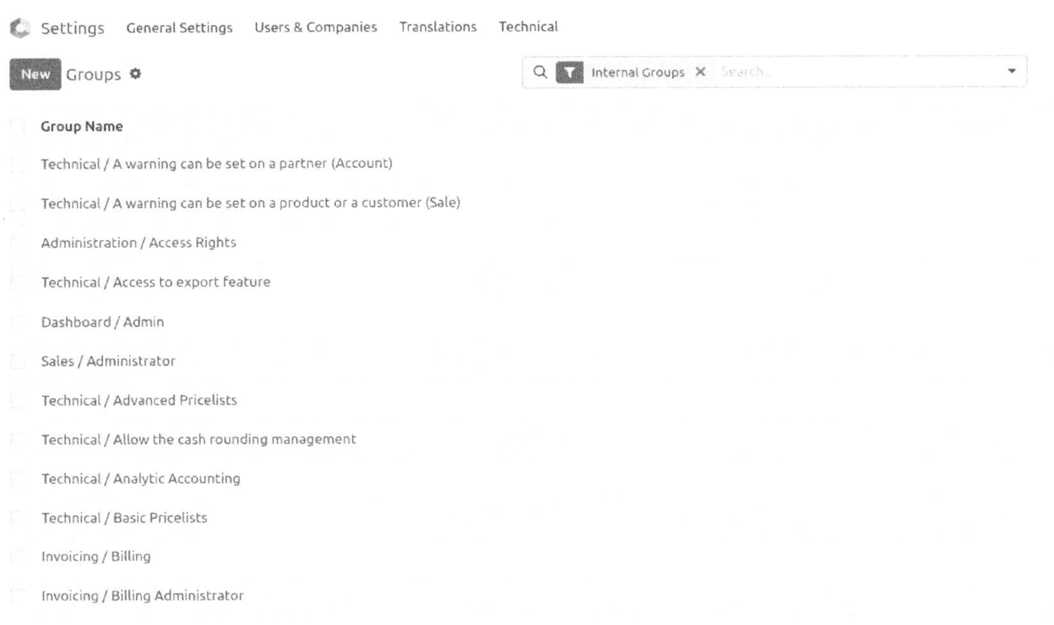

Figure 11.7 – Translation for groups

2. Open one of the group records in the form view, and click on **Edit**:

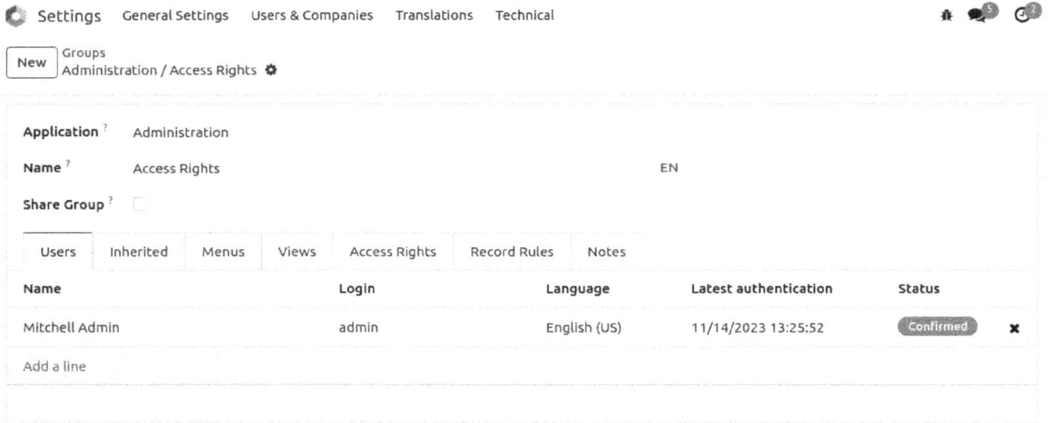

Figure 11.8 – Translation for the field values

3. Note that the **Name** field has a special icon on the far right. This indicates that it is a translatable field. Clicking on this icon opens a **Translate** list with the different installed languages. This allows us to set the translation for each of those languages:

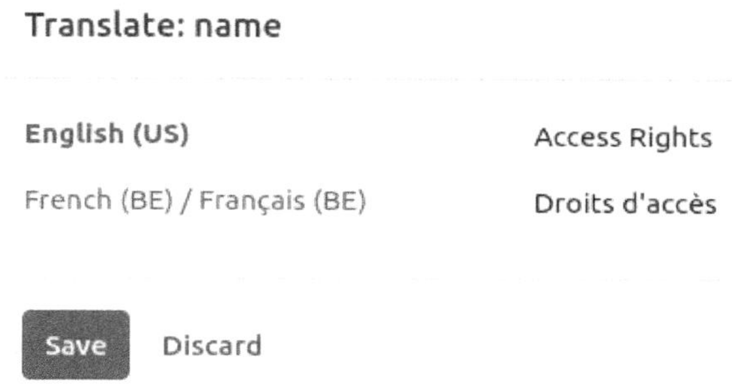

Figure 11.9 – Translation for the field values

How it works...

All translated terms are saved in the name field of any mode/table. In our example, **Access Rights** belongs to the `res_groups` table; when you check the information stored in the name field, it will be saved as a dictionary, where the key is the language code and the value is the translated phrases:

```
odoo_16_ee=# select id,name from res_groups;
 id |                                        name
----+------------------------------------------------------------------------------------------
  2 | {"en_US": "Access Rights", "fr_BE": "Droits d'accès "}
 10 | {"en_US": "Portal", "fr_BE": "Portail"}
 39 | {"en_US": "Librarians"}
 49 | {"en_US": "Hostel Manager"}
 12 | {"en_US": "Access to Private Addresses", "fr_BE": "Accès aux adresses privées"}
  8 | {"en_US": "Access to export feature", "fr_BE": "Accès à la fonctionnalité d'export"}
  3 | {"en_US": "Bypass HTML Field Sanitize", "fr_BE": "Contourner l'assainissement du champ HTML"}
```

Figure 11.10 – Translation for the field values

Exporting translation strings to a file

Translation strings can be exported with or without the translated texts for a selected language. This can either be to include `i18n` data in a module or to later perform translations with a text editor or perhaps with a specialized tool.

We will demonstrate how to do this using our custom `My Hostel` module, so feel free to replace `My Hostel` with your own module.

Getting ready

We will need to have developer mode activated. If it's not already activated, activate it, as demonstrated in the *Activating the Odoo developer tools* recipe in *Chapter 1*, *Installing the Odoo Development Environment*.

How to do it...

To export the translation terms for the `my_hostel` module, follow these steps:

1. In the web client user interface, from the **Settings** top menu, select the **Translations** | **Import/ Export** | **Export Translation** menu option.
2. In the **Export Translations** dialogue box, choose the language translation to export, the file format, and the modules to export. To export a translation template file, select **New Language (Empty translation template)** from the **Language** selection list. It's recommended to use the `.po` format and to export only one add-on module at a time – the **My Hostel** module (`my_hostel` is the technical name for the Discuss app), in our example:

Figure 11.11 – Dialogue to export translation terms

3. In Odoo version 17, you will find a new option in the export settings called **Export Type**, which contains two options: **Module** and **Model**.
4. setting Module Type to Model will provide the new option to select a specific model with a filter option, using which the user can export only specific filter-based records:

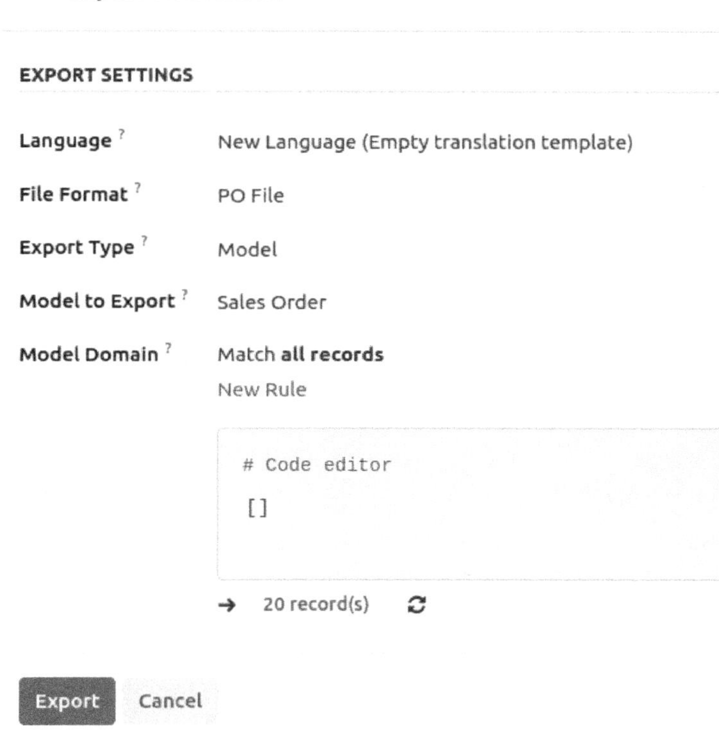

Figure 11.12 – Dialogue to export translation terms

5. Once the export process is complete, a new window will be displayed, with a link to download the file and some additional advice.
6. To export a translation template file for the `my_hostel` add-on module from the Odoo command-line interface, enter the following command:

```
$ ./odoo-bin -d mydb --i18n-export=my_hostel.pot --modules=my_
hostel
$ mv my_hostel.po ./addons/my_hostel/i18n/my_hostel .pot
```

7. To export the translation template file for a language – `es_ES` for Spanish, for example – from the Odoo command-line interface, enter the following command:

```
$ ./odoo-bin -d mydb --i18n-export=es_ES.po --modules=my_hostel
--language=es_ES
$ mv es_ES.po ./addons/my_hostel/i18n
```

How it works...

The **Export Translation** feature extracts the translatable strings from the target modules and then creates a file with the translation terms. This can be done both from the web client and the command-line interface.

When exporting from the web client, we can choose to either export an empty translation template – that is, a file with the strings to translate, along with empty translations – or export a language, resulting in a file with the strings to translate, along with the translation for the selected language.

The available file formats are CSV, PO, and TGZ. The TGZ file format exports a compressed file that contains a `<name>/i18n/` directory structure with the PO or POT file.

The CSV format can be useful for performing translations using a spreadsheet, but the format to use in the add-on modules is PO files. These are expected to be placed inside the `i18n` subdirectory. They are then automatically loaded once the corresponding language is installed. When exporting these PO files, we should export only one module at a time. The PO file is also a popular format supported by translation tools, such as Poedit.

Translations can also be exported directly from the command line by using the `--i18n-export` option. This recipe shows how to extract both the template files and the translated language files.

In *Step 4* of this recipe, we exported a template file. The `--i18n-export` option expects the path and the filename to export. Bear in mind that the file extension is required to be either CSV, PO, or TGZ. This option requires the `-d` option, which specifies the database to use. The `--modules` option is also needed to indicate the add-on modules to export. Note that the `--stop-after-init` option is not needed since the `export` command automatically returns to the command line when finished.

This exports a template file. The Odoo module expects this exported template in the `i18n` folder with the `.pot` extension. When working on a module, after the export operation, we usually want to move the exported PO file to the module's `i18n` directory with a `<module>.pot` name.

In *Step 5*, the `-language` option was also used. With it, instead of an empty translation file, the translated terms for the selected language were also exported. One use case for this is to perform some translations through the web client user interface using the **Technical Translation** feature, and then export and include them in the module.

There's more...

Text strings in view and model definitions are automatically extracted for translation. For models, the `_description` attribute, the field names (the `string` attribute), help text, and selection field options are extracted, as well as the user texts for model constraints (`_constraints` and `_sql_constraints`).

Text strings to translate inside Python or JavaScript code can't be automatically detected, so the code should identify those strings, wrapping them inside the underscore function.

In Python's module file, we should ensure that the file is imported with the following:

```
from odoo import _
```

This file can then be used wherever a translatable text is used with something like this:

```
_('Hello World')
```

For strings that use additional context information, we should use Python string interpolation, as shown here:

```
_('Hello %s') % 'World'
```

Note that the interpolation should go outside the translation function. For example, `_("Hello %s" % 'World')` is wrong. String interpolations should also be preferred to string concatenation so that each interface text is just one translation string.

Be careful with the `Selection` fields! If you pass an explicit list of values to the field definition, the displayed strings are automatically flagged for translation. On the other hand, if you pass a method that returns the list of values, the display strings must be explicitly marked for translation.

Regarding manual translation work, any text file editor will do so, but using an editor that specifically supports the PO file syntax makes this work easier by reducing the risk of formatting errors. Such editors include those listed here:

- **POEDIT**: `https://poedit.net/`
- **Emacs (PO-mode)**: `https://www.gnu.org/software/gettext/manual/html_node/PO-Mode.html`
- **Lokalize**: `https://l10n.kde.org/tools/`
- **Gtranslator**: `https://wiki.gnome.org/Apps/Gtranslator`

Using gettext tools to make translations easier

The PO file format is part of the `gettext` i18n and localization system that's commonly used in Unix-like systems. This system includes tools to ease translation work.

This recipe demonstrates how to use these tools to help translate our add-on modules. We want to use it on a custom module, so the `my_hostel` module we created in *Chapter 3, Creating Odoo Add-On Modules*, is a good candidate. However, feel free to replace it with some other custom module you have at hand, replacing the tutorials' `my_hostel` references as appropriate.

How to do it...

To manage translations from the command line, assuming that your Odoo installation is at `~/odoo-work/odoo`, follow these steps:

1. Create a compendium of translation terms for the target language – for example, Spanish. If we name our compendium file `odoo_es.po`, we should write the following code:

```
$ cd ~/odoo-work/odoo  # Use the path to your Odoo installation
$ find ./ -name es_ES.po | xargs msgcat --use-first | msgattrib
--translated  --no-fuzzy \ -o ./odoo_es.po
```

2. Export the translation template file for the add-on module from the Odoo command-line interface and place it in the module's expected location:

```
$ ./odoo-bin -d mydb --i18n-export=my_module.po --modules=my_
module
$ mv my_module.po ./addons/my_module/i18n/my_module.pot
```

3. If no translation file is available yet for the target language, create the PO translation file, reusing the terms that have been already found and translated in the compendium:

```
$ msgmerge --compendium ./odoo_es.po -o
./addons/my_module/i18n/es_ES.po \
/dev/null ./addons/my_module/i18n/my_module.pot
```

4. If a translation file exists, add the translations that can be found in the compendium:

```
$ mv ./addons/my_module/i18n/es_ES.po /tmp/my_module_es_old.po
$ msgmerge --compendium ./odoo_es.po -o./addons/my_module/i18n/
es_ES.po
\ /tmp/my_module_es_old.po ./addons/my_module/i18n/my_module.pot
$ rm /tmp/my_module_es_old.po
```

5. To take a peek at the untranslated terms in a PO file, use the following command:

```
$ msgattrib --untranslated ./addons/my_module/i18n/es_ES.po
```

6. Use your favorite editor to complete the translation.

How it works...

Step 1 uses commands from the `gettext` toolbox to create a translation compendium for the chosen language – Spanish, in our case. It works by finding all the `es_ES.po` files in the Odoo code base and passing them to the `msgcat` command. We use the `--use-first` flag to avoid conflicting translations (there are a few in the Odoo code base). The result is passed to the `msgattrib` filter. We use the `--translated` option to filter out the untranslated entries and the `--no-fuzzy` option to remove fuzzy translations. We then save the result in `odoo_es.po`.

Step 2 calls `odoo.py` with the `--i18n-export` option. You need to specify a database on the command line, even if one is specified in the configuration file and the `--modules` option, with a comma-separated list of modules to export the translation.

In the `gettext` world, fuzzy translations are those created automatically by the `msgmerge` command (or other tools) using a proximity match on the source string. We want to avoid these in the compendium.

Step 3 creates a new translation file by using existing translated values found in the compendium. The `msgmerge` command is used with the `--compendium` option to find the `msgid` lines in the compendium files, matching those in the translation template file generated in *Step 2*. The result is saved in the `es_ES.po` file.

If you have a preexisting `.po` file for your add-on with translations that you would like to preserve, you should rename it and replace the `/dev/null` argument with this file. The renaming procedure is required to avoid using the same file for input and output.

There's more...

This tutorial only skims the surface of the rich tools that are available with the GNU `gettext` toolbox. Full coverage is well beyond the scope of this book. If you are interested, the GNU `gettext` documentation contains a wealth of precious information about PO file manipulation and is available at `http://www.gnu.org/software/gettext/manual/gettext.html`.

Importing translation files into Odoo

The standard method for loading translations is to store PO files in the module's `i18n` subfolder. The translation files are loaded and additional translated strings are added whenever the add-on module is installed or updated.

However, there may be cases where we want to directly import a translation file. In this recipe, we will learn how to load a translation file, either from the web client or from the command line.

Getting ready

We need to have developer mode activated. If it's not activated already, activate it, as demonstrated in the *Activating the Odoo developer tools* recipe in *Chapter 1, Installing the Odoo Development Environment*. We will also need a translation po file, which we are going import in this tutorial – for example, the `myfile.po` file.

How to do it...

To import the translation terms, follow these steps:

1. In the web client user interface, from the **Settings** top menu, select the **Translations** | **Import/ Export** | **Import Translation** menu option.
2. In the **Import Translations** dialogue box, fill out the language name and the language code, and select the file to import. Finally, click on the **Import** button:

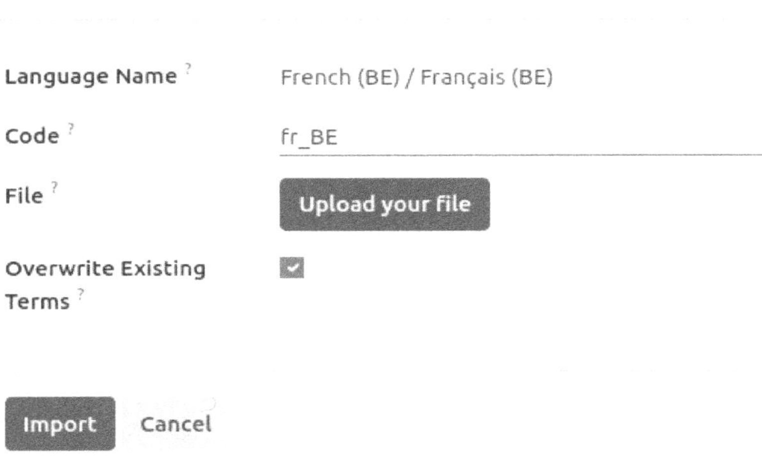

Figure 11.13 – Dialogue to import a translation file

3. To import a translation file from the Odoo command-line interface, we must place it inside the server add-ons path and then perform the import:

```
$ mv myfile.po ./addons/
$ ./odoo.py -d mydb --i18n-import="myfile.po" --lang=fr_BE
```

How it works...

Import Translation reads a PO or CSV file and inserts the translation strings into the appropriate table. Previously, it was stored in the `ir.translation` table, but in newer versions of Odoo, that table is no longer present. As a result, all view-level translations for `arch_db`, such as button strings and selection field values, will be stored in the `ir_ui_view` table, and all field-level translations, such as field labels, will be stored in the `ir_model_fields` table under the `field_description` field.

For example, our hostel table has a `room_number` model. Its field translation of "Room Number" will be stored at the database level as `{"en_US": "Room Number", "fr_BE": "Numéro de chambre"}`.

The web client feature asks for the language name, but this is not used in the import process. It also has an overwrite option. If selected, it forces all the translation strings to be imported, even the ones that already exist, overwriting them in the process.

On the command line, the import can be done using the `--i18n-import` option. It must be provided with the path to the file, relative to an add-on's path directory; `-d` and `--language` (or `-l`) are mandatory. Overwriting can also be achieved by adding the `--i18n-overwrite` option to the command. Note that we didn't use the `--stop-after-init` option here. It isn't needed since the import action stops the server when it finishes.

Altering a website's custom language URL code

Odoo also supports multiple languages for websites. On a website, the current language is identified as a language string. In this recipe, you will learn how to change the language code in a URL.

Getting ready

Before following this recipe, make sure you have installed the `website` module and enabled multiple languages for the website.

How to do it...

To modify a language's URL code, follow these steps:

1. Open the language list from the **Settings** | **Translations** | **Languages** menu option. Clicking on one of the installed languages will open a form that looks like this:

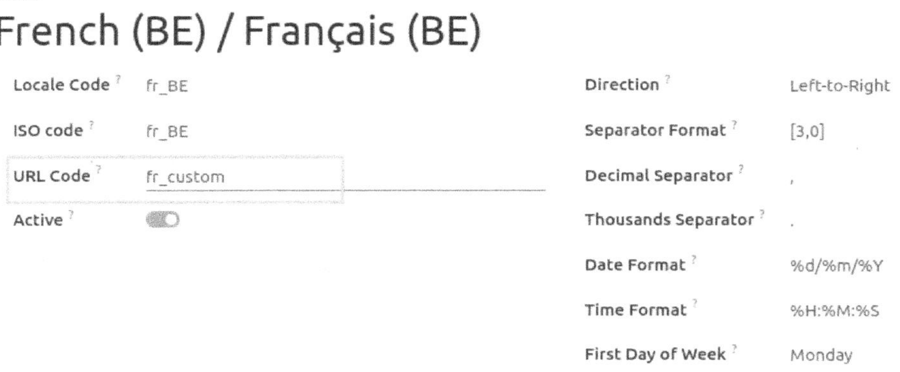

Figure 11.14 – Language URL code for a website

2. Here, you will see the **URL Code** field. Set the value that you want. Make sure you don't add spaces or special characters here.

After configuring this, you can test the results on your website. Open the home page and change the language; you will see the custom language code in the URL.

How it works...

Odoo identifies the languages for a website via the URL path. For example, `www.odoo.com/fr_FR` is used for the French language and `www.odoo.com/es_ES` is used for the Spanish language. Here, the `fr_FR` and `es_ES` parts of the URL are the language ISO codes, and they are used by Odoo to detect the requested language. But sometimes, you want to set the language in a more user-friendly way. In that case, you can update the **URL Code** field. Once you have changed that, the Odoo website will use the **URL Code** value to identify the language. For example, you could set **URL Code** to `fr` for the French language. In this case, `www.odoo.com/fr_FR` would be converted into `www.odoo.com/fr`.

> **Note**
>
> Changing the URL code in production is not a problem; Odoo will automatically redirect the URL containing the language ISO code to your custom URL.

12 Automation, Workflows, Emails, and Printing

Business applications are expected not only to store records but also to manage business workflows. Some objects, such as leads or project tasks, have a lot of records that run in parallel. Having too many records for an object makes it harder to have a clear picture of the business. Odoo has several techniques that can deal with this problem. In this chapter, we will look at how we can set a business workflow with dynamic stages and Kanban groups. This will help the user understand how their business is running.

We will also look at techniques, such as server actions and automated actions, that can be used by power users or functional consultants to add simpler process automation without the need to create custom add-ons. Finally, we will create **QWeb-based PDF reports** and print them out.

In this chapter, we will cover the following recipes:

- Managing dynamic record stages
- Managing Kanban stages
- Adding a quick create form to a Kanban card
- Creating interactive Kanban cards
- Adding a progress bar to Kanban views
- Creating server actions
- Using Python code server actions
- Using automated actions on time conditions
- Using automated actions on event conditions
- Creating QWeb-based PDF reports

- Managing activities from a Kanban card
- Adding a stat button to a form view
- Enabling the archive option for records

Technical requirements

The technical requirement for this chapter is having an online Odoo platform.

All the code that will be used in this chapter can be downloaded from this book's GitHub repository at `https://github.com/PacktPublishing/Odoo-17-Development-Cookbook-Fifth-Edition/tree/main/Chapter12`.

Managing dynamic record stages

In `my_hostel`, we have a `state` field to indicate the current status of a hostel room record. This `state` field is limited to the `Draft` or `Available` statuses and it is not possible to add a new state to the business process. To avoid this, we can use the `many2one` field to give flexibility when designing the Kanban workflow of a user's choice, and you can add/remove a new state at any time.

Getting ready

For this recipe, we will be using the `my_hostel` module from *Chapter 8, Advanced Server-Side Development Techniques*. This module manages the hostel and students. It also records rooms. We added an initial module, `Chapter12/00_initial_module/my_hostel`, to the GitHub repository for this book to help you get started: `https://github.com/PacktPublishing/Odoo-17-Development-Cookbook-Fifth-Edition/tree/main/Chapter12`.

How to do it...

Follow these simple steps to add stages to the `hostel.room` model:

1. Add a new model called `hostel.room.stage`, as follows:

```
class HostelRoomStage(models.Model):
    _name = 'hostel.room.stage'
    _order = 'sequence,name'

    name = fields.Char("Name")
    sequence = fields.Integer("Sequence")
    fold = fields.Boolean("Fold?")
```

2. Add access rights for this new module in the `security/ir.model.access.csv` file, as follows:

```
my_hostel.access_hostel_room_stage,access_hostel_room_stage,my_
hostel.model_hostel_room_stage,base.group_user,1,1,1,1
```

3. Remove the `state` field from the `hostel.room` model and replace it with a new `stage_id` field, which is a `many2one` field, and its methods, as shown in the following example:

```
@api.model
def _default_room_stage(self):
    Stage = self.env['hostel.room.stage']
    return Stage.search([], limit=1)
```

4. Replace the `state` field in the form view with the `stage_id` field, as shown in the following example:

```
<header>
    <field name="stage_id" widget="statusbar"
          options="{'clickable': '1', 'fold_field': 'fold'}"/>
</header>
```

5. Replace the `state` field in the tree view with the `stage_id` field, as follows:

```
<tree string="Room">
    <field name="name"/>
    <field name="room_no"/>
    <field name="floor_no"/>
    <field name="stage_id"/>
</tree>
```

6. Add some initial stages from the `data/room_stages.xml` file. Don't forget to add this file to the manifest, as shown in the following example:

```
<?xml version="1.0" encoding="utf-8"?>
<odoo noupdate="1">
    <record id="stage_draft" model="hostel.room.stage">
        <field name="name">Draft</field>
        <field name="sequence">1</field>
    </record>
    <record id="stage_available" model="hostel.room.stage">
        <field name="name">Available</field>
        <field name="sequence">15</field>
    </record>
    <record id="stage_reserved" model="hostel.room.stage">
        <field name="name">Reserved</field>
```

```
            <field name="sequence">5</field>
            <field name="fold">True</field>
        </record>
    </odoo>
```

After installing the module, you will see stages in the form view, as shown in the following screenshot:

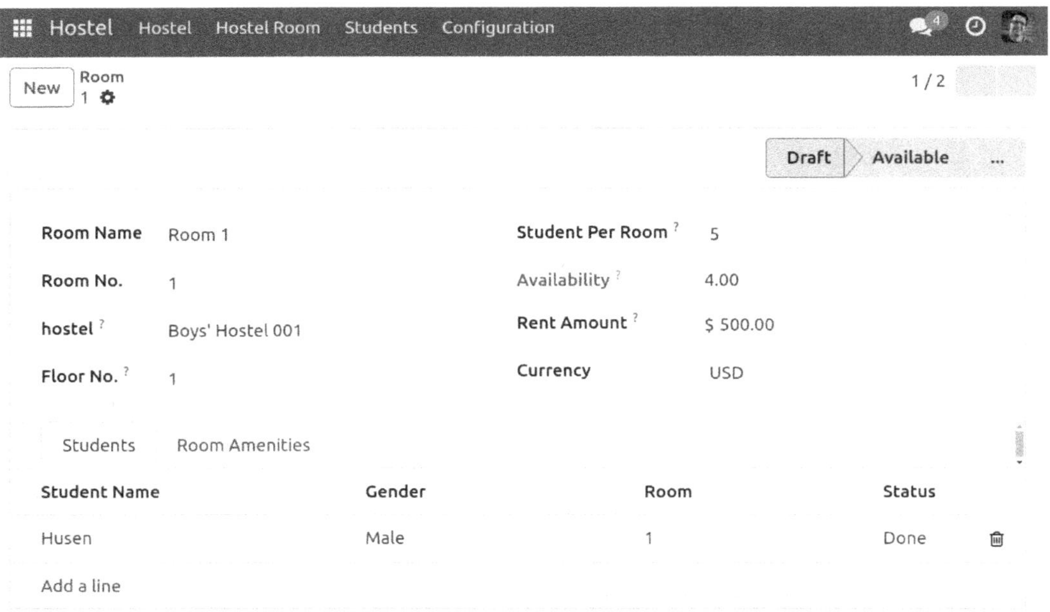

Figure 12.1 – Stage selector in the form view

Here, you'll notice the stages outlined on the hostel record. These stages are clickable, so you will be able to change the stage by clicking on it. Folded stages will be displayed under the **More** dropdown.

How it works...

Since we want to manage the record stages dynamically, we need to create a new model. In *Step 1*, we created a new model called `hostel.room.stage` to store the dynamic stages. In this model, we added a few fields. One of these was the `sequence` field, which is used to determine the order of the stages. We also added the `fold` Boolean field, which is used to collapse the stages and put them in a drop-down list. This is very helpful when your business process has lots of stages because it means that you can hide insignificant stages in the drop-down menu by setting this field.

The `fold` field is also used in Kanban views to display folded Kanban columns. Usually, **Reserved** items are expected to be in the **Unfolded** stage, and terminated items that are marked as either **Done** or **Cancelled** should be in the **Folded** stage.

By default, `fold` is the name of the field that is used to hold the value of the stage fold. You can change this by adding the `_fold_name = 'is_fold'` class attribute.

In *Step 2*, we added the basic access rights rules for the new model.

In *Step 3*, we added the `stage_id` `many2one` field to the `hostel.room` model. While creating a new room record, we wanted to set the default stage value to `Draft`. To accomplish this, we added a `_default_room_stage()` method. This method will fetch the record of the `hostel.room.stage` model with the lowest sequence number, so, while creating a new record, the stage with the lowest sequence will be displayed as active in the form view.

In *Step 4*, we added the `stage_id` field to the form view. By adding the `clickable` option, we made the status bar clickable. We also added an option for the `fold` field, which will allow us to display insignificant stages in the drop-down menu.

In *Step 5*, we added `stage_id` to the tree view.

In *Step 6*, we added the default data for the stages. Users will see these basic stages after installing our module. If you want to learn more about XML data syntax, refer to the *Loading data using XML files* recipe in *Chapter 6*, *Managing Module Data*.

> **Important note**
>
> With this implementation, the user can define new stages on the fly. You will need to add views and menus for `hostel.room.stage` so that you can add new stages from the user interface. Refer to *Chapter 9*, *Backend Views*, if you don't know how to add views and menus.
>
> If you don't want to do this, the Kanban view provides inbuilt features for adding, removing, or modifying stages from the Kanban view itself. We'll look at this in the next recipe.

See also

- Refer to *Chapter 9*, *Backend Views*, to learn about adding views and menus.

Managing Kanban stages

Using a **Kanban board** is a simple way to manage workflows. It is organized into columns, each corresponding to stages, and the work items progress from left to right until they are finished. A Kanban view, with these stages, provides flexibility because it allows users to choose their own workflows. It provides a full overview of the records on a single screen.

Getting started

For this recipe, we will be using the `my_hostel` module from the previous recipe. We will add Kanban to the `hostel.room` model and we will group Kanban cards by stage.

How to do it...

Perform the following steps to enable workflows such as Kanban for the `hostel.room` model:

1. Add a Kanban view for `hostel.room`, as follows:

```
<record id="hostel_room_view_kanban" model="ir.ui.view">
    <field name="name">Hostel room Kanban</field>
    <field name="model">hostel.room</field>
    <field name="arch" type="xml">
        <kanban default_group_by="stage_id">
            <field name="stage_id" />
            <templates>
                <t t-name="kanban-box">
                    <div class="oe_kanban_global_click">
                        <div class="oe_kanban_content">
                            <div class="oe_kanban_card">
                                <div>
                                    <b>
                                        <field name="name" />
                                    </b>
                                </div>
                                <div class="text-muted">
                                    <i class="fa fa-building"/>
                                    <field name="hostel_id" />
                                </div>
                            </div>
                        </div>
                    </div>
                </t>
            </templates>
        </kanban>
    </field>
</record>
```

2. Add Kanban to the `action_hostel_room` action, as follows:

```
...
<field name="view_mode">kanban,tree,form</field>
...
```

3. Add the `_group_expand_stages()` method and the `group_expand` attribute to the `stage_id` field, as follows:

```
@api.model
def _group_expand_stages(self, stages, domain, order):
    return stages.search([], order=order)

stage_id = fields.Many2one(
        'hostel.room.stage',
        default=_default_room_stage,
        group_expand='_group_expand_stages'
    )
```

Restart the server and update the module to apply the changes. This will enable a Kanban board, as shown in the following screenshot:

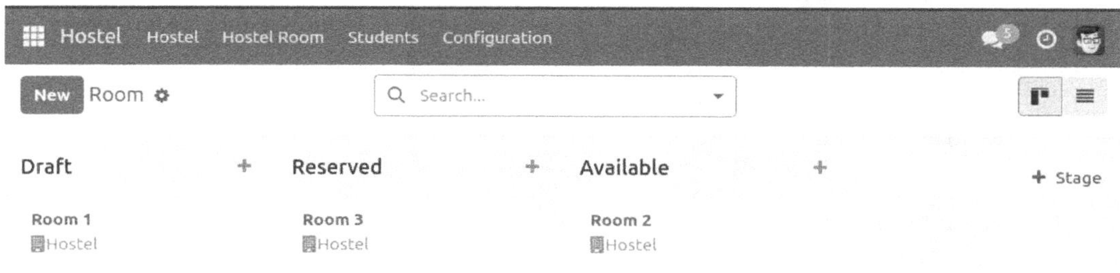

Figure 12.2 – Kanban view with groups by stage

As shown in the preceding screenshot, the Kanban view will show the room records grouped by stage. You will be able to drag and drop cards to another stage column. Moving cards to another column will change the stage value in the database too.

How it works...

In *Step 1*, we added a Kanban view for the `hostel.room.stage` model. Note that we used `stage_id` as the default group for Kanban so that when the user opens Kanban, the Kanban cards will be grouped by stage. To find out more about Kanban, please refer to *Chapter 9*, *Backend Views*.

In *Step 2*, we added the `kanban` keyword to the existing action.

In *Step 3*, we added the `group_expand` attribute to the `stage_id` field. We also added a new `_group_expand_stages()` method. `group_expand` changes the behavior of the field grouping. By default, field grouping shows the stages that are being used. For example, if no rooms record has the `Reserved` stage, the grouping will not return that stage, so Kanban will not display the `Reserved` column. But in our case, we want to display all of the stages, regardless of whether or not they are being used.

The `_group_expand_stages()` function is used to return all the records for the stages. Because of this, the Kanban view will display all the stages and you will be able to use workflows by dragging and dropping them.

There's more...

If you play around with the Kanban board you created in this recipe, you will find lots of different features. Some of these are as follows:

- You can create a new stage by clicking on the **Add new column** option. The `group_create` option can be used to disable the **Add column** option from the Kanban board.
- You can arrange columns in a different order by dragging them by their headers. This will update the sequence field of the `hostel.room.stage` model.
- You can edit or delete columns with the gear icon in the header of a Kanban column. The `group_edit` and `group_delete` options can be used to disable this feature.
- The stages that have a `true` value in the `fold` field will collapse and the column will be displayed as a slim bar. If you click on this slim bar, it will expand and display the Kanban cards.
- If the model has an `active` Boolean field, it will display the option to archive and unarchive records in the Kanban column. The `archivable` option can be used to disable this feature.
- The plus icon on the Kanban column can be used to create records directly from the Kanban view. The `quick_create` option can be used to disable this feature. For the moment, this feature will not work in our example. This will be solved in the next recipe.

See also

- To learn more about Kanban, please refer to *Chapter 9, Backend Views.*

Adding a quick create form to a Kanban card

Grouped Kanban views provide the quick create feature, which allows us to generate records directly from the Kanban view. The plus icon on a column will display an editable Kanban card on the column, using which you can create a record. In this recipe, we will learn how to design a quick create Kanban form of our choice.

Getting started

For this recipe, we will be using the `my_hostel` module from the previous recipe. We will use the quick create option in Kanban for the `hostel.room` model.

How to do it...

Follow these steps to add a custom quick create form for Kanban:

1. Create a new minimal form view for the `hostel.room` model, as follows:

```
<record id="hostel_room_view_form_minimal" model="ir.ui.view">
    <field name="name">Hostel room Form</field>
    <field name="model">hostel.room</field>
    <field name="arch" type="xml">
        <form>
            <group>
                <field name="name"/>
                <field name="room_no"/>
                <field name="hostel_id" required="1"/>
                <field name="floor_no"/>
                <field name="student_per_room"/>
            </group>
        </form>
    </field>
</record>
```

2. Add quick create options to the `<kanban>` tag, as follows:

```
<kanban default_group_by="stage_id" on_create="quick_create"
quick_create_view="my_hostel.hostel_room_view_form_minimal">
```

3. Restart the server and update the module to apply the changes. Then, click on the plus icon in the column. This will enable Kanban forms, as shown in the following screenshot:

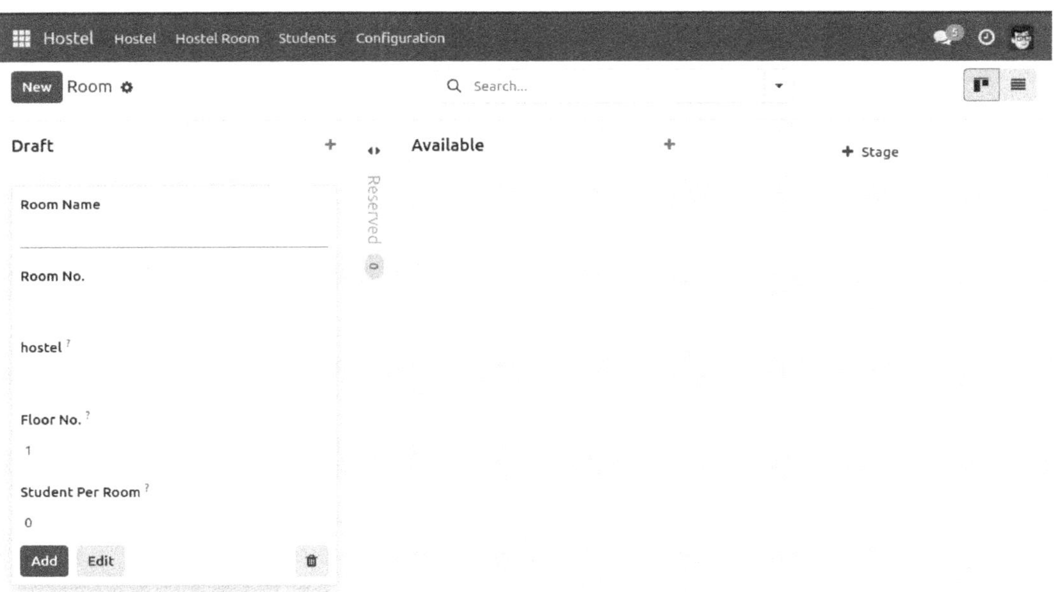

Figure 12.3 – Quickly creating a record directly from the Kanban view

When you click on the **Create** button in the Kanban view, you will see a small card with input instead of being redirected to the form view. You can fill in the values and click on **Add**, which will create a room record.

How it works...

To create a custom quick create option, we need to create a minimal form view. We did this in *Step 1*. We added two required fields because you cannot create a record without filling in the required fields. If you do so, Odoo will generate an error and open the default form view in the dialogue so that you can enter all the required values.

In *Step 2*, we added this new form view to the Kanban view. Using the `quick_create_view` option, you can map the custom form view to the Kanban view. We also added one extra option – `on_create="quick_create"`. This option will display a quick create form in the first column when you click on the **Create** button in the control panel. Without this option, the **Create** button will open a form view in edit mode.

You can disable the quick create feature by adding `quick_create="false"` to the Kanban tag.

Creating interactive Kanban cards

Kanban cards support all HTML tags, which means you can design them however you like. Odoo provides some built-in ways to make Kanban cards more interactive. In this recipe, we will add color options, the star widget, and many2many tags to the Kanban card.

Getting started

For this recipe, we will be using the my_hostel module from the previous recipe.

How to do it...

Follow these steps to create an attractive Kanban card:

1. Add a new model to manage the tags for the hostel.room model, as follows:

```
class HostelAmenities(models.Model):
    _name = "hostel.amenities"
    _description = "Hostel Amenities"

    name = fields.Char("Name", help="Provided Hostel Amenity")
    active = fields.Boolean("Active", default=True,
        help="Activate/Deactivate whether the amenity should be
given or not")
    color = fields.Integer()
```

2. Add basic access rights for the hostel.amenities model, as follows:

```
access_hostel_amenities_manager_id,access.hostel.amenities.
manager,my_hostel.model_hostel_amenities,my_hostel.group_hostel_
manager,1,1,1,1
access_hostel_amenities_user_id,access.hostel.amenities.
user,my_hostel.model_hostel_amenities,my_hostel.group_hostel_
user,1,0,0,0
```

3. Add new fields to the hostel.room model, as follows:

```
color = fields.Integer()
    popularity = fields.Selection([('no', 'No Demand'), ('low',
'Low Demand'), ('medium', 'Average Demand'), ('high', 'High
Demand'),])
hostel_amenities_ids = fields.Many2many(
    "hostel.amenities",
    "hostel_room_amenities_rel", "room_id", "amenitiy_id",
    string="Amenities", domain="[('active', '=', True)]",
    help="Select hostel room amenities")
```

4. Add fields to the form view, as follows:

```
<field name="popularity" widget="priority"/>
<field name="hostel_amenities_ids" widget="many2many_tags"
    options="{'color_field': 'color', 'no_create_edit': True}"/>
```

In the next few steps, we will update an existing Kanban view. The code in bold text is newly added code.

5. Add a `color` field to the Kanban view:

```
<field name="stage_id" />
<field name="color" />
```

6. Add a dropdown to choose a color on the Kanban view:

```
<t t-name="kanban-box">
    <div t-attf-class="#{kanban_color(record.color)} oe_kanban_
global_click">
        <div class="o_dropdown_kanban dropdown">
            <a class="dropdown-toggle o-no-caret btn"
role="button" data-toggle="dropdown">
                <span class="fa fa-ellipsis-v"> </span>
            </a>
            <div class="dropdown-menu" role="menu">
                <t t-if="widget.editable">
                    <a role="menuitem" type="edit"
class="dropdown-item">Edit</a>
                </t>
                <t t-if="widget.deletable">
                    <a role="menuitem" type="delete"
class="dropdown-item">Delete</a>
                </t>
                <ul class="oe_kanban_colorpicker" data-
field="color"/>
            </div>
        </div>
        <div class="oe_kanban_content">
            <div class="oe_kanban_card oe_kanban_global_click">
                <div>
                    <i class="fa fa-bed">  </i>
                    <b>
                        <field name="name" />
                    </b>
                </div>
                <div class="text-muted">
```

```
                    <i class="fa fa-building">    </i>
                    <field name="hostel_id" />
                </div>
                <span class="oe_kanban_list_many2many">
                    <field name="hostel_amenities_ids"
widget="many2many_tags" options="{'color_field': 'color'}"/>
                </span>
                <div>
                    <field name="popularity" widget="priority"/>
                </div>
            </div>
        </div>
    </div>
</t>
...
```

7. Add tags and a `popularity` field to the Kanban view:

```
...
<div class="text-muted">
    <i class="fa fa-building"/>
    <field name="hostel_id" />
</div>
<span class="oe_kanban_list_many2many">
    <field name="hostel_amenities_ids" widget="many2many_tags"
options="{'color_field': 'color'}"/>
</span>
<div>
    <field name="popularity" widget="priority"/>
</div>
```

Important note

The code in bold should be added to the existing Kanban view.

Restart the server and update the module to apply the changes. Then, click on the plus icon on a column. It will display the Kanban, as shown in the following figure:

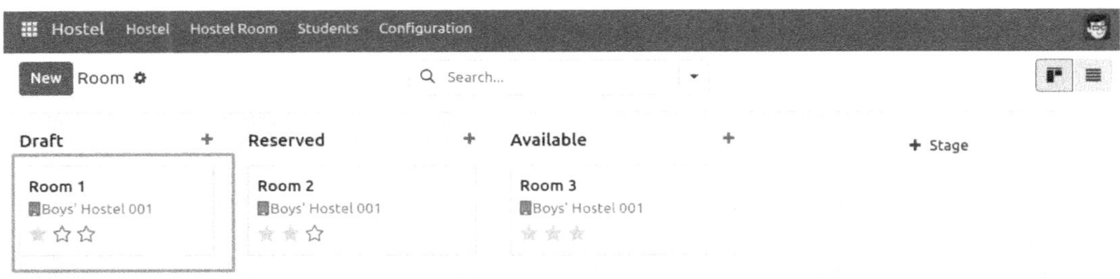

Figure 12.4 – Kanban cards with new options

Our changes in the Kanban structure will enable extra options in the Kanban card. Now, you will be able to choose the color on the Kanban itself. You will also be able to prioritize cards with stars.

How it works...

In *Steps 1* and *2*, we added a new model and security rules for tags. In *Step 3*, we added a few fields to the rooms model.

In *Step 4*, we added those fields to the form view. Note that we used the `priority` widget on the `popularity` field, which displays the selection field with star icons. In the `hostel_amenities_ids` field, we used the `many2many_tags` widget, which displays the `many2many` field in the form of tags. The `color_field` option is passed to enable the color feature on tags. The value of this option is the field name where the color index is stored. The `no_create_edit` option will disable the feature of creating new tags via the form view.

In *Step 5*, we improved lots of things. Fisrt, we added `t-attf-class="#{kanban_color(record.color.raw_value)}` to the Kanban card. This will be used to display the color of the Kanban card. It uses the value of the `color` field and generates a class based on that value. For example, if a Kanban record has a value of `2` in the `color` field, it will add `kanban_color_2` to the class. After that, we added a drop-down menu to add options such as **Edit**, **Delete**, and the Kanban color picker. The **Edit** and **Delete** options are only displayed if the user has proper access rights.

Finally, we added tags and priority to the Kanban card. After adding all of this, the Kanban card will look as follows:

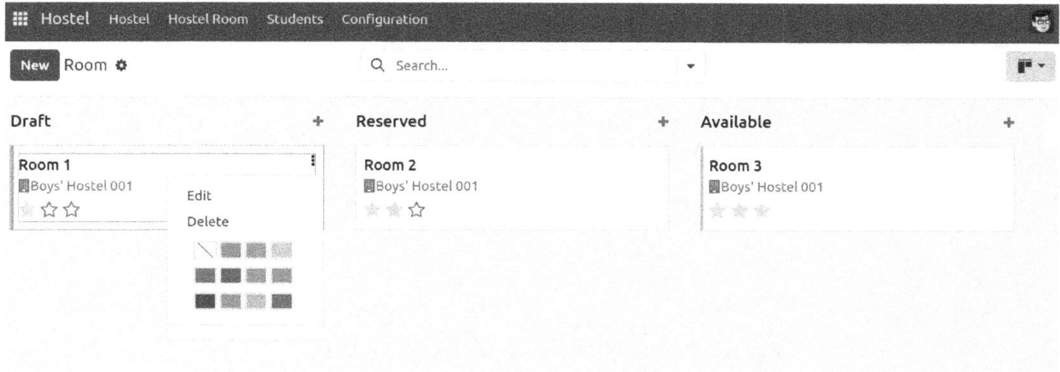

Figure 12.5 – Kanban card options

With this card design, you will be able to set popularity stars and colors directly from the Kanban card.

Adding a progress bar to Kanban views

Sometimes, you have tons of records in columns and it is very difficult to get a clear picture of the particular stages. A progress bar can be used to display the status of any column. In this recipe, we will display a progress bar on Kanban based on the `popularity` field.

Getting started

For this recipe, we will be using the `my_hostel` module from the previous recipe.

How to do it...

To add a progress bar to the Kanban columns, you will need to add a `progressbar` tag to the Kanban view definition, as follows:

```
<progressbar
    field="popularity"
    colors='{"low": "success", "medium": "warning", "high":
"danger"}'/>
```

Note that Kanban column progress bars were introduced in Odoo version 11. Versions before that will not display column progress bars.

Restart the server and update the module to apply the changes. Then, click on the plus icon on a column. This will display the progress bar on the Kanban columns, as shown in the following screenshot:

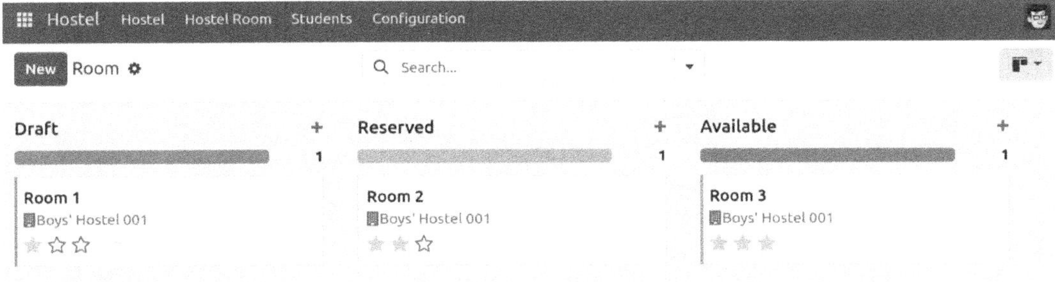

Figure 12.6 – Kanban view with a progress bar

Upon updating the module, you will have added a progress bar to the Kanban columns. The color of the progress bar shows the number of records based on the record state. You will be able to click on one of the progress bars to filter records based on that state.

How it works...

Progress bars on Kanban columns are displayed based on the values of the field. Progress bars support four colors, so you cannot display more than four states. The available colors are green (success), blue (information), red (danger), and yellow (warning). Then, you need to map colors to the field states. In our example, we mapped three states of the `priority` field because we didn't want any progress bars for the rooms that were not in demand.

By default, progress bars show a count of the records on the side. You can see the total of a particular state by clicking on it in the progress bar. Clicking on the progress bar will also highlight the cards for that state. Instead of the count of records, you can also display the sum of the integer or float field. To do this, you need to add the `sum_field` attribute with the field value, such as `sum_field="field_name"`.

Creating server actions

Server actions underpin Odoo's automation tools. They allow us to describe the actions to perform. These actions are then available to be called by **event triggers** or to be triggered automatically when certain time conditions are met.

The simplest case is to let the end user perform an action on a document by selecting it from the **More** button. We will create this kind of action for project tasks so that we can **Set Priority** by starring the currently selected task and setting a deadline on it for 3 days from now.

Getting ready

We will need an Odoo instance with the Project app installed. We will also need **Developer Mode** activated. If it's not already activated, activate it in the Odoo **Settings** dashboard.

How to do it...

To create a server action and use it from the **More** menu, follow these steps:

1. From the **Settings** top menu, select the **Technical** | **Actions** | **Server Actions** menu item and click on the **Create Contextual Action** button at the top of the record list, as shown in the following screenshot:

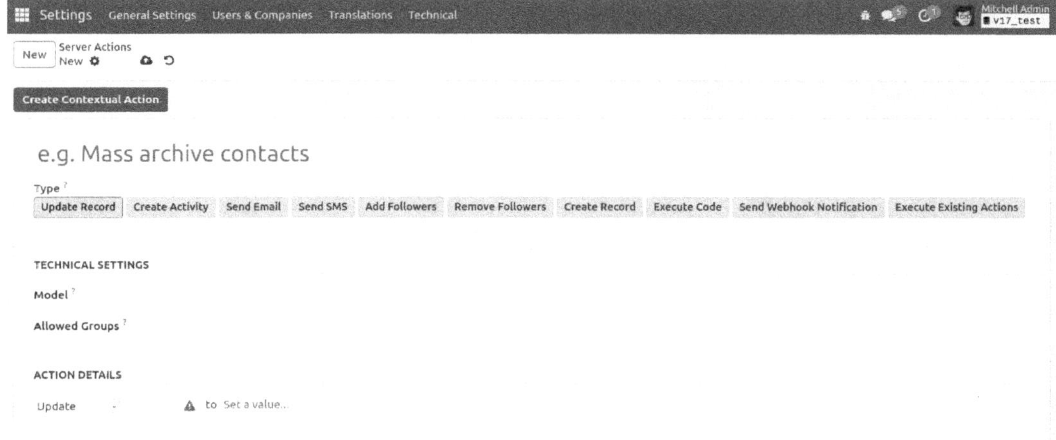

Figure 12.7 – Server action form view

2. Fill out the server action form with these values:

   - **Action Name**: **Set as Priority**
   - **Model**: **Task**
   - **Type**: **Update Record**

3. In the server action, under the **Data to Write** tab, go to the **Update** section. Enter the following parameters:

   - **Field**: `Priority`
   - **Evaluation Type**: `Value`
   - **Value**: `Low`

The following screenshot shows the entered values:

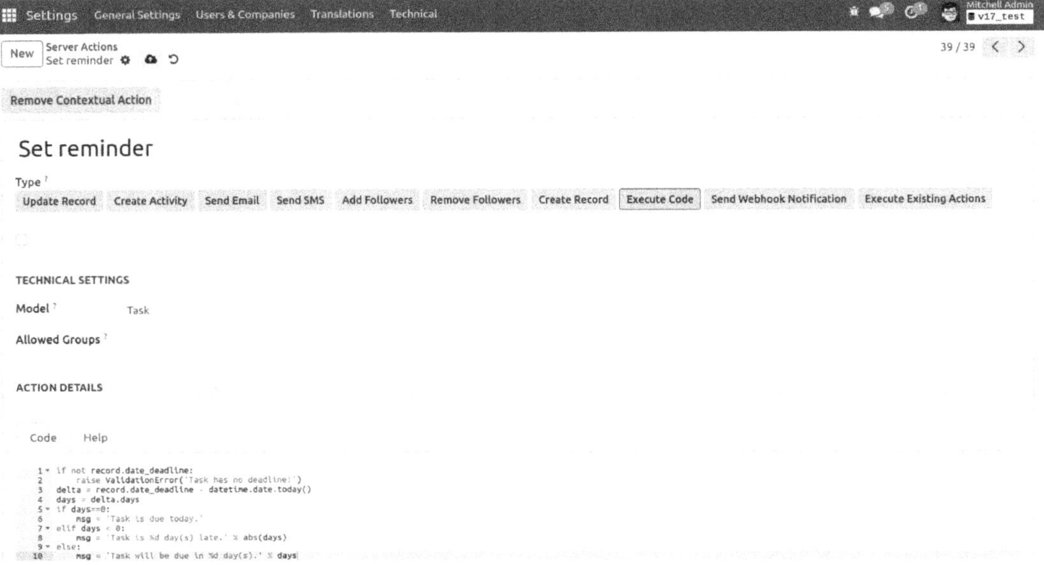

Figure 12.8 – Set lines to write

1. Save the server action and click on the **Create Contextual Action** button at the top left to make it available under the project task's **More** button.
2. To try it out, go to the **Projects** top menu, open the **Project**, and open a random task. By clicking on the action, we should see the **Set Priority** option, as shown in the following screenshot. Selecting this will star the task and change the deadline date to 3 days from now:

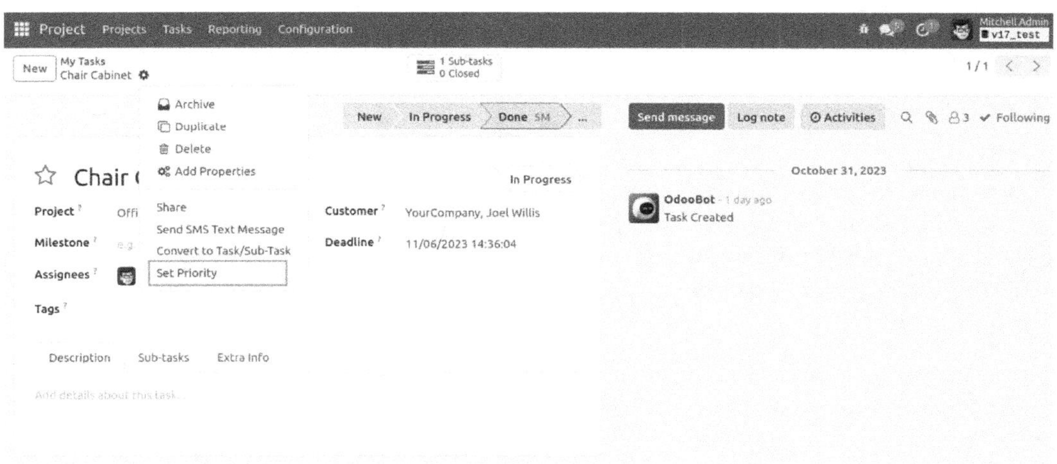

Figure 12.9 – The Set Priority server action

Once you add the server action, you will have set the priority option on the task. Upon clicking on it, the server action star will turn yellow, meaning the priority of the task has increased. Also, the server action will change the deadline.

How it works...

Server actions work on a model, so one of the first things we must do is pick the model we want to work with. In our example, we used project tasks.

Next, we should select the type of action to perform. There are a few options available:

- **Update Record** allows you to set values on the current record or another record.
- **Create activity** allows you to create activity on the selected records
- **Execute Code** allows you to write arbitrary code to execute when none of the other options are flexible enough for what we need.
- **Create Record** allows you to create a new record on the current model or another model.
- **Send Email** allows you to choose an email template. This will be used to send out an email when the action is triggered.
- **Execute Existing actions** can be used to trigger a client or window action, just like when a menu item is clicked on.
- **Add Followers** allows users or channels to subscribe to the record.
- **Create Next Activity** allows you to create a new activity. This will be displayed in the chatter.
- **Send SMS** allows you to send an SMS. You need to select the SMS template.

> **Note**
>
> **Send SMS Text Message** is a chargeable service from Odoo. You need to purchase credit for SMS if you want to send an SMS.

For our example, we used **Update the Record** to set some values on the current record. We set **Priority** to `1` to star the task and set a value on the **Deadline** field. This one is more interesting because the value to use is evaluated from a Python expression. Our example makes use of the `datetime` Python module (`https://docs.python.org/2/library/datetime.html`) to compute the date 3 days from today.

Arbitrary Python expressions can be used there, as well as in several of the other action types available. For security reasons, the code is checked by the `safe_eval` function implemented in the `odoo/tools/safe_eval.py` file. This means that some Python operations may not be allowed, but this rarely proves to be a problem.

When you add a drop-down option to the server action, usually, it is available for all internal users. But if you just want to show this option to selected users, you can assign a group to the server action. This is available under the **Security** tab in the server action form view.

There's more...

The Python code is evaluated in a restricted context, where the following objects are available to use:

- `env`: This is a reference for the `Environment` object, just like `self.env` in a class method.
- `model`: This is a reference to the `model` class that the server action acts upon. In our example, it is equivalent to `self.env['project.task]`.
- `ValidationError`: This is a reference to `from odoo.exceptions import ValidationError`, allowing validations that block unintended actions. It can be used as `raise Warning('Message!')`.
- `Record` or `records`: This provides references to the current record or records, allowing you to access their field values and methods.
- `log`: This is a function that's used to log messages in the `ir.logging` model, allowing database-side logging-on actions.
- `datetime`, `dateutil`, and `time`: These provide access to the Python libraries.

Using Python code server actions

Server actions have several types available, but executing arbitrary Python code is the most flexible option. When used wisely, it empowers users with the capability to implement advanced business rules from the user interface, without the need to create specific add-on modules to install that code.

We will demonstrate using this type of server action by implementing a server action that sends reminder notifications to the followers of a project task.

Getting ready

We will need an Odoo instance with the Project app installed.

How to do it...

To create a Python code server action, follow these steps:

1. Create a new server action. In the **Settings** menu, select the **Technical | Actions | Server Actions** menu item, and click on the **Create** button at the top of the record list.
2. Fill out the **Server Action** form with the following values:
   - **Action Name**: **Send Reminder**
   - **Base Model**: **Task**
   - **Action To Do**: **Execute Code**
3. In the **Python Code** text area, remove the default text and replace it with the following code:

```
if not record.date_deadline:
    raise ValidationError('Task has no deadline!')
delta = record.date_deadline - datetime.date.today()
days = delta.days
if days==0:
    msg = 'Task is due today.'
elif days < 0:
    msg = 'Task is %d day(s) late.' % abs(days)
else:
    msg = 'Task will be due in %d day(s).' % days

record.message_post(body=msg, subject='Reminder', subtype_
xmlid='mail.mt_comment')
```

The following screenshot shows the entered values:

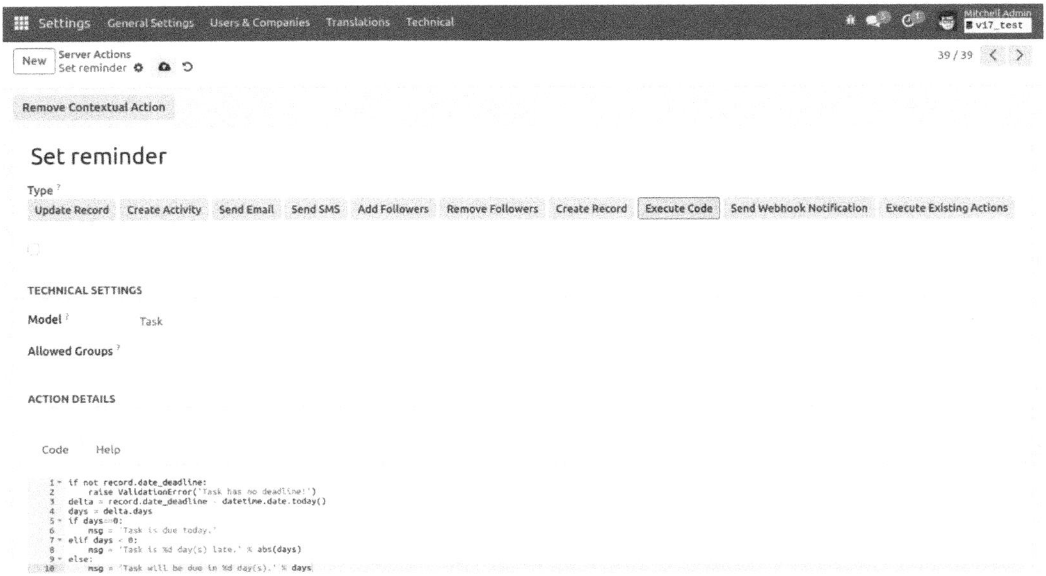

Figure 12.10 – Python code with the values entered

4. Save the server action and click on **Create Contextual Action** at the top left to make it available under the project task's **More** button.
5. Now, click on the **Project** top menu and select the **Search** | **Tasks** menu item. Pick a random task, set a deadline date on it, and then try the **Send Reminder** option under the **More** button.

This works just like the previous recipe; the only difference is that this server action will run your Python code. Once you run the server action on a task, it will put a message in the chatter.

How it works...

The *Creating server actions* recipe of this chapter provides a detailed explanation of how to create a server action in general. For this particular type of action, we need to pick the **Execute Code** option and then write the code to run the text area.

The code can have multiple lines, as is the case in our recipe, and it runs in a context that has references to objects such as the current record object or the session user. The available references were described in the *Creating server actions* recipe.

The code we used computes the number of days from the current date until the deadline date and uses that to prepare an appropriate notification message. The last line does the actual posting of the message in the task's message wall. The `subtype='mt_comment'` argument is needed for email notifications to be sent to the followers, just like when we use the **New Message** button. If no subtype is given, `mt_note` is used by default, posting an internal note without notification, as if we had used the **Log an internal note** button. Refer to *Chapter 23*, *Managing Emails in Odoo*, to learn more about mailing in Odoo.

There's more...

Python code server actions are a powerful and flexible resource, but they do have some limitations compared to the custom add-on modules.

Because the Python code is evaluated at runtime, if an error occurs, the stack trace is not as informative and can be harder to debug. It is also not possible to insert a breakpoint in the code of a server action using the techniques shown in *Chapter 7*, *Debugging Modules*, so debugging needs to be done using logging statements. Another concern is that, when trying to track down the cause of behavior in the module code, you may not find anything relevant. In this case, it's probably caused by a server action.

When carrying out a more intensive use of server actions, the interactions can be quite complex, so it is advisable to plan properly and keep them organized.

See also

- Refer to *Chapter 23*, *Managing Emails in Odoo*, to learn more about mailing in Odoo.

Using automated actions on time conditions

Automated actions can be used to automatically trigger actions based on time conditions. We can use them to automatically perform some operations on records that meet certain criteria and time conditions.

As an example, we can trigger a reminder notification for project tasks one day before their deadline, if they have one. Let's see how this can be done.

Getting ready

To follow this recipe, we will need to have both the *project management* app (which has the technical name `project`) and the **Automated Action Rules** add-on (which has the technical name `base_automation`) already installed, and have **Developer Mode** activated. We will also need the server action we created in the *Using Python code server actions* recipe of this chapter.

How to do it...

To create an automated action with a timed condition on tasks, follow these steps:

1. In the **Settings** menu, select the **Technical** | **Automation** | **Automated Actions** menu item, and click on the **Create** button.

2. Fill out the basic information on the **Automated Actions** form:

   - **Rule Name**: `Send notification near deadline`
   - **Model**: **Task**
   - Select **Based on Time Condition** in the **Trigger** field
   - For **Action To Do**, select **Execute Existing actions**

3. To set the record criteria, click on the **Edit Domain** button in the **Apply on** section. In the pop-up dialogue, set a valid domain expression in the code editor, `["&",["date_deadline","!=",False],["stage_id.fold","=",False]]`, and click on the **Save** button. When changing to another field, the information on the number of records meeting the criteria will be updated and display **Record(s)** buttons. By clicking on the **Records** button, we can check the records list of the records meeting the domain expression.
4. To set the time condition for **Trigger Date**, select the field to use, which is **Deadline**, and set **Delay after trigger date** to `-1` **Days**.
5. On the **Actions** tab, under **Server actions to run**, click on **Add an item** and pick **Send Reminder** from the list; this should have been created previously. Refer to the following screenshot:

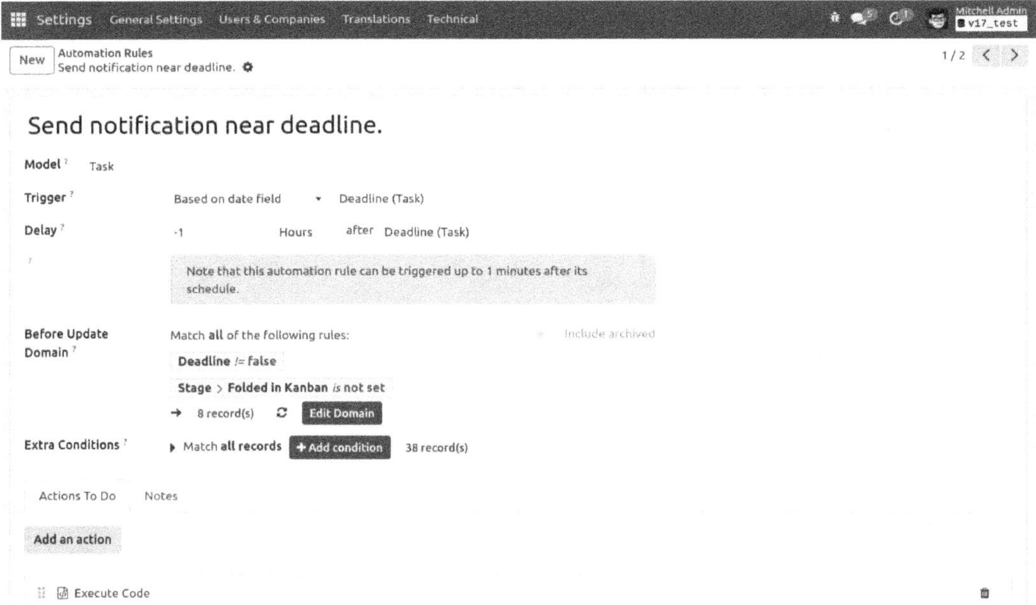

Figure 12.11 – Automated action form view

   If not, we can still create the server action to run using the **Create** button.

6. Click on **Save** to save the automated action.

7. Perform the following steps to try it out:

    i. Go to the **Project** menu, go to **Search** | **Tasks**, and set a deadline on a task with the date in the past.

    ii. Go to the **Settings** menu, click on the **Technical** | **Automation** | **Scheduled Actions** menu item, find the **Base Action Rule: check and execute** action in the list, open its form view, and press on the **Run Manually** button at the top left. This forces timed automated actions to be checked. This is shown in the following screenshot. Note that this should work on a newly created demo database, but it might not work this way in an existing database:

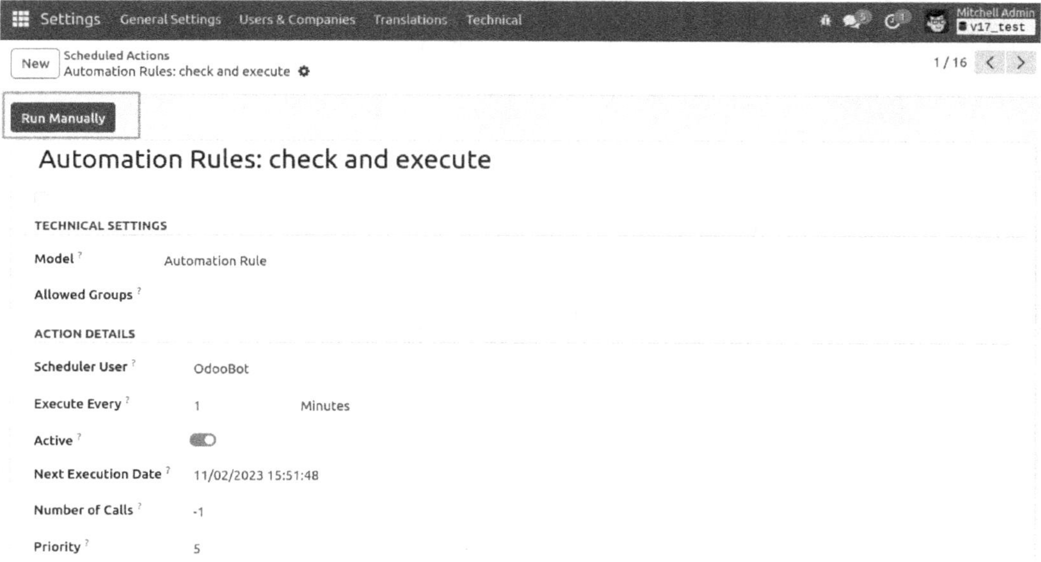

Figure 12.12 – Running an automated action (for test)

8. Again, go to the **Project** menu and open the same task you previously set a deadline date on. Check the message board; you should see the notification generated by the server action triggered by our automated action.

    After adding the time-based automated action for the deadline, a reminder message will be added to the task 1 day before the deadline.

How it works...

Automated actions act on a model, and can be triggered either by events or time conditions. First, we must set the **Model** and **When to Run** values.

Both methods can use a filter to narrow down the records that we can perform the action on. We can use a domain expression for this. You can find further information about writing domain expressions in *Chapter 9, Backend Views*. Alternatively, you can create and save a filter on project tasks by using the user interface features and then copy the automatically generated domain expression, selecting it from the **Set** selection based on a search filter list.

The domain expression we used selects all the records with a non-empty **Deadline** date, in a stage where the `Fold` flag is not checked. Stages without the `Fold` flag are considered to be work-in-progress. This way, we avoid triggering notifications on tasks that are in the **Done**, **Canceled**, or **Closed** stages.

Then, we should define the time condition – the date field to use and when the action should be triggered. The period can be in minutes, hours, days, or months, and the number set for the period can be positive, indicating the time after the date, or negative, indicating the time before the date. When using a period in days, we can provide a resource calendar that defines the working days and that can be used by the day count.

These actions are checked by the **Check Action Rules** scheduled job. Note that, by default, this is run every 4 hours. This is appropriate for actions that work on a day or month scale, but if you need actions that work on smaller timescales, you need to change the running interval to a smaller value.

Actions will be triggered for records that meet all the criteria and whose triggering date condition (the field date plus the interval) is after the last action execution. This is to avoid repeatedly triggering the same action. Also, this is why manually running the preceding action will work in a database in which the scheduled action has not yet been triggered, but why it might not work immediately in a database where it has already been run by the scheduler.

Once an automated action is triggered, the **Actions** tab tells you what should happen. This might be a list of server actions that do things such as changing values on the record, posting notifications, or sending out emails.

There's more...

These types of automated actions are triggered once a certain time condition is reached. This is not the same as regularly repeating an action while a condition is still true. For example, an automated action will not be capable of posting a reminder for every day after the deadline has been exceeded.

This type of action can, instead, be performed by scheduled actions, which are stored in the `ir.cron` model. However, scheduled actions do not support server actions; they can only call an existing method of a model object. So, to implement a custom action, we need to write an add-on module, adding the underlying Python method.

For reference, the technical name for the model is `base.action.rule`.

See also

- For further details about writing domain expressions, refer to *Chapter 9, Backend Views*.

Using automated actions on event conditions

Business applications provide systems with records for business operations but are also expected to support dynamic business rules that are specific to the organization's use cases.

Carving these rules into custom add-on modules can be inflexible and out of the reach of functional users. Automated actions triggered by event conditions can bridge this gap and provide a powerful tool to automate or enforce the organization's procedures. As an example, we will enforce validation on project tasks so that only the project manager can change tasks to the **Done** stage.

Getting ready

To follow this recipe, you will need to have the project management app already installed. You also need to have **Developer Mode** activated. If it's not activated already, activate it in the Odoo **About** dialogue.

How to do it...

To create an automated action with an event condition on tasks, follow these steps:

1. In the **Settings** menu, select the **Technical** | **Automation** |**Automated Actions** menu item, and click on the **Create** button.
2. Fill out the basic information in the **Automated Actions** form:
   - **Action Name**: `Validate Closing Tasks`
   - **Model**: **Task**
   - **Trigger**: **On Save**
   - **Action To Do**: **Execute Existing actions**
   - **Watched fields**: `Stage id`
3. The **When updating** rules allow you to set two record filters, before and after the update operation:
   - For the **Before Update Filter** field, click on the **Edit Domain** button, set a valid domain expression – `[('stage_id.name', '!=', 'Done')]` – in the code editor, and save
   - For the **Apply on** field, click on the **Edit Domain** button, set the `[('stage_id.name', '=', 'Done')]` domain in the code editor, and save, as shown in the following screenshot:

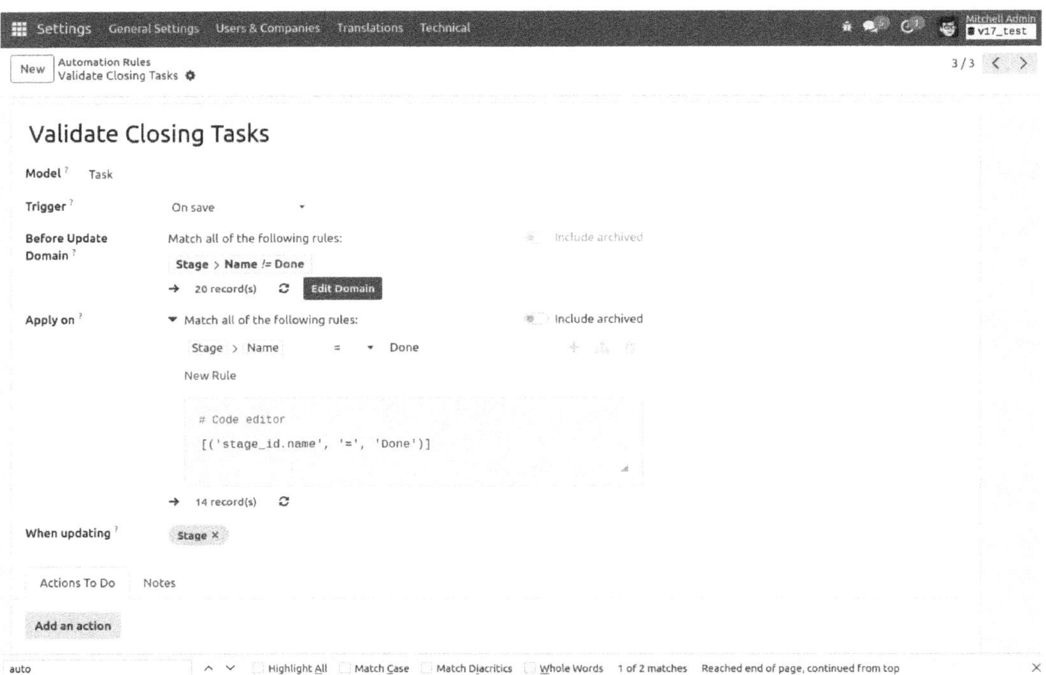

Figure 12.13 – Automated action form view

4. In the **Actions** tab, click on **Add an item**. In the list dialogue, click on the **Create** button to create a new server action.
5. Fill out the server action form with the following values, and then click on the **Save** button:
   - **Action Name**: `Validate Closing tasks`
   - **Model**: **Task**
   - **Action To Do**: **Execute Code**
   - **Python Code**: Enter the following code:

```
if user != record.project_id.user_id:
    raise Warning('Only the Project Manager can close Tasks')
```

The following screenshot shows the entered values:

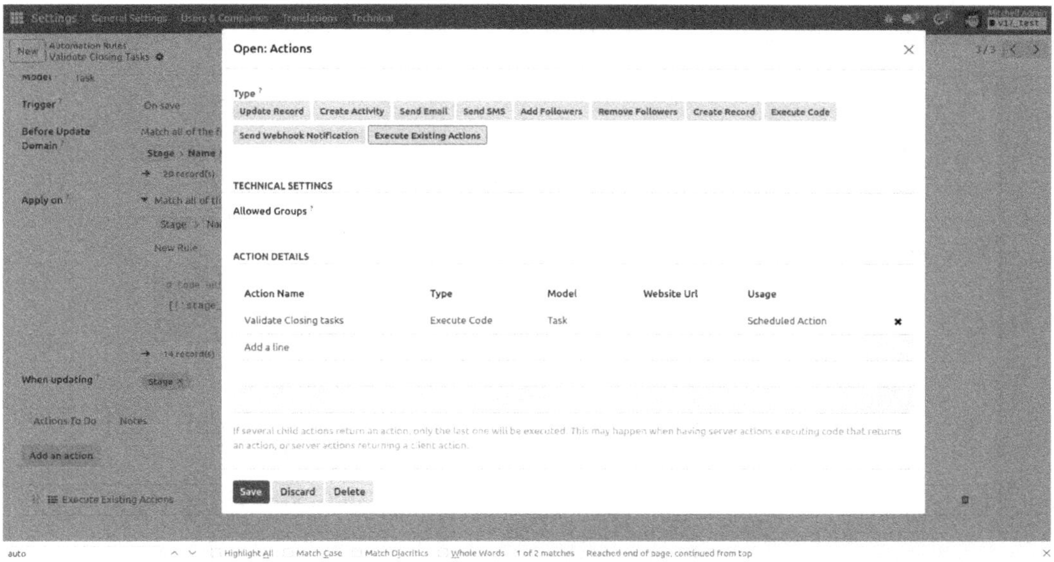

Figure 12.14 – Adding a child action

6. Click on **Save & Close** to save the automated action and try it out:

    i. On a database with demo data and where you're logged in as an administrator, go to the **Project** menu and click on the project to open the Kanban view of the tasks.

    ii. Then, try dragging one of the tasks into the **Done** stage column. Since this project's manager is the `Demo` user and we are working with the `Administrator` user, our automated action should be triggered, and our warning message should block the change.

How it works...

We start by giving a name to our automated actions and setting the model it should work with. For the type of action we require, we should choose **On Save**, but the **On Creation**, **On Creation & Update**, **On Deletion**, and **Based On Form Modification** options are also available.

Next, we define the filters to determine when our action should be triggered. The **On Save** actions allow us to define two filters – one to check before and the other after the changes are made to the record. This can be used to express transitions – to detect when a record changes from *state A* to *state B*. In our example, we want to trigger the action when a *not-done* task changes to the *done* stage. The **On Save** action is the only one that allows these two filters; the other action types only allow one filter.

Important note

It is important to note that our example condition will only work correctly for English language users. This is because **Stage Name** is a translatable field that can have different values for different languages. So, the filters on the translatable fields should be avoided or used with care.

Finally, we create and add one (or more) server actions with whatever we want to be done when the automated action is triggered. In this case, we chose to demonstrate how to implement custom validation, making use of a Python code server action that used the `Warning` exception to block the user's changes.

There's more...

In *Chapter 5, Basic Server-Side Development*, we saw how to redefine the `write()` methods of a model to perform actions on record updates. Automated actions on record updates provide another way to achieve this, with some benefits and drawbacks.

Among the benefits, it is easy to define an action that's triggered by the update of a stored computed field, which is tricky to do in pure code. It is also possible to define filters on records and have different rules for different records or for records matching different conditions that can be expressed with search domains.

However, automated actions can have disadvantages compared to Python business logic code inside modules. With poor planning, the flexibility provided can lead to complex interactions that are difficult to maintain and debug. Also, the before-and-after write filter operations bring some overhead, which can be an issue if you are performing sensitive actions.

Creating QWeb-based PDF reports

When communicating with the outside world, it is often necessary to produce a PDF document from a record in the database. Odoo uses the same template language that's used for form views: QWeb.

In this recipe, we will create a QWeb report to print information about a room that is currently being borrowed by a student. This recipe will reuse the models presented in the *Adding a progress bar in Kanban views* recipe from earlier in this chapter.

Getting ready

If you haven't done so already, install `wkhtmltopdf`, as described in *Chapter 1, Installing the Odoo Development Environment*; otherwise, you won't get shiny PDFs as a result of your efforts.

Also, double-check that the `web.base.url` configuration parameter (or `report.url`) is a URL that is accessible from your Odoo instance; otherwise, the report will take a long time to generate and the result will look strange.

How to do it...

Follow these steps:

1. In this recipe, we will add a report to `hostel.student` that prints a list of students that the student borrowed. We need to add a `one2many` field to the student model concerning the `hostel.room` model, as shown in the following example:

```
class HostelStudent(models.Model):
    _name = "hostel.student"
    _description = "Hostel Student Information"
    name = fields.Char("Student Name")
    gender = fields.Selection([("male", "Male"),
        ("female", "Female"), ("other", "Other")],
        string="Gender", help="Student gender")
    active = fields.Boolean("Active", default=True,
        help="Activate/Deactivate hostel record")
```

2. Define a view for your report in `reports/hostel_room_detail_report_template.xml`, as follows:

```
<?xml version="1.0" encoding="utf-8"?>
<odoo>
<template id="hostel_room_detail_reports_template">
    <t t-call="web.html_container">
        <t t-foreach="docs" t-as="doc">
            <t t-call="web.internal_layout">
                <div class="page">
                    <h1>Room name: <t t-esc="doc.name"/></h1>
                    <h1>Room No: <t t-esc="doc.name"/></h1>
                    <table class="table table-condensed">
                        <thead>
                            <tr>
                                <th>Student Name</th>
                                <th>Gender</th>
                            </tr>
                        </thead>
                        <tbody>
                            <tr t-foreach="doc.student_ids"
t-as="student" >
                                <td><t t-esc="student.name" /></
td>
                                <td><t t-esc="student.gender"
/></td>
                            </tr>
```

```
                        </tbody>
                    </table>
                </div>
            </t>
        </t>
    </t>
</template>
</odoo>
```

3. Add a tag in `reports/hostel_room_detail_report.xml`, as shown in the following example:

```
<?xml version="1.0" encoding="utf-8"?>
<odoo>

    <record id="report_hostel_room_detail" model="ir.actions.
report">
        <field name="name">Room detail report</field>
        <field name="model">hostel.room</field>
        <field name="report_type">qweb-pdf</field>
        <field name="binding_model_id" ref="model_hostel_room"/>
        <field name="report_name">my_hostel.hostel_room_detail_
reports_template</field>
        <field name="report_file">my_hostel.hostel_room_detail_
reports_template</field>
    </record>

</odoo>
```

4. Add both files to the manifest of the add-on, as shown in the following example:

```
...

"depends": ["base"],
    "data": [
        "security/hostel_security.xml",
        "security/ir.model.access.csv",
        "data/room_stages.xml",
        "views/hostel.xml",
        "views/hostel_amenities.xml",
        "views/hostel_room.xml",
        "views/hostel_room_stages_views.xml",
        "views/hostel_student.xml",
        "views/hostel_categ.xml",
        "reports/hostel_room_detail_report_template.xml",
```

```
                "reports/hostel_room_detail_report.xml",
            ],
```

Now, when opening the room form view, or when selecting students in the list view, you should be offered the option to print the rooms detail report in a drop-down menu, as shown in the following screenshot:

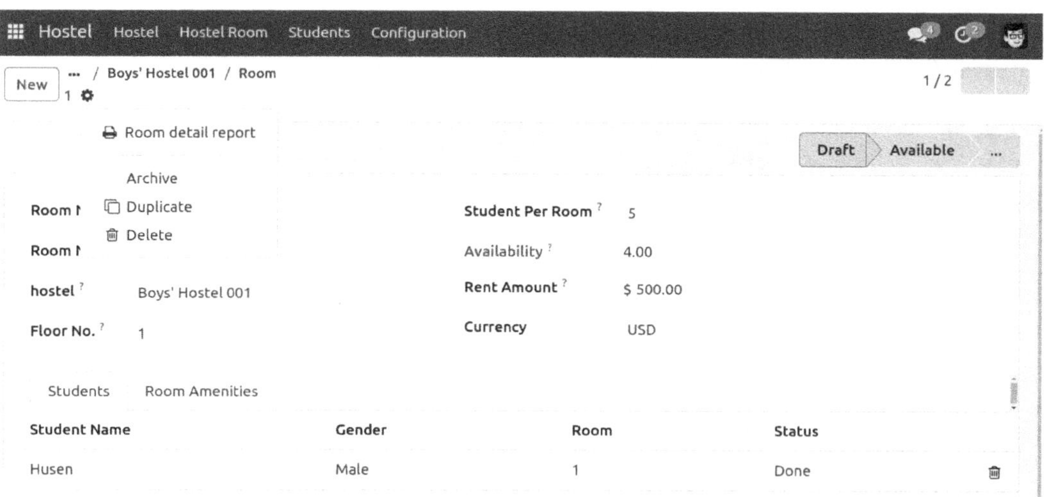

Figure 12.15 – Print action for report

How it works...

In *Step 1*, we added a `one2many` `hostel_student_ids` field. This field will contain rooms records for the student. We will use it in the QWeb report to list the rooms that the student has reserved.

In *Step 2*, we defined the QWeb template. The content of the template will be used to generate the PDF. In our example, we used some basic HTML structure. We also used some attributes such as `t-esc` and `t-foreach`, which are used to generate dynamic content in the report. Don't worry about this syntax within the `template` element for now. This topic will be addressed extensively in the *Creating or modifying templates – QWeb* recipe in *Chapter 14*, *CMS Website Development*. Another important thing to notice in the template is the layout. In our example, we have used `web.internal_layout` in our template, which will generate the final PDF with a minimal header and footer. If you want an informative header and footer that uses the company logo and company information, use the `web.external_layout` layout. We also added one `for` loop to the `docs` parameter that will be used to generate a report for multiple records when the user prints it from the list view.

In *Step 3*, we declared the report in another XML file via the `<record>` tag. It will register the report's `ir.actions.report` model. The crucial part here is that you set the `report_name` field to the complete XML ID (that is, `modulename.record_id`) of the template you defined; otherwise, the report generation process will fail. The `model` field determines which type of record the report operates, and the `name` field is the name shown to the user in the print menu.

> **Note**
> In previous versions of Odoo, a `<report>` tag was used to register a report. But from version v14, it is deprecated and you need to create a record of `ir.actions.report` with the `<record>` tag. The `<report>` tag is still supported in Odoo v14 for backward compatibility but using it will show a warning in the log.

By setting `report_type` to `qweb-pdf`, we requested that the HTML generated by our view is run through `wkhtmltopdf` to deliver a PDF to the user. In some cases, you may want to use `qweb-html` to render the HTML within the browser.

There's more...

There are some marker classes in a report's HTML that are crucial for the layout. Ensure that you wrap all your content in an element with the `page` class set. If you forget that, you'll see nothing at all. To add a header or footer to your record, use the `header` or `footer` class.

Also, remember that this is HTML, so make use of *CSS attributes* such as `page-break-before`, `page-break-after`, and `page-break-inside`.

You'll have noted that all of our template body is wrapped in two elements with the `t-call` attribute set. We'll examine the mechanics of this attribute later in *Chapter 14, CMS Website Development*, but you must do the same in your reports. These elements ensure that the HTML generates links to all the necessary CSS files and contains some other data that is needed for report generation. While `web.html_container` doesn't have an alternative, the second `t-call` can be `web.external_layout`. The difference is that the external layout already comes with a header and footer displaying the company logo, the company's name, and some other information you expect from a company's external communication, while the internal layout just gives you a header with pagination, the print date, and the company's name. For the sake of consistency, always use one of the two.

> **Important note**
> Note that `web.internal_layout`, `web.external_layout`, `web.external_layout_header`, and `web.external_layout_footer` (the last two are called by the external layout) are just views by themselves, and you already know how to change them via inheritance. To inherit with the template element, use the `inherit_id` attribute.

Managing activities from a Kanban card

Odoo uses activities to schedule actions on records. These activities can be managed in the form view and the Kanban view. In this recipe, we will learn how to manage activities from the Kanban view card. We will add an activity widget to the cards of the rooms Kanban.

Getting started

For this recipe, we will be using the `my_hostel` module from the previous recipe.

How to do it...

Follow these steps to add and manage activity from the Kanban view:

1. Add mail dependencies to the `manifest` file:

   ```
   'depends': ['base', 'mail'],
   ```

2. Inherit an activity mixin in the `hostel.room` model:

   ```
   class HostelRoom(models.Model):

       _name = "hostel.room"
       _description = "Hostel Room Information"
       _rec_name = "room_no"
       _inherit = ['mail.thread', 'mail.activity.mixin']
   ```

3. Add the `activity_state` field to the Kanban view under the `color` field:

   ```
   <field name="color" />
   <field name="activity_state"/>
   ```

4. Add the `activity_ids` field inside the Kanban template. Add this field under the `popularity` field, as shown here:

   ```
   <div>
       <field name="popularity" widget="priority"/>
   </div>
   <div>
       <field name="activity_ids" widget="kanban_activity"/>
   </div>
   ```

   Update the `my_hostel` module to apply the change. Open the **Rooms** Kanban view; you will see the activity manager on the Kanban card, as shown in the following screenshot:

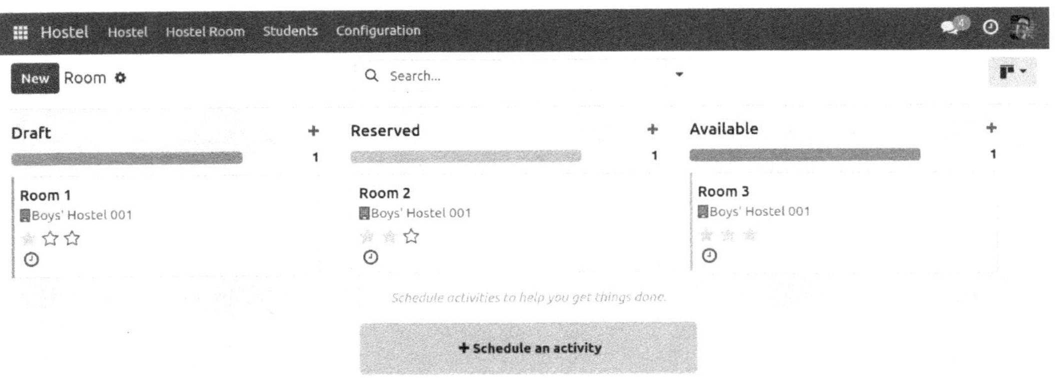

Figure 12.16 – Activity manager in a Kanban card

As you can see, after applying the code from this recipe, you will be able to manage activity from a Kanban card. You can now also process or create an activity from a Kanban card.

How it works...

In *Step 1*, we added a dependency to the manifest of our module. We did this because all the implementation associated with the activity is part of the `mail` module. Without installing `mail`, we cannot use activities in our model.

In *Step 2*, we added `activity mixin` to the `hostel.room` model. This will enable activities for the rooms records. Adding `mail.activity.mixin` will add all the fields and methods required for activities. We also added the `mail.thread` mixin because the activity logs the message when the user processes the activity. If you want to learn more about this activity, please refer to the *Managing activities on documents* recipe of *Chapter 23, Managing Emails in Odoo*.

In *Step 3*, we added the `activity_state` field to the Kanban view. This field is used by the activity widget to display the color widget. The color will represent the current state of the upcoming activity.

In *Step 4*, we added the activity widget itself. It uses the `activity_ids` field. In our example, we added the activity widget in a separate `<div>` tag, but you can put it anywhere according to your design requirements. With the activity widget, you can schedule, edit, and process the activity directly from the Kanban card.

There's more...

In the *Adding a progress bar in Kanban views* recipe of this chapter, we displayed a Kanban progress bar based on the `popularity` field. But we can also show a progress bar based on the state of the upcoming activity:

```
<progressbar field="activity_state"
    colors='{"planned": "success",
```

```
            "today": "warning",
            "overdue": "danger"}'/>
```

This will show the progress bar based on the state of the upcoming activity. A state-based progress bar is used in several views in Odoo.

See also

- If you want to learn more about the mail thread, refer to the *Managing chatter on documents* recipe of *Chapter 23, Managing Emails in Odoo*.
- If you want to learn more about activities, refer to the *Managing activities on documents* recipe of *Chapter 23, Managing Emails in Odoo*.

Adding a stat button to a form view

Odoo uses a stat button to relate two different objects visually on the form view. It is used to show some basic KPIs for related records. It is also used to redirect and open another view. In this recipe, we will add a stat button to the form view of a room. This stat button will display the count of room records and on clicking it, we will be redirected to the list of Kanban views.

Getting started

For this recipe, we will be using the `my_hostel` module from the previous recipe.

How to do it...

Follow these steps to add a stat button to the hostel's form view:

1. Add the `rooms_count` compute field to the `hostel.hostel` model. This field will count the number of active rooms in the hostel:

```
rooms_count = fields.Integer(compute="_compute_rooms_count")
def _compute_rooms_count(self):
        room_obj = self.env['hostel.room']
        for hostel in self:
            hostel.rooms_count = room_obj.search_
count([('hostel_id', '=', hostel.id)])
```

2. Add a stat button to the form view of the `hostel.hostel` model. Prepend it just inside the `<sheet>` tag:

```
<div class="oe_button_box" name="button_box">
    <button class="oe_stat_button" name="%(action_hostel_room)
d" type="action" icon="fa-building" context="{'search_default_
hostel_id': active_id}">
```

```
            <field string="Rooms" name="rooms_count"
    widget="statinfo"/>
        </button>
    </div>
```

Update the `my_hostel` module to apply the changes. Open the form view of any hostel; you will find the stat button, as shown in the following screenshot:

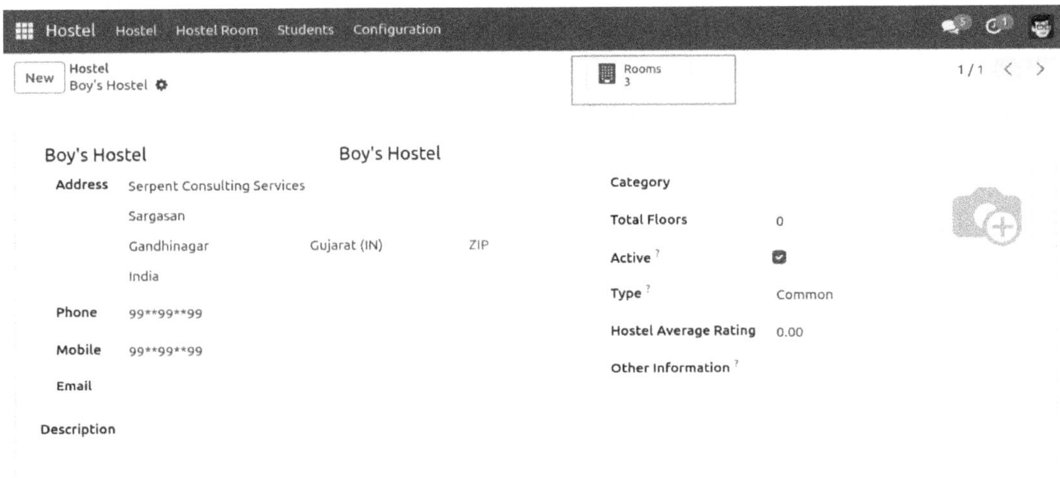

Figure 12.17 – Stat button in a hostel's form view

On clicking the stat button, you will be redirected to the **Rooms** Kanban view. Here, you will see orders only from the current hostel.

How it works...

In *Step 1*, we added a compute field that calculates the number of rooms records for the current hostel. The value of this field will be used for a stat button to show the count. If you want to learn more about compute, refer to the *Adding computed fields to a model* recipe in *Chapter 4, Application Models*.

In *Step 2*, we added the stat button in the form view of the `hostel.hostel` model. There are a specific syntax and location for the stat button. All the stat button needs to do is wrap under `<div>` with the `oe_button_box` class. The stat button box needs to be placed inside the `<sheet>` tag. Note that we used a `name` attribute on the button box. This `name` attribute is useful when you want to add a new stat, but then you will need to add a stat button with the `<button>` tag with the `oe_stat_button` class. Internally, the stat button is just a form view button with a different user interface. This means it supports all of the attributes that are supported by a normal button, such as an action, icon, and context.

In our example, we used the action of rooms orders, which means that when the user clicks on the stat button, they will be redirected to the rooms records but it will show all the rooms records. We only want to show the rooms records for the current room. To do so, we have to pass `search_default_hostel_id`. This will apply a default filter for the current room. Note that `hostel_id` is the `many2one` field on the `hostel.room` model. If you want to filter by another field, use it by prefixing it with `search_default_` in `context`.

Stat buttons are used often as they are very useful and show the overall statistics related to a record. You could use them to show all the information that relates to the current record. For example, on the contact record, Odoo shows stat buttons that show information related to the current contact total of the invoice, the number of leads, the number of orders, and so on.

See also

- To learn more about buttons, refer to the *Adding buttons to forms* recipe in *Chapter 9, Backend Views*.
- To learn more about actions, refer to the *Adding a menu item and window action* recipe in *Chapter 9, Backend Views*.

Enabling the archive option for records

Odoo provides inbuilt features to enable archive and unarchive options for records. This will help the user hide records that are no longer important. In this recipe, we will add an archive/unarchive option for a room. We can archive a room once it is not available.

Getting started

For this recipe, we will be using the `my_hostel` module from the previous recipe.

How to do it...

Archive and unarchive mostly work automatically. The options are available on a record if the model has a Boolean field named `active`. We already have an `active` field in the `hostel.room` model. But if you have not added it, follow these steps to add the `active` field:

1. Add an `active` Boolean field to the `hostel.room` model, like this:

```
active = fields.Boolean(default=True)
```

2. Add an `active` field to the form view:

```
<field name="active" invisible="1"/>
```

Update the `my_hostel` module to apply the changes. Now, you will be able to archive rooms. The **Archive** option is available in the **Action** dropdown, as shown in the following screenshot:

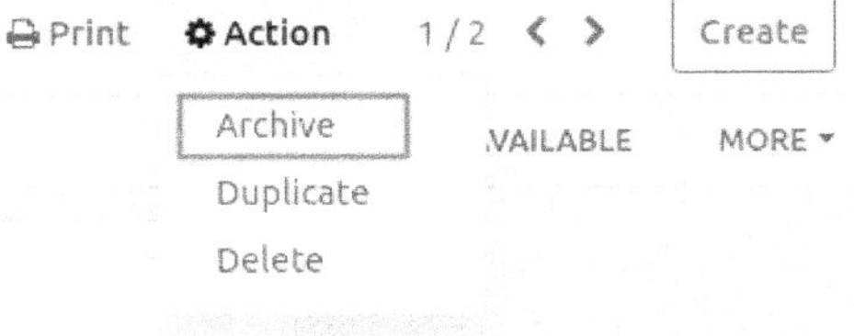

Figure 12.18 – Archive option on the form view

Once you archive a record, you'll want to see that record anywhere in Odoo. To see it, you need to apply a filter from the search view.

How it works...

A Boolean field named `active` has a special purpose in Odoo. If you add an `active` field to your model, records with a `false` value in the `active` field won't be displayed anywhere in Odoo.

In *Step 1*, we added an `active` field to the `hostel.room` model. Note that we kept the default value of `True` here. If we don't add this default value, the new records will be created in archive mode by default and won't be displayed in views, even if we have recently created them.

In *Step 2*, we added the `active` field in the form view. If you don't add an `active` field in the form view, the archive/unarchive option won't be displayed in the **Action** drop-down menu. If you don't want to show the field in the form view, you can use the `invisible` attribute to hide it from the form view.

In our example, once you archive a room, that room will not be displayed in the tree view or any other view. The room won't even be displayed in the `many2one` dropdown in the hostel record. If you want to unarchive that room, then you need to apply a filter to display archived records from the search view, and then restore the room.

There's more...

If your model has an `active` Boolean field, the `search` method will not return an archived record. If you want to search all the records, whether they are archived or not, then pass `active_test` in a context, like this:

```
self.env['hostel.room'].with_context(active_test=False).search([])
```

Note that if the archive record is linked to another record, it will be displayed in the related form view. For example, say you have *Room 1*. Then, you archive *Room 1*, which means from now on, you cannot select *Room 1* in the room. But if you open *Order 1*, you will see the archived *Room 1*.

13

Web Server Development

We'll introduce the basics of the web server part of Odoo in this chapter. Note that this will cover the fundamental aspects; for high-level functionality, you should refer to *Chapter 14, CMS Website Development*.

The Odoo web server is a crucial component of the Odoo framework, responsible for handling web requests and serving the web interface to users.

Here are key aspects of the Odoo web server:

- **The web interface and modules**: The web server provides a user-friendly web interface to access and interact with Odoo applications. Users can navigate through different modules, access data, and perform various business operations using this interface.
- **An HTTP server**: Odoo uses an HTTP server to handle web requests. It can be configured to work with popular web servers such as Nginx or Apache or can run its own built-in HTTP server.
- **Werkzeug**: Werkzeug is a **WSGI** (**Web Server Gateway Interface**) library for Python, and Odoo uses it to handle HTTP requests and responses. Werkzeug helps in routing requests, handling sessions, and managing other web-related tasks.
- **Controllers and routing**: Odoo uses controllers to handle different web requests and routes them to the appropriate controllers and methods. The routing mechanism ensures that requests are directed to the correct modules and functionalities.
- **Views and templates**: Odoo uses views and templates to define how data should be presented in the web interface. Views determine the structure of pages, and templates provide the HTML and presentation logic to render data.
- **Business logic**: The web server is tightly integrated with the business logic of Odoo. It communicates with the backend to fetch and update data, ensuring that the web interface reflects the most current state of the business applications.
- **Security**: Security is a critical aspect of the Odoo web server. It includes features such as authentication, authorization, and session management to ensure that users have appropriate access levels and that their interactions with the system are secure.

- **JavaScript and CSS**: The Odoo web interface relies on JavaScript and CSS to enhance user experience and provide dynamic and responsive features. This includes form validation, interactive elements, and real-time updates.
- **A RESTful API**: The web server also provides a RESTful API, allowing external applications to interact with Odoo programmatically. This enables integration with third-party systems and the development of custom applications.
- **Customization and extensions**: Developers can extend and customize the Odoo web server to meet specific business requirements. This includes creating custom modules, views, and controllers.

Understanding the Odoo web server is essential for developers and administrators working with Odoo to deploy, configure, and customize the system based on the unique needs of a business.

Werkzeug (`https://werkzeug.palletsprojects.com/en/2.3.x`) is a WSGI library for Python and is used by Odoo to handle HTTP requests and responses. WSGI is a specification for how web servers and web applications communicate in Python. Werkzeug provides a set of utilities and classes that make it easier to work with WSGI applications. Here are some details about how Werkzeug is used in the context of Odoo:

- **Request handling**: Werkzeug provides a `Request` object that represents an incoming HTTP request. In Odoo, this object is used to extract information from the incoming HTTP request, such as form data, query parameters, and headers.
- **Response generation**: The `Response` object in Werkzeug is used to create HTTP responses. Odoo utilizes this to construct and send responses back to the client, including rendering web pages or providing data in response to AJAX requests.
- **Routing**: Werkzeug enables easy URL routing. In Odoo, the routing mechanism is used to map incoming requests to the appropriate controller methods or views. This helps in directing requests to the correct functionality within the Odoo application.
- **Middleware**: Middleware components can be added to the Odoo application using Werkzeug. Middleware sits between the web server and the Odoo application and can perform tasks such as authentication, logging, or modifying requests and responses.
- **URL building**: Werkzeug provides a URL building facility that helps to generate URLs for different routes within the Odoo application. This is essential for creating links and redirects dynamically in the web interface.
- **Session management**: Werkzeug supports session management, which Odoo utilizes to handle user sessions. This is important for maintaining user state across multiple requests and ensuring security features such as user authentication.
- **Utilities for common tasks**: Werkzeug includes various utilities that simplify common web development tasks. Odoo leverages these utilities for tasks such as parsing form data, handling file uploads, and managing cookies.

- **Error handling**: Werkzeug provides mechanisms to handle errors, including HTTP error responses. Odoo uses this to ensure that appropriate error messages are returned to the client when needed.

To work with Werkzeug in the context of Odoo, developers often interact with these features through the controllers and views defined in Odoo modules. Understanding Werkzeug is beneficial for developers who want to extend or customize Odoo, as it provides insights into the underlying mechanisms to handle HTTP requests and responses within the application. However, in day-to-day Odoo development, developers often work at a higher level using the Odoo framework itself, without directly interacting with Werkzeug.

In this chapter, we'll cover the following topics:

- Making a path accessible from a network
- Restricting access to web-accessible paths
- Consuming parameters passed to your handlers
- Modifying an existing handler
- Serving static

Technical requirements

The technical requirements for this chapter include the online Odoo platform.

All the code used in this chapter can be downloaded from the GitHub repository at `https://github.com/PacktPublishing/Odoo-17-Development-Cookbook-Fifth-Edition/tree/main/Chapter13`.

Making a path accessible from a network

Making a path accessible from a network means defining the entry points or URLs through which users can access the application. This is fundamental to any web development project, as it determines how users will interact with the system. In this recipe, we'll look at how to make a URL of the `http://yourserver/path1/path2` form accessible to users. This can be either a web page or a path returning arbitrary data to be consumed by other programs. In the latter case, you would usually use the JSON format to consume parameters and offer your data.

Getting ready

We'll make use of the `hostel.student` model, which we looked at in *Chapter 4, Application Models*; therefore, if you haven't done so yet, grab the initial module from `https://github.com/PacktPublishing/Odoo-17-Development-Cookbook-Fifth-Edition/tree/main/Chapter13/00_initial_module` so that you will be able to follow the examples.

We want to allow any user to query the full list of students in the hostel. Furthermore, we want to provide the same information to programs through a JSON request. Let's check out how to do it.

How to do it...

We'll need to add controllers, which go into a folder called `controllers` by convention:

1. Add a `controllers/main.py` file with the HTML version of our page, as follows:

```
from odoo import http
from odoo.http import request
    class Main(http.Controller):
        @http.route('/my_hostel/students', type='http',
auth='none')
    def students(self):
        students = request.env['hostel.student'].sudo().
search([])
        html_result = '<html><body><ul>'
        for student in students:
            html_result += "<li> %s </li>" % student.name
        html_result += '</ul></body></html>'
        return html_result
```

2. Add a function to serve the same information in the JSON format, as shown in the following example:

```
@http.route('/my_hostel/students/json', type='json',
auth='none')
    def students_json(self):
        records = request.env['hostel.student'].sudo().
search([])
        return records.read(['name'])
```

3. Add the `controllers/__init__.py` file, as follows:

```
from . import main
```

4. Import `controllers` into your `my_hostel/__init__.py` file, as follows:

```
from . import controllers
```

After restarting your server, you can visit /my_hostel/students in your browser, and you'll be presented with a flat list of the student names.

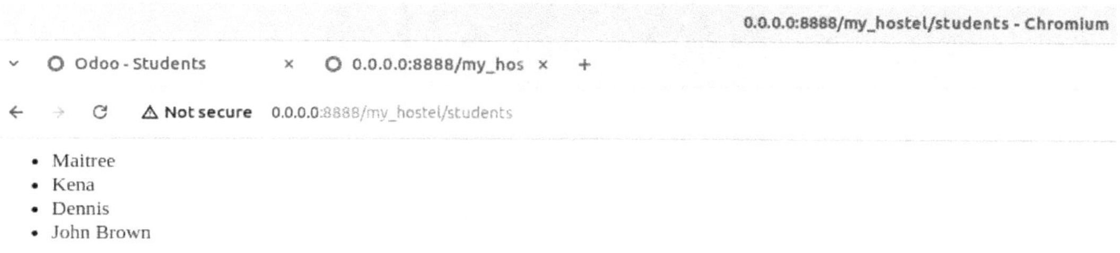

Figure 13.1 – A list of students

To test the JSON-RPC part, you'll have to craft a JSON request. A simple way to do that is by using the following command to receive the output on the command line:

```
curl -i -X POST -H "Content-Type: application/json" -d "{}"
localhost:8069/my_hostel/students/json
```

If you get 404 errors at this point, you probably have more than one database available on your instance. If so, it's impossible for Odoo to determine which database is meant to serve the request.

Use the --db-filter='^yourdatabasename$' parameter to force Odoo to use the exact database you installed the module in. The path should now be accessible.

How it works...

The two crucial parts here are that our controller is derived from odoo.http.Controller, and the methods we use to serve content are decorated with odoo.http.route. Inheriting from odoo.http.Controller registers the controller with Odoo's routing system in a similar way to how the models are registered – by inheriting from odoo.models.Model. Also, Controller has a metaclass that takes care of this.

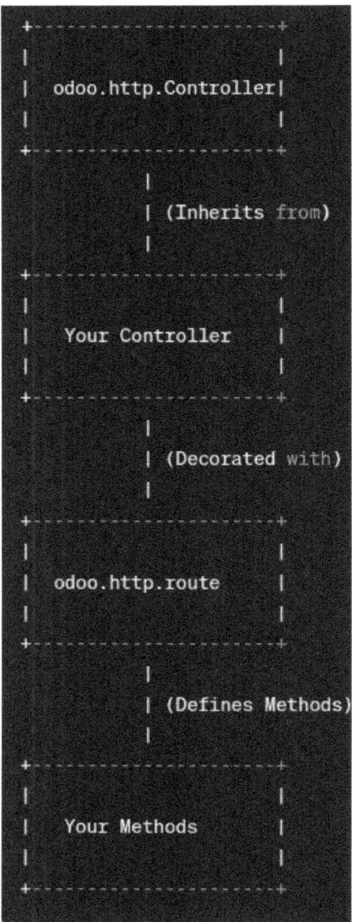

Figure 13.2 – A diagram of controllers

In general, paths handled by your add-on should start with your add-on's name, to avoid name clashes. Of course, if you extend some of the add-on's functionality, you'll use this add-on's name.

odoo.http.route

The `route` decorator allows us to tell Odoo that a method should be web-accessible in the first place, and the first parameter determines on which path it is accessible. Instead of a string, you can also pass a list of strings, if you use the same function to serve multiple paths.

The `type` argument defaults to `http` and determines what type of request will be served. While, strictly speaking, JSON is HTTP, declaring the second function as `type='json'` makes life a lot easier because Odoo then handles type conversions for us.

Don't worry about the `auth` parameter for now; it will be addressed in the *Restricting access to web-accessible paths* recipe in this chapter.

Return values

Odoo's treatment of the functions' return values is determined by the `type` argument of the `route` decorator. For `type='http'`, we usually want to deliver some HTML, so the first function simply returns a string containing it. An alternative is to use `request.make_response()`, which gives you control over the headers to send in the response. So, to indicate when our page was last updated, we might change the last line in `students()` to the following code:

```
return request.make_response(
  html_result, headers=[
    ('Last-modified', email.utils.formatdate(
      (
        fields.Datetime.from_string(
        request.env['hostel.student'].sudo()
        .search([], order='write_date desc', limit=1)
        .write_date) -
        datetime.datetime(1970, 1, 1)
      ).total_seconds(),
      usegmt=True)),
])
```

This code sends a `Last-modified` header along with the HTML we generated, telling the browser when the list was modified for the last time. We can extract this information from the `write_date` field of the `hostel.student` model.

In order for the preceding snippet to work, you'll have to add some imports at the top of the file, as follows:

```
import email
import datetime
from odoo import fields
```

You can also create a `Response` object of `werkzeug` manually and return that, but there's little to gain for the effort.

> **Important information**
>
> Generating HTML manually is nice for demonstration purposes, but you should never do this in production code. Always use templates, as demonstrated in the *Creating or modifying templates – QWeb* recipe in *Chapter 15, Web Client Development*, and return them by calling `request.render()`. This will give you localization for free and will make your code better by separating business logic from the presentation layer. Also, templates provide you with functions to escape data before outputting HTML. The preceding code is vulnerable to cross-site scripting attacks (if a user manages to slip a `script` tag into the book name, for example).

For a JSON request, simply return the data structure you want to hand over to the client; Odoo takes care of serialization. For this to work, you should restrict yourself to data types that are JSON-serializable, which generally means dictionaries, lists, strings, floats, and integers.

odoo.http.request

The `request` object is a static object referring to the currently handled request, which contains everything you need in order to take action. The most important aspect here is the `request.env` property, which contains an `Environment` object that is just the same as `self.env` for models. This environment is bound to the current user, which is not in the preceding example, because we used `auth='none'`. The lack of a user is also why we have to `sudo()` all our calls to model methods in the example code.

If you're used to web development, you'll expect session handling, which is perfectly correct. Use `request.session` for an `OpenERPSession` object (which is quite a thin wrapper around the `Session` object of `werkzeug`) and `request.session.sid` to access the session ID. To store session values, just treat `request.session` as a dictionary, as shown in the following example:

```
request.session['hello'] = 'world'
request.session.get('hello')
```

> **Important note**
>
> Note that storing data in the session is no different from using global variables. Only do so if you must. This is usually the case for multi-request actions, such as a checkout in the `website_sale` module.

There's more...

The `route` decorator can have some extra parameters in order to customize its behavior further. By default, all HTTP methods are allowed, and Odoo intermingles the parameters passed. Using the `methods` parameter, you can pass a list of methods to accept, which would usually be one of either `['GET']` or `['POST']`.

To allow cross-origin requests (browsers block AJAX and some other types of requests to domains other than where the script was loaded from, for security and privacy reasons), set the `cors` parameter to `*` to allow requests from all origins, or a URI to restrict requests to ones originating from that URI. If this parameter is unset, which is the default, the `Access-Control-Allow-Origin` header is not set, leaving you with the browser's standard behavior. In our example, we might want to set it on

`/my_module/students/json` in order to allow scripts pulled from other websites to access the list of students.

By default, Odoo protects certain types of requests from an attack known as cross-site request forgery, by passing a token along on every request. If you want to turn that off, set the `csrf` parameter to `False`, but note that this is a bad idea, in general.

See also

Refer to the following points to learn more about the HTTP routes:

- If you host multiple Odoo databases on the same instance, then different databases might be running on different domains. If so, you can use the `--db-filter` options, or you can use the `dbfilter_from_header` module from `https://github.com/OCA/server-tools`, which helps you filter databases based on the domain.
- To see how using templates makes modularity possible, check out the *Modifying an existing handler* recipe later in the chapter.

Restricting access to web-accessible paths

As an Odoo developer, one of your primary concerns is ensuring the security of the application. Restricting access to web-accessible paths is a crucial aspect of access control. It involves determining who can or cannot access specific routes or functionalities within the Odoo application. Odoo provides different authentication mechanisms to control user access. Understanding and implementing these mechanisms is essential for ensuring that only authorized users can interact with sensitive or protected parts of the application. For instance, you might want to restrict certain routes to authenticated users only.

We'll explore the three authentication mechanisms Odoo provides for routes in this recipe. We'll define routes with different authentication mechanisms in order to show their differences.

Getting ready

As we extend the code from the previous recipe, we'll also depend on the `hostel.student` model of *Chapter 4*, *Application Models*, so you should retrieve its code in order to proceed.

How to do it...

Define the handlers in `controllers/main.py`:

1. Add a path that shows all the students, as shown in the following example:

```
    @http.route('/my_hostel/all-students', type='http',
auth='none')
    def all_students(self):
        students = request.env['hostel.student'].sudo().
search([])
        html_result = '<html><body><ul>'
        for student in students:
            html_result += "<li> %s </li>" % student.name
        html_result += '</ul></body></html>'
        return html_result
```

2. Add a path that shows all the students and indicates which ones belong to the current user, if any. This is shown in the following example:

```
@http.route('/my_hostel/all-students/mark-mine', type='http',
auth='public')
    def all_students_mark_mine(self):
        students = request.env['hostel.student'].sudo().
search([])
        hostels = request.env['hostel.hostel'].sudo().
search([('rector', '=', request.env.user.partner_id.id)])
        hostel_rooms = request.env['hostel.room'].sudo().
search([('hostel_id', 'in', hostels.ids)])
        html_result = '<html><body><ul>'
        for student in students:
            if student.id in hostel_rooms.student_ids.ids:
                html_result += "<li> <b>%s</b> </li>" % student.
name
            else:
                html_result += "<li> %s </li>" % student.name
        html_result += '</ul></body></html>'
        return html_result
```

3. Add a path that shows the current user's students, as follows:

```
@http.route('/my_hostel/all-students/mine', type='http',
auth='user')
    def all_students_mine(self):
        hostels = request.env['hostel.hostel'].sudo().
search([('rector', '=', request.env.user.partner_id.id)])
        students = request.env['hostel.room'].sudo().
search([('hostel_id', 'in', hostels.ids)]).student_ids
        html_result = '<html><body><ul>'
        for student in students:
            html_result += "<li> %s </li>" % student.name
        html_result += '</ul></body></html>'
        return html_result
```

With this code, the `/my_hostel/all-students` and `/my_hostel/all-students/mark-mine` paths look the same for unauthenticated users, while a logged-in user sees their students in a bold font on the latter path.

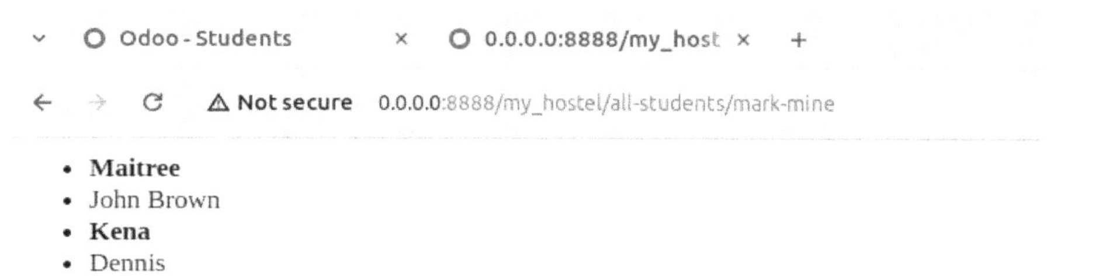

Figure 13.3 – Mark as mine students – with login

The following screenshot shows the results without login:

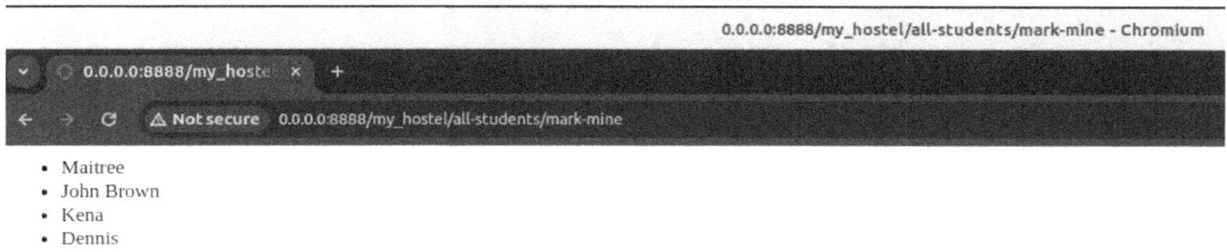

Figure 13.4 – Mark as mine students – without login

The `/my_hostel/all-students/mine` path is not accessible at all for unauthenticated users. If you try to access it without being authenticated, you'll be redirected to the login screen in order to do so.

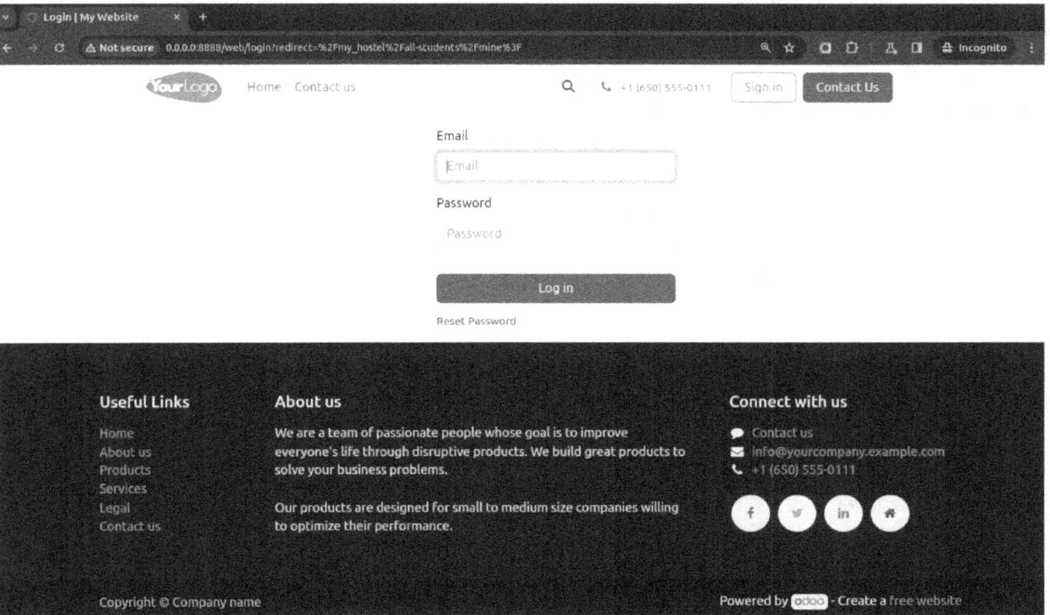

Figure 13.5 – Access via unauthenticated users

How it works...

The difference between authentication methods is basically what you can expect from the content of `request.env.user`.

For `auth='none'`, the user record is always empty, even if an authenticated user accesses the path. Use this if you want to serve content that has no dependencies on users, or if you want to provide database-agnostic functionality in a server-wide module.

The `auth='public'` value sets the user record as a special user with an XML ID of `base.public_user` for unauthenticated users, and to the user's record for authenticated ones. This is the right choice if you want to offer functionality to both unauthenticated and authenticated users, while the authenticated ones get some extras, as demonstrated in the preceding code.

Use `auth='user'` to ensure that only authenticated users have access to what you've got to offer. With this method, you can be sure that `request.env.user` points to an existing user.

There's more...

The magic of authentication methods happens in the `ir.http` model from the base add-on. For whatever value you pass to the `auth` parameter in your route, Odoo searches for a function called `_auth_method_<yourvalue>` on this model, so you can easily customize it by inheriting it and declaring a method that takes care of your authentication method of choice.

As an example, we will provide an authentication method called `base_group_user`, which will only authorize the user if the currently logged-in user is part of the `base.group_user` group, as shown in the following example:

```
from odoo import exceptions, http, models
from odoo.http import request
class IrHttp(models.Model):
  _inherit = 'ir.http'
  def _auth_method_base_group_user(self):
    self._auth_method_user()
    if not request.env.user.has_group('base.group_user'):
      raise exceptions.AccessDenied()
```

Now, you can say `auth='base_group_user'` in your decorator and be sure that users running this route's handler are members of the group. With a little trickery, you can extend this to `auth='groups(xmlid1,...)'`; its implementation is left as an exercise for you but is included in the GitHub repository example code at `Chapter13/02_paths_auth/my_hostel/models/sample_auth_http.py`.

Consuming parameters passed to your handlers

It's nice to be able to show content, but it's better to show content as a result of user input. This recipe will demonstrate the different ways to receive this input and react to it. As in the previous recipes, we'll make use of the `hostel.student` model.

How to do it...

First, we'll add a route that expects a traditional parameter with a student's ID to show some details about them. Then, we'll do the same again, but we'll incorporate our parameter into the path itself:

1. Add a path that expects a student's ID as a parameter, as shown in the following example:

```
        @http.route('/my_hostel/student_details', type='http',
auth='none')
        def student_details(self, student_id):
                record = request.env['hostel.student'].sudo().
browse(int(student_id))
                return u'<html><body><h1>%s</h1>Room No: %s' % (
                        record.name, str(record.room_id.room_no)
or 'none')
```

2. Add a path where we can pass the student's ID in the path, as follows:

```
@http.route("/my_hostel/student_details/<model('hostel.
student'):student>",
            type='http', auth='none')
    def student_details_in_path(self, book):
        return self.student_details(student.id)
```

If you point your browser to `/my_hostel/student_details?student_id=1`, you should see a detailed page of the student with ID 1.

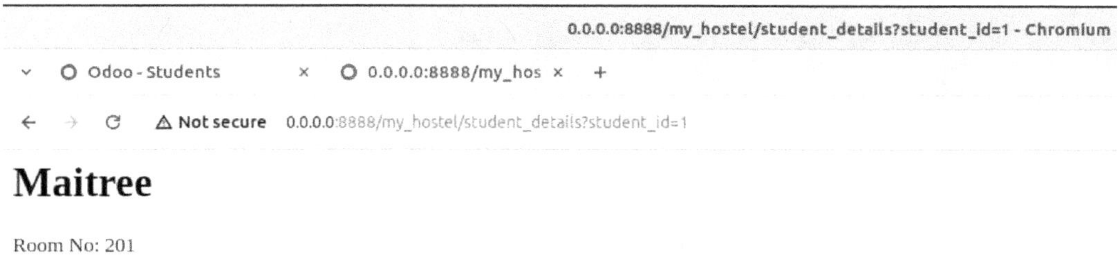

Figure 13.6 – The student details web page

If this doesn't exist, you'll receive an error page.

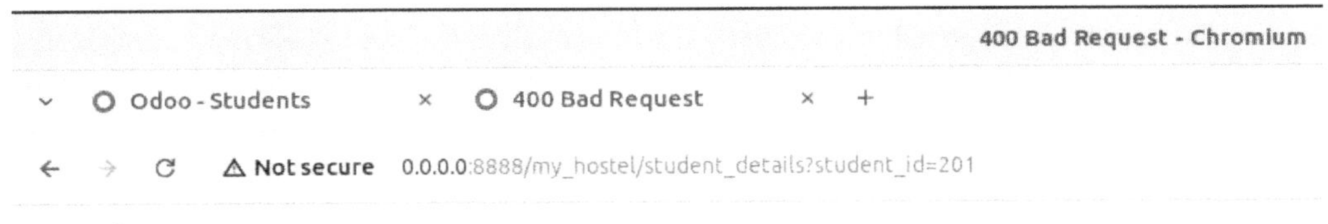

Bad Request

Record does not exist or has been deleted.
(Record: hostel.student(201,), User: None)

Figure 13.7 – The student not found:Error page

How it works...

By default, Odoo (actually, `werkzeug`) intermingles the `GET` and `POST` parameters and passes them as keyword arguments to your handler. So, by simply declaring your function as expecting a parameter called `student_id`, you introduce this parameter as either `GET` (the parameter in the URL) or `POST` (usually passed by the `<form>` element with your handler as the `action` attribute). Given that we didn't add a default value for this parameter, the runtime will raise an error if you try to access this path without setting the parameter.

The second example makes use of the fact that, in a `werkzeug` environment, most paths are virtual anyway. So, we can simply define our path as containing some input. In this case, we say that we expect the ID of a `hostel.student` instance as the last component of the path. The name after the colon is the name of a keyword argument. Our function will be called, with this parameter passed as a keyword argument. Here, Odoo takes care of looking up this ID and delivering a browse record, which, of course, only works if the user accessing this path has appropriate permissions. Given that `student` is a browse record, we can simply recycle the first example's function by passing `student.id` as a `student_id` parameter, outputing the same content.

There's more...

Defining parameters within a path is a functionality delivered by `werkzeug`, called `converters`. The `model` converter is added by Odoo, which also defines the converter models that accept a comma-separated list of IDs and pass a recordset containing those IDs to your handler.

The beauty of converters is that the runtime coerces parameters to the expected type, whereas you're on your own with normal keyword parameters. These are delivered as strings, and you have to take care of the necessary type conversions yourself, as seen in the first example.

Built-in `werkzeug` converters include not only `int`, `float`, and `string` but also more intricate ones, such as `path`, `any`, and `uuid`. You can look up their semantics at `https://werkzeug.palletsprojects.com/en/2.3.x/`.

See also

If you want to learn more about the HTTP routes, refer to the following points:

- Odoo's custom converters are defined in `ir_http.py` in the base module and registered in the `_get_converters` class method of `ir.http`
- If you want to learn more about the form submission on the route, refer to the *Getting input from users* recipe from *Chapter 14, CMS Website Development*

Modifying an existing handler

When you install the website module, the `/website/info` path displays some information about your Odoo instance. In this recipe, we will override this in order to change this information page's layout, as well as to change what is displayed.

Getting ready

Install the `website` module and inspect the `/website/info` path. In this recipe, we will update the `/website/info` route to provide more information.

How to do it...

We'll have to adapt the existing template and override the existing handler. We can do this as follows:

1. Override the qweb template in a file called `views/templates.xml`, as follows:

```
<?xml version="1.0"?>
<odoo>
  <template id="show_website_info"
            inherit_id="website.show_website_info">
   <xpath expr="//dl[@t-foreach='apps']" position="replace">
            <table class="table">
                <tr t-foreach="apps" t-as="app">
                    <th>
                        <a t-att-href="app.website"
groups='base.group_no_one'>
                            <t t-out="app.name" />
                        </a>
                    </th>
                    <td>
                        <span t-out="app.summary" />
                    </td>
                </tr>
            </table>
```

```
        </xpath>
    </template>
</odoo>
```

2. Override the handler in a file called `controllers/main.py`, as shown in the following example:

```
from odoo import http
from odoo.addons.website.controllers.main import Website
class WebsiteInfo(Website):
    @http.route()
    def website_info(self):
        result = super(WebsiteInfo, self).website_info()
        result.qcontext['apps'] = result.qcontext['apps'].
filtered(
            lambda x: x.name != 'website'
        )
        return result
```

3. Now, when visiting the info page, we'll only see a filtered list of installed applications in a table, as opposed to the original definition list.

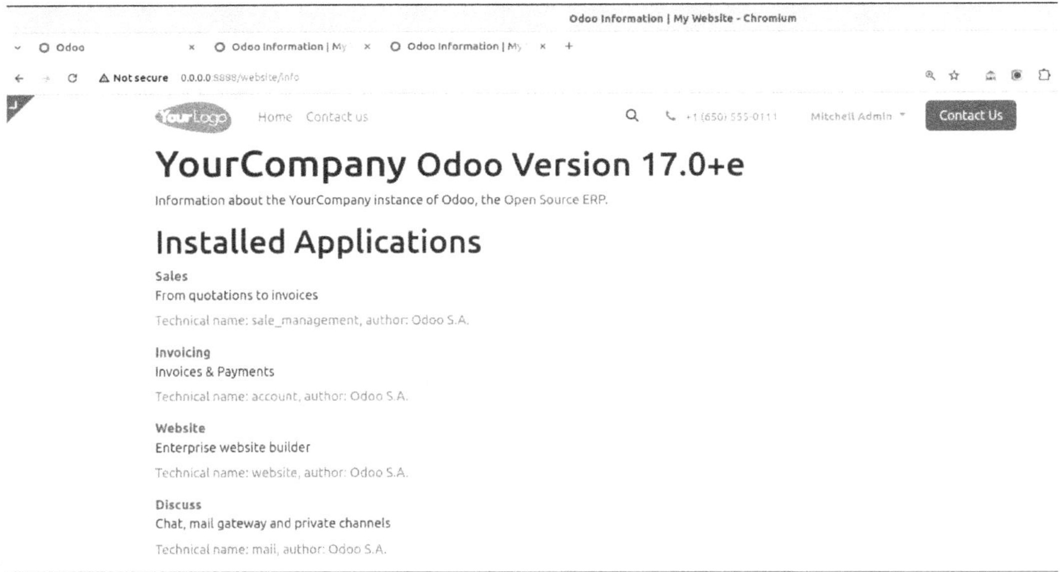

Figure 13.8 – The website info page (original)

The following screenshot shows the customized page:

Figure 13.9 – The website info page (customized)

How it works...

In the first step, we overrode an existing QWeb template. In order to find out which one it is, you'll have to consult the code of the original handler. Usually, this will give you something similar to the following line, which tells you that you need to override `template.name`:

```
return request.render('template.name', values)
```

In our case, the handler used a template called `website_info`, but this one was immediately extended by another template called `website.show_website_info`, so it's more convenient to override this one. Here, we replaced the definition list showing installed apps with a table. For details on how QWeb inheritance works, consult *Chapter 15, Web Client Development*.

In order to override the handler method, we must identify the class that defines the handler, which is `odoo.addons.website.controllers.main.Website` in this case. We need to import the class to be able to inherit from it. Now, we can override the method and change the data passed to the response. Note that what the overridden handler returns here is a `Response` object and not a string of HTML, as the previous recipes did, for the sake of brevity. This object contains a reference to the template to be used and the values accessible to the template, but it is only evaluated at the very end of the request.

In general, there are three ways to change an existing handler:

- If it uses a QWeb template, the simplest way to change it is to override the template. This is the right choice for layout changes and small logic changes.

- QWeb templates get a context passed, which is available in the response as the `qcontext` member. This is usually a dictionary where you can add or remove values to suit your needs. In the preceding example, we filtered the list of apps to the website only.
- If the handler receives parameters, you can also preprocess those in order to make the overridden handler behave in the way you want.

There's more...

As seen in the preceding section, inheritance with controllers works slightly differently than model inheritance; you actually need a reference to the base class and to use Python inheritance on it.

Don't forget to decorate your new handler with the `@http.route` decorator; Odoo uses it as a marker, for which methods are exposed to the network layer. If you omit the decorator, you actually make the handler's path inaccessible.

The `@http.route` decorator itself behaves similarly to field declarations – every value you don't set will be derived from the decorator of the function you're overriding, so we don't have to repeat values we don't want to change.

After receiving a `response` object from the function you override, you can do a lot more than just change the QWeb context:

- You can add or remove HTTP headers by manipulating `response.headers`.
- If you want to render an entirely different template, you can overwrite `response.template`.
- To detect whether a response is based on QWeb in the first place, query `response.is_qweb`.
- The resulting HTML code is available by calling `response.render()`.

See also

- Details on QWeb templates will be given in *Chapter 15*, *Web Client Development*.

Serving static resources

Web pages contain several types of static resources, such as images, videos, CSS, and so on. In this recipe, we will see how you can manage such static resources for your module.

Getting ready

For this recipe, we will display an image on the page. Grab the `my_hostel` module from the previous recipe. Select any image from your system and put that image inside the `/my_hostel/static/src/img` directory.

How to do it...

Follow these steps to show an image on the web page:

1. Add your image to the `/my_hostel/static/src/img` directory.
2. Define the new route in `controller`. In the code, replace the image URL with the URL of your image:

```
@http.route('/demo_page', type='http', auth='none')
def students(self):
        image_url = '/my_hostel/static/src/image/odoo.
png'
        html_result = """<html>
                <body>
                <img src="%s"/>
                </body>
            </html>""" % image_url
return html_result
```

3. Restart the server and update the module to apply the changes. Now, visit `/demo_page` to see the image on the page.

Figure 13.10 – The static image on the web page

How it works...

All the files placed under the `/static` folder are considered static resources and are publicly accessible. In our example, we have put our image in the `/static/src/img` directory. You can place the static resource anywhere under the `static` directory, but there is a recommended directory structure based on the type of file:

- `/static/src/img` is the directory used for images
- `/static/src/css` is the directory used for CSS files
- `/static/src/scss` is the directory used for SCSS files
- `/static/src/fonts` is the directory used for font files
- `/static/src/js` is the directory used for JavaScript files
- `/static/src/xml` is the directory used for XML files for client-side QWeb templates
- `/static/lib` is the directory used for files of external libraries

In our example, we displayed an image on the page. You can also access the image directly from `/my_hostel/static/src/image/odoo.png`.

In this recipe, we displayed a static resource (an image) on the web page, and we saw the recommended directories for different static resources. There are more simple ways to present page content and static resources, which we will see in the next chapter.

14

CMS Website Development

Odoo has a built-in feature called Website Builder, which is a powerful tool that allows you to create and manage websites within the Odoo ERP ecosystem. It offers a user-friendly and visual approach to web design, making it accessible to users without extensive technical knowledge.

Here are some key features and aspects of Odoo Website Builder:

- **Drag-and-drop interface**: Website Builder provides a drag-and-drop interface, allowing you to easily add and arrange various content elements on your web pages. This includes text, images, videos, forms, buttons, and more.
- **Pre-designed templates**: Odoo offers a selection of pre-designed website templates that you can use as a starting point. These templates are customizable and can be adapted to your brand's identity.
- **Responsive design**: Websites created with Odoo are designed to be responsive, which means they automatically adapt to different screen sizes and devices, ensuring a consistent user experience on desktops, tablets, and smartphones.
- **Content management**: You can create and manage web pages, blogs, product listings, and other types of content easily. Website Builder provides a **content management system** (**CMS**) to organize and update your content.
- **Search engine optimization (SEO)**: Odoo includes tools for SEO, allowing you to set metadata, define SEO-friendly URLs, and manage sitemaps to improve your website's visibility in search engines.
- **Multilingual support**: Odoo supports multiple languages, making it suitable for businesses with international audiences. You can translate content and adapt your website for different regions.
- **Integration with other Odoo modules**: One of the advantages of using Odoo Website Builder is its seamless integration with other Odoo modules, such as CRM, sales, inventory, and more. This means you can manage various aspects of your business within a unified system.
- **Analytics and reporting**: Odoo provides built-in analytics and reporting tools to track the performance of your website, including visitor statistics, conversion rates, and more.

- **Custom development**: For businesses with unique requirements, Odoo's modular architecture allows for custom development to extend the platform's functionality.

In this chapter, you will explore the developments of the Odoo website's custom features and learn how to create web pages. You will also learn how to create building blocks that users can drag and drop on a page. Advanced things such as **Urchin Tracking Modules** (**UTMs**), SEO, multi-websites, GeoIP, and sitemaps are also covered in this chapter.

In this chapter, we will cover the following recipes:

- Managing assets
- Adding CSS and JavaScript for a website
- Creating or modifying templates
- Managing dynamic routes
- Offering static snippets to the user
- Offering dynamic snippets to the user
- Getting input from website users
- Managing SEO options
- Managing sitemaps for the website
- Getting a visitor's country information
- Tracking a marketing campaign
- Managing multiple websites
- Redirecting old URLs
- Publish management for website-related records

Managing assets

In the context of Odoo's website, assets refer to various types of resources, such as **Cascading Style Sheets** (**CSS**), JavaScript files, fonts, and images, that are used to enhance the appearance and functionality of your website. Managing assets in Odoo is important for maintaining a well-structured and efficient website. When a page is loaded in the browser, these static files make a separate request to the server. The higher the number of requests, the lower the website speed. To avoid this issue, most websites serve static assets by combining multiple files. There are several tools on the market to manage these sorts of things, but Odoo has its own implementation for managing static assets.

What are asset bundles and different assets in Odoo?

In Odoo, asset bundles are collections of different assets, such as CSS, JavaScript files, and other resources, grouped together for efficient and organized loading on your website. Asset bundles help manage the loading of these resources by allowing you to define which assets should be loaded together to improve performance and ensure that your website functions properly. The job of an asset bundle is to combine all the JavaScript and CSS in a single file and reduce its size by minimizing it.

Here are the different asset bundles used in Odoo:

- `web._assets_primary_variables`
- `web._assets_secondary_variables`
- `web.assets_backend`
- `web.assets_frontend`
- `web.assets_frontend_minimal`
- `web.assets_frontend_lazy`
- `web.report_assets_common`
- `web.report_assets_pdf`
- `web.assets_web_dark`
- `web._assets_frontend_helpers`
- `web_editor.assets_wysiwyg`
- `website.assets_wysiwyg`
- `website.assets_editor`

> **Important information**
> There are some other asset bundles used for specific applications;
>
> for example, `point_of_sale.assets`, `survey.survey_assets`, `mass_mailing.layout`, and `website_slides.slide_embed_assets`.

Odoo manages its static assets through the `AssetBundle` class, which is located at `/odoo/addons/base/models/assetsbundle.py`.

Now, `AssetBundle` not only combines multiple files; it is also packed with more features. Here is the list of features it provides:

- It combines multiple JavaScript and CSS files.
- It minifies the JavaScript and CSS files by removing comments, extra spaces, and carriage returns from the file content. Removing this extra data will reduce the size of static assets and improve the page loading speed.

- It has built-in support for CSS preprocessors, such as **Sassy CSS** (**SCSS**) and **Leaner Style Sheets** (**LESS**). This means you can add SCSS and LESS files and they will automatically be compiled and added to the bundle.

Custom assets

As we have seen, Odoo has different assets for different code bases. To get the right result, you will need to choose the right asset bundle in which to place your custom JavaScript and CSS files. For example, if you are designing a website, you need to put your file in `web.assets_frontend`. Although it is rare, sometimes, you need to create a whole new asset bundle. You can create your own asset bundle, as we will describe in the following section.

How to do it...

To load assets, you can use the `web.assets_frontend` template in your module's `__manifest__.py` file; for example:

```
'assets': {
    'web.assets_backend': [
        'my_hostel/static/src/xml/**/*',
    ],
    'web.assets_frontend: [
        'my_hostel/static/lib/bootstrap/**/*',
        'my_hostel/static/src/js/**',
        'my_hostel/static/src/scss/**',
    ],
},
```

Here are some of the most important bundles:

- `web.assets_common`
- `web.assets_backend`
- `web.assets_frontend`
- `web.qunit_suite_tests`
- `web.qunit_mobile_suite_tests`

Operations

Here are all directives targeting a certain asset file:

- `before`
- `after`

- `replace`
- `remove`

append

An appending assets operation refers to adding additional CSS or JavaScript files to existing bundles or templates provided by other modules or the Odoo core. This allows you to extend the functionality or appearance without modifying the original code directly; for example:

```
'web.assets_common': [
    'my_hostel/static/src/js/**/*',
],
```

Always consider the sequence in which your assets are loaded. If your code depends on any specific libraries or functionalities defined in other assets, ensure they are loaded in the correct order to avoid conflicts or errors.

prepend

Prepending assets in Odoo involves adding your own CSS or JavaScript files at the beginning of existing bundles or templates provided by other modules or the Odoo core. This helps ensure that your customizations take precedence over existing styles or scripts; for example:

```
'web.assets_common': [
    ('prepend','my_hostel/static/src/css/bootstrap_overridden.scss'),
],
```

Determine the sequence in which your assets are loaded. Prepending assets means they'll be loaded before other styles or scripts, potentially impacting functionality or design. Be cautious with overriding core functionalities.

before

In Odoo, organizing assets such as CSS or JavaScript files before other modules' assets involves controlling the loading order of resources to ensure your module's files are loaded before those of other modules; for example:

```
'web.assets_common': [
    ('before', 'web/static/src/css/bootstrap_overridden.scss',    'my_
hostel/static/src/css/bootstrap_overridden.scss'),
],
```

Ensure that you're referring to the correct assets or templates of modules you want to load your resources beforehand. Incorrect referencing might lead to errors or unexpected behavior.

after

In Odoo, organizing assets such as CSS or JavaScript files to load after other modules' assets involves controlling the loading order to ensure your module's files are loaded after those of other modules. This is useful when you need your assets to rely on or override styles or scripts from other modules; for example:

```
'web.assets_common': [
    ('after', 'web/static/src/css/list_view.scss', 'my_hostel/static/
src/css/list_view.scss'),
],
```

Controlling the loading sequence using the `after` attribute or Python code helps ensure your module's assets are loaded after other modules, enabling you to manage dependencies and customizations effectively.

include

In Odoo, including assets such as CSS or JavaScript files involves linking these resources to your module or theme to enhance its functionality or appearance; for example:

```
'web.assets_common': [
    ('include', 'web._primary_variables'),
],
```

Including assets in Odoo allows you to extend your module's capabilities by adding custom styles or scripts, enhancing the user experience and functionality.

remove

Remove one or multiple file(s).

Removing assets, such as CSS or JavaScript files, in Odoo involves excluding them from your module's assets; for example:

```
'web.assets_common': [
    ('remove', 'web/static/src/js/boot.js'),
],
```

Removing assets in Odoo enables you to customize your module by excluding specific styles or scripts that are not needed or conflict with your module's functionalities.

replace

In Odoo, replacing assets involves substituting existing CSS or JavaScript files with new ones in your module or theme; for example:

```
'web.assets_common': [
    ('replace', 'web/static/src/js/boot.js', 'my_addon/static/src/js/
```

```
boot.js'),
],
```

Replacing assets in Odoo allows you to update and customize the appearance or functionality of your module by substituting existing files with new ones. Be cautious when replacing assets to maintain the stability and functionality of your application.

Loading order

In Odoo, managing the loading order of assets (CSS, JavaScript, and so on) is crucial for ensuring that dependencies are resolved correctly and that the user interface renders properly. The loading order can be controlled to determine which assets are loaded first or after others; for example:

```
'web.assets_common': [
    'my_addon/static/lib/jquery/jquery.js',
    'my_addon/static/lib/jquery/**/*',
],
```

When an asset bundle is called (for example, `t-call-assets="web.assets_common"`), an empty list of assets is generated.

All records of type `ir.asset` matching the bundle are fetched and sorted by sequence number. Then, all records with a sequence strictly less than 16 are processed and applied to the current list of assets.

All modules declaring assets for the said bundle in their manifest apply their assets' operations to this list. This is done following the order of module dependencies (for example, web assets are processed before the website). If a directive tries to add a file already present in the list, nothing is done for that file. In other words, only the first occurrence of a file is kept in the list.

The remaining `ir.asset` records (those with a sequence greater than or equal to 16) are then processed and applied as well.

Assets declared in the manifest may need to be loaded in a particular order; for example, `jquery.js` must be loaded before all other `jquery` scripts when loading the `lib` folder. One solution would be to create an `ir.asset` record with a lower sequence or a `prepend` directive, but there is another, simpler way to do so.

There's more...

The following are a few things you need to know if you are working with assets in Odoo.

Debugging JavaScript can be very hard in Odoo because `AssetBundle` merges multiple JavaScript files into a single file and also minifies them. By enabling developer mode with assets, you can skip asset bundling, and the page will load static assets separately so that you can debug easily.

Combined assets are generated once and stored in the `ir.attachment` model. After that, they are served from the attachment. If you want to regenerate assets, you can do so from the debug options, as shown in the following screenshot:

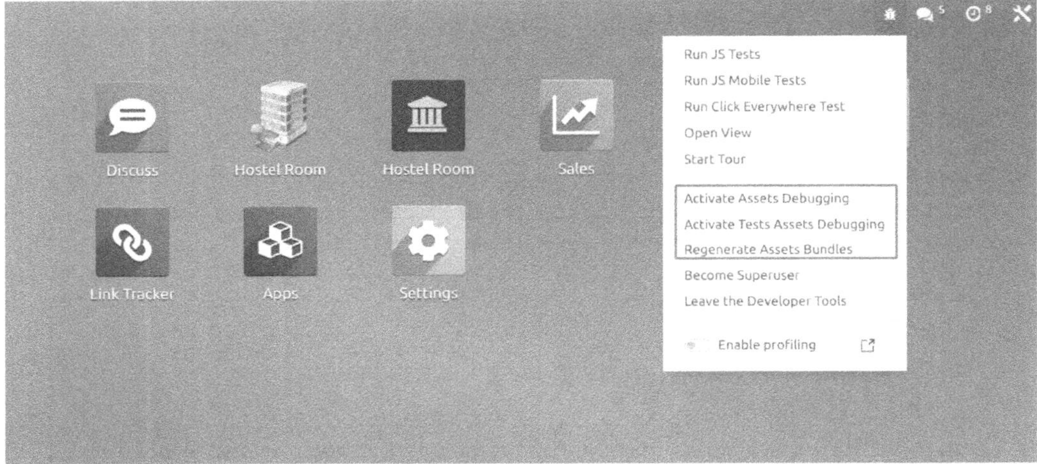

Figure 14.1 – Screenshot of assets' activation options

> **Tip**
>
> As you know, Odoo will generate an asset only once. This behavior can be a headache during development, as it requires frequent server restarts. To overcome this issue, you can use `dev=xml` in the command line, which will load assets directly, so there will be no need for a server restart.

Lazy loading

Lazy loading is a technique that defers loading non-essential resources until they're needed, often used for images, scripts, or other assets to improve performance:

```
await loadAssets({
    jsLibs: ["/web/static/lib/stacktrace-js/stacktrace.js"],
});
```

However, implementing lazy loading for specific assets or components in Odoo can be achieved through custom development or by utilizing third-party libraries. Here are some approaches you might consider.

Lazy loading images

You can implement lazy loading for images using JavaScript libraries such as Intersection Observer. This library allows you to load images only when they enter the user's viewport.

1. Using Intersection Observer, JavaScript code could look like this:

```
document.addEventListener("DOMContentLoaded", function () {
    var lazyImages = [].slice.call(document.
querySelectorAll("img.lazy"));

    if ("IntersectionObserver" in window) {
        let lazyImageObserver = new \
        IntersectionObserver(function (entries, \
        observer) {
            entries.forEach(function (entry) {
                if (entry.isIntersecting) {
                    let lazyImage = entry.target;
                    lazyImage.src = \
                    lazyImage.dataset.src;
                    lazyImage.classList.remove("lazy");
                    lazyImageObserver.unobserve(lazyImage);
                }
            });
        });

        lazyImages.forEach(function (lazyImage) {
            lazyImageObserver.observe(lazyImage);
        });
    }
});
```

   You would then need to assign the `lazy` class to your `<img>` tags and use the `data-src` attribute for the actual image source.

2. **Integrating with Odoo views**: To integrate this with Odoo views, you'd need to add this JavaScript code to your Odoo module's assets and update your XML templates to apply the `lazy` class and `data-src` attributes to the image tags.

Adding CSS and JavaScript for a website

Managing assets such as CSS, JavaScript, and other static files can be done through the module's asset management system. You can control the loading of these assets in your module by defining them in your manifest file and linking them to views or templates.

Here's an overview of how to manage CSS and JavaScript in Odoo.

Defining assets in module manifest (__manifest__.py)

In the manifest file, specify the assets your module requires:

```
'assets': {
        'web.assets_frontend': [
             'my_hostel/static/src/scss/hostel.scss',
             'my_hostel/static/src/js/hostel.js',
        ],
    },
```

We will add CSS, SCSS, and JavaScript files, which will modify the website. As we are modifying the website, we will need to add the website as a dependency. Modify the manifest file like this:

```
'depends': ['base', 'website'],
```

Add some SCSS code to `static/src/scss/hostel.scss`, as follows:

```
$my-bg-color: #1C2529;
$my-text-color: #D3F4FF;
nav.navbar {
    background-color: $my-bg-color !important;
    .navbar-nav .nav-link span{
        color: darken($my-text-color, 15);
        font-weight: 600;
    }
}
footer.o_footer {
    background-color: $my-bg-color !important;
    color: $my-text-color;
}
```

Add some JavaScript code to `static/src/js/my_library.js`, as follows:

```
/** @odoo-module **/

import { _t } from "@web/core/l10n/translation";
import publicWidget from "@web/legacy/js/public/public_widget";


publicWidget.registry.MyHostel = publicWidget.Widget.extend({
     selector: '#wrapwrap',
```

```
    init() {
        this._super(...arguments);
        this.orm = this.bindService("orm");
        alert(_t('Hello world'));
    },
});
```

After updating your module, you should see that the Odoo website has custom colors in the menu, body, and footer, and a somewhat annoying Hello World popup on each page load, as shown in the following screenshot:

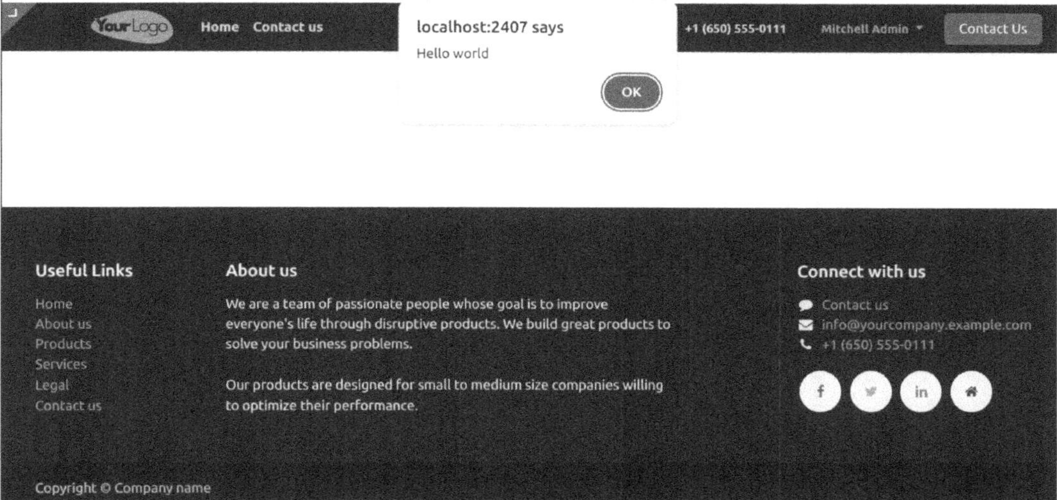

Figure 14.2 – Screenshot of Hello World popup from JavaScript code

> **Tip**
>
> For CSS/SCSS files, sometimes, order matters. So, if you need to override a style defined in another add-on, you will have to ensure that your file is loaded after the original file you want to modify. This can be done by either adjusting your view's priority field or directly inheriting from the add-on's view that injects the reference to the CSS file.

We have added basic SCSS. Odoo has built-in support for the SCSS preprocessor. Odoo will automatically compile SCSS files into CSS. In our example, we have used basic SCSS with some variables and the `darken` function to make `$my-text-color` darker by 15%. The SCSS preprocessor has tons of other features; if you want to learn more about SCSS, refer to `http://sass-lang.com/`.

Creating or modifying templates

Website templates are created using QWeb, a templating language that integrates seamlessly with the Odoo framework. These templates are used to define the structure and appearance of web pages within the Odoo website module.

Here's an overview of how to work with website templates in Odoo.

Understanding QWeb templates

QWeb templates in Odoo allow you to create dynamic web pages using XML-like syntax mixed with control structures and placeholders. They enable you to define the structure, content, and presentation of web pages.

Creating a basic website template

To create a simple website template, follow these steps:

Create a view file: Create a new XML file within your module's `views` directory:

```
<!-- Example: custom_template.xml -->
<template id="custom_template" name="Custom Template">
    <t t-call="website.layout">
        <t t-set="page_title">Custom Page</t>
        <!-- Your content here -->
        <div class="custom-content">
            <h1>Welcome to my custom page!</h1>
            <p>This is a custom template created in Odoo.</p>
        </div>
    </t>
</template>
```

Here's an explanation of the code:

- `<template>`: Defines the QWeb template
- `id`: Unique identifier for the template
- `name`: Name of the template
- `<t t-call="website.layout">`: Indicates that this template inherits from the website's main layout
- `<t t-set="page_title">`: Sets the page title dynamically
- `<div class="custom-content">`: Example of content within the template

Include in the manifest file: Include your view file in the module's manifest file:

```
{
    # Other manifest information
    'data': [
        'views/custom_template.xml',
        # Other XML or CSV files
    ],
    # Other manifest information
}
```

Using Odoo Website Builder

You can also use the Odoo Website Builder interface to create and customize web pages using predefined blocks and templates. This allows for a more visual and interactive way to design web pages without directly editing XML templates.

Styling and customization

For styling and customization, you can use CSS, which can be included within your QWeb templates or as separate files linked to your templates.

Remember—the structure and styling of your website templates can vary based on your specific needs and the complexity of the web pages you're creating. Additionally, consider exploring existing Odoo website modules and official documentation for more detailed and advanced usage of QWeb templates within the Odoo framework.

Loops

To work on recordsets or iterable data types, you need a construct to loop through lists. In the QWeb template, this can be done with the `t-foreach` element. Iteration can happen in a `t` element, in which case its contents are repeated for every member of the iterable that was passed in the `t-foreach` attribute, as follows:

```
<t t-foreach="[1, 2, 3, 4, 5]" t-as="num">
    <p><t t-esc="num"/></p>
</t>
```

This will be rendered as follows:

```
<p>1</p>
<p>2</p>
<p>3</p>
```

```
<p>4</p>
<p>5</p>
```

You can also place the `t-foreach` and `t-as` attributes in some arbitrary element, at which point this element and its contents will be repeated for every item in the iterable. Take a look at the following code block. This will generate exactly the same result as the previous example:

```
<p t-foreach="[1, 2, 3, 4, 5]" t-as="num">
    <t t-esc="num"/>
</p>
```

In our example, take a look at the inside of the `t-call` element, where the actual content generation happens. The template expects to be rendered with a context that has a variable called `hostel` set that iterates through it in the `t-foreach` element. The `t-as` attribute is mandatory and will be used as the name of the iterator variable to access the iterated data. While the most common use for this construction is to iterate over recordsets, you can use it on any iterable Python object.

Dynamic attributes

QWeb templates can set attribute values dynamically. This can be done in the following three ways.

The first way is through `t-att-$attr_name`. At the time of template rendering, an attribute, `$attr_name`, is created; its value can be any valid Python expression. This is computed with the current context and the result is set as the value of the attribute, like this:

```
<div t-att-total="10 + 5 + 5"/>
```

It will be rendered like this:

```
<div total="20"></div>
```

The second way is through `t-attf-$attr_name`. This is similar to the previous option. The only difference is that only strings between `{{ ..}}` and `#{..}` are evaluated. This is helpful when values are mixed with the strings. It is mostly used to evaluate classes, as in this example:

```
<t t-foreach="['info', 'danger', 'warning']" t-as="color">
    <div t-attf-class="alert alert-#{color}">
        Simple bootstrap alert
    </div>
</t>
```

It will be rendered like this:

```
<div class="alert alert-info">
    Simple bootstrap alert
</div>
```

```
<div class="alert alert-danger">
    Simple bootstrap alert
</div>
<div class="alert alert-warning">
    Simple bootstrap alert
</div>
```

The third way is through the `t-att=mapping` option. This option accepts the dictionary after the template rendering the dictionary's data is converted into attributes and values. Take a look at the following example:

```
<div t-att="{'id': 'my_el_id', 'class': 'alert alert-danger'}"/>
```

After rendering this template, it will be converted into the following:

```
<div id="my_el_id" class="alert alert-danger"/>
```

In our example, we have used `t-attf-class` to get a dynamic background based on index values.

Fields

The `h3` and `div` tags use the `t-field` attribute. The value of the `t-field` attribute must be used with the recordset with a length of one; this allows the user to change the content of the web page when they open the website in edit mode. When you save the page, updated values will be stored in the database. Of course, this is subject to a permission check and is only allowed if the current user has write permissions for the displayed record. With an optional `t-options` attribute, you can give a dictionary option to be passed to the field renderer, including the widget to be used. Currently, there is not a vast collection of widgets for the backend, so the choices are a bit limited here. For example, if you want to display an image from the binary field, then you can use the `image` widget like this:

```
<span t-field="author.image_small" t-options="{'widget': 'image'}"/>
```

`t-field` has some limitations. It only works on recordsets, and it cannot work on the `<t>` element. For this, you need to use some HTML elements, such as `<span>` or `<div>`. There is an alternative to the `t-field` attribute, which is `t-esc`. The `t-esc` attribute is not limited to recordsets; it can also be used on any data type, but it is not editable on a website.

Another difference between `t-esc` and `t-field` is that `t-field` shows values based on the user's language, while `t-esc` shows raw values from the database. For example, for users who configured the English language in their preferences and set the `datetime` field as used with `t-field`, the result will be rendered in `12/15/2023 14:17:15` format. In contrast, if the `t-esc` attribute is used, then the result will be in a rendered format like this: `2023-12-15 21:12:07`.

Conditionals

Note that the division showing the publication date is wrapped by a `t` element with the `t-if` attribute set. This attribute is evaluated as Python code, and the element is only rendered if the result is a truthy value. In the following example, we only show the `div` class if there is actually a publication date set. However, in complex cases, you can use `t-elif` and `t-else`, as in the following example:

```
<div t-if="state == 'new'">
    Text will be added if the state is new.
</div>
<div t-elif="state == 'progress'">
    Text will be added if the state is progress.
</div>
<div t-else="">
    Text will be added for all other stages.
</div>
```

Setting variables

The QWeb template is also capable of defining the variable in the template itself. After defining the template, you can use the variable in the subsequent template. You can set the variable like this:

```
<t t-set="my_var" t-value="5 + 1"/>
<t t-esc="my_var"/>
```

Subtemplates

If you are developing a big application, managing large templates can be difficult. The QWeb template supports subtemplates, so you can divide large templates into smaller subtemplates and you can reuse them in multiple templates. For subtemplates, you can use a `t-call` attribute, as in this example:

```
<template id="first_template">
    <div> Test Template </div>
</template>
<template id="second_template">
    <t t-call="first_template"/>
</template>
```

Inline editing

The user will be able to modify records directly from the website in edit mode. The data loaded with the `t-field` node will be editable by default. If the user changes the value in such a node and saves the page, the values will also be updated in the backend. Don't worry; in order to update the record, a user will need write permissions on the record. Note that `t-field` only works on a recordset.

To display other types of data, you can use `t-esc`. This works exactly like `t-field`, but the only difference is that `t-esc` is not editable and can be used with any type of data.

If you want to enable snippet drag-and-drop support on the page, you can use the `oe_structure` class. In our example, we have added this at the top of the template. Using `oe_structure` will enable editing and snippet drag-and-drop support.

If you want to disable the website editing feature on a block, you can use the `contenteditable=False` attribute. This makes an element read-only. We have used this attribute in the last `<section>` tag.

> **Note**
>
> To make the page multi-website compatible, when you edit a page/view through the website editor, Odoo will create a separate copy of the page for that website. This means that subsequent code updates will never make it to the edited website page. In order to also get the ease of use of inline editing and the possibility of updating your HTML code in subsequent releases, create one view that contains semantic HTML elements and a second one that injects editable elements. Then, only the latter view will be copied, and you can still have updates for the parent view.

For the other CSS classes used here, consult Bootstrap's documentation.

In *step 1*, we have declared the route to render the template. If you noticed, we have used the `website=True` parameter in `route()`, which will pass some extra context in the template, such as menus, user language, company, and so on. This will be used in `website.layout` to render the menus and footers. The `website=True` parameter also enables multi-language support in a website and displays exceptions in a better way.

Managing dynamic routes

In website development projects, it is often the case that we need to create pages with dynamic URLs. For example, in e-commerce, each product has a detailed page linked with a different URL. In this recipe, we will create a web page to display hostel details.

Getting ready

Add basic fields in the `hostel` model:

```
from odoo import fields, models


class Hostel(models.Model):
    _name = 'hostel.hostel'
    _description = "Information about hostel"
    _order = "id desc, name"
```

```
    _rec_name = 'hostel_code'

    name = fields.Char(string="hostel Name", required=True)
    hostel_code = fields.Char(string="Code", required=True)
    street = fields.Char('Street')
    street2 = fields.Char('Street2')
    zip = fields.Char('Zip', change_default=True)
    city = fields.Char('City')
    state_id = fields.Many2one("res.country.state", string='State')
    country_id = fields.Many2one('res.country', string='Country')
    phone = fields.Char('Phone',required=True)
    mobile = fields.Char('Mobile',required=True)
    email = fields.Char('Email')
    hostel_floors = fields.Integer(string="Total Floors")
    image = fields.Binary('Hostel Image')
    active = fields.Boolean("Active", default=True,
        help="Activate/Deactivate hostel record")
    type = fields.Selection([("male", "Boys"), ("female", "Girls"),
        ("common", "Common")], "Type", help="Type of Hostel",
        required=True, default="common")
    other_info = fields.Text("Other Information",
        help="Enter more information")
    description = fields.Html('Description')
    hostel_rating = fields.Float('Hostel Average Rating', digits=(14,
4))
```

How to do it...

Follow these steps to generate a details page for the hostel:

1. Add a new route for hostel details in `main.py`, as follows:

```
@http.route('/hostel/<model("hostel.hostel"):hostel>',
type='http', auth="user", website=True)
def hostel_room_detail(self, hostel):
    return request.render(
        'my_hostel.hostel_detail', {
            'hostel': hostel,
        })
```

2. Add a new template for hostel details in `hostel_templates.xml`, as follows:

```
<template id="hostel_detail" name="Hostel Detail">
   <t t-call="website.layout">
       <div class="container">
```

```
            <div class="row mt16">
                <div class="col-5">
                    <span t-field="hostel.image" t-options="{
                        'widget': 'image',
                        'class': 'mx-auto d-block
img-thumbnail'}"/>
                </div>
                <div class="offset-1 col-6">
                    <h1 t-field="hostel.name"/>
                    <p t-esc="hostel.hostel_rating"></p>
                    <t t-if="hostel.hostel_code">
                        <div t-field=
                        "hostel.hostel_code"
                        class="text-muted"/>
                    </t>
                    <b class="mt8"> State: </b>
                    <ul>
                        <li t-foreach="hostel.state_id"
t-as="state">
                            <span t-esc="state.name" />
                        </li>
                    </ul>
                </div>
            </div>
        </div>
        <div t-field="hostel.description"/>
    </t>
</template>
```

3. Add a button in the hostel list template, as follows. This button will redirect to the hostel details web page:

```
...
<div t-attf-class="card mt24 #{'bg-light' if hostel_rating else
''}">
    <div class="card-body">
        <h3 t-field="hostel.name"/>
        <t t-if="hostel.hostel_rating">
            <div t-field="hostel.hostel_rating"
            class="text-muted"/>
        </t>
        <b class="mt8"> Authors </b>
        <ul>
            <li t-foreach="hostel.state_id"
            t-as="state">
```

```
                    <span t-esc="state.name" />
                </li>
            </ul>
            <a t-attf-href="/hostel/#{hostel.id}"
            class="btn btn-primary btn-sm">
                <i class="fa fa-building"/> Hostel Detail
            </a>
        </div>
    </div>
    ...
```

Update the `my_hostel` module to apply changes. After the update, you will see hostel details page links on the hostel card. Upon clicking those links, the hostel details pages will open.

How it works...

In *step 1*, we created a dynamic route for the hostel details page. In this route, we added `<model("hostel.hostel"):hostel>`. This accepts URLs with integers, as in `/hostel/1`. Odoo considers this integer as the ID of the `hostel.hostel` model, and when this URL is accessed, Odoo fetches a recordset and passes it to the function as the argument. So, when `/hostel/1` is accessed from the browser, the `hostel` parameter in the `hostel_detail()` function will have a recordset of the `hostel.hostel` model with the ID `1`. We passed this `hostel` recordset and rendered a new template called `my_hostel.hostel2_detail`.

In *step 2*, we created a new QWeb template called `hostel_detail` to render a hostel details page. This is simple and is created using the Bootstrap structure. If you check, we have added `html_description` in the details page. The `html_description` field has a field type of HTML, so you can store HTML data in the field. Odoo automatically adds snippet drag-and-drop support to HTML types of fields. So, we are now able to use snippets in the hostel details page. The snippets dropped in the HTML fields are stored in a hostel's records, so you can design different content for different records.

In *step 3*, we added a link with the anchor tag so that a visitor can be redirected to the hostel details page.

> **Note**
>
> The model route also supports domain filtering. For example, if you want to restrict some records based on a condition, you can do so by passing the domain to the route as follows:
>
> `/hostel/<model("hostel.hostel", "[(name','!=', 'Hostel 1')]"):hostel>`
>
> This will restrict access to the hostel that has the name `Hostel 1`.

There's more...

Odoo uses `werkzeug` to handle HTTP requests. Odoo adds a thin wrapper around `werkzeug` to easily handle routes. You saw the `<model("hostel.hostel"):hostel>` route in the last example. This is Odoo's own implementation, but it also supports all features from the `werkzeug` routing. Consequently, you can use routing like this:

- `/page/<int:page>` accepts integer values
- `/page/<any(about, help):page_name>` accepts selected values
- `/pages/<page>` accepts strings
- `/pages/<category>/<int:page>` accepts multiple values

There are lots of variations available for routes, which you can read about at `http://werkzeug.pocoo.org/docs/0.14/routing/`.

Offering static snippets to the user

Static snippets are reusable components or blocks of HTML, CSS, and JavaScript that can be inserted into website pages using Website Builder. These snippets allow for easy customization and construction of web pages without needing to write code from scratch.

Odoo's website editor offers several editing building blocks, which can be dragged onto the page and edited according to your needs. This recipe will cover how to offer your own building blocks. These blocks are referred to as snippets. There are several types of snippets, but in general, we can categorize them into two types: static and dynamic. A static snippet is fixed and does not change until the user changes it. Dynamic snippets depend on database records and are changed based on record values. In this recipe, we will see how to create a static snippet.

How to do it...

A snippet is actually just a QWeb view that gets injected into the **Insert blocks** bar. We will create a small snippet that will show the hostel's image and its title. You will be able to drag and drop the snippet on the page, and you will be able to edit the image and text. Follow these steps to add a new static snippet:

1. Add a file called `views/snippets.xml`, as follows (do not forget to register the file in the manifest):
2. Add a QWeb template for the snippet in `views/snippets.xml`, as follows:

```
<template id="snippet_hostel_card" name="Hostel Card">
       <section class="pt-3 pb-3">
           <div class="container">
```

```
                <div class="row align-items-center">
                    <div class="col-lg-6 pt16 pb16">
                        <h1>This is Hostel Card Block</h1>
                        <p>
                            Learn snippet development
                            quickly with examples
                         </p>
                        <a class="btn btn-primary"
                        href="#" >Hostel Details</a>
                    </div>
                    <div class="col-lg-6 pt16 pb16">
                        <img
                          src="/my_hostel/static/src/img/cover.
jpeg"
                          class="mx-auto img-thumbnail w-50 img
img-fluid shadow"/>
                    </div>
                </div>
            </div>
    </section>
</template>
```

3. List the template in the snippet list like this:

```
<template id="hostel_snippets_options" inherit_id="website.
snippets">
    <xpath    expr="//div[@id='snippet_structure']/
div[hasclass('o_panel_body')]" position="inside">
        <t t-snippet="my_hostel.snippet_hostel_card"
         t-thumbnail="/my_hostel/static/src/img/s_hostel_thumb.
png"/>
    </xpath>
</template>
```

4. Add the cover image and snippet thumbnail image in the `/my_hostel/static/src/img` directory.

Restart the server and update the `my_hostel` module to apply the changes. When you open the website page in edit mode, you will be able to see our snippet in the snippets blocks panel:

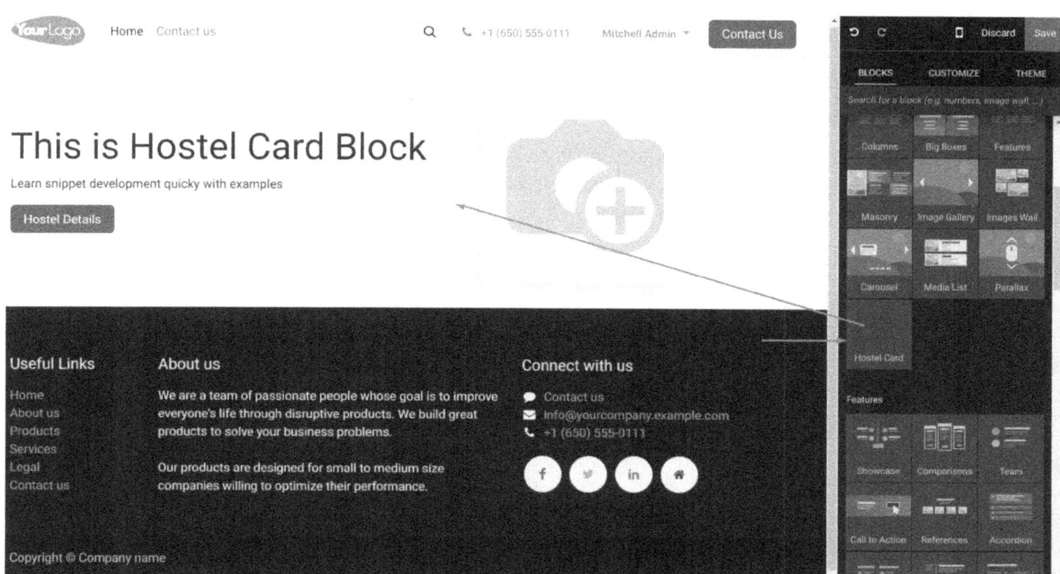

Figure 14.3 – Screenshot of static snippet

How it works...

A static snippet is nothing but a block of HTML code. In *step 1*, we created a QWeb template with our HTML for the hostel block. In this HTML, we have just used a Bootstrap column structure, but you can use any HTML code. Note that the HTML code you add in the snippet's QWeb template will be added to the page when you drag and drop. In general, it is a good idea to use `section` elements and Bootstrap classes for snippets, because for them, Odoo's editor offers edit, background, and resize controls out of the box.

In *step 2*, we registered our snippet in the snippet list. You will need to inherit `website.snippets` to register a snippet. In the website editor GUI, snippets are divided into different sections based on their usage. In our example, we have registered our snippet in the `Structure` section via `xpath`. To list your snippet, you need to use a `<t>` tag with the `t-snippet` attribute. The `t-snippet` attribute will have the XML ID of the QWeb template, which is `my_hostel.snippet_hostel_card` in our example. You will also need to use the `t-thumbnail` attribute, which is used to show a small snippet image in the website editor.

> **Note**
>
> The `website.snippets` template contains all the default snippets, and you can learn more about it by exploring the `/addons/website/views/snippets/snippets.xml` file.
>
> Odoo will add some default options to your snippets when you have a proper Bootstrap structure. For example, in our snippet, you would be able to set a background color, a background image, width, height, and so on. Explore the `/addons/website/views/snippets/snippets.xml` file to see all the snippet options. In the next recipe, we will see how to add our own options.

In *step 3*, we listed our snippet under the `structure` block. Once you update the module, you will be able to drag and drop the snippet. In *step 4*, we just added an image for the snippet thumbnail.

There's more...

In such cases, there will be no need for extra JavaScript. Odoo's editor offers lots of options and controls out of the box, and they are more than enough for static snippets. You will find all existing snippets and options at `website/views/snippets.xml`.

Snippet options also support the `data-exclude`, `data-drop-near`, and `data-drop-in` attributes, which determine where a snippet can be placed when dragging it out of the snippet bar. These are also jQuery selectors, but in *step 3* of this recipe, we didn't use them, because we allow putting the snippet basically anywhere that content can go.

Offering dynamic snippets to the user

Dynamic snippets refer to reusable components or blocks that are capable of displaying dynamic content pulled from various sources such as databases, models, or external services. These snippets enable the creation of versatile and adaptable web pages that display real-time or context-specific information.

Identify data sources:

- Determine the data sources you want to use in your dynamic snippet. This can include Odoo models, databases, APIs, and so on.
- Implement dynamic placeholders using QWeb templating tags (`{% %}`) or Odoo-specific directives (`<t t-foreach="..." t-as="...">`).

We will see how we can create dynamic snippets for Odoo. We will generate content based on database values.

How to do it...

Perform the following steps to add a dynamic snippet that shows a list of hostel data:

1. Add a given QWeb template for the snippet in `views/snippets.xml`:

```
<template id="snippet_hostel_dynamic" name="Hostel Dynamic">
    <section class="hostel_list">
        <div class="container">
            <h2>Hostel</h2>
            <table class="table hostel_snippet table-striped"
                   data-number-of-hostel="5">
                <tr>
                    <th>Name</th>
                    <th>Available date</th>
                </tr>
            </table>
        </div>
    </section>
</template>
```

2. Register the snippet and add an option to change the snippet behavior:

```
<template id="hostel_snippets_options"
           inherit_id="website.snippets">
    <!-- register snippet -->
    <xpath expr="//div[@id='snippet_structure']/
                 div[hasclass('o_panel_body')]"
           position="inside">
        <t t-snippet="my_hostel.snippet_hostel_dynamic"
           t-thumbnail="/my_hostel/static/src/img/s_list.png"/>
    </xpath>
    <xpath expr="//div[@id='snippet_options']"
position="inside">
  <!--Add step 3 here -->
    </xpath>
</template>
```

3. Then, add snippet options for the hostel snippet:

```
<div data-selector=".hostel_snippet">
    <we-select string="Table Style">
        <we-button data-select-class="table-striped">
                Striped
        </we-button>
        <we-button data-select-class="table-dark">
```

```
                Dark
            </we-button>
            <we-button data-select-class="table-bordered">
                Bordered
            </we-button>
        </we-select>
        <we-button-group string="No of Rooms"
            data-attribute-name="numberOfRooms">
            <we-button data-select-data-attribute="5">
                5
            </we-button>
            <we-button data-select-data-attribute="10">
                10
            </we-button>
            <we-button data-select-data-attribute="15">
                15
            </we-button>
        </we-button-group>
    </div>
```

4. Add a new `/static/src/snippets.js` file and add code to render a dynamic snippet.
5. Add a `public` widget to render the hostel snippet dynamically:

```
/** @odoo-module **/
import { _t } from "@web/core/l10n/translation";
import publicWidget from "@web/legacy/js/public/public_widget";
publicWidget.registry.HostelSnippet = publicWidget.Widget.
extend({
   selector: '.hostel_snippet',
   disabledInEditableMode: false,

   start: function () {
       var self = this;
       var rows = this.$el[0].dataset.numberOfRooms || '5';
       this.$el.find('td').parents('tr').remove();
       this._rpc({
           model: 'hostel.hostel',
           method: 'search_read',
           domain: [],
           fields: ['name', 'hostel_code'],
           orderBy: [{ name: 'hostel_code', asc: false }],
           limit: parseInt(rows)
       }).then(function (data) {
```

```
                _.each(data, function (hostel) {
                    self.$el.append(
                        $('<tr />').append(
                            $('<td />').text(hostel.name),
                            $('<td />').text(hostel.hostel_code)
                        ));
                });
            });
        },
    });
```

6. Load the JavaScript file to the `__manifest__.py` module:

```
'assets': {
        'web.assets_frontend': [
            'my_hostel/static/src/js/snippets.js',
        ],
    },
```

After updating the module, you will be offered a new snippet called `Hostels`, which has an option to change the number of recently added rooms. We have also added the option to change the table design, which can be displayed when you click on the table.

How it works...

In *step 1*, we added a QWeb template for the new snippet (it is just like the previous recipe). Note that we added a basic structure for the table. We will dynamically add lines for the hostel in the table.

In *step 2*, we registered our dynamic snippet and added custom options to change the behavior of our dynamic snippet. The first option we added is `Table Style`. It will be used to change the style of the table. The second option we added is `No of Rooms`. We used the `<we-select>` and `<we-button-group>` tags for our options. These tags will provide different GUIs to the snippet option. The `<we-select>` tag will show the options as a dropdown, while the `<we-button-group>` tag will show the options as a button group. There are several other GUI options, such as `<we-checkbox>` and `<we-colorpicker>`. You can explore more GUI options in the `/addons/website/views/snippets/snippets.xml` file.

If you look at the options closely, you will see we have `data-select-class` and `data-select-data-attribute` attributes for the option buttons. This will let Odoo know which attribute to change when the user chooses an option. `data-select-class` will set the class attribute on the element when the user chooses this option, while `data-select-data-attribute` will set the custom attribute and value on the element. Note that it will use the value of `data-attribute-name` to set the attribute.

Now, we have added the snippet options. If you drag and drop the snippet at this point, you will only see the table header and the snippet options. Changing the snippet options will change the table style, but there is no hostel data yet. For that, we need to write some JavaScript code that will fetch the data and display it in the table. In *step 3*, we added JavaScript code that will render the hostel data in the table. To map a JavaScript object to an HTML element, Odoo uses `PublicWidget`. Now, `PublicWidget` is available through `import publicWidget from "@web/legacy/js/public/public_widget";`. The key attribute in using `PublicWidget` is the `selector` attribute. In the `selector` attribute, you will need to use the CSS selector of the element, and Odoo will automatically bind the element with `PublicWidget`. You can access the related element in the `$el` attribute. The rest of the code is basic JavaScript and jQuery except `_rpc`.

There's more...

If you want to create your own snippet option, you can use the `t-js` option on the snippet option. After that, you will need to define your own option in the JavaScript code. Explore the `addons/website/static/src/js/editor/snippets.options.js` file to learn more about snippet options.

Getting input from website users

In Odoo, you can collect input from website users through forms, surveys, or interactive elements integrated into your website. Odoo provides functionalities to create forms easily and manage the data collected from these forms. Here's how you can set up input collection.

- Submitted form data is typically stored in the database as records of a specific model associated with the form
- Access the collected data either through the website backend or by configuring views to display the form submissions
- Optionally, you can link the form submissions to specific models in Odoo, allowing you to manage and process the data within the Odoo backend
- Define models and fields to store the form data securely

Getting ready

For this recipe, we will be using the `my_hostel` module. We will need a new model to store hostel booking inquiries submitted by users.

So, before starting this recipe, modify the previous code and create one new model for booking inquiries, `my_hostel/models/inquiries.py`:

```
from odoo import fields, models
```

```
class Inquiries(models.Model):
    _name = 'hostel.inquiries'
    _description = "Inquiries about hostel"
    _order = "id desc,"

    name = fields.Char(string="Student Name", required=True)
    phone = fields.Char(string="Phone", required=True)
    email = fields.Char(string="Email")
    book_fy = fields.Char(string="Book for Year")
    queries = fields.Html(string="Your Question", required=True)
```

Now, create menus, actions, and views in the backend to store the submitted data from the website inquiries form.

For that, create an XML file in `my_hostel/views/inquiries_view.xml`, then add menus, actions, and its basic tree and form view:

```
<?xml version="1.0" encoding="utf-8"?>
<odoo>
    <data>
        <record id="view_hostel_inquiry_tree" model="ir.ui.view">
            <field name="name">hostel.inquiry.tree</field>
            <field name="model">hostel.inquiries</field>
            <field name="arch" type="xml">
                <tree string="Inquiries">
                    <field name="name"/>
                    <field name="phone"/>
                    <field name="email"/>
                    <field name="book_fy"/>
                </tree>
            </field>
        </record>

        <record id="view_hostel_inquiry_form" model="ir.ui.view">
            <field name="name">hostel.inquiry.form</field>
            <field name="model">hostel.inquiries</field>
            <field name="arch" type="xml">
                <form string="Inquiries">
                    <sheet>
                        <div class="oe_title">
                            <h3>
                                <table>
                                    <tr>
```

```
                                        <td style="padding-
right:10px;"><field name="name" required="1"
                                        placeholder="Name" /></td>
                                    </tr>
                                </table>
                            </h3>
                        </div>
                        <group>
                            <group>
                                <field name="phone"/>
                                <field name="email"/>
                                <field name="book_fy"/>
                            </group>
                        </group>
                        <group>
                            <field name="queries"/>
                        </group>
                    </sheet>
                </form>
            </field>
        </record>

        <record model="ir.actions.act_window"
        id="action_inquiry">
            <field name="name">Inquiries</field>
            <field name="type">
            ir.actions.act_window</field>
            <field name="res_model">
            hostel.inquiries</field>
            <field name="view_mode">tree,form</field>
            <field name="help" type="html">
                <p class="oe_view_nocontent_create">
                    Create Inquiries.
                </p>
            </field>
        </record>

        <menuitem id="hostel_inquiry_main_menu" name="Inquiries"
        parent="hostel_main_menu" sequence="2" />

        <menuitem id="hostel_inquiry_menu" name="Inquiries"
        parent="hostel_inquiry_main_menu"
            action="my_hostel.action_inquiry"
            groups="my_hostel.group_hostel_manager"
```

```
            sequence="1"/>

    </data>
</odoo>
```

Now, create a basic form to get the details from the customers, which is published on the website page. Once a user has submitted that form, all the filled data will be stored in the `Inquiries` table.

For that, create a new folder in the module, `my_hostel/controllers/main.py`:

```
# -*- coding: utf-8 -*-
from odoo import http, tools, _
from odoo.http import request


class InquiryForm(http.Controller):


    @http.route('/inquiry/form', type='http', auth="public",
website=True)
    def inquiry_form_template(self, **kw):
        return request.render("my_hostel.hostel_inquiry_form")


    @http.route('/inquiry/submit', type='http', auth="public",
website=True)
    def inquiry_form(self, **kwargs):
        inquiry_obj = request.env['hostel.inquiries']
        form_vals = {
            'name': kwargs.get('name') or '',
            'email': kwargs.get('email') or '',
            'phone': kwargs.get('phone') or '',
            'book_fy': kwargs.get('book_fy') or '',
            'queries': kwargs.get('queries') or '',
            }
        submit_success = inquiry_obj.sudo().create(form_vals)
        return request.redirect('/contactus-thank-you')
```

Now, design a form for the website called `my_hostel/views/form_template.xml`:

```
<odoo>
  <template id="hostel_inquiry_form"
    name="Hostel Inquiry Form">
    <t t-call="website.layout">
      <section class="s_website_form" data-snippet="s_website_form">
```

```
<div class="container">
  <div class="row">
    <div class="col-md-12 mb64">
      <div class="aboutus-section pl-5 pr-5 p-t-100 p-b-50">
        <div class="wrapper wrapper--w900">
          <div class="card">
            <div class="card-body mt8">
              <form action="/inquiry/submit" method="POST"
class="o_mark_required" id="inquiry_form" enctype="multipart/form-
data">
                <input type="hidden" name="csrf_token" t-att-
value="request.csrf_token()"/>
                  <div class="row">
                    <div class="form-group col-md-12">
                      <label for="name"> Your Name </label>
                        <input type="text"
                        class="form-control" name="name"
                        id="name" required="True" />
                    </div>
                  </div>
                  <div class="row">
                    <div class="form-group col-md-12">
                      <label for="phone"> Phone </label>
                        <input type="text" class="form-
control"

                        name="phone" id="phone"

required="True" />
                    </div>
                  </div>
                  <div class="row">
                    <div class="form-group col-md-12">
                      <label for="email"> Email ID </label>
                          <input type="text" class="form-
control"

                          name="email" id="email"/>
                      </div>
                    </div>
                    <div class="row">
                      <div class="form-group col-md-12">
                        <label for="book_fy"> Booking for the
Year </label>

                          <input type="text" class="form-
control"

                          name="book_fy" id="book_fy"/>
                      </div>
```

```
                                </div>
                                <div class="row">
                                  <div class="form-group col-md-12">
                                    <label for="queries"> Your Question </
label>
                                      <input type="text" class="form-
control"
                                      name="queries" id="queries"/>
                                  </div>
                                </div>
                                <div class="form-group row">
                                  <div class="col-sm-12">
                                    <button type="submit"
                                    class="btn btn-primary btn-lg
a-submit">
                                      <span>Submit</span>
                                    </button>
                                  </div>
                                </div>
                              </form>
                            </div>
                          </div>
                        </div>
                      </div>
                    </div>
                  </div>
                </div>
              </section>
          </t>
      </template>
</odoo>
```

How to do it...

Update the module and open the `/inquiry/form` URL. From this page, you will be able to submit queries for the hostel. After submission, you can check them into the respective inquiries form view in the backend.

Managing SEO options

Odoo has built-in SEO features for templates (pages). However, some templates are used for multiple URLs. For example, in an online shop, each product page uses the same template but different product data. For these cases, we need different SEO options for each URL.

Getting ready

For this recipe, we will be using the `my_hostel` module. We will store separate SEO data for each hostel details page. Before following this recipe, you should test the SEO options in the different hostel pages. You can get an SEO dialog from the **Promote** drop-down menu on the top, as shown in the following screenshot:

Figure 14.4 – Opening the SEO configuration for a page

If you test SEO options on different hostel details pages, you will notice that changing the SEO data in one book page will reflect on all hostel pages. We will fix this issue in this recipe.

How to do it...

To manage separate SEO options for each hostel, follow these steps:

1. Inherit the `website.seo.metadata` mixin in the `hostel.hostel` model, as follows:

```
class Hostel(models.Model):
_name = 'hostel.hostel'
_description = "Information about hostel"
_inherit = ['website.seo.metadata']
_order = "id desc, name"
_rec_name = 'hostel_code'
```

2. Pass the `hostel` object in the hostel details route as `main_object`, as follows:

```
@http.route('/hostels/<model("hostel.hostel"):hostel>',
type='http', auth='public', website = True)
def hostel_detail(self, hostel):
        return request.render(
        'my_hostel.hostel_detail', {
            'hostel': hostel,
```

```
                'main_object': hostel
            }
    ...
```

Update the module and change the SEO on the different hostel pages. It can be changed through the **Optimize SEO** option. Now, you will be able to manage separate SEO details per hostel.

How it works...

To enable SEO on each record of the model, you will need to inherit the `website.seo.metadata` mixin in your model. This will add a few fields and methods to the `hostel.hostel` model. These fields and methods will be used from the website to store separate data for each book.

> **Tip**
>
> If you want to see fields and methods for the SEO mixin, search for the `website.seo.metadata` model in the `/addons/website/models/website.py` file.

All SEO-related code is written in `website.layout`, and it gets all the SEO meta-information from the recordset passed as `main_object`. Consequently, in *step 2*, we passed a `hostel` object with the `main_object` key so that the website layout will get all SEO information from the hostel. If you don't pass `main_object` from the controller, then the template recordset will be passed as `main_object`, and that's why you were getting the same SEO data in all hostels.

There's more...

In Odoo, you can add custom metatags for Open Graph and Twitter sharing. If you want to add your custom metatags to a page, you can override `_default_website_meta()` after adding the SEO mixin. For example, if we want to use the hostel cover as the social sharing image, then we can use the following code in our `hostel` model:

```
    def _default_website_meta(self):
        res = super(Hostel, self)._default_website_meta()
        res['default_opengraph']['og:image'] = self.env['website'].
image_url(self, 'image')
        res['default_twitter']['twitter:image'] = self.
env['website'].image_url(self, 'image')
        return res
```

After this, the hostel cover will be displayed on social media when you share the hostel's URL. Additionally, you can also set the page title and the description from the same method.

Managing sitemaps for the website

A website's sitemaps are crucial for any website. The search engine will use website sitemaps to index the pages of a website. In this recipe, we will add hostel details pages to the sitemap.

Getting ready

For this recipe, we will be using the `my_hostel` module from the previous recipe. If you want to check the current sitemap in Odoo, open `<your_odoo_server_url>/sitemap.xml` in your browser.

How to do it...

Follow these steps to modify a hostel's page to `sitemap.xml`:

1. Import the methods in `main.py`, as follows:

```
from odoo.addons.http_routing.models.ir_http import slug
from odoo.addons.website.models.ir_http import sitemap_qs2dom
```

2. Add the `sitemap_hostels` method to `main.py`, as follows:

```
    def sitemap_hostels(env, rule, qs):
        Hostels = env['hostel.hostel']
        dom = sitemap_qs2dom(qs, '/hostels', Hostels._rec_name)
    #Ex. to filter urls
        #dom += [('name', 'ilike', 'abc')]
        for f in Hostels.search(dom):
            loc = '/hostels/%s' % slug(f)
            if not qs or qs.lower() in loc:
                yield {'loc': loc}
```

3. Add the `sitemap_hostels` function reference in a hostel's detail routes as follows:

```
@http.route('/hostels/<model("hostel.hostel"):hostel>',
type='http', auth='public', website = True,
sitemap=sitemap_hostels)
    def hostel_detail(self, hostel):
```

Update the module to apply the changes. A `sitemap.xml` file is generated and stored in `Attachments`. Then, it is regenerated every few hours. To see our changes, you will need to remove the sitemap file from the attachment. To do this, visit **Settings** | **Technical** | **Database Structure** | **Attachments**, search for the sitemap, and delete the file. Now, access the `/sitemap.xml` URL in a browser, and you will see the hostel's pages in the sitemap.

How it works...

In *step 1*, we imported a few required functions. `slug` is used to generate a clean, user-friendly URL, based on a record name. `sitemap_qs2dom` is used to generate a domain based on route and query strings.

In *step 2*, we created a Python generator function,`sitemap_hostels()`. This function will be called whenever a sitemap is generated. During the call, it will receive three arguments—the `env` Odoo environment, the `rule` route rule, and the `qs` query string. In the function, we generated a domain with `sitemap_qs2dom`. Then, we used the generated domain to search the hostel records, which are used to generate the location through the `slug()` method. With `slug`, you will get a user-friendly URL, such as `/hostels/cambridge-1`. If you do not want to list all the hostels on the sitemap, you can just use a valid domain in the search to filter the hostel.

In *step 3*, we passed the `sitemap_hostels()` function reference to the route with a `sitemap` keyword.

There's more...

In this recipe, we have seen how you can use a custom method to generate a URL for a sitemap. But if you do not want to filter hostels and you want to list all hostels in a sitemap, then instead of the function reference, just pass `True` as follows:

```
...
@http.route('/hostels/<model("hostel.hostel"):hostel>', type='http',
auth='public', website = True, sitemap=True)
...
```

In the same way, if you don't want any URL to display in the sitemap, just pass `False` as follows:

```
...
@http.route('/hostels/<model("hostel.hostel"):hostel>', type='http',
auth='public', website = True, sitemap=False)
...
```

Getting a visitor's country information

The Odoo CMS has built-in support for **GeoIP**. In a live environment, you can track a visitor's country based on the IP address. In this recipe, we will get the country of the visitor based on the visitor's IP address.

Getting ready

For this recipe, we will be using the `my_hostel` module from the previous recipe. In this recipe, we will hide some hostels on the web page based on the visitor's country. You will need to download the GeoIP database for this recipe. After that, you will need to pass the database location from the `cli` option, like this:

```
./odoo-bin -c config_file --geoip-db=location_of_geoip_DB
```

Or, follow the steps from this document sheet:

`https://www.odoo.com/documentation/17.0/applications/websites/website/configuration/on-premise_geo-ip-installation.html`.

If you don't want to locate the GeoIP database in `/usr/share/GeoIP/`, use the `--geoip-city-db` and `--geoip-country-db` options of the Odoo command-line interface. These options take the absolute path to the GeoIP database file and use it as the GeoIP database.

How to do it...

Follow these steps to restrict books based on country:

1. Add the `restrict_country_ids` Many2many field in the `hostel.hostel` model, as follows:

```
class Hostel(models.Model):
    _name = 'library.book'
    _inherit = ['website.seo.metadata']
    ...
    restrict_country_ids = fields.Many2many('res.country')
    ...
```

2. Add a `restrict_country_ids` field in the form view of the `hostel.hostel` model, as follows:

```
...
    <field name="restrict_country_ids" widget="many2many_tags"/>

...
```

3. Update the `/hostel` controller to restrict books based on country, as follows:

```
@http.route('/hostels', type='http', auth='public', website =
True)
    def hostel(self):
        country_id = False
        country_code = request.geoip and request.geoip.
get('country_code') or False
        if country_code:
            country_ids = request.env['res.country'].sudo().
search([('code', '=', country_code)])
```

```
            if country_ids:
                country_id = country_ids[0].id
        domain = ['|', ('restrict_country_ids', '=', False),
('restrict_country_ids', 'not in', [country_id])]
        hostels = request.env['hostel.hostel'].sudo().
search(domain)
        return request.render(
        'my_hostel.hostels', {
            'hostels': hostels,})
```

> **Warning**
>
> This recipe does not work with the local server. It will require a hosted server because, with the local machine, you will get the local IP, which is not related to any country. You will also need to configure NGINX properly.

How it works...

In *step 1*, we added a new `restricted_country_ids` many2many-type field in the `hostel.hostel` model. We will hide the book if the website visitor is from a restricted country.

In *step 2*, we just added a `restricted_country_ids` field in the book's form view. If GeoIP and NGINX are configured properly, Odoo will add GeoIP information to `request.geoip`, and then you can get the country code from that.

In *step 3*, we fetched the country code from GeoIP, followed by the recordset of the country, based on `country_code`. After getting a visitor's country information, we filtered hostels with domains based on a restricted country.

> **Important information**
>
> If you don't have a real server and you want to test this anyway, you can add a default country code in the controller, like this: `country_code = request.geoip and request.geoip.get('country_code') or 'IN'`.

The GeoIP database gets updated from time to time, so you will need to update your copy to get up-to-date country information.

Tracking a marketing campaign

In any business or service, it is really important to be familiar with the **return on investment** (**ROI**). The ROI is used to evaluate the effectiveness of an investment. Investments in ads can be tracked through UTM codes. A UTM code is a small string that you can add to a URL. This UTM code will help you to track campaigns, sources, and media.

Getting ready

For this recipe, we will be using the `my_library` module. Odoo has built-in support for UTMs. With our hostel application, we don't have any practical cases where UTMs can be used. However, in this recipe, we will add a UTM in the issues generated by `/books/submit_issues` in `my_library`.

How to do it...

Follow these steps to link UTMs in a book issue generated from our web page to the `/books/submit_issues` URL:

1. Add a `utm` module in the `depends` section of `manifest.py`, as follows:

```
'depends': ['base', 'website', 'utm'],
```

2. Inherit `utm.mixin` in the `book.issue` model, as follows:

```
class LibraryBookIssues(models.Model):
    _name = 'book.issue'
    _inherit = ['utm.mixin']
    book_id = fields.Many2one('library.book', required=True)
    submitted_by = fields.Many2one('res.users')
    issue_description = fields.Text()
```

3. Add a `campaign_id` field in the tree view of the `book_issue_ids` field, as follows:

```
...
<group string="Book Issues">
    <field name="book_issue_ids" nolabel="1">
        <tree name="Book issues">
            <field name="create_date"/>
            <field name="submitted_by"/>
            <field name="issue_description"/>
            <field name="campaign_id"/>
        </tree>
    </field>
</group>
...
```

Update the module to apply the changes. To test the UTM, you need to perform the following steps:

1. In Odoo, a UTM is processed based on cookies, and some browsers do not support cookies in the localhost, so if you are testing it with the localhost, access the instance with `http://127.0.0.1:8069`.

By default, UTM tracking is blocked for salespeople. Consequently, to test the UTM feature, you need to log in with a portal user.

2. Now, open the `http://127.0.0.1:8069/books/submit_issues?utm_campaign=sale` URL.
3. Submit the book issue and check the book issue in the backend. This will display the campaign in the book's form view.

How it works...

In the first step, we inherited `utm.mixin` in the `book.issue` model. This will add the following fields to the `book.issue` model:

- `campaign_id`: The `Many2one` field with the `utm.campaign` model. This is used to track different campaigns, such as the *Summer* and *Christmas Special* campaigns.
- `source_id`: The `Many2one` field with the `utm.source` model. This is used to track different sources, such as search engines and other domains.
- `medium_id`: The `Many2one` field with the `utm.medium` model. This is used to track different media, such as postcards, emails, and banner ads.

To track the campaign, medium, and source, you need to share a URL in the marketing media like this: `your_url?utm_campaign=campaign_name&utm_medium=medium_name&utm_source=source_name`.

If a visitor visits your website from any marketing media, then the `campaign_id`, `source_id`, and `medium_id` fields are automatically filled when records are created on the website page.

In our example, we just tracked `campaign_id`, but you can also add `source_id` and `medium_id`.

> **Important note**
>
> In our test example, we have used `campaign_id=sale`. Now, `sale` is the name of the record in the `utm.campaign` model. By default, the `utm` module adds a few records of the campaign, medium, and source. The `sale` record is one of them. If you want to create a new campaign, medium, and source, you can do this by visiting the `Link Tracker > UTMs` menu in developer mode.

Managing multiple websites

Odoo has built-in support for multiple websites. This means that the same Odoo instance can be run on multiple domains as well as when displaying different records.

Getting ready

For this recipe, we will be using the `my_hostel` module from the previous recipe. In this recipe, we will hide hostels based on the website.

How to do it...

Follow these steps to make the online website multi-website compatible:

1. Add `website.multi.mixin` in the `hostel.hostel` model, as follows:

```
class Hostel(models.Model):
     _name = 'hostel.hostel'
     _description = "Information about hostel"
     _inherit = ['website.seo.metadata', 'website.multi.mixin']
...
```

2. Add `website_id` in the hostel form view, as follows:

```
...
<group>
    <field name="website_id"/>
</group>
...
```

3. Modify the domain in the `/hostels` controller, as follows:

```
@http.route('/hostels', type='http', auth='public', website =
True)
     def hostel(self, **post):
...
   domain = ['|', ('restrict_country_ids', '=', False),
('restrict_country_ids', 'not in', [country_id])]
          domain += request.website.website_domain()
     hostels = request.env['hostel.hostel'].sudo().search(domain)
          return request.render(
          'my_hostel.hostels', {
              'hostels': hostels,
          })
...
```

4. Import `werkzeug` and modify a hostel details controller to restrict hostel access from another website, as follows:

```
import werkzeug
...
```

```
@http.route('/hostels/<model("hostel.hostel"):hostel>',
type='http', auth='public', website = True, sitemap=sitemap_
hostels)
def hostel_detail(self, hostel, **post):
    if not hostel.can_access_from_current_website():
        raise werkzeug.exceptions.NotFound()
    return request.render(
         'my_hostel.hostel_detail', {
             'hostel': hostel,
             'main_object': hostel
         })
...
```

Update the module to apply the changes. To test this module, set up different websites in some hostels. Now, open the /hostels URL and check the list of books. After this, change the website and check the list of books. For testing, you can change the website from the website switcher drop-down menu. Refer to the following screenshot to do that:

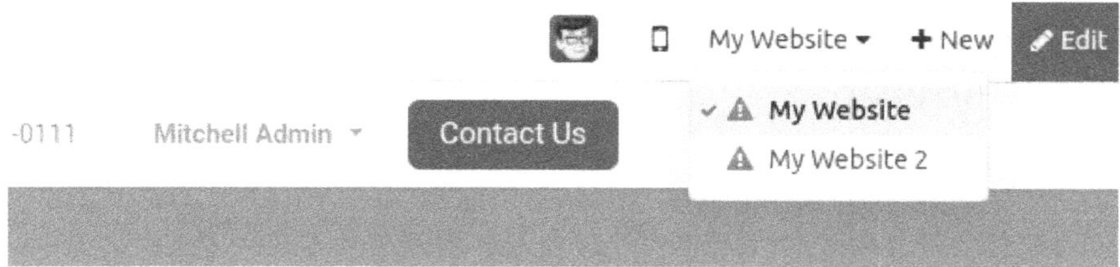

Figure 14.5 – Website switcher

You can also try to access the book details directly from the URL, such as for /hostels/1. If a hostel is not from that website, it will show as 404.

How it works...

In *step 1*, we added website.multi.mixin. This mixin adds a basic utility to handle multiple websites in the model. This mixin adds the website_id field in the model. This field is used to determine which website a record is meant for.

In *step 2*, we added the website_id field in the form view of the hostel so that the hostels would be filtered based on the website.

In *step 3*, we modified the domain used to find a list of hostels. request.website.website_domain() will return the domain that filters out hostels that are not from the website.

> **Important note**
>
> Notice that there are records that do not have any `website_id` field set. Such records will be displayed on all websites. This means that if you don't have a `website_id` field on a particular hostel, then that hostel will be displayed on all websites.

Then, we added the domain in the web search, as follows:

- In *step 4*, we restricted book access. If the book is not meant for the current website, then we will raise a `Not found` error. The `can_access_from_current_website()` method will return a `True` value if a hostel record is meant for the currently active website and `False` if a hostel record is meant for another website.
- If you noticed, we added `**post` in both controllers. This is because, without it, `**post` `/hostels` and `/hostels/<model("hostel.hostel"):hostel>` will not accept a query parameter. They will also generate an error while switching the website from the website switcher, so we added it. Normally, it is a good practice to add `**post` in every controller so that they can handle query parameters.

Redirecting old URLs

When you move to the Odoo website from an existing system or website, you must redirect your old URLs to new URLs. With proper redirection, all of your SEO rankings will be moved to new pages. In this recipe, we will see how to redirect old URLs to new URLs in Odoo.

Getting ready

For this recipe, we will be using the `my_hostels` module from the previous recipe. For this recipe, we are assuming that you used to have a website and have just moved to Odoo.

How to do it...

Imagine that, in your old website, books were listed at the `/my-hostels` URL; as you know, the `my_hostel` module lists hostels on the `/hostels` URL as well. So, we will now add a **redirection** rule in Odoo that will redirect your old `/my-hostels` URL to the new `/hostels` URL. Perform the following steps to add the redirection rule:

1. Activate developer mode.
2. Open **Website** | **Configuration** | **Redirects**.
3. Click on **New** to add a new rule.
4. Enter values in the form, as shown in the following screenshot. In **URL from**, enter `/my-hostels`, and in **URL to**, enter `/hostels`.

5. Select the **Action** value of **301 Moved permanently**.
6. Save the record. Once you have filled in the data, your form will look like this:

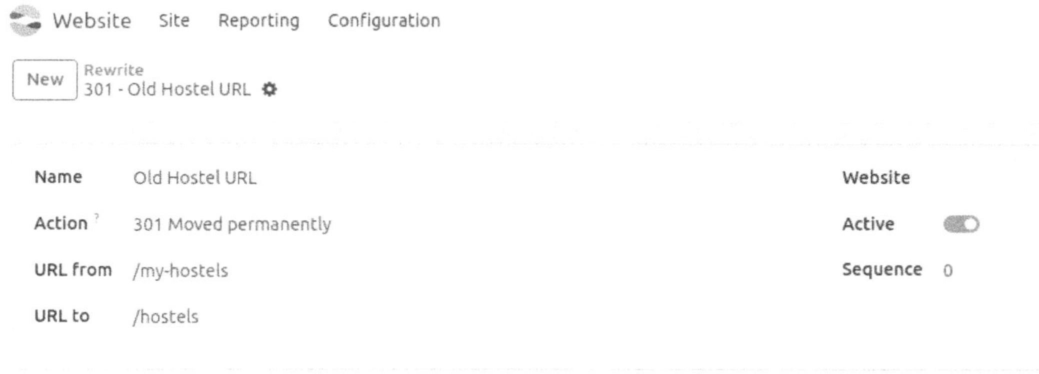

Figure 14.6 – Redirection rule

Once you have added this rule, open the `/my-hostels` page. You will notice that the page gets redirected automatically to the `/hostels` page.

How it works...

Page redirection is simple; it's just part of the HTTP protocol. In our example, we moved `/my-hostels` to `/hostels`. We used a **301 Moved permanently** redirect for redirection. Here are all the redirection options that are available in Odoo:

- **404 Not Found**: This option is used if you want to give a `404 Not Found` response for a page. Note that Odoo will display the default `404` page for such requests.
- **301 Moved permanently**: This option redirects old URLs to new ones permanently. This type of redirection will move SEO rankings to a new page.
- **302 Moved temporarily**: This option redirects old URLs to new ones temporarily. Use this option when you need to redirect a URL for a limited time. This type of redirection will not move SEO rankings to a new page.
- **308 Redirect/Rewrite**: An interesting option – with this, you will be able to change/rewrite existing Odoo URLs to new ones.

In this recipe, this would allow us to rewrite the old `/my-hostels` URL to the new `/hostels` URL. Hence, we would have no need to redirect the old URL by using the **301 Moved permanently** rule for `/my-hostels`.

There are a few more fields on the redirection rule form. One of them is the **Active** field, which can used if you want to enable/disable rules from time to time. A second important field is **Website**. The **Website** field is used when you are using the multi-website feature and you want to limit the redirection rule to one website only. By default, however, the rule will be applied to all websites.

Publish management for website-related records

In business flows, there are some cases where you need to allow or revoke page access to public users. One such case is e-commerce products, where you need to publish or unpublish products based on availability. In this recipe, we will see how you can publish and unpublish hostel records for public users.

Getting ready

For this recipe, we will be using the `my_hostel` module from the previous recipe.

> **Important note**
> If you notice, we have put `auth='user'` on the `/hostels` and `/hostels/<model("hostel.hostel"):hostel>` routes. Please change this to `auth='public'` to make those URLs accessible to public users.

How to do it...

Perform the following steps to enable a publish/unpublish option for hostel details pages:

1. Add `website.published.mixin` to the `hostel.hostel` model like this:

```
class Hostel(models.Model):
      _name = 'hostel.hostel'
      _description = "Information about hostel"
      _inherit = ['website.seo.metadata', 'website.multi.mixin',
' website.published.mixin']
      _order = "id desc, name"
     ...
```

2. Add a new file to `my_hostel/security/rules.xml` and add a record rule for hostels like this (make sure you register the file in the manifest):

```
<?xml version="1.0" encoding="utf-8"?>
<odoo noupdate="1">

     <record id="hostels_rule_portal_public" model="ir.rule">
         <field name="name">Portal/Public user: View published
Hostels</field>
```

```
            <field name="model_id" ref="my_hostel.model_hostel_
hostel"/>
            <field name="groups" eval="[(4, ref('base.group_
portal')), (
            4, ref('base.group_public'))]"/>
            <field name="domain_force">
            [('website_published','=', True)]</field>
            <field name="perm_read" eval="True"/>
        </record>

</odoo>
```

- Update the `my_hostel` module to apply the changes. Now, you can publish and unpublish hostel pages:

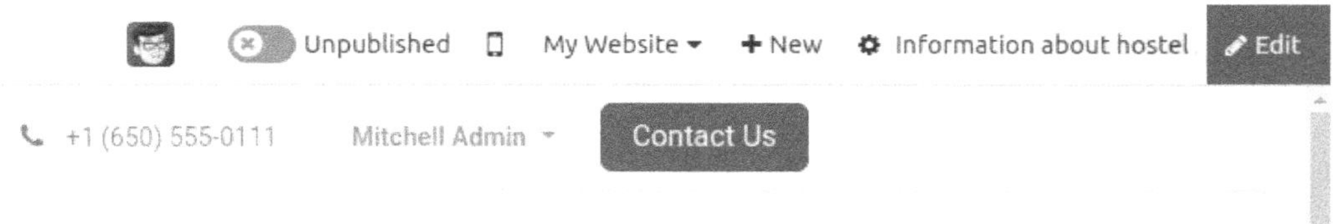

Figure 14.7 – The Publish/Unpublish toggle

To publish/unpublish hostels, you can use the toggle shown in the preceding screenshot of a hostel details page.

How it works...

Odoo provides a ready-made mixin to handle publish management for your records. It does most of the job for you. All you need to do is add `website.published.mixin` to your model. In *step 1*, we added `website.published.mixin` to our `hostel` model. This will add all the fields and methods required to publish and unpublish hostels. Once you add this mixin to the book model, you will be able to see the button to toggle the state on the book details page, as shown in the preceding screenshot.

> **Note**
>
> We are sending a hostel record as `main_object` from our hostel details route. Without this, you will not be able to see the publish/unpublish button on the hostel details page.

Adding the mixin will show the publish/unpublish button on the hostel's details page, but it will not restrict a public user from accessing it. To do this, we need to add a record rule. In *step 2*, we added a record rule to restrict access to unpublished hostels. If you want to learn more about record rules, refer to *Chapter 10, Security Access*.

There's more...

The `publish` mixin will enable the publish/unpublish button on the website. But if you want to show a redirect button on the backend form view, the publish mixin can provide a means for that too. The following steps show how to add a redirect button to a hostel's form view:

1. Add a method in the `hostel.hostel` model to compute the URL for a hostel:

```
@api.depends('name')
def _compute_website_url(self):
    for hostel in self:
        hostel.website_url = '/hostels/%s' % (slug(hostel))
```

2. Add a button in the form view to redirect to the website:

```
...
<sheet>
    <div class="oe_button_box" name="button_box">
        <field name="is_published"
               widget="website_redirect_button"/>
    </div>
...
```

Once you add the button, you will be able to see the button in the hostel's form view, and by clicking on it, you will be redirected to the hostel's details page.

15
Web Client Development

Odoo's web client, or backend, is where employees spend most of their time.

In *Chapter 9, Backend Views*, you saw how to use the existing functionality that backends provide. Here, we'll take a look at how to extend and customize those functionalities.

The `web` module contains everything related to the user interface in Odoo.

All of the code in this chapter will depend on the `web` module. As you know, Odoo has two different editions (Enterprise and Community).

The Community version uses the `web` module for user interfaces, while the Enterprise version uses an extended version of the Community `web` module, which is the `web_enterprise` module.

The Enterprise version provides more features than the Community version, including mobile compatibility, searchable menus, and material design. We'll work on the Community version here. Don't worry—the modules developed in Community work perfectly in Enterprise because, internally, `web_enterprise` depends on the Community `web` module and just adds some features to it.

> **Important information**
>
> Odoo 17 is a bit different for the backend web client compared to other Odoo versions. It contains two different frameworks to maintain the GUI of the Odoo backend. The first one is the widget-based legacy framework, and the second one is the component-based modern framework called the **Odoo Web Library** (**OWL**). OWL is the new UI framework introduced in Odoo v16. Both use QWeb templates for structure, but there are significant changes in the syntax and the way those frameworks work.
>
> Although Odoo 17 has a new framework OWL, Odoo does not use this new framework everywhere. Most of the web client is still written with the old widget-based framework. In this chapter, we will see how to customize the web client using a widget-based framework. In the next chapter, we will look at the OWL framework.

In this chapter, you will learn how to create new field widgets to get input from users. We will also be creating a new view from scratch. After reading this chapter, you will be able to create your own UI elements in the Odoo backend.

> **Note**
> Odoo's user interface heavily depends on JavaScript. Throughout this chapter, we will assume you have a basic knowledge of JavaScript, jQuery, and SCSS.

In this chapter, we will cover the following recipes:

- Creating custom widgets
- Using client-side QWeb templates
- Making RPC calls to the server
- Creating a new view
- Debugging your client-side code
- Improving onboarding with tours
- Mobile app JavaScript

Technical requirements

The technical requirement for this chapter is the online Odoo platform.

All the code used in this chapter can be downloaded from the GitHub repository at `https://github.com/PacktPublishing/Odoo-17-Development-Cookbook-Fifth-Edition/tree/main/Chapter15`

Creating custom widgets

As you saw in *Chapter 9, Backend Views*, we can use widgets to display certain data in different formats. For example, we used `widget='image'` to display a binary field as an image. To demonstrate how to create your own widget, we'll write one widget that lets the user choose an integer field, but we will display it differently. Instead of an input box, we will display a color picker so that we can select a color number. Here, each number will be mapped to its related color.

Getting ready

For this recipe, we will be using the `my_hostel` module with basic fields and views. You will find the basic `my_hostel` module in the `Chapter15/00_initial_module` directory in the GitHub repository.

How to do it...

We'll add a JavaScript file that contains our widget's logic, an XML file that contains design logic, and an SCSS file to do some styling. Then, we will add one integer field to the books form to use our new widget.

Follow these steps to add a new field widget:

1. This widget can be written with a very small amount of JavaScript. Let's create a file called `static/src/js/field_widget.js` with this:

```
/** @odoo-module */
import { Component} from "@odoo/owl";
import { registry } from "@web/core/registry";
```

2. Create your widget by extending `Component`:

```
export class CategColorField extends Component {
```

3. Capture the JavaScript color widget code:

```
export class CategColorField extends Component {
    setup() {
      this.totalColors = [1,2,3,4,5,6];
      super.setup();
   }
   clickPill(value) {
     this.props.record.update({ [this.props.name]: value });
   }
}
```

4. Set the `template` and the supported field types for the widget:

```
CategColorField.template = "CategColorField";
CategColorField.supportedTypes = ["integer"];
```

5. In the same file, register the component to the `fields` registry:

```
registry.category("fields").add("category_color", {
  component: CategColorField,
});
```

6. Add QWeb template design code in `static/src/xml/field_widget.xml`:

```
<templates xml:space="preserve">
  <t t-name="CategColorField" owl="1">
    <div>
      <t t-foreach="totalColors" t-as="color" t-key="color">
```

```
            <span t-attf-class="o_color_pill o_color_#{color}
               {{props.record.data[props.name] == color ? 'active':
''}}"
               t-att-data-value="color"
               t-on-click="() => this.clickPill(color)"/>
         </t>
         <div class="categInformationPanel
"/>
      </div>
   </t>
</templates>
```

7. Add SCSS in `static/src/scss/field_widget.scss`:

```
.o_field_category_color {
  .o_color_pill {
    display: inline-block;
    height: 25px;
    width: 25px;
    margin: 4px;
    border-radius: 15px;
    position: relative;
    @for $size from 1 through length($o-colors) {
      &.o_color_#{$size - 1} {
        background-color: nth($o-colors, $size);
        &:not(.readonly):hover {
          transform: scale(1.2);
          transition: 0.3s;
          cursor: pointer;
        }
        &.active:after{
          content: "\f00c";
          display: inline-block;
          font: normal 14px/1 FontAwesome;
          font-size: inherit;
          color: #fff;
          position: absolute;
          padding: 4px;
          font-size: 16px;
        }
      }
    }
  }
}
```

8. Register files in the manifest file:

```
'assets': {
  'web.assets_backend': [
    'my_hostel/static/src/scss/field_widget.scss',
    'my_hostel/static/src/js/field_widget.js',
    'my_hostel/static/src/xml/field_widget.xml',
  ],
}
```

9. Then, add the `Category` integer field to the `hostel.room` model:

```
category = fields.Integer('Category')
```

10. Add the category field to the hostel form view, and then add `widget="category_color"`:

```
<field name="category" widget="category_color"/>
```

Update the module to apply the changes. After the update, open the hostel form view and you will see the category color picker, as shown in the following screenshot:

Figure 15.1 – How the custom widget is displayed

How it works...

In *step 1*, we imported the `Component` and registry.

In *step 2*, we created a `CategColorField` by extending the `Component`. Through this, `CategColorField` will get all the properties and methods from the `Component`.

In *step 3*, we inherited the `setup` method and set the value of the `this.totalColors` attribute. We will use this variable to decide on the number of color pills. We want to display six color pills, so we assigned `[1,2,3,4,5,6]`.

In *step 4*, we added the `clickPill` handler method to manage pill clicks. To set the field value, we used `this. props.update` method. This method is added from the `Component` class.

In *step 5*, we added a template name where we rendered the `CategColorField` design and set supported types.

`supportedTypes` has been used to decide which types of field are supported by this widget. In our case, we want to create a widget for integer fields.

In *step 6*, after registering the component to the fields registry.

Finally, we exported our `widget` class so that other add-ons can extend it or inherit from it. Then, we added a new integer field called category to the `hostel.room` model. We also added the same field to the form view with the `widget="category_color"` attribute. This will display our widget in the form instead of the default integer widget.

Using client-side QWeb templates

Just as it's a bad habit to programmatically create HTML code in JavaScript, you should only create minimal DOM elements in your client-side JavaScript code. Fortunately, there's a templating engine available for the client side.

A client-side template engine is also available in Odoo. This template engine is known as **Qweb Templates** and is carried out completely in JavaScript code and rendered inside the browser.

Getting ready

For this recipe, we will be using the `my_hostel` module from the previous recipe and add `informationPanel` below the category color icon.

Using `renderToElement`, we render the category information element and set it on `informationPanel`.

How to do it...

We need to add the QWeb definition to the manifest and change the JavaScript code so that we can use it. Perform the following steps to get started:

1. Import `@web/core/utils/render` and extract the `renderToElement` reference to a variable, as shown in the following code:

```
import { renderToElement } from "@web/core/utils/render";
```

2. Add the template file to `static/src/xml/field_widget.xml`:

```
<t t-name="CategColorField">
  <div>
    <t t-foreach="totalColors" t-as="color" t-key="color">
       <span t-attf-class="o_color_pill o_color_#{color}
            {{props.record.data[props.name] == color ? 'active':
''}}"
            t-att-data-value="color"
            t-on-click="() => this.clickPill(color)"
            t-on-mouseover.prevent="categInfo"/>
    </t>
    <div class="categInformationPanel"/>
```

```
    </div>
  </t>

  <t t-name="CategInformation">
    <div t-attf-class="categ_info o_color_pill o_color_#{value}">
      <t t-if="value == 1">
        Single Room With AC<br/>
        <ul>
          <li>
            Small Dressing Table
          </li>
          <li>
            Small Bedside Table
          </li>
          <li>
            Small Writing Table
          </li>
          <li>
            Attached Bathroom
          </li>
        </ul>
      </t>
      <t t-if="value == 2">
        Single Room With None AC<br/>
        <ul>
          <li>
            Small Dressing Table
          </li>
          <li>
            Small Bedside Table
          </li>
          <li>
            Small Writing Table
          </li>
          <li>
            Attached Bathroom
          </li>
        </ul>
      </t>
      <t t-if="value == 3">
        King Double Room With AC<br/>
        <ul>
          <li>
```

```
          King Size Double Bed
        </li>
        <li>
          Small Dressing Table
        </li>
        <li>
          Small Bedside Table
        </li>
        <li>
          Small Writing Table
        </li>
        <li>
          TV
        </li>
        <li>
          Small Fridge
        </li>
        <li>
          Attached Bathroom
        </li>
      </ul>
    </t>
    <t t-if="value == 4">
      King Double Room With None AC<br/>
      <ul>
        <li>
          King Size Double Bed
        </li>
        <li>
          Small Dressing Table
        </li>
        <li>
          Small Bedside Table
        </li>
        <li>
          Small Writing Table
        </li>
        <li>
          TV
        </li>
        <li>
          Small Fridge
        </li>
```

```
        <li>
          Attached Bathroom
        </li>
      </ul>
    </t>
    <t t-if="value == 5">
      Queen Double Room With AC<br/>
      <ul>
        <li>
          Queen Size Double Bed
        </li>
        <li>
          Small Dressing Table
        </li>
        <li>
          Small Bedside Table
        </li>
        <li>
          Small Writing Table
        </li>
        <li>
          TV
        </li>
        <li>
          Small Fridge
        </li>
        <li>
          Attached Bathroom
        </li>
      </ul>
    </t>
    <t t-if="value == 6">
      Queen Double Room With None AC<br/>
      <ul>
        <li>
          Queen Size Double Bed
        </li>
        <li>
          Small Dressing Table
        </li>
        <li>
          Small Bedside Table
        </li>
```

```
          <li>
            Small Writing Table
          </li>
          <li>
            TV
          </li>
          <li>
            Small Fridge
          </li>
          <li>
            Attached Bathroom
          </li>
        </ul>
      </t>
    </div>
  </t>
```

3. Add the mouseover function and, in this function, simply render the category information element using `renderToElement.render` and append the `categ` information element to `categInformationPanel`:

```
categInfo(ev){
  var $target = $(ev.target);
  var data = $target.data();
  $target.parent().find(".categInformationPanel").html(
    $(renderToElement('CategInformation',{
      'value': data.value,
      'widget': this
    }))
  )
}
```

4. Add SCSS in `static/src/scss/field_widget.scss` to set the category information style:

```
.categInformationPanel .categ_info{
  padding: 10px;
  height: 100%;
  width: 100%;
  color: white;
  font-weight: bold;
}
```

5. Register the QWeb file in your manifest:

```
'assets': {
  'web.assets_backend': [
    'my_hostel/static/src/scss/field_widget.scss',
    'my_hostel/static/src/js/field_widget.js',
    'my_hostel/static/src/xml/field_widget.xml',
  ],
}
```

Restart the server to apply the changes. After the restart, open the hotel form view and you will see the category information panel, as shown in the following screenshot:

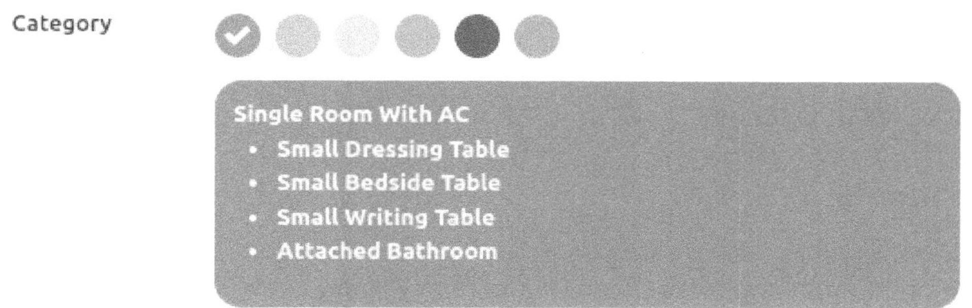

Figure 15.2 – Category information panel

When we hover over the category color icon:

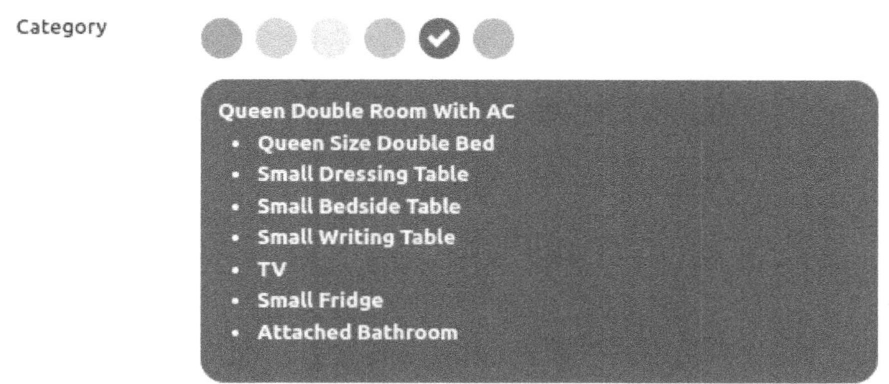

Figure 15.3 – Hovering over the category icon displays category information

How it works...

As there is already a comprehensive discussion on the basics of QWeb in the *Creating or Modifying Templates – QWeb* recipe in *Chapter 14, CMS Website Development*, we'll focus on what is different here. First of all, you need to realize that we're dealing with the JavaScript QWeb implementation, as opposed to the Python implementation on the server side. This means that you don't have access to browsing records or the environment; you only have access to the parameters you have passed from the `renderToElement` function.

In our case, we have passed the current object via the `widget` key. This means that you should have all the intelligence in the widget's JavaScript code and have your template only access properties, or possibly functions. Given that we can access all the available properties on the widget, we can simply check the value in the template by checking the hover category color property.

As client-side QWeb has nothing to do with QWeb views, there's a different mechanism to make those templates known to the web client—add them via the QWeb key to your add-on's manifest in a list of filenames relative to the add-on's root.

There's more...

The reason for going to the effort of using QWeb here was extensibility. If, for example, we want to add info icons to our widget from another module, we'll use the following code to have an icon in each pill:

```
<t t-name="CategInformationCustom" t-inherit="my_hostel.
CategInformation"
    t-inherit-mode="extension">
    <xpath expr='//t[@t-if="value == 1"]' position="before">
      <i class="fa fa-info-circle" aria-hidden="true"></i>
    </xpath>
  </t>
```

Category

King Double Room With None AC

- King Size Double Bed
- Small Dressing Table
- Small Bedside Table
- Small Writing Table
- TV
- Small Fridge
- Attached Bathroom

Figure 15.4 – Info icon on the category information panel

> **Note**
>
> If you want to learn more about the QWeb templates, refer to the following points:
>
> - The client-side QWeb engine has less convenient error messages and handling than other parts of Odoo. A small error often means that nothing happens, and it's hard for beginners to continue from there.
>
> - Fortunately, there are some debug statements for client-side QWeb templates that will be described later in this chapter in the *Debugging your client-side code* recipe.

Making RPC calls to the server

Sooner or later, your widget will need to look up some data from the server. In this recipe, we will add a category booked panel to the category information panel. When the user hovers their cursor over the category color pill element, the booked panel will show the number of booked rooms related to that category color. We will make an RPC call to the server to fetch a book count of the data associated with that particular category.

Getting ready

For this recipe, we will be using the `my_hostel` module from the previous recipe.

How to do it...

Follow these steps to make an RPC call to the server and display the result in a `colorPreviewPanel`:

1. Import `@odoo/owl` and extract the `onWillStart`, `onWillUpdateProps` reference to a variable, as shown in the following code:

```
import { Component, onWillStart , onWillUpdateProps} from "@
odoo/owl";
```

2. Add the `onWillStart` method to the `setup` method and call our custom `load ColorData` method:

```
onWillStart(() => {
  this.loadCategInformation();
});
onWillUpdateProps(() => {
  this.loadCategInformation();
});
```

3. Add the `loadCategInformation` method and set `categInfoData` in the RPC call:

```
async loadCategInformation() {
  var self = this;
```

```
    self.categoryInfo = {};
    var resModel = self.env.model.root.resModel;
    var domain = [];
    var fields = ['category'];
    var groupby = ['category'];
    const categInfoPromise = await self.env.services.orm.
readGroup(
    resModel,
    domain,
    fields,
    groupby
);
    categInfoPromise.map((info) => {
      self.categoryInfo[info.category] = info.category_count;
    });
}
```

4. Update `CategoryInformation` template and add the count data:

```
<t t-name="CategInformationCustom" t-inherit="my_hostel.
CategInformation"
    t-inherit-mode="extension">
    <xpath expr='//t[@t-if="value == 1"]' position="before">
      <i class="fa fa-info-circle" aria-hidden="true"></i>
    </xpath>
    <xpath expr="//div" position="inside">
      <div class="text-center"
         style="color:gray;background:
white;padding:3px;padding: 5px;border-radius: 5px;">
        Total Booked Rooms: <t t-esc="widget.categoryInfo[value]
or 0"/>
      </div>
    </xpath>
  </t>
```

Update the module to apply the changes. After the update, you will see a count of category information, as shown in the following screenshot:

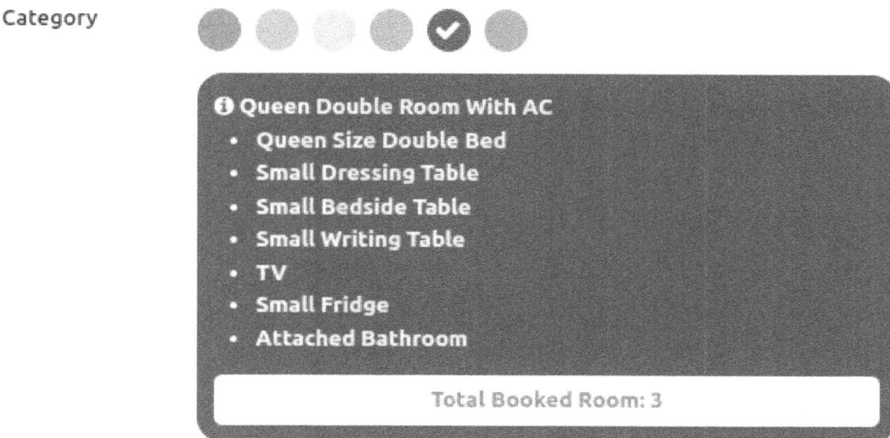

Figure 15.5 – Data Fetch Using RPC

How it works...

The `onWillStart` hook will be called just before the component is rendered for the first time. It will be useful if we need to do some actions before the component is rendered to the view, such as loading some initial data.

`onWillUpdateProps` is also an asynchronous hook that is called whenever an update is made to a related component. The reactive nature of the OWL framework can be maintained using this amazing hook.

When dealing with data access, we rely on the `_rpc` function provided by the `ORM` class, as we explained earlier. This function allows you to call any public function on models such as `search`, `read`, `write`, or, in this case, `read_group`.

In *step 1*, we made an RPC call and invoked the `read_group` method on the current model, which is `hostel.room` in our case. We grouped data based on the `category` field so that the RPC call will return book data that were grouped by `category` and add an aggregate in the `category_count` key. We also mapped the `category_count` and `category` index in the `categoryInfo` so that we could use it in the QWeb template.

Step 2 is nothing special. We just initialized the bootstrap tooltip.

In *step 3*, we used `categoryInfo` to set the attributes that are needed to display the category information. In the `loadCategInformation` method, we assigned a color map via `this.categoryInfo` so that you can access them in the QWeb template via `widget.categoryInfo`. This is because we passed the widget reference; this is the `renderToElement` method.

See also

Odoo's RPC returns JavaScript's native Promise object. You will get the requested data once the Promise is resolved. You can learn more about Promise here: https://developer.mozilla.org/en-US/docs/Web/JavaScript/Reference/Global_Objects/Promise

Creating a new view

As you saw in *Chapter 9, Backend Views*, there are different kinds of views, such as form, list, and kanban. In this recipe, we will create a new view. This view will display the list of rooms, along with their students.

Getting ready

For this recipe, we will be using the my_hostel module from the previous recipe. Note that views are very complex structures, and each view has a different purpose and implementation. The purpose of this recipe is to make you aware of the MVC pattern view and how to create simple views. In this recipe, we will create a view called m2m_group, the purpose of which is to display records in groups. To divide records into different groups, the view will use the many2x field data. In the my_hostel module, we have the room_id field. Here, we will group students based on room and display them in cards.

In addition, we will add a new button to the control panel. With the help of this button, you will be able to add a new student record. We will also add a button to the room's card so that we can redirect users to another view.

How to do it...

Follow these steps to add a new view called m2m_group:

1. Add a new view type in ir.ui.view:

```
class View(models.Model):
  _inherit = 'ir.ui.view'

  type = fields.Selection(selection_add=[('m2m_group', 'M2m
Group')])
```

2. Add a new view mode in ir.actions.act_window.view:

```
class ActWindowView(models.Model):
  _inherit = 'ir.actions.act_window.view'

  view_mode = fields.Selection(selection_add=[
 ('m2m_group', 'M2m group')], ondelete={'m2m_group': 'cascade'})
```

3. Add a new method by inheriting from the base model. This method will be called from the JavaScript model (see *step 4* for more details):

```
class Base(models.AbstractModel):
  _inherit = 'base'

  @api.model
  def get_m2m_group_data(self, domain, m2m_field):
    records = self.search(domain)
    result_dict = {}
    for record in records:
      for m2m_record in record[m2m_field]:
        if m2m_record.id not in result_dict:
          result_dict[m2m_record.id] = {
            'name': m2m_record.name,
            'children': [],
            'model': m2m_record._name
          }
        result_dict[m2m_record.id]['children'].append({
          'name': record.display_name,
          'id': record.id,
        })
    return result_dict
```

4. Add a new file called `/static/src/js/m2m_group_model.js` and add the following content to it:

```
/** @odoo-module **/

import { Model } from "@web/model/model";

export class M2mGroupModel extends Model {
  setup(params) {
    const metaData = Object.assign({}, params.metaData, {});
    this.data = params.data || {};
    this.metaData = this._buildMetaData(metaData);
    this.m2m_field = this.metaData.m2m_field;
  }
  _buildMetaData(params) {
    const metaData = Object.assign({}, this.metaData, params);
    return metaData;
  }
  async load(searchParams) {
    var self = this;
```

```
        const model = self.metaData.resModel;
        const method = 'get_m2m_group_data'
        const m2m_field = self.m2m_field
        const result = await this.orm.call(
          model,
          method,
          [searchParams.domain, m2m_field]
        )
        self.data = result;
        return result;
      }
    }
```

5. Add a new file called `/static/src/js/m2m_group_controller.js` and add the following content to it:

```
/** @odoo-module **/

import { useService } from "@web/core/utils/hooks";
import { Layout } from "@web/search/layout";
import { useModelWithSampleData } from "@web/model/model";
import { standardViewProps } from "@web/views/standard_view_
props";
import { Component } from "@odoo/owl";

export class M2mGroupController extends Component {
  setup() {
    this.actionService = useService("action");
    this.model = useModelWithSampleData(this.props.Model, this.
props.modelParams);
  }
  _onBtnClicked(domain) {
    this.actionService.doAction({
      type: 'ir.actions.act_window',
      name: this.model.metaData.title,
      res_model: this.props.resModel,
      views: [[false, 'list'], [false, 'form']],
      domain: domain,
    });
  }
  _onAddButtonClick(ev) {
    this.actionService.doAction({
      type: 'ir.actions.act_window',
```

```
        name: this.model.metaData.title,
        res_model: this.props.resModel,
        views: [[false, 'form']],
        target: 'new'
      });
    }
  }
M2mGroupController.template = "M2mGroupView";
M2mGroupController.components = { Layout };

M2mGroupController.props = {
  ...standardViewProps,
  Model: Function,
  modelParams: Object,
  Renderer: Function,
  buttonTemplate: String,
};
```

6. Add a new file called `/static/src/js/m2m_group_renderer.js` and add the following content to it:

```
/** @odoo-module **/

import { Component } from "@odoo/owl";
export class M2mGroupRenderer extends Component {
  onClickViewButton(group) {
     var children_ids = group.children.map((group_id) => {
      return group_id.id;
    });
    const domain = [['id', 'in', children_ids]]
    this.props.onClickViewButton(domain);
  }
  get groups() {
    return this.props.model.data
  }
}
M2mGroupRenderer.template = "M2mGroupRenderer";
M2mGroupRenderer.props = ["model", "onClickViewButton"];
```

7. Add a new file called `/static/src/js/m2m_group_arch_parser.js` and add the following content to it:

```
/** @odoo-module **/

import { visitXML } from "@web/core/utils/xml";
```

```
export class M2mGroupArchParser {
  parse(arch, fields = {}) {
    const archInfo = { fields, fieldAttrs: {} };
    visitXML(arch, (node) => {
      switch (node.tagName) {
        case "m2m_group": {
          const m2m_field = node.getAttribute("m2m_field");
          if (m2m_field) {
            archInfo.m2m_field = m2m_field;
          }
          const title = node.getAttribute("string");
          if (title) {
            archInfo.title = title;
          }
          break;
        }
        case "field": {
          const fieldName = node.getAttribute("name"); // exists
(rng validation)
          if (fieldName === "id") {
            break;
          }
          const string = node.getAttribute("string");
          if (string) {
            if (!archInfo.fieldAttrs[fieldName]) {
              archInfo.fieldAttrs[fieldName] = {};
            }
            archInfo.fieldAttrs[fieldName].string = string;
          }
          const modifiers = JSON.parse(node.
getAttribute("modifiers") || "{}");
          if (modifiers.invisible === true) {
            if (!archInfo.fieldAttrs[fieldName]) {
              archInfo.fieldAttrs[fieldName] = {};
            }
            archInfo.fieldAttrs[fieldName].isInvisible = true;
            break;
          }
          break;
        }
      }
    });
    return archInfo;
```

```
    }
  }
```

8. Add a new file called `/static/src/js/m2m_group_view.js` and add the following content to it:

```
/** @odoo-module **/

import { _lt } from "@web/core/l10n/translation";
import { registry } from "@web/core/registry";
import { M2mGroupArchParser } from "./m2m_group_arch_parser";
import { M2mGroupController } from "./m2m_group_controller";
import {  M2mGroupModel } from "./m2m_group_model";
import { M2mGroupRenderer } from "./m2m_group_renderer";

const viewRegistry = registry.category("views");

export const M2mGroupView = {
  type: "m2m_group",
  display_name: _lt("Author"),
  icon: "fa fa-id-card-o",
  multiRecord: true,
  Controller: M2mGroupController,
  Renderer: M2mGroupRenderer,
  Model: M2mGroupModel,
  ArchParser: M2mGroupArchParser,
  searchMenuTypes: ["filter", "favorite"],
  buttonTemplate: "ViewM2mGroup.buttons",
  props: (genericProps, view) => {
    const modelParams = {};
    const { arch, fields, resModel } = genericProps;
    // parse arch
    const archInfo = new view.ArchParser().parse(arch);
    modelParams.metaData = {
      m2m_field: archInfo.m2m_field,
      fields: fields,
      fieldAttrs: archInfo.fieldAttrs,
      resModel: resModel,
      title: archInfo.title || _lt("Untitled"),
      widgets: archInfo.widgets,
    };
    return {
      ...genericProps,
      Model: view.Model,
```

```
            modelParams,
            Renderer: view.Renderer,
            buttonTemplate: view.buttonTemplate,
        };
    },
};
viewRegistry.add("m2m_group", M2mGroupView);
```

9. Add the QWeb template for the view to the `/static/src/xml/m2m_group_controller.xml` file:

```
<?xml version="1.0" encoding="UTF-8"?>
<templates xml:space="preserve">
  <t t-name="M2mGroupView" owl="1">
    <div t-att-class="props.className" t-ref="root">
      <Layout display="props.display">
        <t t-set-slot="layout-buttons">
          <t t-call="{{ props.buttonTemplate }}"/>
        </t>
         <div>
            <t t-component="props.Renderer"
             model="model"
             onClickViewButton="group => this._
onBtnClicked(group)"/>
         </div>
      </Layout>
    </div>
  </t>
</templates>
```

10. Add the QWeb template for the view to the `/static/src/xml/m2m_group_renderer.xml` file:

```
<?xml version="1.0" encoding="UTF-8"?>
<templates xml:space="preserve">
  <t t-name="M2mGroupRenderer" owl="1">
    <div class="row ml16 mr16">
      <div t-foreach="groups" t-as="group" class="col-3"
t-key="group">
        <t t-set="group_data" t-value="groups[group]" />
        <div class="card mt16">
          <img class="card-img-top"
          t-attf-src="/web/image/#{group_data.model}/#{group}/
image"
          style="height: 300px;"/>
```

```
            <div class="card-body">
              <h5 class="card-title mt8">
                <t t-esc="group_data['name']"/>
              </h5>
            </div>
            <ul class="list-group list-group-flush">
              <t t-foreach="group_data['children']" t-as="child"
t-key="child.id">
                <li class="list-group-item">
                  <i class="fa fa-user"/><t t-esc="child.name"/>
                </li>
              </t>
            </ul>
            <div class="card-body">
              <a href="#" class="btn btn-sm btn-primary o_primay_
button"
               t-att-data-group="group"
               t-on-click="() => this.onClickViewButton(group_
data)">View</a>
            </div>
          </div>
        </div>
      </div>
    </t>
</templates>
```

11. Add the QWeb template for the view to the `/static/src/xml/m2m_group_view.xml` file:

```
<?xml version="1.0" encoding="UTF-8"?>
<templates xml:space="preserve">
  <t t-name="ViewM2mGroup.buttons" owl="1">
    <button type="button" class="btn btn-primary"
        t-on-click="() => this._onAddButtonClick()">
      Add Record
    </button>
  </t>
</templates>
```

12. Add all of the JavaScript and XML files to the backend assets:

```
'assets': {
  'web.assets_backend': [
    'my_hostel/static/src/js/m2m_group_arch_parser.js',
    'my_hostel/static/src/js/my_hostel_tour.js',
```

```
        'my_hostel/static/src/js/m2m_group_view.js',
        'my_hostel/static/src/js/m2m_group_renderer.js',
        'my_hostel/static/src/js/m2m_group_model.js',
        'my_hostel/static/src/js/m2m_group_controller.js',
        'my_hostel/static/src/xml/m2m_group_controller.xml',
        'my_hostel/static/src/xml/m2m_group_renderer.xml',
        'my_hostel/static/src/xml/m2m_group_view.xml',
        'my_hostel/static/src/xml/field_widget.xml',
    ],
},
```

13. Finally, add our new view for the `hostel.student` model:

```
<record id="view_hostel_student_m2m_group" model="ir.ui.view">
  <field name="name">Students</field>
  <field name="model">hostel.student</field>
  <field name="arch" type="xml">
    <m2m_group m2m_field="room_id">
    </m2m_group>
  </field>
</record>
```

14. Add `m2m_group` to the action:

```
<field name="view_mode">tree,m2m_group,form</field>
```

Update the `my_hostel` module to open the **Students** view, and then, from the view switcher, open the new view that we just added. This will look as follows:

Figure 15.6 – Many2many group view

Figure 15.7 – Many2many group view

> **Important information**
>
> Odoo views are very easy to use and are very flexible. However, it is often the case that easy and flexible things have complex implementations under the hood.
>
> This is true of Odoo JavaScript views: they are easy to use, but complex to implement. They consist of lots of components, including the model, renderer, controller, view, and QWeb template. In the next section, we have added all of the required components for the views and have also used a new view for the `my_hostel` model. If you don't want to add everything manually, grab a module from the example file in this book's GitHub repository.

How it works...

In *steps 1* and *2*, we registered a new type of view, called `m2m_group`, in `ir.ui.view` and `ir.actions.act_window.view`.

In *step 3*, we added the `get_m2m_group_data` method to the base. Adding this method to the base will make that method available in every model. This method will be called via an RPC call from the JavaScript view. The view will pass two parameters—the `domain` and `m2m_field`. In the `domain` argument, the value of the domain will be the domain generated with a combination of the search view domain and the action domain. `m2m_field` is the field name by which we want to group the records. This field will be set on the view definition.

In the next few steps, we added the JavaScript files that are required to new the view. An Odoo JavaScript view consists of the view, model, renderer, and controller. The word view has historical meaning in the Odoo code base, so **model, view, controller** (**MVC**) becomes **model, renderer, controller** (**MRC**) in Odoo. In general, the view sets up the model, renderer, and controller, and sets the MVC hierarchy so that it looks similar to the following:

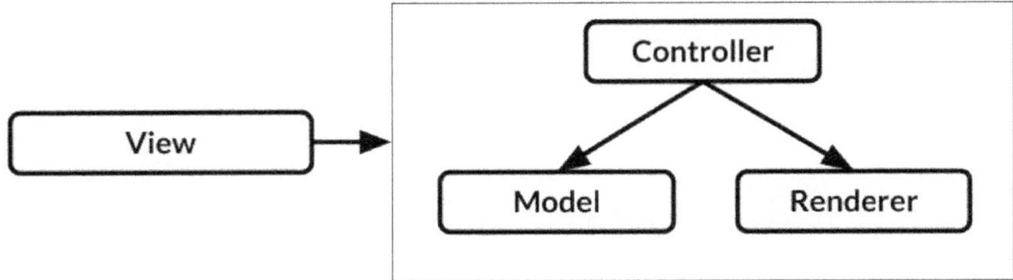

Figure 15.8 – View components

Its job is to get a set of fields, arch, context, and some other parameters, then to construct a controller/renderer/model triplet:

- The view's role is to properly set up each piece of the MVC pattern with the correct information. Usually, it has to process the arch string and extract the data necessary for each other parts of the view.

  Note that the view is a class, not a widget. Once its job has been done, it can be discarded.

- The renderer has one job: representing the data being viewed in a DOM element. Each view can render the data in a different way. Also, it should listen to appropriate user actions and notify its parent (the controller) if necessary. The renderer is the V in the MVC pattern.
- The model: its job is to fetch and hold the state of the view. Usually, it represents in some way a set of records in the database. The model is the owner of the business data. It is the M in the MVC pattern.
- The controller: Its job is to coordinate the renderer and the model. Also, it is the main entry point for the rest of the web client. For example, when the user changes something in the search view, the `update` method of the controller will be called with the appropriate information. It is the C in the MVC pattern.

> **Note**
>
> The JavaScript code for the views has been designed to be usable outside of the context of a view manager/action manager. It could be used in a client action, or it could be displayed on the public website (with some work on the assets).

In *step 8*, we added JavaScript and XML files to the assets.

Finally, in the last two steps, we added a view definition for the `hostel.student` model.

In *step 9*, we used the `<m2m_group>` tag for the view, and we also passed the `m2m_field` attribute as the option. This will be passed to the model to fetch the data from the server.

Debugging your client-side code

This book contains a whole chapter for debugging server-side code, *Chapter 7, Debugging Modules*. For the client-side part, you'll get a kick-start in this recipe.

Getting ready

This recipe doesn't rely on specific code, but if you want to be able to reproduce exactly what's going on, grab the previous recipe's code.

How to do it...

What makes debugging client-side scripts difficult is that the web client relies heavily on jQuery's asynchronous events. Given that breakpoints halt the execution, there is a high chance that a bug caused by timing issues will not occur when debugging. We'll discuss some strategies for this later:

1. For the client-side debugging, you will need to activate debug mode with the assets. If you don't know how to activate debug mode with the assets, read the *Activating the Odoo developer tools* recipe in *Chapter 1, Installing the Odoo Development Environment*.
2. In the JavaScript function you're interested in, call `debugger`:

   ```
   debugger;
   ```

3. If you have timing problems, log in to the console through a JavaScript function:

   ```
   console.log("Debugging call……");
   ```

4. If you want to debug during template rendering, call the debugger from QWeb:

   ```
   <t t-debug="" />
   ```

5. You can also have QWeb log in to the console, as follows:

   ```
   <t t-log="myvalue" />
   ```

All of this relies on your browser offering the appropriate functionality for debugging. While all major browsers do that, we'll only look at Chromium here, for demonstration purposes. To be able to use the debug tools, open them by clicking on the top-right menu button and selecting **More tools | Developer tools**:

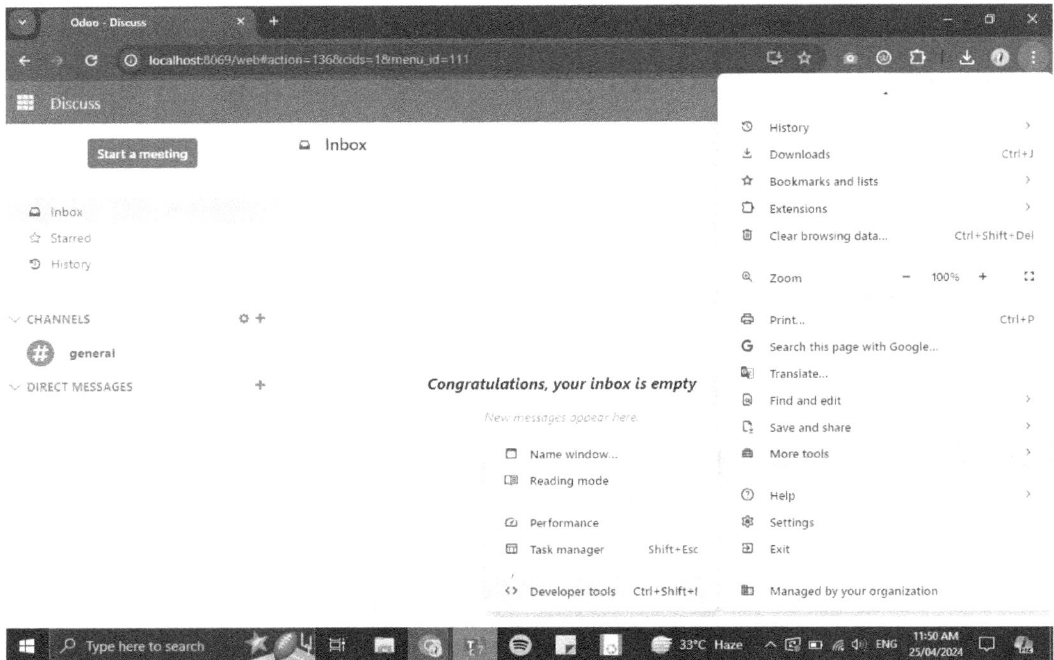

Figure 15.9 – Opening Developer Tools in Chrome

How it works...

When the debugger is open, you should see something similar to the following screenshot:

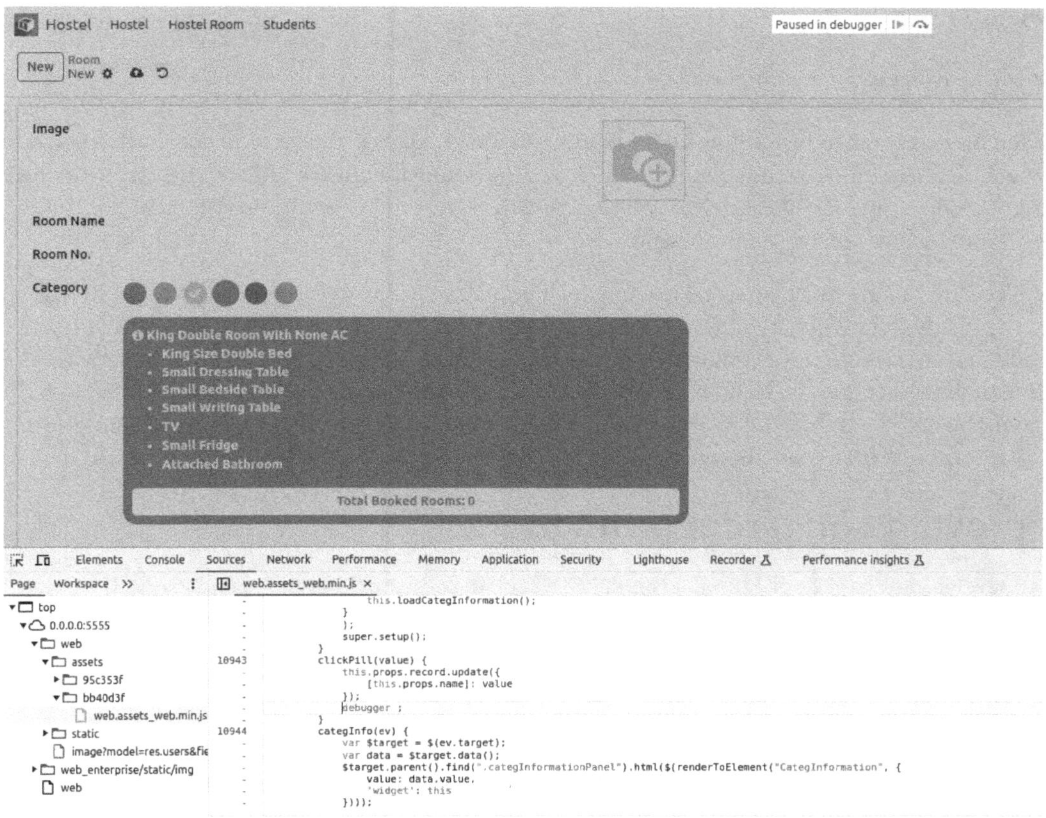

Figure 15.10 – Opening Developer Tools in Chrome

Here, you have access to a lot of different tools in separate tabs. The currently active tab in the preceding screenshot is the JavaScript debugger, and we have set a breakpoint in line 31 by clicking on the line number. Every time our widget fetches the list of users, the execution should stop at this line, and the debugger will allow you to inspect variables or change their values. Within the watch list to the right, you can also call functions to try out their effects without having to continuously save your script file and reload the page.

The debugger statements we described earlier will behave the same as soon as you have the developer tools open. The execution will then stop, and the browser will switch to the **Sources** tab, with the file in question opened and the line with the debugger statement highlighted.

The two logging possibilities from earlier will end up on the **Console** tab. This is the first tab you should inspect in case of problems in any case because, if some JavaScript code doesn't load at all because of syntax errors or similar fundamental problems, you'll see an error message there explaining what's going on.

There's more...

Use the **Elements** tab to inspect the DOM representation of the page the browser currently displays. This will prove helpful when it comes to familiarizing yourself with the HTML code the existing widgets produce, and it will also allow you to play with classes and CSS attributes, in general. This is a great resource for testing layout changes.

The **Network** tab gives you an overview of which requests the current page made and how long it took. This is helpful when it comes to debugging slow page loads as, in the **Network** tab, you will usually find the details of the requests. If you select a request, you can inspect the payload that was passed to the server and the result returned, which helps you to figure out the reason for unexpected behavior on the client side. You'll also see the status codes of requests made—for example, 404—in case a resource can't be found because you misspelled a filename, for instance.

Improving onboarding with tours

After developing a large application, it is crucial to explain software flows to the end users. The Odoo framework includes a built-in tour manager. With this tour manager, you can guide an end user through learning specific flows. In this recipe, we will create a tour so that we can create a book in the library.

Getting ready

We will be using the `my_hostel` module from the previous recipe. Tours are only displayed in the database without demo data, so if you are using a database with demo data, create a new database without demo data for this recipe.

How to do it...

To add a tour to a hostel, follow these steps:

1. Add a new `/static/src/js/my_hostel_tour.js` file with the following code:

```
/** @odoo-module **/

import { _t } from "@web/core/l10n/translation";
import { registry } from "@web/core/registry";
import { markup } from "@odoo/owl"
import { stepUtils } from "@web_tour/tour_service/tour_utils";
```

```
registry.category("web_tour.tours").add('hostel_tour',  {
  url: "/web",
  rainbowManMessage: _t("Congrats, best of luck catching such
big fish! :)"),
  sequence: 5,
  steps: () => [stepUtils.showAppsMenuItem(), {
    trigger: '.o_app[data-menu-xmlid="my_hostel.hostel_main_
menu"]',
    content: markup(_t("Ready to launch your <b>hostel</b>?")),
    position: 'bottom',
  }, {
    trigger: '.o_list_button_add',
    content: markup(_t("Let's create new room.")),
    position: "bottom",
  },{
    trigger: '.o_form_button_save',
    content: markup(_t('Save this room record')),
    position: "bottom",
  }]
});
```

2. Add the tour JavaScript file in the backend assets:

```
'assets': {
  'web.assets_backend': [
    'my_hostel/static/src/scss/field_widget.scss',
    'my_hostel/static/src/js/field_widget.js',
    'my_hostel/static/src/js/my_hostel_tour.js',
    'my_hostel/static/src/xml/field_widget.xml',
  ],
},
```

Update the module and open the Odoo backend. At this point, you will see the tour, as shown in the following screenshot:

Or you can click on debug icon and click on **Start Tour**.

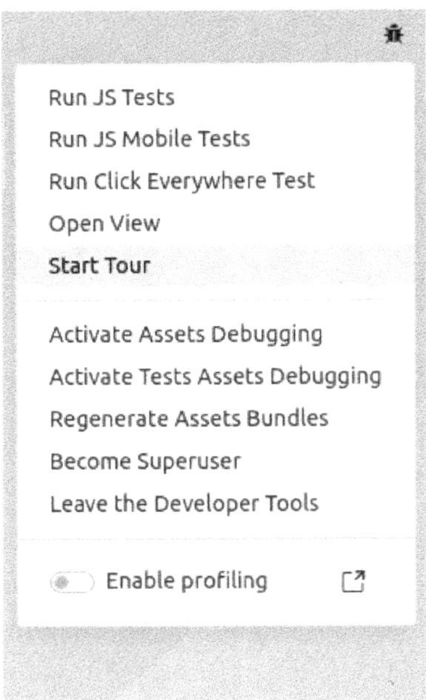

Figure 15.11 – Tour step for user onboarding

It displays below the **Tours** popup.

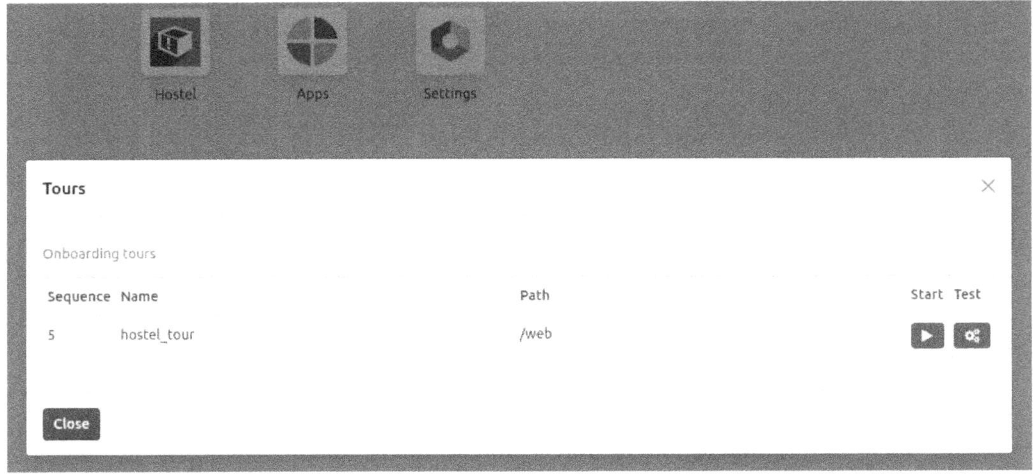

Figure 15.12 – Tour step for user onboarding

Click the start icon button to see the **Ready to lunch your hostel?** Tour content:

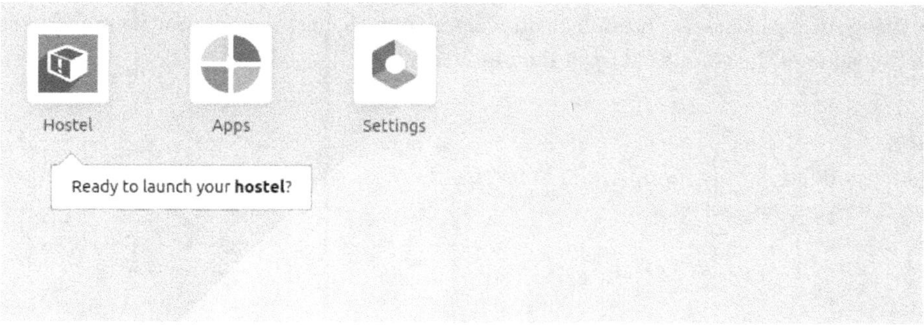

Figure 15.13 – Tour step for user onboarding

How it works...

The tour manager is available in the **web_tour.tours** category.

In the first step, we imported `registry`. We can then added a new tour with `registry.category("web_tour.tours")`. We registered our tour with the `hostel_tour` name and passed the URL on which this tour should run.

The next parameter is a list of these tour stops. A tour step requires three values. The trigger is used to select the element on which the tour should be displayed. This is a JavaScript selector. We used the XML ID of the menu because it is available in the DOM.

The first step, `stepUtils.showAppsMenuItem()`, is the predefined step from the tour for the main menu. The next key is the content, and this is displayed when the user hovers over the tour drop. We used the `markup(_t())` function because we want to translate the string, while the position key is used to decide on the position of the tour drop. Possible values are top, right, left, or bottom.

> **Important information**
>
> Tours improve the onboarding experience of the user and manage the integration tests. When you run Odoo with test mode internally, it also runs the tours and causes the test case to fail if a tour has not finished.

Mobile app JavaScript

Odoo v10 introduced the Odoo mobile application. It provides a few small utilities to perform mobile actions, such as vibrating the phone, showing a toast message, and scanning QR codes.

Getting ready

We will be using the my_hostel module from the previous recipe. We will show you the toast when we change the value of the color field from the mobile app.

> **Warning**
> The Odoo mobile app only supports the Enterprise Edition, so if you don't have the Enterprise Edition, then you cannot test it.

How to do it...

Follow these steps to show a toast message in the Odoo mobile app:

```
import mobile from "@web_mobile/js/services/core";

clickPill(value) {
    if (mobile.methods.showToast) {
      mobile.methods.showToast({ 'message': 'Color changed' });
    }
    this.props.record.update({ [this.props.name]: value });
  }
```

Update the module and open the form view of the hostel.room model in the mobile app. When you change the color, you will see the toast, as shown in the following screenshot:

Color changed

Figure 15.14 – Toast on color change

How it works...

`@web_mobile/js/services/core` provides the bridge between a mobile device and Odoo JavaScript. It exposes a few basic mobile utilities. In our example, we used the `showToast` method to display a toast in the mobile app. We also need to check the availability of the function. The reason for this is that some mobile phones might not support a few features. For example, if devices don't have a camera, then you can't use the `scanBarcode()` method. In such cases, to avoid tracebacks, we need to wrap them with an `if` condition.

There's more...

The mobile utilities that are to be found in Odoo are as follows:

- `showToast()`: To display a toast message
- `vibrate()`: To make a phone vibrate
- `showSnackBar()`: To display a snack bar with a button
- `showNotification()`: To display a mobile notification
- `addContact()`: To add a new contact to the phonebook
- `scanBarcode()`: To scan QR codes
- `switchAccount()`: To open the account switcher in Android

To learn more about mobile JavaScript, refer to `https://www.odoo.com/documentation/16.0/developer/reference/frontend/mobile.html`.

16

The Odoo Web Library (OWL)

The Odoo V17 Javascript framework uses a custom component framework called **OWL** (short for **Odoo Web Library**). It is a declarative component system loosely inspired by **Vue** and **React**. **OWL** is a component-based UI framework and uses QWeb templates for structure. OWL is very fast compared to Odoo's legacy widget system and introduces tons of new features, including **hooks**, **reactivity**, the **auto instantiation** of **subcomponents**, and more besides.

In this chapter, we will learn how to use an OWL component to generate interactive UI elements. We will start with a minimal OWL component and then we will learn about the component's life cycle. Finally, we will create a new field widget for the form view. In this chapter, we will cover the following recipes:

- Creating an OWL component
- Managing user actions in an OWL component
- Making OWL components with hooks
- Understanding the OWL component life cycle
- Adding an OWL field to the form view

> **Note**
>
> The following question may occur to you: why is Odoo not using some well-known JavaScript frameworks, such as React.js or Vue.js? Please check out the following link for more information: `https://github.com/odoo/owl/blob/master/doc/miscellaneous/comparison.md`.
>
> You can refer to `https://github.com/odoo/owl` to learn more about the OWL framework.

Technical requirements

OWL components are defined with ES6 classes. In this chapter, we will be using some ES6 syntax. Also, some ES6 syntaxes are not supported by old browsers, so make sure you are using the latest version of Chrome or Firefox. You will find the code for this chapter at `https://github.com/PacktPublishing/Odoo-17-Development-Cookbook-Fifth-Edition/tree/main/Chapter16`.

Creating an OWL component

The main building blocks of OWL are components and templates.

In OWL, every part of the UI is managed by a component: they hold the logic and define the templates that are used to render the user interface

The goal of this recipe is to learn the basics of an OWL component. We will create a minimal OWL component and append it to the Odoo web client. In this recipe, we will create a component for a small horizontal bar with some text.

Getting ready

For this recipe, we will be using the `my_hostel` module with basic fields and views. You will find the basic `my_hostel` module in the `https://github.com/PacktPublishing/Odoo-17-Development-Cookbook-Fifth-Edition/tree/main/Chapter16/00_initial_module/my_hostel` directory in the GitHub repository.

How to do it...

We will add a small horizontal bar component to the Odoo web client. Follow these steps to add your first component to the Odoo web client:

1. Add a `my_hostel/static/src/js/component.js` JavaScript file and define the new module's namespace:

```
    odoo.define('my_hostel.component', [], function (require) {
    "use strict";
    console.log("Load component......");
});
```

2. Add the component JavaScript to `assets`:

```
'assets': {
    'web.assets_backend': [
        'my_hostel/static/src/js/component.js',
    ],
},
```

3. Define the OWL utilities to the `component.js` file created in *step 1*:

```
const { Component, mount, xml , whenReady } = owl;
```

4. Add the OWL component and its basic template to the `component.js` file created in *step 1*:

```
class MyComponent extends Component {
    static template = xml`
        <div class="bg-info text-white text-center p-3">
            <b> Welcome To Odoo </b>
        </div>`
}
```

5. Initialize and append the component to the web client. Add this to the `component.js` file added in *step 1*:

```
whenReady().then(() => {
    mount(MyComponent, document.body);
});
```

Install/upgrade the `my_hostel` module to apply our changes. Once our module is loaded in Odoo, you will see the horizontal bar, as shown in the following screenshot:

Figure 16.1 – OWL component

This is just a simple component. Right now, it will not handle any user events and you cannot remove it.

How it works...

In *step 1* and *step 2*, we added a JavaScript file and listed it in the backend assets. If you want to learn more about assets, refer to the *Static assets management* recipe in *Chapter 14, CMS Website Development*.

In *step 3*, we initialized a variable from OWL. All the utilities from OWL are available under a single global variable, `owl`. In our example, we pulled an OWL utility. we declared `Component`, `mount`, `xml` , `whenReady`. `Component` is the main class for the `OWL` component and, by extending it, we will create our own components.

In *step 4*, we created our component, `MyComponent`, by extending OWL's `Component` class. For the sake of simplicity, we have just added the QWeb template to the definition of the `MyComponent` class. Here, as you may have noticed, we have used `xml`...`` to declare our template. This syntax is known as an inline template.

However, you can load QWeb templates via separate files, which is usually the case. We will see examples of external QWeb templates in the upcoming recipes.

> **Note**
>
> Inline QWeb templates do not support translations or modifications via inheritance. So, always endeavor to load QWeb templates from a separate file.

In *step 5*, we instantiated the `MyComponent` component and appended it to the body. The OWL component is an ES6 class, so you can create an object via the `new` keyword. Then you can use the `mount()` method to add the component to the page. If you notice, we have placed our code inside the `whenReady()` callback. This will ensure that all OWL functionality is properly loaded before we start using OWL components.

There's more...

OWL is a separate library and is loaded in Odoo as an external JavaScript library. You can use OWL in your other projects, too. The OWL library is listed at `https://github.com/odoo/owl`. There is also an online playground available in case you just want to test OWL without setting it in your local machine. You can play with OWL at `https://odoo.github.io/owl/playground/`.

Managing user actions in an OWL component

To make the UI interactive, components need to handle `user actions` such as `click`, `hover`, and `form submission`. In this recipe, we will add a button to our component, and we will handle a click event.

Getting ready

For this recipe, we will continue using the `my_hostel` module from the previous recipe.

How to do it...

In this recipe, we will add a delete button to the component. Upon clicking the delete button, the component gets removed. Perform the following steps to add a delete button and its event in the component:

1. Update the QWeb template and add an icon to remove the bar:

```
class MyComponent extends Component {
    static template = xml`
```

```
            <div class="bg-info text-white text-center p-3">
                <b> Welcome To Odoo </b>
                <i class="fa fa-close p-1 float-end"
                   style="cursor: pointer;"
                   t-on-click="onRemove"> </i>
            </div>`
    }
```

2. To remove the component, add the `onRemove` method to the `MyComponent` class, as follows:

```
    class MyComponent extends Component {
        static template = xml`
            <div class="bg-info text-white text-center p-3">
                <b> Welcome To Odoo </b>
                <i class="fa fa-close p-1 float-end"
                   style="cursor: pointer;"
                   t-on-click="onRemove"> </i>
            </div>`
        onRemove(ev) {
            $(ev.target).parent().remove();
        }
    }
```

Update the module to apply the changes. Following the update, you will see a little cross icon on the right side of the bar, as in the following screenshot:

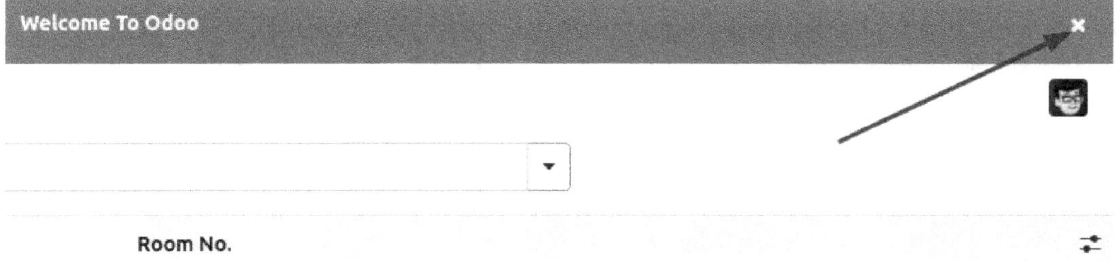

Figure 16.2 – The remove button on the top bar component

Upon clicking the remove icon, our OWL component will be removed. The bar will reappear when you reload the page.

How it works...

In *step 1*, we added a remove icon to the component. We have added a `t-on-click` attribute. This will be used to bind a click event. The value of the attribute will be the method in the component. In our example, we have used `t-on-click="onRemove"`. This implies that when the user clicks on the remove icon, the `onRemove` method in the component will be called. The syntax to define the event is simple:

```
t-on-<name of event>="<method name in component>"
```

For example, if you want to call the method when the user moves the mouse over the component, you can do so by adding the following code:

```
t-on-mouseover="onMouseover"
```

After adding the preceding code, whenever the user moves the mouse cursor over the component, OWL will call the `onMouseover` method specified in the component.

In *step 2*, we have added the `onRemove` method. This method will be called when the user clicks on the remove icon. In the method, we have called the `remove()` method, which will remove the component from the DOM. We will be seeing several default methods in the upcoming recipes.

There's more...

Event handling is not limited to the DOM events. You can use your custom events as well. For instance, if you are manually triggering the event called `my-custom-event`, you can use `t-on-my-custom-event` to catch custom-triggered events.

Making OWL components with hooks

OWL is a powerful framework and supports automatic updates for the UI based on **hooks**. With update hooks, a component's UI will be automatically updated when the internal state of the component is changed. In this recipe, we will update the message in the component based on user actions.

Getting ready

For this recipe, we will continue using the `my_hostel` module from the previous recipe.

How to do it...

In this recipe, we will add arrows around the text in the component. When we click on the arrow, we will change the message. Follow these steps to make the OWL component reactive:

1. Update the XML template of the component. Add two buttons with an event directive around the text. Also, retrieve the message dynamically from the list:

```
static template = xml`
    <div class="bg-info text-white text-center p-3">
        <i class="fa fa-arrow-left p-1"
           style="cursor: pointer;"
```

```
            t-on-click="onPrevious"> </i>
        <b t-esc="messageList[Math.abs(state.currentIndex%4)]"/>
        <i class="fa fa-arrow-right p-1"
            style="cursor: pointer;"
            t-on-click="onNext"> </i>
        <i class="fa fa-close p-1 float-end"
            style="cursor: pointer;"
            t-on-click="onRemove"> </i>
    </div>`
```

2. In the JavaScript file of the component, import the `useState` hook as follows:

```
const { Component, mount, xml , whenReady, useState } = owl;
```

3. Add the `setup` method to the component and initialize some variables as follows:

```
setup() {
    this.messageList = [
        'Hello World',
        'Welcome to Odoo',
        'Odoo is awesome',
        'You are awesome too'
    ];
    this.state = useState({ currentIndex: 0 });
}
```

4. In the `Component` class, add methods to handle the user's click event:

```
onNext(ev) {
    this.state.currentIndex++;
}
onPrevious(ev) {
    this.state.currentIndex--;
}
```

Restart and update the module to apply the changes to the module. Following the update, you will see the two arrow icons around the text like this:

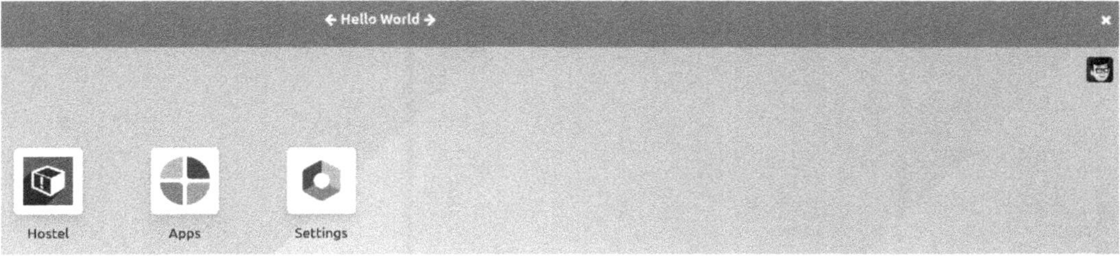

Figure 16.3 – Arrows around the text

If you click on the arrow, the message text will be changed based on the list of messages in the constructor.

How it works...

In *step 1*, we updated the XML template of our component. Basically, we made two changes to the template. We rendered the text message from the list of messages, and we selected the message based on the value of `currentIndex` in the state variable. We added two arrow icons around the text block. In the arrow icons, we added `the t-on-click` attribute to bind the click event to the arrow.

In *step 2*, we imported the `useState` hook from OWL. This hook is used to handle the state of the component.

In *step 3*, we added a `setup`. This will be called when you create an instance of the object. In the `setup`, we added a list of messages that we want to show, and then we added the `state` variable using the `useState` hook. This will make the component reactive. When the `state` is `changed`, the `UI` will be updated based on the new state. In our example, we used `currentIndex` in the `useState` hook. This implies that whenever the value of `currentIndex` changes, the UI will be updated as well.

> **Important information**
>
> There is only one rule for defining hooks, which is that the hooks will only work if you have declared them in `setup`. Several other types of hooks are available, which you can find here: `https://github.com/odoo/owl/blob/master/doc/reference/hooks.md`.

In *step 4*, we added methods to handle the click events of the arrow. Upon clicking the arrow, we are changing the state of the component. As we are using a hook on the state, the UI of the component will be automatically updated.

Understanding the OWL component life cycle

OWL components have several methods that help developers to create powerful and interactive components. Some of the important methods of the OWL components are as follows:

- `setup()`
- `onWillStart()`
- `onWillRender()`
- `onRendered()`
- `onMounted()`
- `onWillUpdateProps()`
- `onWillPatch()`

- `onPatched()`
- `onMounted()`
- `onWillUnmount()`
- `onWillDestroy()`
- `onError()`

In this recipe, we will log the message in the console to help us understand the life cycle of the OWL component.

Getting ready

For this recipe, we will continue using the `my_hostel` module from the previous recipe.

How to do it...

To add methods of the component to show the life cycle of an OWL component, you need to carry out the following steps:

1. First, you need to import the `all` hook, as follows:

```
const {
        Component,
        mount,
        whenReady,
        onWillStart,
        onMounted,
        onWillUnmount,
        onWillUpdateProps,
        onPatched,
        onWillPatch,
        onWillRender,
        onRendered,
        onError,
        onWillDestroy,
    } = owl;
```

2. As we already have `setup` in the component, let's add a message to the console like this:

```
setup() {
    console.log('CALLED:> setup');
}
```

3. Add the `willStart` method to the component:

```
setup() {
    onWillStart(async () => {
        console.log('CALLED:> willStart');
    });
}
```

4. Add the `willrender` method to the component:

```
setup() {
    onWillRender(() => {
        console.log('CALLED:> willRender');
    });
}
```

5. Add the `render` method to the component:

```
setup() {
    onRendered(() => {
        console.log('CALLED:> Rendered');
    });
}
```

6. Add the `mounted` method to the component:

```
setup() {
    onMounted(() => {
        console.log('CALLED:> Mounted');
    });
}
```

7. Add the `willUpdateProps` method to the component:

```
setup() {
    onWillUpdateProps(() => {
        console.log('CALLED:> WillUpdateProps');
    });
}
```

8. Add the `willPatch` method to the component:

```
setup() {
    onWillPatch(() => {
        console.log('CALLED:> WillPatch');
    });
}
```

9. Add the `patched` method to the component:

```
setup() {
    onPatched(() => {
        console.log('CALLED:> Patch');
    });
}
```

10. Add the `willUnmount` method to the component:

```
setup() {
    onWillUnmount(() => {
        console.log('CALLED:> WillUnmount');
    });
}
```

11. Add the `willDestroy` method to the component:

```
setup() {
    onWillDestroy(() => {
        console.log('CALLED:> WillDestroy');
    });
}
```

12. Add the `Error` method to the component:

```
setup() {
    onError(() => {
        console.log('CALLED:> Error');
    });
}
```

Restart and update the module to apply the module changes. Following the update, perform some operations, such as changing the message via arrows and removing the component. In the browser console, you will see the logs like this:

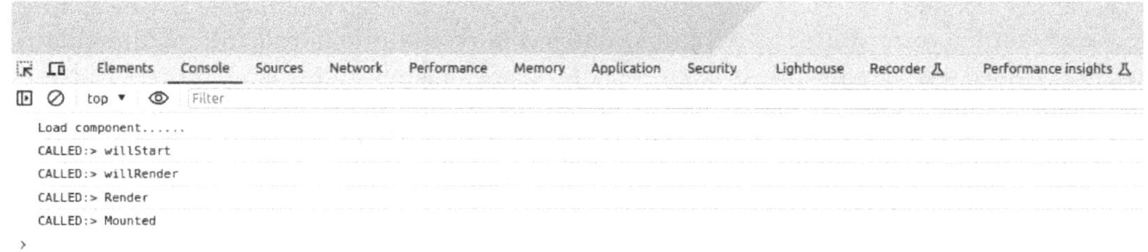

Figure 16.4 – Logs in the browser console

You may have different logs based on the operation you have performed on the component.

How it works...

In this recipe, we have added several methods and added logged messages to the method. You can use these methods based on your requirements. Let's see the life cycle of the component and when these methods are called.

setup

`setup` is run just after the component is constructed. It is a life cycle method that's very similar to the `constructor`, except that it does not receive any arguments.

It is the proper place to call hook functions. Note that one of the main reasons to have the setup hook in the component lifecycle is to make it possible to **monkey patch** it. It is a common need in the Odoo ecosystem.

willStart

`willStart` is an asynchronous hook that can be implemented to perform some (most of the time asynchronous) action before the initial rendering of a component.

It will be called exactly once before the initial rendering. It is useful in some cases, for example, to load external assets (such as a JavaScript library) before the component is rendered. Another use case is to load data from a server.

The `onWillStart` hook is used to register a function that will be executed:

```
setup() {
    onWillStart(async () => {
        this.data = await this.loadData()
    });
}
```

willRender

It is uncommon, but you may need to execute code just before a component is rendered (more precisely, when its compiled template function is executed). To do that, we can use the `onWillRender` hook.

`willRender` hooks are called just before rendering templates, parent first, then children.

rendered

Similarly, it is uncommon, but you may need to execute code just after a component is rendered (more precisely, when its compiled template function is executed). To do that, we can use the `onRendered` hook.

`rendered` hooks are called just after rendering templates, parent first, then children. Note that at this moment, the actual DOM may not exist yet (if it is the first rendering), or is not updated yet. This will be done in the next animation frame, as soon as all the components are ready.

mounted

The `mounted` hook is called each time a component is attached to the DOM, after the initial rendering. At this point, the component is considered active. This is a good place to add some listeners, or to interact with the DOM, if the component needs to perform some measure for example.

It is the opposite of `willUnmount`. If a component has been mounted, it will always be unmounted at some point in the future.

The `mounted` method will be called recursively on each of its children. First children, then parents.

It is allowed (but not encouraged) to modify the state in the mounted hook. Doing so will cause a rerender, which will not be perceptible by the user, but will slightly slow down the component.

The `onMounted` hook is used to register a function that will be executed at this moment.

willUpdateProps

`willUpdateProps` is an asynchronous hook that is called just before new props are set. This is useful if the component needs to perform an asynchronous task, depending on the props (for example, assuming that the props are some record ID, fetching the record data).

The `onWillUpdateProps` hook is used to register a function that will be executed at this moment.

Notice that it receives the next props for the component.

This hook is not called during the first render (but `willStart` is called and performs a similar job). Also, like most hooks, it is called in the usual order: parents first, then children.

willPatch

The `willPatch` hook is called just before the DOM patching process starts. It is not called on the initial render. This is useful to read information from the DOM, such as the current position of the scrollbar.

Note that modifying the state is not allowed here. This method is called just before an actual DOM patch, and is only intended to be used to save some local DOM state. Also, it will not be called if the component is not in the DOM.

The `onWillPatch` hook is used to register a function that will be executed at this moment. `willPatch` is called in the usual parent/children order.

patched

This hook is called whenever a component actually updates its DOM (most likely via a change in its state/props or environment).

This method is not called on the initial render. It is useful to interact with the DOM (for example, through an external library) whenever the component is patched. Note that this hook will not be called if the component is not in the DOM.

The `onPatched` hook is used to register a function that will be executed at this moment.

Updating the component state in this hook is possible, but not recommended. We need to be careful because updates here will create additional rendering, which in turn will cause other calls to the `patched` method. So, we need to be particularly careful to prevent endless cycles.

Like `mounted`, the `patched` hook is called in the order: children first, then parent.

willUnmount

`willUnmount` is a hook that is called just before a component is unmounted from the DOM. This is a good place to remove listeners, for example.

The `onWillUnmount` hook is used to register a function that will be executed at this moment.

This is the opposite method of `mounted`. Note that if a component is destroyed before being mounted, the `willUnmount` method may not be called.

Parent `willUnmount` hooks will be called before children.

willDestroy

Sometimes, components need to do some action in the setup and clean it up when they are inactive. However, the `willUnmount` hook is not appropriate for the cleaning operation, since the component may be destroyed before it has even been mounted. The `willDestroy` hook is useful in this situation since it is always called.

The `onWillUnmount` hook is used to register a function that will be executed at this moment.

`willDestroy` hooks are first called on children, then on parents.

onError

Sadly, components may crash at runtime. This is an unfortunate reality, and this is why OWL needs to provide a way to handle these errors.

The `onError` hook is useful when we need to intercept and properly react to errors that occur in some sub-components.

There's more...

There is one more method in the component life cycle, but it is used when you are using subcomponents. OWL passes the parent component state via the `props` parameter, and when `props` is changed, the `willUpdateProps` method is called. This is an asynchronous method, which means you can perform an asynchronous operation such as RPC here.

Adding an OWL field to the form view

Up to this point, we have learned about all the basics of OWL. Now we will move on to more advanced aspects and create a field widget that can be used in the form view, just like the field widget recipe from the previous chapter.

Odoo has many widgets in the UI for different functionalities, such as a status bar, checkboxes, and radio buttons. which makes the operations in Odoo simpler and run with ease. For example, we used `widget='image'` to display a binary field as an image. To demonstrate how to create your own widget, we'll write one widget that lets the user choose an integer field, but we will display it differently. Instead of an input box, we will display a color picker so that we can select a color number. Here, each number will be mapped to a color.

In this recipe, we will create a color picker widget that will save integer values based on the color selected.

To make the example more informative, we will use some advanced concepts of OWL.

Getting ready

For this recipe, we will be using the `my_hostel` module.

How to do it...

We'll add a JavaScript file that contains our widget's logic, an XML file that contains design logic, and an SCSS file to do some styling. Then, we will add one integer field to the books form to use our new widget.

Perform the following steps to add a new field widget:

1. Add the category integer field to the `hostel.room` model as follows:

```
category = fields.Integer('Category')
```

2. Add the same field to the form view, with a `widget` attribute as well:

```
<field name="category" widget="category_color"/>
```

3. Add the QWeb templates for the field at `static/src/xml/field_widget.xml`:

```
<t t-name="OWLColorPill">
    <span t-attff-class="o_color_pill o_color_#{props.color}
#{props.value == props.color ? 'active': ''}"
            t-att-data-val="props.color"
            t-on-click="() => this.pillClicked()"
            t-attf-title="#{props.category_count or 0 } Room
booked in this category" />
</t>

<span t-name="OWLFieldColorPills">
    <t t-foreach="totalColors" t-as='color' t-key="color">
        <ColorPill onClickColorUpdated="data => this.
colorUpdated(data)"
                    color='color'
                    value="props.value"
                    category_count="categoryInfo[color]"/>
    </t>
</span>
```

4. List the QWeb file in the module's `manifest` file:

```
'assets': {
    'web.assets_backend': [
        'my_hostel/static/src/js/field_widget.js',
    ],
},
```

5. Now we want to add some SCSS for the field at `static/src/scss/field_widget.scss`. As the content of SCSS is too long, please find the content of the SCSS file in this book's GitHub repository at `https://github.com/PacktPublishing/Odoo-17-Development-Cookbook-Fifth-Edition/tree/main/Chapter16/05_owl_field/my_hostel/static/src/scss`.

6. Add the static `/src/js/field_widget.js` JavaScript file with the following basic content:

```
/** @odoo-module */
import { Component, onWillStart , onWillUpdateProps} from "@
odoo/owl";
import { registry } from "@web/core/registry";

class ColorPill extends Component {
    static template = 'OWLColorPill';
    pillClicked() {
        this.props.onClickColorUpdated(this.props.color);
    }
```

```
}

export class OWLCategColorField extends Component {
    static supportedFieldTypes = ['integer'];
    static template = 'OWLFieldColorPills';
    static components = { ColorPill };
    setup() {
        this.totalColors = [1,2,3,4,5,6];
        onWillStart(async() => {
            await this.loadCategInformation();
        });
        onWillUpdateProps(async() => {
            await this.loadCategInformation();
        });
        super.setup();
    }

    colorUpdated(value) {
        this.props.record.update({ [this.props.name]: value });
    }

    async loadCategInformation() {
        var self = this;
        self.categoryInfo = {};
        var resModel = self.env.model.root.resModel;
        var domain = [];
        var fields = ['category'];
        var groupby = ['category'];
        const categInfoPromise = await self.env.services.orm.
readGroup(
            resModel,
            domain,
            fields,
            groupby
        );
        categInfoPromise.map((info) => {
            self.categoryInfo[info.category] = info.category_
count;
        });
    }
}
registry.category("fields").add("category_color",{
    component: OWLCategColorField
});
```

7. Add JavaScript and an SCSS file to the backend assets as follows:

```
'assets': {
    'web.assets_backend': [
        'my_hostel/static/src/scss/field_widget.scss',
        'my_hostel/static/src/js/field_widget.js',
        'my_hostel/static/src/xml/field_widget.xml',
    ],
},
```

8. Restart and update the module to apply the module changes. Open the room form view. You will be able to see the color picker widget, as shown in the following screenshot:

Category

Figure 16.5 – Color picker OWL widget

9. This field looks just like the color widget from the last chapter, but the actual difference lies under the hood. This new field is built with OWL components and subcomponents, while the previous one was built with widgets.
10. The benefit of this subcomponent is to provide a comprehensive framework for building modern, responsive, and interactive UIs in OWL. By modularizing functionality into small, reusable units, developers can create more maintainable and extensible applications while reducing code duplication and improving development efficiency.

How it works...

In *step 1*, we added an integer field to the `hostel.room` model.

In *step 2*, we added the field to the form view of the room.

In *step 3*, we added the QWeb template file. If you notice, we added two templates to the file, one for the color pill and the other for the field itself. We used two templates because we want to see the concept of the `subcomponent`. If you observe the template closely, you will find that we have used the `<ColorPill>` tag. This will be used to instantiate the subcomponent. On the `<ColorPill>` tag, we have passed the active and color attributes. These attributes will be received as `props` in the template of the subcomponent. Also note that the `onClickColorUpdated` attribute is used to listen to the custom event triggered from the subcomponent.

> **Important information**
> Odoo v17 uses both the widget system and the OWL framework.

In *step 4*, we listed our QWeb template in the manifest. This will automatically load our template in the browser.

In *step 5*, we added SCSS for the color. This will help us to have a beautiful UI for the color picker.

In *step 6*, we added JavaScript for the field component.

We imported the OWL utility and we also imported the component and `fieldRegistry`.

`fieldRegistry` is used to list the OWL component as a field component.

In *step 7*, we created the `ColorPill` component. The `template` variable on the component is the name of the template that is loaded from the external XML file. The `ColorPill` component has the `pillClicked` method, which is called when the user clicks on the color pill. Inside the method body, we have triggered the `onClickColorUpdated` event, which will be captured by the parent `OWLCategColorField` component as we used `colorUpdated` on the `OWLCategColorField` component.

In *step 8* and *step 9*, we created the `OWLCategColorField` component by extending `Component`. We used the `Component` because it will have all the utilities that are required to create the field widget.

If you notice, we used the `components` static variable at the start. You need to list the components via the `components` static variable when you are using subcomponents in the template. We also added the `onWillStart` method in our example. The `willStart` method is an asynchronous method, so we have called RPC (network call) to fetch data regarding the number of the room booked for a particular color. Toward the end, we added the `colorUpdated` method, which will be called when the user clicks on the pill. So, we are changing the values of the field. The `this.props.record.update` method is used to set the field values (which will be saved in the database). Note here that the data triggered from the child component is available under the `detail` attribute in the `event` parameter. Finally, we registered our widget in `fieldRegistry`, implying that henceforth, we will be able to use our field via the `widget` attribute in the form view.

In *step 10*, we loaded JavaScript and SCSS files into the backend assets.

There's more...

Understanding QWeb

QWeb is the primary templating engine used by Odoo. It is an XML templating engine and is used mostly to generate HTML fragments and pages. Template directives are specified as XML attributes prefixed with `t-`, for instance, `t-if` for conditionals, with elements and other attributes being rendered directly. The following are the different operations of the QWeb template:

- **Data output**:

  QWeb's output directive, `out`, will automatically HTML-escape its input, limiting XSS risks when displaying user-provided content. `out` takes an expression, evaluates it, and injects the result into the document:

  ```
  <p><t t-out="value"/></p>
  ```

Setting `value` set to `42` yields the following:

```
<p>42</p>
```

- **Setting variables**:

  QWeb allows us to create variables from within the template, memorize a computation (to use it multiple times), and give a piece of data a clearer name.

  This is done via the `set` directive, which takes the name of the variable to create. The value of `set` can be provided in two ways:

  - A `t-value` attribute containing an expression, and the result of its evaluation will be set:

  ```
  <t t-set="foo" t-value="2 + 1"/>
  <t t-out="foo"/>
  ```

  This will print `3`.

  - If there is no `t-value` attribute, the node's body is rendered and set as the variable's value:

  ```
  <t t-set="foo">
      <li>ok</li>
  </t>
  <t t-out="foo"/>
  ```

- **Conditionals**:

  QWeb has a conditional directive, `if`, which evaluates an expression given as an attribute value:

  ```
  <div>
      <t t-if="condition">
          <p>ok</p>
      </t>
  </div>
  ```

  The element is rendered if the condition is `true`:

  ```
  <div>
      <p>ok</p>
  </div>
  ```

  But if the condition is `false`, it is removed from the result:

  ```
  <div>
  </div>
  ```

  Extra conditional branching directives, `t-elif` and `t-else`, are also available:

  ```
  <div>
      <p t-if="user.birthday == today()">Happy birthday!</p>
      <p t-elif="user.login == 'root'">Welcome master!</p>
  ```

```
    <p t-else="">Welcome!</p>
</div>
```

- **Loops**:

  QWeb has an iteration directive, `foreach`, which takes an expression that returns the collection to iterate on, and a second parameter, `t-as`, providing the name to use for the current item of the iteration:

```
<t t-foreach="[1, 2, 3]" t-as="i">
    <p><t t-out="i"/></p>
</t>
```

  This will be rendered as follows:

```
<p>1</p>
<p>2</p>
<p>3</p>
```

- **Attributes**:

  QWeb can compute attributes on the fly and set the result of the computation on the output node. This is done via the `t-att` (attribute) directive, which exists in three different forms:

```
t-att-$name
```

  Attribute `t-att` is `$name`, then the set will be rendered the attribute value as shown in the example below.

```
< div t-att-a="42"/>
```

  This will be rendered as follows:

```
<div a="42"></div>
    t-attf-$name
```

  This is the same as previous, but the parameter is a format string instead of just an expression. This is often useful for mixing literal and non-literal strings (e.g., classes):

```
<t t-foreach="[1, 2, 3]" t-as="item">
    <li t-attf-class="row {{ (item_index % 2 === 0) ? 'even' :
'odd' }}">
        <t t-out="item"/>
    </li>
</t>
```

  This will be rendered as follows:

```
< li class="row even">1</li>
<li class="row odd">2</li>
<li class="row even">3</li>
      t-att=mapping
```

If the parameter is a mapping, each key/value pair generates a new attribute and its value:

```
<div t-att="{'a': 1, 'b': 2}"/>
```

This will be rendered as follows:

```
<div a="1" b="2"></div>
t-att=pair
```

If the parameter is a pair (a tuple or array of two elements), the first item of the pair is the name of the attribute and the second item is the value:

```
<div t-att="['a', 'b']"/>
```

This will be rendered as follows:

```
<div a="b">
</div>
```

- **Calling sub-templates**

  QWeb templates can be used for top-level rendering, but they can also be used from within another template (to avoid duplication or to give names to parts of templates) using the `t-call` directive:

  ```
  <t t-call="other-template"/>
  ```

  This calls the named template with the execution context of the parent, if `other_template` is defined as follows:

  ```
  <p><t t-value="var"/></p>
  ```

  The preceding call will be rendered as `<p/> (no content).`

  ```
  <t t-set="var" t-value="1"/>
  <t t-call="other-template"/>
  ```

  The preceding code will be rendered as `<p>1</p>`.

  However, this has the problem of being visible from outside the `t-call.` Alternatively, content set in the body of the `call` directive will be evaluated before calling the sub-template, and can alter a local context:

  ```
  <t t-call="other-template">
            <t t-set="var" t-value="1"/>
      </t>
      <!-- "var" does not exist here -->
  ```

  The body of the `call` directive can be arbitrarily complex (not just set directives), and its rendered form will be available within the called template as a magical `0` variable:

  ```
  <div>
      This template was called with content:
  ```

```
        <t t-out="0"/>
    </div>
```

being called thus:

```
    <t t-call="other-template">
        <em>content</em>
    </t>
```

This will result in the following:

```
    <div>
        This template was called with content:
        <em>content</em>
    </div>
```

Understanding subcomponents

In the context of OWL, subcomponents refer to small, modular units of functionality that can be integrated into larger components to enhance their capabilities or provide additional features.

Subcomponents in OWL can include various elements, such as widgets, utilities, services, and views, which are designed to work together within the OWL framework to create rich, interactive UIs and manage client-side logic efficiently.

These subcomponents work together to provide a comprehensive framework for building modern, responsive, and interactive UIs in OWL. By modularizing functionality into small, reusable units, developers can create more maintainable and extensible applications while reducing code duplication and improving development efficiency.

It is convenient to define a component using other (sub) components. This is called composition and is very powerful in practice. To do that in OWL, we can just use a tag starting with a capital letter in its template, and register the subcomponent class in its static `component` object:

```
class Child extends Component {
  static template = xml`<div>child component <t t-esc="props.
value"/></div>`;
}

class Parent extends Component {
  static template = xml`
    <div>
      <Child value="1"/>
      <Child value="2"/>
    </div>`;

  static components = { Child };
}
```

Here, `<Child>` has `subcomponent`. This example also shows how we can pass information from the parent component to the child component as props. In OWL, `props` (short for properties) is an object that contains every piece of data given to a component by its parent. Note that `props` is an object that only makes sense from the perspective of the child component.

The `props` object is made of every attribute defined on the template, with the following exceptions: every attribute starting with `t-` is not a prop (they are QWeb directives).

In the following example:

```
<div>
    <Child value="string"/>
    <Child t-if="condition" model="model"/>
</div>
```

The `props` object contains the following keys:

```
for Child: value,
for Child: model,
```

17
In-App Purchasing with Odoo

Odoo has had built-in support for **in-app purchasing** (**IAP**) since version 11. IAP is used to provide recurring services without any complex configurations. Usually, apps purchased from the app store only require a one-time payment from the customer, because they are normal modules and once the user has purchased and started using the module, it won't cost the developer anything. In contrast to this, IAP apps are used to provide services to users, and so there is an operational cost to providing continuous service. In such cases, it is not possible to provide a service with just a single initial purchase. The service provider needs something that charges the user in a recurring manner, based on usage. Odoo's IAP fixes these issues and provides a way to charge based on usage.

In-app purchases typically refer to the ability to buy additional features, content, or services within an application. However, Odoo is highly customizable, and while it might not have a dedicated IAP module, you can create similar functionalities using custom development or by leveraging existing modules.

This feature allows users to expand the functionality of their Odoo instance by acquiring additional apps, features, or services without leaving the Odoo environment. Here's an overview of Odoo IAP:

- **App marketplace integration**: Odoo's IAP is tightly integrated with the Odoo App Store or marketplace. Users can browse, select, and purchase additional apps or modules from a wide range of options.
- **Easy access to extensions**: Users can access and evaluate the available apps and extensions directly from their Odoo dashboard. This makes it convenient for businesses to extend the capabilities of their Odoo instance without the need for extensive manual installation.
- **Trial versions**: Some apps in the marketplace may offer trial versions or limited-time trials, allowing users to test the functionality of the app before making a purchase. This helps users make informed decisions.
- **Simplified licensing**: Odoo IAP simplifies the licensing and subscription management for the purchased apps. Users can easily subscribe, renew, or manage their licenses without external processes.
- **One-click installation**: After purchasing an app, users can typically install it with just one click from within their Odoo instance. This streamlined process reduces the complexity of app installation.

- **Centralized billing**: Billing and payment for the purchased apps are typically managed through Odoo's central billing system, simplifying the financial aspects of app acquisition.
- **App updates**: Odoo IAP often includes automatic updates for purchased apps, ensuring that users have access to the latest features and security updates.
- **Support and documentation**: Many apps available through Odoo IAP come with documentation and support options, making it easier for users to get assistance if needed.
- **Integration with core Odoo**: The purchased apps are seamlessly integrated with the core Odoo system, ensuring compatibility and a unified user experience.

There are several use cases where you can use IAP, such as a fax service for sending documents or an SMS service. In this chapter, we will explain the Partner autocomplete service that will be provided by Odoo.

In this short chapter, we will cover the following topics:

- IAP concepts
- Buying credits
- IAP accounts
- The IAP portal
- Getting notifications for low credits

IAP concepts

IAP involves several key concepts and elements that are essential to understand when using this feature within the Odoo ERP system. We will explore the different entities that are a part of the IAP process and also look at the role of each entity and how they combine to complete the IAP process.

Odoo IAP is a valuable tool for businesses using the Odoo ERP system, as it simplifies the process of extending and customizing their software environment. It offers a centralized platform for managing additional apps and modules, helping organizations optimize their business processes and operations. This feature adds a layer of flexibility and scalability to the Odoo ecosystem, making it an even more powerful and adaptable solution for a wide range of businesses and industries.

Odoo IAP simplifies the process of discovering, acquiring, and installing additional apps and modules directly from the Odoo environment. Users can extend the functionality of their Odoo system with ease, without leaving the platform.

App developers can offer their products with various pricing models, including one-time purchases, subscription plans, and trial versions. This flexibility caters to diverse customer needs.

Odoo's IAP feature significantly enhances the Odoo ERP system's adaptability and customization capabilities. It streamlines the process of app acquisition, encourages developer innovation, and offers a user-centric experience, ultimately contributing to the platform's versatility and value for businesses across various industries.

Odoo IAP

IAP simplifies the process of acquiring and managing additional applications, modules, and features for your Odoo ERP system.

How it works...

There are three main entities in the IAP process: the customer, the service provider, and Odoo itself. These are described as follows:

- **The customer** is the end user who wants to use the service. In order to use the service, the customer needs to install the application provided by the service provider. The customer then needs to purchase a service plan according to their usage requirements. With that, the customer can start to use the service straight away. This prevents difficulties for the customer, as it is not necessary to carry out complex configurations. Instead, they just pay for the service and start to use it.
- **The service provider** is the developer who wants to sell the service (probably you, as you are the developer). The customer will ask the provider for the service, at which point the service provider will check whether the customer purchased a valid plan and whether there is enough credit in the customer's account. If the customer has enough credit, the service provider will deduct the credit and provide the service to the customer.
- **Odoo** itself is a kind of broker in this. It provides a medium for handling payments, credits, plans, and so on. Customers purchase the service credit from Odoo, and the service provider draws this credit when serving the service. Odoo then bridges the gap between the customer and the service provider, so the customer has no need to do complex configurations and the service provider has no need to set up a payment gateway, customer account management, and so on.

There is also an optional entity in the process, which is the **external service**. In some cases, service providers use some external services. However, we will ignore external services here, as they are the secondary service provider. An example of this could be an SMS service. If you are providing an SMS IAP service to Odoo users, then you (the service provider) will use an SMS service internally.

Buying credits

Each IAP service has its own pricing. Customers have to buy that service from the IAP service provider. To check your services, go to **Settings** | **Odoo IAP** | **View My Services**.

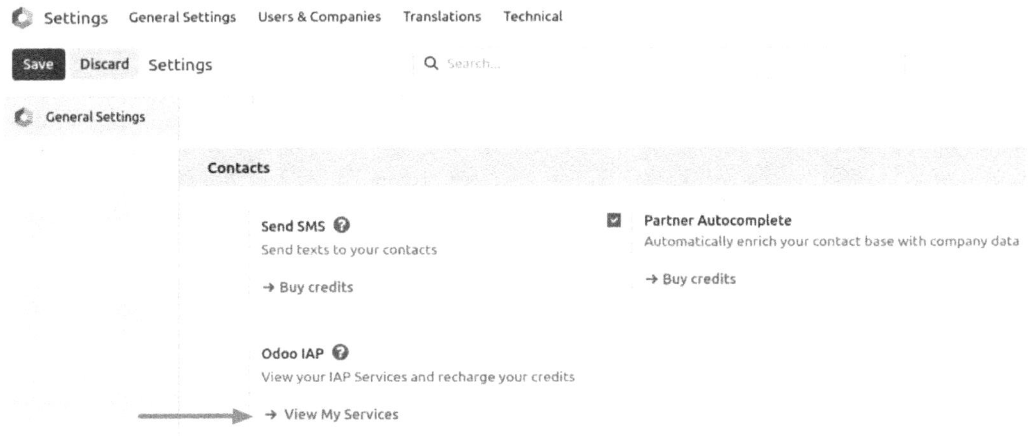

Figure 17.1 – Buy credits

The preceding screenshot shows the screen you see when you want to buy credits.

IAP accounts

Once you buy credits from the provider, it is stored on IAP accounts, which are to be used for each service. By default, IAP accounts are common for all companies but can be configured to be company-specific.

To create a new **IAP Account**, activate the developer mode and go to **Technical Settings | IAP Account**.

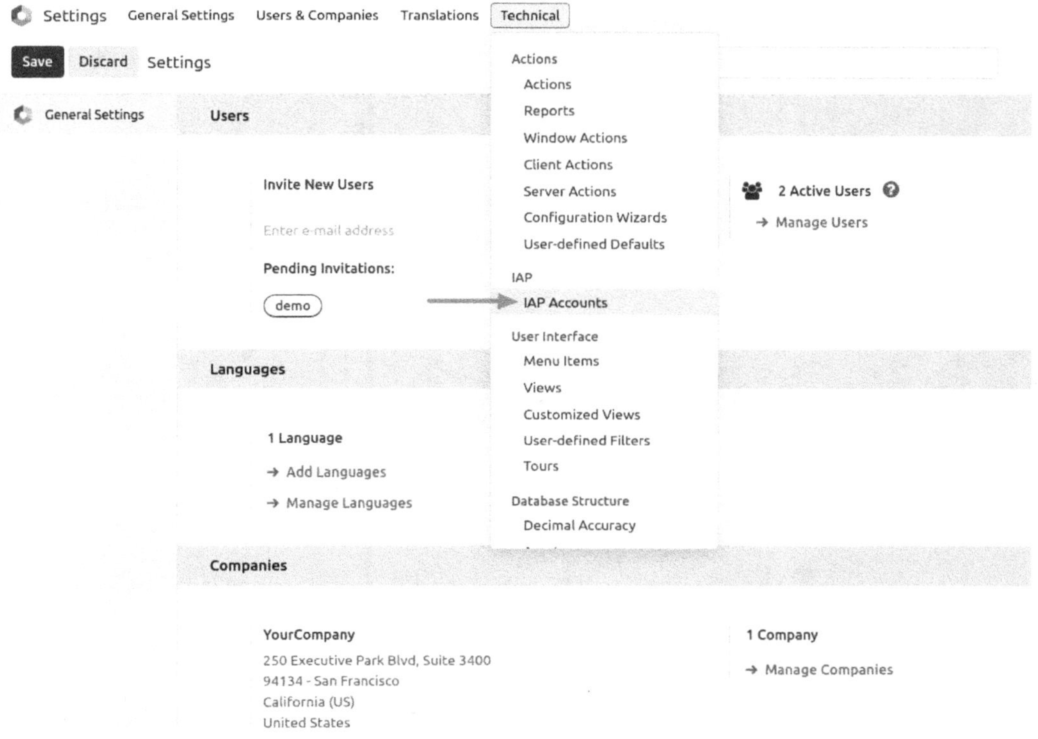

Figure 17.2 – IAP Accounts

The following is the screenshot of the **IAP Account** screen:

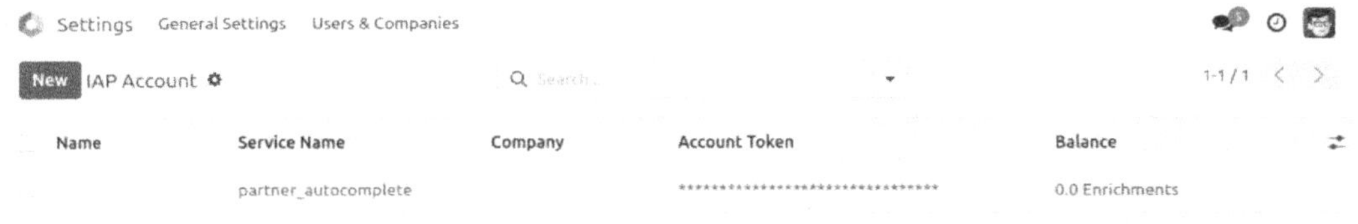

Figure 17.3 – IAP Account

The IAP portal

The IAP portal is a platform where you can see your IAP services and their credits and can recharge them by clicking on the **Buy Credit** button, which will redirect you to the IAP portal. can be set Threshold once it reaches you will get notified via mentioned email ID.

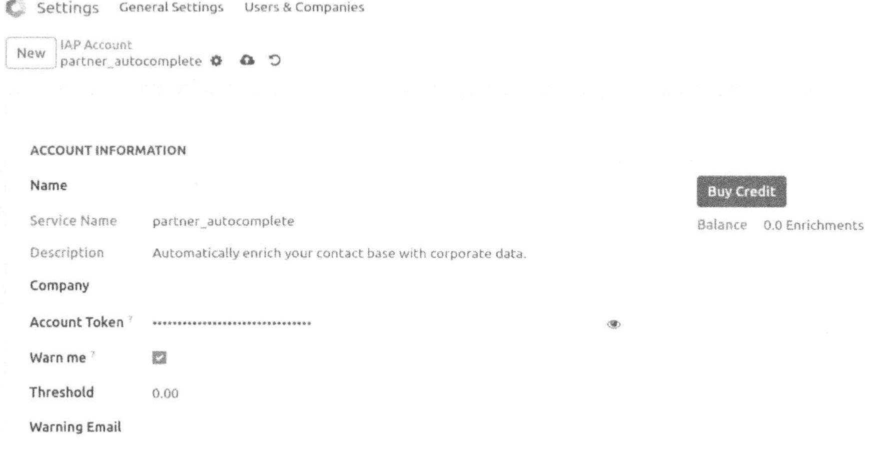

Figure 17.4 – IAP Account

Get low credits notification

Here, we can set credits in **Threshold**, which means we have to set a minimum credit limit and email address. Once it reaches the limit, an automatic reminder will be sent to the mentioned email ID go to **Settings** | **Odoo IAP** | **View My Services**. Next, unfold the service, check the credits, and configure it accordingly.

Figure 17.5 – Low credits notification

This is how in-app purchasing in Odoo works. In the next chapter, we'll see automated test cases.

18
Automated Test Cases

When it comes to developing large applications, using automated test cases is good practice to improve the reliability of your module. This makes your module more robust. Every year, Odoo releases a new version of its software, and automated test cases are very helpful in detecting regression in your application, which may have been caused by a version upgrade. Luckily, any Odoo framework comes with different automated testing utilities. Odoo includes the following three main types of tests:

- **A Python test case**: Used to test Python business logic
- **A JavaScript QUnit test**: Used to test JavaScript implementation in Odoo
- **Tours**: An integration test to check that Python and JavaScript work with each other properly

In this chapter, we will cover the following recipes:

- Adding Python test cases
- Running tagged Python test cases
- Setting up Headless Chrome for client-side test cases
- Adding client-side QUnit test cases
- Adding tour test cases
- Running client-side test cases from the UI
- Debugging client-side test cases
- Generating videos/screenshots for failed test cases
- Populating random data for testing

Technical requirements

In this chapter, we will look at all the test cases in detail. In order to cover all of the test cases in a single module, we have created a small module. Its Python definition is as follows:

```
import logging

from odoo import api, fields, models
from odoo.exceptions import UserError
from odoo.tools.translate import _

_logger = logging.getLogger(__name__)


class HostelRoom(models.Model):

    _name = 'hostel.room'
    _description = "Information about hostel Room"

    name = fields.Char(string="Hostel Name", required=True)
    room_no = fields.Char(string="Room Number", required=True)
    other_info = fields.Text("Other Information",
                              help="Enter more information")
    description = fields.Html('Description')
    room_rating = fields.Float('Hostel Average Rating', digits=(14,
4))
    member_ids = fields.Many2many('hostel.room.member',
string='Members')
    state = fields.Selection([
        ('draft', 'Unavailable'),
        ('available', 'Available'),
        ('closed', 'Closed')],
        'State', default="draft")

    @api.model
    def is_allowed_transition(self, old_state, new_state):
        allowed = [('draft', 'available'),
                   ('available', 'closed'),
                   ('closed', 'draft')]
        return (old_state, new_state) in allowed

    def change_state(self, new_state):
        for room in self:
            if room.is_allowed_transition(room.state, new_state):
```

```
                room.state = new_state
            else:
                message = _('Moving from %s to %s is not allowed') %
(room.state, new_state)
                raise UserError(message)

    def make_available(self):
        self.change_state('available')
        return True

    def make_closed(self):
        self.change_state('closed')


class HostelRoomMember(models.Model):

    _name = 'hostel.room.member'
    _inherits = {'res.partner': 'partner_id'}
    _description = "Hostel Room member"

    partner_id = fields.Many2one('res.partner', ondelete='cascade')
    date_start = fields.Date('Member Since')
    date_end = fields.Date('Termination Date')
    member_number = fields.Char()
    date_of_birth = fields.Date('Date of birth')
```

The Python code given here will help us to write test cases for Python business cases.

For JavaScript test cases, we have added the `int_color` widget from the *Creating custom widgets* recipe in *Chapter 15, Web Client Development*.

You can grab this initial module from the GitHub repository of this room at the following link: `https://github.com/PacktPublishing/Odoo-17-Development-Cookbook-Fifth-Edition/tree/main/Chapter18/00_initial_module`.

Adding Python test cases

Python test cases are used to check the correctness of business logic. In *Chapter 5, Basic Server-Side Development*, you saw how you can modify the business logic of our existing app. This makes it even more important, as customization might break the app's functionality. In this chapter, we will write a test case to validate the business logic to change a hostel room's state.

Getting ready

We will use the my_hostel module from the Chapter18/00_initial_module directory of the GitHub repository.

How to do it...

Follow these steps to add Python test cases to the my_hostel module:

1. Add a new file, tests/__init__.py, as follows:

```
from . import test_hostel_room_state
```

2. Add a tests/test_hostel_room_state.py file, and then add the test case, as follows:

```
from odoo.tests.common import TransactionCase

class TestHostelRoomState(TransactionCase):

    def setUp(self, *args, **kwargs):
        super(TestHostelRoomState, self).setUp(*args, **kwargs)
        self.partner_nikul = self.env['res.partner'].
create({'name': 'Nikul Chaudhary'})
        self.partner_deepak = self.env['res.partner'].
create({'name': 'Deepak Ahir'})
        self.member_ids = self.env['hostel.room.member'].
create([
            {'partner_id': self.partner_nikul.id, 'member_
number': '007'},
            {'partner_id': self.partner_deepak.id, 'member_
number': '357'}])
        self.test_hostel_room = self.env['hostel.room'].create({
            'name': 'Hostel Room 01',
            'room_no': '1',
            'member_ids': [(6, 0, self.member_ids.ids)]
        })

    def test_button_available(self):
        """Make available button"""
        self.test_hostel_room.make_available()
        self.assertIn(self.partner_nikul, self.test_hostel_room.
mapped('member_ids.partner_id'))
        self.assertEqual(
            self.test_hostel_room.state, 'available', 'Hostel
Room state should changed to available')
```

```
    def test_button_closed(self):
        """Make closed button"""
        self.test_hostel_room.make_available()
        self.test_hostel_room.make_closed()
        self.assertEqual(
            self.test_hostel_room.state, 'closed', 'Hostel Room
state should changed to closed')
```

3. To run the test cases, start the Odoo server with the following option:

```
./odoo-bin -c server.conf -d db_name -i my_hostel --test-enable
```

4. Now, check the server log. You will find the following logs if our test cases ran successfully:

```
INFO test odoo.addons.my_hostel.tests.test_hostel_room_state:
Starting TestHostelRoomState.test_button_available ...
INFO test odoo.addons.my_hostel.tests.test_hostel_room_state:
Starting TestHostelRoomState.test_button_closed ...
INFO test odoo.modules.loading: Module my_hostel loaded in 0.31s
(incl. 0.05s test), 240 queries (+33 test, +240 other)
```

You will see the `ERROR` log instead of `INFO` if a test case fails or there is an error.

How it works...

In Odoo, Python test cases are added to the `tests/` directory of the module. Odoo will automatically identify this directory and run the test under the folder.

> **Note**
>
> *You also need to list your test case files in* `tests/__init__.py`. *If you don't do that, that test case will not execute.*

Odoo uses Python's `unittest` for Python test cases. To learn more about `unittest`, refer to `https://docs.python.org/3.5/library/unittest.html`. Odoo provides the following helper classes:

- The `Common` class: This class provides common methods and setup for test cases. It includes functionalities such as creating and managing database transactions during the tests.
- The `SavepointCase` class: This extends the Common class.
- The `SavepointCase` provides additional features to handle savepoints during tests. This is useful when you want to roll back the changes made to the database during a test, ensuring that each test starts with a clean state,

- The `TransactionCase` class: This class extends `SavepointCase` and provides transaction-related functionality. It helps to manage database transactions during the tests.
- The `HttpCase` class: This class is used to test HTTP requests and responses. It allows you to simulate HTTP requests and test the responses.
- The `BaseCase` class: This is a base class for various test cases in Odoo. It provides common functionality that can be reused in different test scenarios,
- The `SingleTransactionCase` class: This class extends `TransactionCase` and ensures that each test case is executed within a single database transaction. This can be useful in scenarios where you want to isolate tests completely from each other.
- The `FormCase` class: This class is used to test form views and their interactions. It provides methods to simulate user interactions with form views.
- The `FunctionCase` class: This class is designed to test server-side Python functions. It helps in testing various functions and methods within the Odoo framework, wrapped over `unittest`.

These classes simplify the process of developing test cases. In our case, we have used `TransactionCase`. Now, `TransactionCase` runs each test case method in a different transaction. Once a test case method runs successfully, a transaction is automatically rolled back. This means the next test case will not have any modification made by the previous test case.

The class method starts from `test_` and is considered a test case. In our example, we have added two test cases. This checks the methods that change the hostel room's state. The `self.assertEqual` method (`assertEqual()` in Python) is a `unittest` library function that is used in unit testing to check the equality of two values. This function will take three parameters as input and return a Boolean value, depending upon the `assert` condition. If both input values are equal `assertEqual()` will return `true` else return `false`) is used to check whether the test case runs successfully. We have checked the hostel room state after performing operations on the hostel room's record. So, if the developer makes a mistake and the method does not change states as expected, the test case will fail.

> **Important information**
>
> *Note that the `setUp()` method will automatically call for every test case we run, so, in this recipe, we have added two test cases so that `setUp()` will call twice. As per the code in this recipe, there will only be one record of the hostel room present during testing because, with `TransactionCase`, the transaction is rolled back with every test case.*

In Python, a docstring is a string literal that occurs as the first statement in a module, function, class, or method definition. Docstrings are used to provide documentation about what a piece of code does. They serve as a form of inline documentation that can be accessed using various tools, such as the `help()` function.. This can be very helpful to check the status of a particular test case.

There's more...

The test suite provides the following additional test utility classes:

- `SingleTransactionCase`: Test cases generated through this class will run all cases in a single transaction, so changes made from one test case will be available in a second test case. In this way, the transaction begins with the first test method and is only rolled back at the end of the last test case.
- `SavepointCase`: This is the same as `SingleTransactionCase`, but in this case, test methods run inside a rolled-back save point, instead of having all test methods in a single transaction. This is used to create large test cases and make them faster, by generating test data only once. Here, we use the `setUpClass()` method to generate the initial test data.

Running tagged Python test cases

When you run the Odoo server with the `--test-enabled` module name, the test cases run immediately after the module is installed. If you want to run a test case after the installation of all the modules, or if you just want to run a test case for only one module, a `tagged()` decorator is the answer.

In this recipe, we'll show you how to utilize this decorator specifically for shaping test cases. It's important to note that this decorator only applies to classes; it doesn't affect functions or methods. Tags can be modified by adding a minus (-) sign as a prefix, which removes them instead of adding or selecting them. For example, if you want to prevent your test from being executed by default, you can remove the standard tag.

Getting ready

For this recipe, we will use the `my_hostel` module from the last recipe. We will modify the sequence of the test case.

How to do it...

Follow these steps to add tags to the Python test cases:

1. Add a `tagged()` decorator (such as the following) to the test class to run it after the installation of all modules:

```
from odoo.tests.common import TransactionCase, tagged

@tagged('-at_install', 'post_install')
class TestHostelRoomState(TransactionCase):

    def setUp(self, *args, **kwargs):
```

```
        super(TestHostelRoomState, self).setUp(*args, **kwargs)
        self.partner_nikul = self.env['res.partner'].
create({'name': 'Nikul Chaudhary'})
        self.partner_deepak = self.env['res.partner'].
create({'name': 'Deepak Ahir'})
        self.member_ids = self.env['hostel.room.member'].
create([
            {'partner_id': self.partner_nikul.id, 'member_
number': '007'},
            {'partner_id': self.partner_deepak.id, 'member_
number': '357'}])
        self.test_hostel_room = self.env['hostel.room'].create({
            'name': 'Hostel Room 01',
            'room_no': '1',
            'member_ids': [(6, 0, self.member_ids.ids)]
        })

    def test_button_available(self):
        """Make available button"""
        self.test_hostel_room.make_available()
        self.assertIn(self.partner_nikul, self.test_hostel_room.
mapped('member_ids.partner_id'))
        self.assertEqual(
            self.test_hostel_room.state, 'available', 'Hostel
Room state should changed to available')

    def test_button_closed(self):
        """Make closed button"""
        self.test_hostel_room.make_available()
        self.test_hostel_room.make_closed()
        self.assertEqual(
            self.test_hostel_room.state, 'closed', 'Hostel Room
state should changed to closed')
```

2. After that, run the test case as follows, just like before:

```
./odoo-bin -c server.conf -d db_name -i my_hostel --test-enable
```

3. Now, check the server log. This time, you will see our test case log after the following logs, meaning that our test cases were run after all of the modules were installed, as follows:

```
INFO test odoo.modules.loading: Module my_hostel loaded in
0.21s, 240 queries (+240 other)
INFO test odoo.modules.loading: Modules loaded
INFO test odoo.service.server: Starting post tests
```

```
INFO test odoo.addons.my_hostel.tests.test_hostel_room_state:
Starting TestHostelRoomState.test_button_available ...
INFO test odoo.addons.my_hostel.tests.test_hostel_room_state:
Starting TestHostelRoomState.test_button_closed ...
INFO test odoo.service.server: 2 post-tests in 0.04s, 36 queries
INFO test odoo.tests.stats: my_hostel: 4 tests 0.04s 36 queries
```

In these logs, the first line shows that nine modules were loaded. The second line shows that all requested modules and their dependencies were installed successfully, and the third line shows that it will start running the test cases that are tagged as `post_install`.

How it works...

By default, all of the test cases are tagged with `standard`, `at_install`, and the current module's technical name (in our case, the technical name is `my_hostel`). Consequently, if you do not use a `tagged()` decorator, your test case will have these three tags.

In our case, we want to run the test case after installing all of the modules. To do so, we have added a `tagged()` decorator to the `TestHostelRoomState` class. By default, the test case has the `at_install` tag. Because of this tag, your test case will run immediately after the module is installed; it will not wait for other modules to be installed. We don't want this, so to remove the `at_install` tag, we have added `-at_install` to the tagged function. The tags that are prefixed by `-` will remove that tag.

By adding `-at_install` to the `tagged()` function, we stopped the test case execution after the module installation. As we haven't specified any other tag in this, the test case won't run.

So, we have added a `post_install` tag. This tag specifies that the test case needs to be run after the installation of all modules is completed.

As you have seen, all test cases are tagged with the `standard` tag, by default. Odoo will run all of the test cases tagged with the `standard` tag, in case you don't want to run the specific test case all of the time and only want to run it when it is requested. To do so, you need to remove the `standard` tag by adding `-standard` to the `tagged()` decorator, and you need to add a custom tag like this:

```
@tagged('-standard', 'my_custom_tag')
class TestClass(TransactionCase):
...
```

All of the non-standard test cases will not run with the `--test-enable` option. To run the preceding test case, you need to use the `--test-tags` option, as follows (note that, here, we do not need to pass the `--test-enable` option explicitly):

```
./odoo-bin -c server.conf -d db_name -i my_hostel --test-tags=my_
custom_tag
```

There's more...

During the development of the test case, it is important to run the test case for just one module. By default, the technical name of the module is added as a tag, so you can use the module's technical name with the `--test-tags` option. For example, if you want to run test cases for the `my_hostel` module, then you can run the server like this:

```
./odoo-bin -c server.conf -d db_name -i my_hostel --test-tags=my_
hostel
```

The command given here will run the test case in the `my_hostel` module, but it will still decide the sequence based on the `at_install` and `post_install` options.

Setting up Headless Chrome for client-side test cases

Odoo employs Headless Chrome to execute JavaScript and tour test cases, facilitating the simulation of end-user environments. Headless Chrome, devoid of the complete UI, enables seamless execution of JavaScript test cases, ensuring a consistent testing environment.

How to do it...

You will need to install Chrome to enable a JavaScript test case. For the development of the modules, we will mostly use the desktop OS. Consequently, if you have a Chrome browser installed on your system, then there is no need to install it separately. You can run client-side test cases with desktop Chrome. Make sure that you have a Chrome version higher than Chrome 59. Odoo also supports the Chromium browser.

> **Note**
>
> Headless Chrome client-side test cases work fine with macOS and Linux, but Odoo does not support Headless Chrome test cases on Windows.

The situation changes slightly when you want to run test cases on the production server or Server OS. Server OS does not have a GUI, so you need to install Chrome differently. If you are using a Debian-based OS, you can install Chromium with the following command:

```
apt-get install chromium-browser
```

> **Important information**
>
> *Ubuntu 22.04 Server Edition has not enabled the* `universe` *repository by default. So, it's possible that installing* `chromium-browser` *will show an installation candidate error. To fix this error, enable the* `universe` *repository with the following command –* `sudo add-apt-repository universe`.

Odoo also uses **WebSockets** for JavaScript test cases. For that, Odoo uses the `websocket-client` Python library. To install it, use the following command:

```
pip3 install websocket-client
```

Now, your system is ready to run client-side test cases.

How it works...

Odoo uses Headless Chrome for JavaScript test cases. The reason behind this is that it runs test cases in the background, so it can be run on Server OS, too. Headless Chrome prefers to run the Chrome browser in the background, without opening a GUI browser. Odoo opens a Chrome tab in the background and starts running the test cases in it. It also uses **jQuery's QUnit** for JavaScript test cases. In the next few recipes, we will create a QUnit test case for our custom JavaScript widgets.

For test cases, Odoo opens Headless Chrome in a separate process, so to find out the status of a test case running in that process, the Odoo server uses WebSockets. The `websocket-client` Python library is used to manage WebSockets to communicate with Chrome from the Odoo server.

Adding client-side QUnit test cases

Building new fields or views is very simple in Odoo. In just a few lines of XML, you can define a new view. However, under the hood, it uses a lot of JavaScript. Modifying/adding new features on the client side is complex, and it might break a few things. Most client-side issues go unnoticed, as most errors are only displayed in the console. So, QUnit test cases are used in Odoo to check the correctness of different JavaScript components.

QUnit is a JavaScript testing framework primarily used for client-side testing. It's commonly associated with testing JavaScript code in web applications, particularly for frontend development. QUnit is often used to test the logic and behavior of JavaScript functions, modules, and components in a web browser environment.

Getting ready

For this recipe, we will continue using the `my_hostel` module from the previous recipe. We will add a QUnit test case for the `int_color` widget.

How to do it...

Follow these steps to add JavaScript test cases to the `int_color` widget:

1. We have already implemented a widget for `int_color` using JavaScript in our module.
2. Add `/static/tests/colorpicker_tests.js` with the following code:
3. Create a `beforeEach` function to load the data field-wise before applying the test case:

```
/** @odoo-module */

import { registry } from "@web/core/registry";
import { session } from "@web/session";
import { uiService } from "@web/core/ui/ui_service";
import { makeView, setupViewRegistries} from "@web/../tests/
views/helpers";
import { click, getFixture, patchWithCleanup } from "@web/../
tests/helpers/utils";

const serviceRegistry = registry.category("services");

QUnit.module("Color Picker Widget Tests", (hooks) => {
    let serverData;
    let target;
    hooks.beforeEach(async function (assert) {
        target = getFixture();
        serverData = {
            models: {
                'hostel.room': {
                    fields: {
                        name: { string: "Hostel Name", type:
"char" },
                        room_no: { string: "Room Number", type:
"char" },
                        color: { string: "color", type:
"integer"},
                    },
                    records: [{
                        id: 1,
                        name: "Hostel Room 01",
                        room_no: 1,
                        color: 1,
                    }, {
                        id: 2,
                        name: "Hostel Room 02",
                        room_no: 2,
                        color: 3
                    }],
```

```
                    },
                },
                views: {

                    "hostel.room,false,form": `<form>
                        <field name="name"/>
                        <field name="room_no"/>
                        <field name="color" widget="int_color"/>
                    </form>`,
                },
            };
            serviceRegistry.add("ui", uiService);
            setupViewRegistries();
        });
```

4. Add a `QUnit` test case for the color picker field, like this:

```
        QUnit.module("IntColorField");
        QUnit.test("factor is applied in IntColorField", async
    function (assert) {
            const form = await makeView({
                serverData,
                type: "form",
                resModel: "hostel.room",
            });
            assert.containsOnce(target, '.o_field_int_color');
            assert.strictEqual(target.querySelectorAll(".o_int_
    color .o_color_pill").length, 10, "Color picker should have 10
    pills");
            await click(target.querySelectorAll(".o_int_color .o_
    color_pill")[3]);
            assert.strictEqual(target.querySelector('.o_int_color
    .o_color_4').classList.contains("active"), true, "Click on pill
    should make pill active");
        });
    });
```

5. Add the following code to `__manifest__.py` to register it in the test suite:

```
'assets': {

'web.qunit_suite_tests': [

    'my_hostel/static/tests/**/*',

],
 },
```

To run this test case, start your server with the following command in the Terminal:

```
./odoo-bin -c server.conf -i my_hostel,web --test-enable
```

To check that the tests have run successfully, search for the following log:

```
... INFO test odoo.addons.web.tests.test_js.WebSuite: console log:
"Color Picker Widget Tests" passed 2 tests.
```

How it works...

In Odoo, JavaScript test cases are added to the `/static/tests/` directory. In *step 1*, we added a `colorpicker_test.js` file for the test case. In that file, we imported the registry for use in `serviceRegistry` and `setupViewRegistries` and `makeView` from test helpers. `makeView` is imported because we created the `int_color` widget for the form view, so to test the widget, we will need the form view.

`@web/../tests/helpers/utils` will provide us with the test utilities we require to build the JavaScript test cases. If you don't know how JavaScript import works, refer to the *Extending CSS and JavaScript for the website* recipe in *Chapter 14, CMS Website Development*.

Odoo client-side test cases are built with the QUnit framework, which is the jQuery framework for the JavaScript unit test case. Refer to `https://qunitjs.com/` to learn more about this. The `beforeEach` function is called before running the test cases, and this helps to initialize the test data. The reference of the `beforeEach` function is provided by the QUnit framework itself.

We initialized some data in the `beforeEach` function. Let's see how that data is used in the test case. The client-side test case runs in an isolated (mock) environment, and it doesn't make a connection to the database, so for these test cases, we need to create test data. Internally, Odoo creates the mock server to mimic the **Remote Procedure Call** (**RPC**) calls and uses the `serverData` property as the database. Consequently, in `beforeEach`, we initialized our test data in the `serverData` property. The keys in the `serverData` property are considered a table, and the values contain information about the fields and the table rows. The `fields` key is used to define table fields, and the `records` key is used for the table rows. In our example, we added a `room` table with three fields – `name(char)`, `room_no(char)`, and `color(integer)`. Note that, here, you can use any Odoo fields, even relational fields – for example, `{string: "M2o Field", type: "many2one", relation: 'partner'}`. We also added two room records with the `records` key.

Then, we added the test cases with the `QUnit.test` function. The first argument in the function is `string` to describe the test case. The second argument is the function to which you need to add code for the test cases. This function is called from the QUnit framework, and it passes the assert utilities as the argument. In our example, we passed the number of expected test cases in the `assert.expect` function. We are adding two test cases, so we passed `2`.

We want to add to the test case the `int_color` widget in the editable form view, so we created the editable form view with `makeView`. The `makeView` function accepts different arguments, as follows:

- `resModel` is the name of the model for which the given view is created. All of the models are listed in the `resModel` as properties. We want to create a view for the room model, so in our example, we used the room as a model.
- `serverData` is the record that we are going to use in the view. The views key from `serverData` is the definition of the view you want to create. Because we want to test the `int_color` widget, we passed the view definition with the widget. Note that you can only use the fields that are defined in the model.
- `Type`: The type of view.

After creating the form view with the `int_color` widget, we added two test cases. The first one is used to check the number of color pills on the UI, and the second test case is used to check that the pill is activated correctly after the click. We have the `strictEqual` function from the asserted utility of the QUnit framework. The `strictEqual` function passes the test case if the first two arguments match. If they do not match, it will fail the test case.

There's more...

There are a few more assert functions available for QUnit test cases, such as `assert.deepEqual`, `assert.ok`, and `assert.notOk`. To learn more about QUnit, refer to its documentation at `https://qunitjs.com/`.

Adding tour test cases

You have now seen the Python and JavaScript test cases. Both of these work in an isolated environment, and they don't interact with each other. To test integration between JavaScript and Python code, tour test cases are used.

Getting ready

For this recipe, we will continue using the `my_hostel` module from the previous recipe. We will add a tour test case to check the flow of the room model. Also, make sure you have installed the `web_tour` module or have added the `web_tour` module dependency to the manifest.

How to do it...

Follow these steps to add a tour test case for `rooms`:

1. Add a `/static/src/js/my_hostel_tour.js` file, and then add a tour as follows:

```
/** @odoo-module **/

import { _t } from "@web/core/l10n/translation";
import { registry } from "@web/core/registry";
import { stepUtils } from "@web_tour/tour_service/tour_utils";
import { markup } from "@odoo/owl";


registry.category("web_tour.tours").add("hostel_tour", {
    url: "/web",
    rainbowMan: false,
    sequence: 20,
    steps: () => [stepUtils.showAppsMenuItem(), {
    trigger: '.o_app[data-menu-xmlid="my_hostel.hostel_base_
menu"]',
    content: markup(_t("Ready to launch your <b>Hostel</b>?")),
    position: 'bottom',
    edition: 'community',
}
```

2. Add steps for the test tour:

```
    trigger: '.o_app[data-menu-xmlid="my_hostel.hostel_base_
menu"]',
    content: markup(_t("Ready to launch your <b>Hostel</b>?")),
    position: 'bottom',
    edition: 'enterprise',
}, {
    trigger: '.o_list_button_add',
    content: markup(_t("Let's create new room.")),
    position: 'bottom',
}, {
    trigger: ".o_form_view .o_field_char[name='name']",
    content: markup(_t('Add a new <b> Hostel Room </b>.')),
    position: "top",
    run: function (actions) {
    actions.text("Hostel Room 01", this.$anchor.find("input"));
},
}, {
```

```
    trigger: ".ui-menu-item > a",

    auto: true,
    in_modal: false,
}, {
    trigger: ".breadcrumb-item:not(.active):first",
    content: _t("Click on the breadcrumb to go back to your
Pipeline. Odoo will save all modifications as you navigate."),
    position: "bottom",
    run: function (actions) {
    actions.auto(".breadcrumb-item:not(.active):last");
},
},]});
```

3. Add the `my_hostel_tour.js` file to the test assets:

```
'web.assets_backend': [

'my_hostel/static/src/js/tours/my_hostel_tour.js',
],
```

4. Add a `/tests/test_tour.py` file, and run the tour through `HttpCase`, as follows:

```
from odoo.tests.common import TransactionCase, tagged
from odoo.tests import HttpCase

@tagged('post_install', '-at_install')
class TestUi(HttpCase):


    def test_01_hostel_tour(self):
        self.start_tour("/web", 'hostel_tour', login="admin")
```

In order to run test cases, start the Odoo server with the following option:

```
./odoo-bin -c server.conf -i my_hostel --test-enable
```

Now, check the server log. Here, you will find the following logs if our test cases ran successfully:

```
...INFO test odoo.addons.my_hostel.tests.test_tour.TestroomUI: console
log: Tour hostel_tour succeeded
```

How it works...

In order to create tour test cases, you need to create the UI tour first. If you want to learn more about UI tours, refer to the *Improve onboarding with tours* recipe in *Chapter 15*, *Web Client Development*.

In *step 1*, we registered a new tour with the name `hostel_tour`. This tour is exactly like the tour we created in the *Improve onboarding with tours* recipe in *Chapter 15*. In *step 2*, we added the steps for the tours.

Here, we have two main changes compared to the onboarding tour. First, we added a `test=true` parameter for the tour definition; second, we added one extra property, `run`. In the `run` function, you have to write the logic to perform the operation that is normally done by the user. For example, in the fourth step of the tour, we ask the user to enter the room title.

To automate this step, we added a `run` function to set the value in the `title` field. The `run` function passes the action utility as the parameter. This provides some shortcuts to perform basic actions. The most important ones are as follows:

- `actions.click(element)` is used to click on a given element.
- `actions.dblclick(element)` is used to double-click on a given element.
- `actions.tripleclick(element)` is used to triple-click on a given element.
- `actions.text(string)` is used to set the input values.
- `actions.drag_and_drop(to, element)` is used to drag and drop an element.
- `actions.keydown(keyCodes, element)` is used to trigger particular keyboard events on an element.
- `actions.auto()` is the default action. When you don't pass the `run` function in the tour step, `actions.auto()` is performed. This usually clicks on the trigger element of the tour step. The only exception here is an input element. If the trigger element is `input`, the tour will set the default value, `Test`, in the input. That is why we don't need to add `run` functions to all of the steps.

Alternatively, you can perform whole actions manually if default actions are not enough. In the next tour step, we want to set a value for the color picker. Note that we used the manual action because default values won't help here. Consequently, we added the `run` method with the basic jQuery code to click on the third pill of the color picker. Here, you will find the trigger element with the `this.$anchor` property.

By default, registered tours are displayed to the end user to improve the onboarding experience. In order to run them as a test case, you need to run them in Headless Chrome. To do so, you need to use the `HttpCase` Python test case. This provides the `browser_js` method, which opens the URL and executes the command passed as the second parameter. You can run the tour manually, like this:

```
odoo.__DEBUG__.services['web_tour.tour'].run('hostel_tour')
```

In our example, we passed the name of the tour as the argument in the `browser_js` method. The next parameter is used to wait for a given object to be ready before performing the first command. The last parameter in the `browser_js()` method is the name of the user. This username will be used to create a new test environment, and all of the test actions will be performed on behalf of this user.

Running client-side test cases from the UI

Odoo provides a way to run client-side test cases from the UI. By running the test case from the UI, you will be able to see each step of the test case in action. This way, you can verify that the UI test case works exactly as you wanted.

How to do it...

You can run both the `QUnit` test case and the tours test case from the UI. It is not possible to run Python test cases from the UI, as it runs on the server side. In order to see the options to run test cases from the UI, you need to enable developer mode.

Running QUnit test cases from the UI

Click on the bug icon to open the drop-down menu, as shown in the following figure. Click on the **Run JS Tests** option:

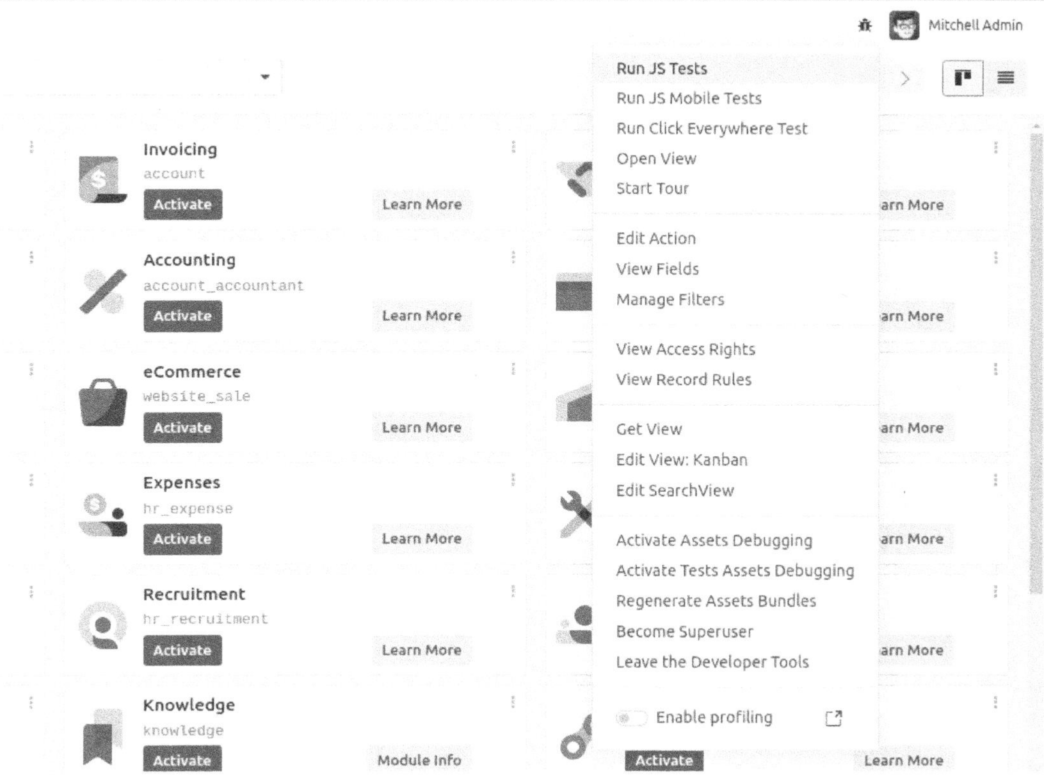

Figure 18.1 – The option to run test cases

This will open the QUnit suite, and it will start running the test cases one by one, as shown in the following screenshot. By default, it will only show the failed test cases. To show all the passed test cases, uncheck the **Hide passed tests** checkbox, as shown in the following screenshot:

Figure 18.2 – The results of the QUnit test cases

Running tours from the UI

Click on the bug icon to open the drop-down menu, as shown in the following screenshot, and then click on **Start Tour**:

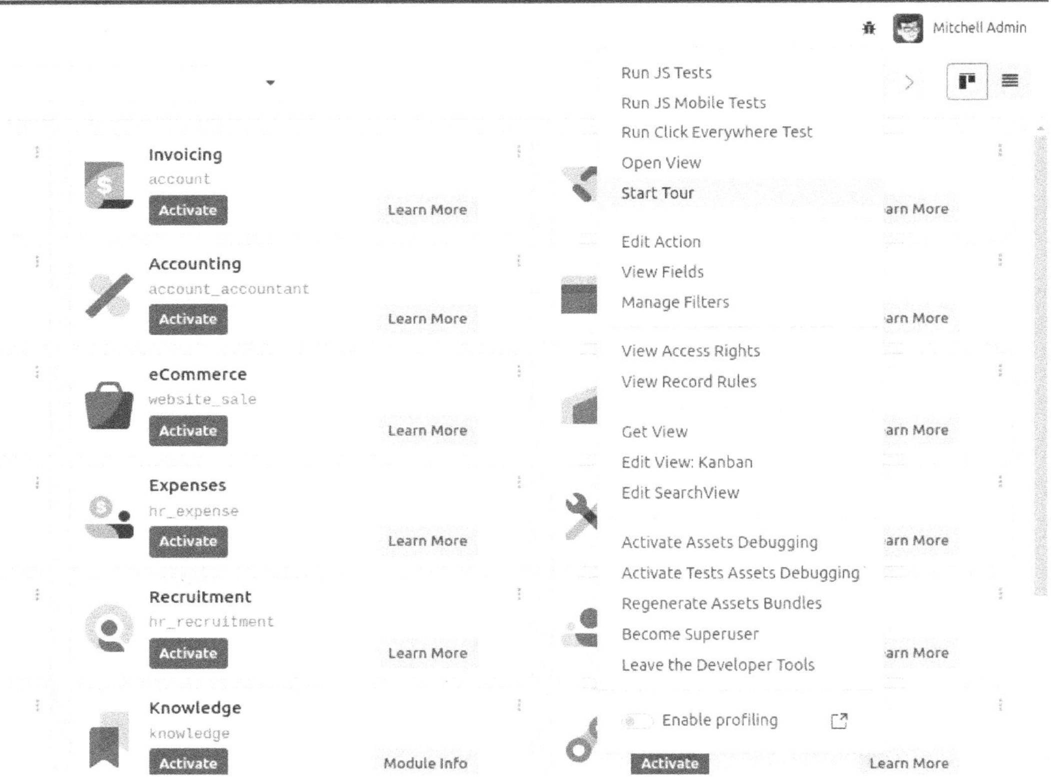

Figure 18.3 – The option to run tour test cases

This will open a dialog with a list of registered tours, as you can see in the following screenshot. Click on the play button on the side to run the tour:

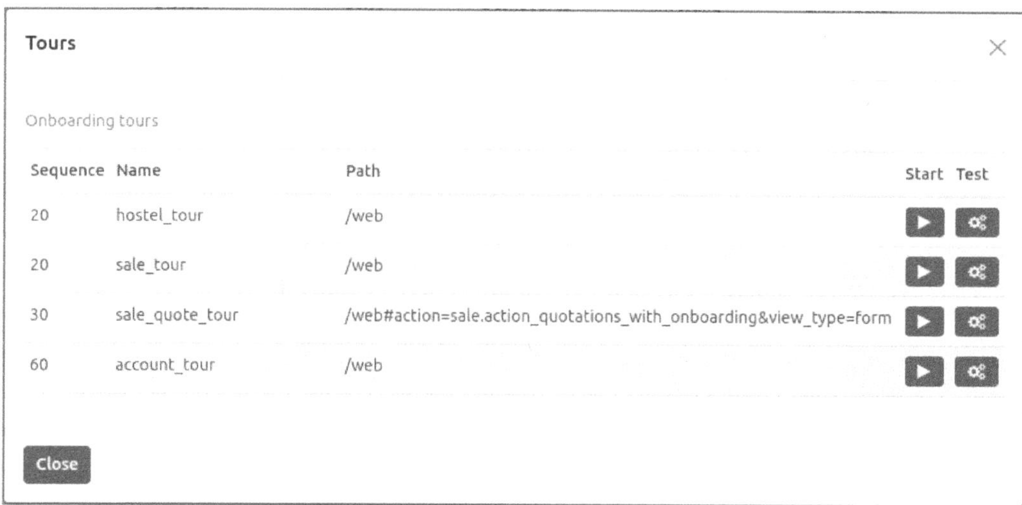

Figure 18.4 – A list of tour test cases

The test tours only display in a list if you have enabled test assets mode. If you can't find the `hostel_tour` tour in the list, make sure you have activated test assets mode.

How it works...

The UI for QUnit is provided by the QUnit framework itself. Here, you can filter the test cases for the modules. You can even run a test case for just one module. With the UI, you can see the progress of each test case, and you can drill down to each step of the test case. Internally, Odoo just opens the same URL in Headless Chrome.

Clicking on the **Run tours** option will display the list of available tours. By clicking on the play button on the list, you can run the tour. Note that when the tour runs via the command-line options, it runs in the rolled-back transaction, so changes made through the tour are rolled back after the tour is successful. However, when the tour runs from the UI, it works just as though a user was operating it, meaning changes made from the tour are not rolled back and stay there, so use this option carefully.

Debugging client-side test cases

Developing complex client-side test cases can be a headache. In this recipe, you will learn how you can debug the client-side test cases in Odoo. Instead of running all of the test cases, we will run just the one. Additionally, we will display the UI of the test case.

Getting ready

For this recipe, we will continue using the `my_hostel` module from the previous recipe.

How to do it...

Follow these steps to run a test case in debug mode:

1. Open the `/static/tests/colorpicker_test.js` file and update and add the `makeView` function, like this:

```
await makeView({
    type: "form",
    resModel: "hostel.room",
    serverData: {
        models: {
            'hostel.room': {
                fields: {
                    name: { string: "Hostel Name", type: "char"
},
                    room_no: { string: "Room Number", type:
"char" },
                    color: { string: "color", type: "integer"},
                },
                records: [
                    {
                        id: 1,
                        name: "Hostel Room 01",
                        room_no: 1,
                        color: 1,
                    },
                    {
                        id: 2,
                        name: "Hostel Room 02",
                        room_no: 2,
                        color: 3
                    }
                ],
            },
        },
        views: { },
    },
    arch: `
    <form>
```

```
            <field name="name"/>
            <field name="room_no"/>
            <field name="color" widget="int_color"/>
        </form>`,
    });
```

2. Check the `target` parameter in the `containtsN` function, as follows:

```
    assert.containsN(
                target,
                ".o_field_int_color",
                1,
                "Both records are rendered"
            );
        });
    });
```

Open developer mode and open the drop-down menu by clicking on the bug icon on the top menu, and then click on **Run JS Tests**. This will open the QUnit suite:

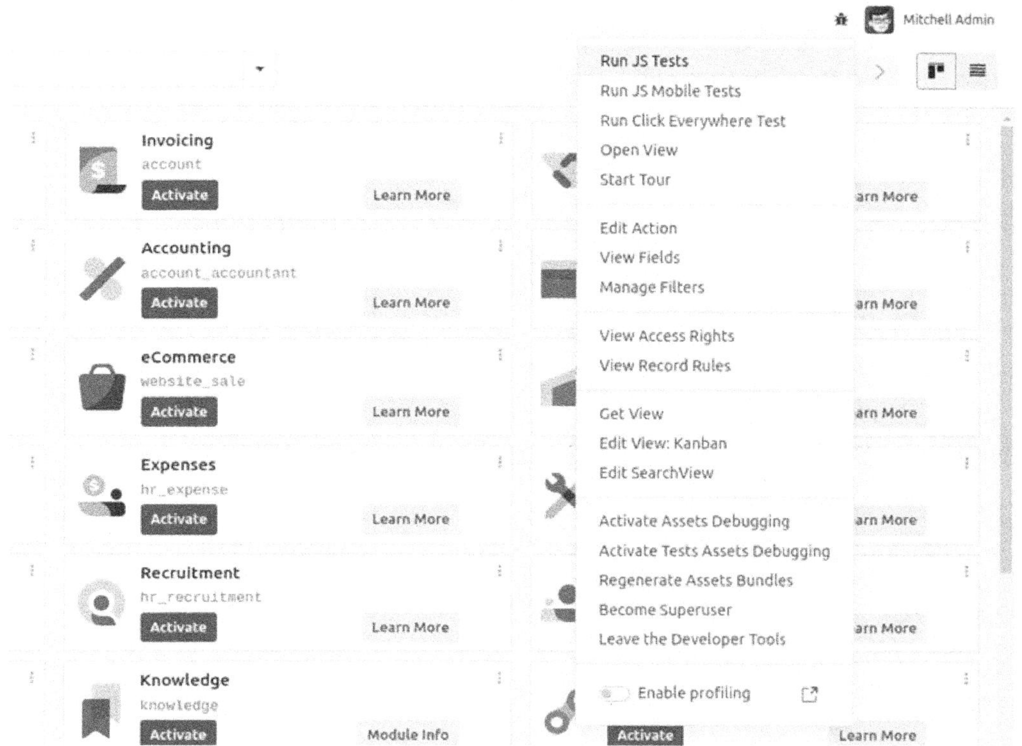

Figure 18.5 – The option to run test cases

This will run only one test case, which is our color picker test case.

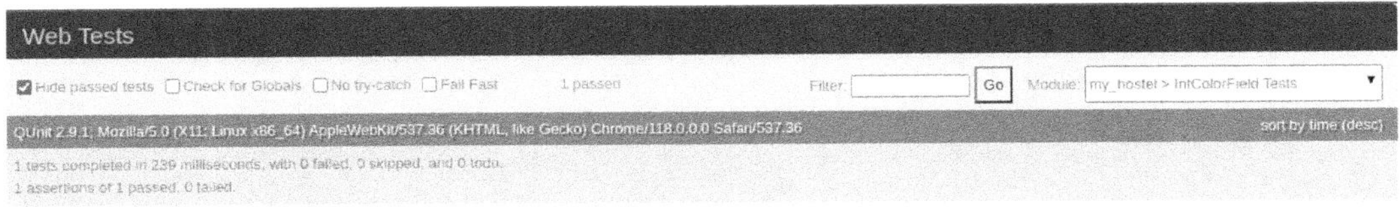

Figure 18.6 – Color picker test case

How it works...

In *step 1*, we replaced `QUnit.test` with `QUnit.only`. This will run this test case only. During the development of the test case, this can be time-saving. Note that using `QUnit.only` will stop the test case from running via the command-line options. This can only be used for debugging or testing, and it can only work when you open the test case from the UI, so don't forget to replace it with `QUnit.test` after the development.

In our QUnit test case example, we created the form view to test the `int_color` widget. If you run the QUnit test cases from the UI, you will find that you are not able to see the created form views in the UI. From the UI of the QUnit suite, you are only able to see the logs. This makes developing a QUnit test case very difficult. To solve this issue, the `debug` parameter is used in the `makeView` function. In *step 2*, we added `debug: true` in the `makeView` function. This will display the test form view in the browser. Here, you will be able to locate **Document Object Model** (**DOM**) elements via the browser debugger.

> **Warning**
>
> *At the end of the test case, we destroy the view through the* `destroy()` *method. If you have destroyed the view, then you won't be able to see the form view in the UI, so in order to see it in the browser, remove that line during development. This will help you debug the test case.*

Running QUnit test cases in debug mode helps you develop test cases very easily and quickly.

Generating videos/screenshots for failed test cases

Odoo uses Headless Chrome, which opens new possibilities. Starting from Odoo 12, you can record videos of the failed test cases, and you can take screenshots of them as well.

How to do it...

Recording a video for a test case requires an `ffmpeg` package:

1. To install this, you need to execute the following command in the terminal (note that this command only works on a Debian-based OS):

   ```
   apt-get install ffmpeg
   ```

2. To generate a video or screenshot, you will need to provide a directory location to store the video or screenshots.
3. If you want to generate a screencast (video) of a test case, use the `--screencasts` command, like this:

   ```
   ./odoo-bin -c server.conf -i my_hostel --test-enable
   --screencasts=/home/pga/odoo_test/
   ```

4. If you want to generate screenshots of a test case, use the `--screenshosts` command, like this:

   ```
   ./odoo-bin -c server.conf -i my_hostel --test-enable
   --screenshots=/home/pga/odoo_test/
   ```

How it works...

In order to generate screenshots/screencasts for failed test cases, you need to run the server with the path to save the video or image files. When you run the test cases, and if a test case fails, Odoo will save a screenshot/video of the failed test case in the given directory.

To generate a video of a test case, Odoo uses the `ffmpeg` package. If you haven't installed this package on the server, then it will only save a screenshot of a failed test case. After installing the package, you will be able to see the `mp4` file of any failed test case.

> **Note**
>
> Generating videos for test cases can consume more space on disks, so use this option with caution and only when it is really necessary.

Keep in mind that screenshots and videos are only generated for failed test cases, so if you want to test them, you need to write a test case that fails.

Populating random data for testing

So far, we have seen test cases that have been used to detect errors or bugs in business logic. However, at times, we need to test our development with large amounts of data. Generating large amounts of data can be a tedious job. Odoo provides a set of tools that helps you generate a lot of random data for your model. In this recipe, we will use the `populate` command to generate test data for the `hostel.room` and `hostel.room.member` models.

Getting ready

For this recipe, we will continue using the `my_hostel` module from the previous recipe. We will add the `_populate_factories` method, which will be used to generate test data.

How to do it...

Follow these steps to generate data for the `hostel.room` model:

1. Add a `populate` folder to the `my_hostel` module. Also, add an `__init__.py` file with this content:

   ```
   from . import hostel_data
   ```

2. Add a `my_hostel/populate/hostel_data.py` file, and then add this code to generate the hostel room's data:

   ```
   import logging
   import random

   from odoo import models
   from odoo.tools import populate

   _logger = logging.getLogger(__name__)


   class RoomData(models.Model):
       _inherit = 'hostel.room.member'
       _populate_sizes = {'small': 10, 'medium': 100, 'large': 500}
       _populate_dependencies = ["res.partner"]

       def _populate_factories(self):
           partner_ids = self.env.registry.populated_models['res.
   partner']
           return [
               ('partner_id', populate.randomize(partner_ids)),
           ]
   ```

```
class HostelData(models.Model):
    _inherit = 'hostel.room'
    _populate_sizes = {'small': 10, 'medium': 100, 'large': 500}
    _populate_dependencies = ["hostel.room.member"]

    def _populate_factories(self):
        member_ids = self.env.registry.populated_models['hostel.
room.member']
        def get_member_ids(values, counter, random):
            return [
                (6, 0, [
                    random.choice(member_ids) for i in
range(random.randint(1, 2))
                ])
            ]
        return [
            ('name', populate.constant('Hostel Room
{counter}')),
            ('room_no', populate.constant('{counter}')),
            ('member_ids', populate.compute(get_member_ids)),
        ]
```

3. Run this command to generate the hostel's data:

```
./odoo-bin -c server.conf -d db_name -i my_hostel
./odoo-bin populate --models=hostel.room --size=medium -c
server.conf -d db_name
```

This will generate 100 units of data for the hostel rooms. After generating the data, the process will be terminated. To see the hostel room's data, run the command without the `populate` parameters.

How it works...

In *step 1*, we added the `populate` folder to the `my_hostel` module. This folder contains the code to populate the test data.

In *step 2*, we added code to populate the room data. To generate random data, the `_populate_factories` method was used. The `_populate_factories` method returns factories for model fields, which will be used to generate random data. The `hostel.room` model has the required name and `room_no` fields, so in our example, we returned the generator for those fields. This generator will be used to generate random data for the `hostel room` record. We used the `populate.constant` generator for the name field; this will generate different names when we iterate during data generation.

Just like `populate.constant`, Odoo provides several other generators to populate data; here is a list of those generators:

- `populate.randomize(list)` will return a random element from the given list.
- `populate.cartesian(list)` is just like `randomize()`, but it will try to include all the values from the list.
- `populate.iterate(list)` will iterate over a given list, and once all the elements are iterated, it will return based on `randomize` or random elements.
- `populate.constant(str)` is used to generate formatted strings. You can also pass the `formatter` parameter to format values. By default, the formatter is a string-format function.
- `populate.compute(function)` is used when you want to compute a value based on your function.
- `populate.randint(a, b)` is used to generate a random number between the `a` and `b` parameters.

These generators can be used to generate test data of your choice.

Another important attribute is `_populate_sizes`. It is used to define the number of records you want to generate based on the `--size` parameter. Its value always depends on the business object.

In *step 3*, we generated a data hostel room model. To populate test data, you will need to use the `--size` and `--model` parameters. Internally, Odoo uses the `_populate` method to generate random records. The `_populate` method itself uses the `_populate_factories` method to get random data for records. The `_populate` method will generate data for the models given in the `--model` parameter, and the amount of test data will be based on the `_populate_sizes` attribute of the model. Based on our example, if we use `--size=medium`, the data for 100 hostel rooms will be generated.

> **Note**
>
> If you run the `populate` command multiple times, the data will be generated multiple times as well. It's important to use this carefully; if you run the command in a production database, it will generate test data in the production database itself. This is something you want to avoid.

There's more...

At times, you might like to generate relational data too. For example, with rooms, you might also want to create member records. To manage such records, you can use the `_populate_dependencies` attribute:

```
class RoomData(models.Model):
    _inherit = 'hostel.room.member'
    _populate_sizes = {'small': 10, 'medium': 100, 'large': 500}
    _populate_dependencies = ["res.partner"]
    . . .
```

This will populate the data for dependencies before populating the current model. Once that is done, you can access the populated data via the `populated_models` registry:

```
partner_ids = self.env.registry.populated_models['res.partner']
```

The preceding line will give you the list of companies that are populated before generating test data for the current model.

19

Managing, Deploying, and Testing with Odoo.sh

In 2017, Odoo released Odoo.sh, a new cloud service. Odoo.sh is a platform that makes the process of testing, deploying, and monitoring Odoo instances as easy as possible. In this chapter, we will look at how Odoo.sh works, when you should use it over other deployment options, and its features.

In this chapter, we will cover the following recipes:

- Exploring some basic concepts of Odoo.sh
- Creating an Odoo.sh account
- Adding and installing custom modules
- Managing branches
- Accessing debugging options
- Getting a backup of your instance
- Checking the status of your builds
- All Odoo.sh options

> **Important note**
>
> This chapter is written under the assumption that you have Odoo.sh access. It is a paid service, and you will need a subscription code to access the platform. If you are an Odoo partner, you will get a free Odoo.sh subscription code. Otherwise, you will need to purchase it from `https://www.odoo.sh/pricing`. You can still go through this chapter even if you don't have a subscription code. It contains enough screenshots to help you understand the platform.

> **A note for print readers**
>
> For the benefit of print readers, there are certain images showing the layout of a window in this chapter that may require zooming to view them clearly. You can access the graphic bundle containing high-quality images at this link: `https://packt.link/gbp/9781805124276`

Exploring some basic concepts of Odoo.sh

In this recipe, we will look at some of the features of the Odoo.sh platform. We will answer some basic questions, such as when you should use it and why it should be used.

What is Odoo.sh?

Odoo.sh is a cloud service that provides the platform with the ability to host Odoo instances with custom modules. Putting it simply, it is Odoo's **platform as a service** (**PaaS**) cloud solution. It is fully integrated with GitHub. Any GitHub repository with valid Odoo modules can be launched on Odoo.sh within minutes. You can examine the ongoing development by testing multiple branches in parallel. Once you have moved your instance to production, you can test some new features with a copy of the production database; this helps to avoid regression. It also takes daily backups. With Odoo.sh, you can deploy Odoo instances efficiently, even if you don't have sound knowledge of DevOps. It automatically sets up an Odoo instance with top-notch configurations. Note that Odoo.sh is the Enterprise edition of Odoo. You cannot use the Odoo Community edition because Odoo.sh will only load the Enterprise edition.

Why was Odoo.sh introduced?

Before Odoo.sh was introduced, there were two ways to host Odoo instances. The first was to use Odoo Online, which is a **software as a service** (**SaaS**) cloud service. The second method was the on-premises option, in which you needed to host an Odoo instance and configure it on your server yourself. Now, both of these options have pros and cons. In the Odoo online option, you don't need to configure or deploy it, as it is a SaaS service. However, you cannot use custom modules on this platform. On the other hand, with the on-premises option, you can use custom modules, but you need to do everything yourself. You need to purchase the server, you need to configure the database and NGINX, and you need to set up the mail server, daily backups, and security.

For this reason, there was a need for a new option that provided the simplicity of Odoo online and the flexibility of the on-premises option. Odoo.sh lets you use custom modules without a complex configuration. It also provides additional features, such as testing branches, staging branches, and automated tests.

> **Important note**
>
> It is not completely true that customization is not possible on Odoo online. With Odoo Studio and other techniques, you can carry out customization. The scope of this customization, however, is very narrow.

When should you use Odoo.sh?

If you don't need customization or you only need a small amount of customization that is possible in Odoo online, you should go for Odoo online. This will save both time and money. If you want a significant amount of customization and you have teamed up with expert DevOps engineers, you can choose the on-premises option. Odoo.sh is suitable for when you have good knowledge of Odoo customization but do not have any expertise in DevOps. With Odoo.sh, there's no need to carry out complex configurations; you can start using it straight away, along with your customization. It even configures the mailing server.

Odoo.sh is very useful when you are developing a large project with agile methodology. This is because on Odoo.sh, you can test multiple development branches in parallel and deploy the stable development in production in minutes. You can even share the test development with the end customer.

What are the features of Odoo.sh?

Odoo has invested a lot of time in the development of the Odoo.sh platform, and it is packed with features as a result. Let's have a look at the features of Odoo.sh. Note that Odoo adds new features from time to time. In this section, I have mentioned the features that are available at the time of writing this book, but you might find some further features as well:

- **GitHub integration**: This platform is fully integrated with GitHub. You can test every branch, pull, or commit here. For every new commit, a new branch will be pulled automatically. It will also run an automated test for the new commits. You can even create/merge branches from the Odoo.sh UI itself.
- **Web shell**: Odoo.sh provides the web shell in the browser for the current build (or production server). Here, you can see all the modules and logs.
- **Web code editor**: Just like the web shell, Odoo.sh provides the code editor in the browser. Here, you can access all of the source code and also get the Odoo interactive shell for the current build.
- **SSH access**: By registering your public keys, you can connect to any container via SSH.
- **External dependencies**: You can install any Python package. To do this, you just need to add `requirement.txt` to the root of your GitHub repository. Right now, you can only install Python packages. It is not possible to install system packages (apt packages).
- **Server logs**: You can access the server log for each build from this browser. These logs are in real time, and you can also filter them from here.

- **Automated tests**: Odoo.sh provides your own runbot, which you can use to perform a series of automated tests for your development. Whenever you add a new commit or a new development branch, Odoo.sh will automatically run all of the test cases and show the status of the tests. You can access the full test log, which will help you find issues if a test case fails.
- **Staging and development branches**: Odoo.sh provides two types of branches: the development branch and the staging branch. In the development branch, you can test ongoing development with demonstration data. The staging branch is used when the development is finished, and you want to test the feature before merging it into production. The staging branch does not load the demonstration data; instead, it uses a copy of the production server.
- **Mail server**: Odoo.sh automatically sets up a mail server for the production server. Just like Odoo online, Odoo.sh does not need any extra configuration for email, although it is possible to use your own mail server.
- **Mail catcher**: The staging branch uses a copy of your production database, so it has information about your real customers. Testing on such a database can make it possible to send emails to real customers. To avoid this issue, the email feature is only activated on production branches. Staging and development branches do not send real emails; instead, they use a mail catcher so that you can test and see emails in the staging and development branches.
- **Share the build**: With Odoo.sh, you can share the development branches with your customers so they can test them before merging the feature into production.
- **Faster deployment**: As Odoo.sh is fully integrated with GitHub, you can merge and deploy the development branches directly from the browser with a simple drag-and-drop procedure.
- **Backup and recovery**: Odoo.sh keeps full backups for the production instance. You can download or restore any of these backups in just a few clicks. Refer to the *Getting a backup of your instance* recipe to learn more about backups. Odoo.sh keeps 14 full backups for up to 3 months: 1 per day for 7 days, 1 per week for 4 weeks, 1 per month for 3 months.
- **Community modules**: You can test and install any community module in a few simple clicks. You can also test free modules directly from the app store.

Creating an Odoo.sh account

In this recipe, we will create an Odoo.sh account and an empty repository for the custom add-ons.

Getting ready

For this recipe, you will need a GitHub account on which you can add custom modules. You will also need an Odoo.sh subscription code. If you are an Odoo partner, you will get a free Odoo.sh subscription code. Otherwise, you will need to purchase it from `https://www.odoo.sh/pricing`.

How to do it...

Follow these steps to create an Odoo.sh account:

1. Open `https://www.odoo.sh` and click on **Sign in** in the top menu. This will redirect you to the GitHub page:

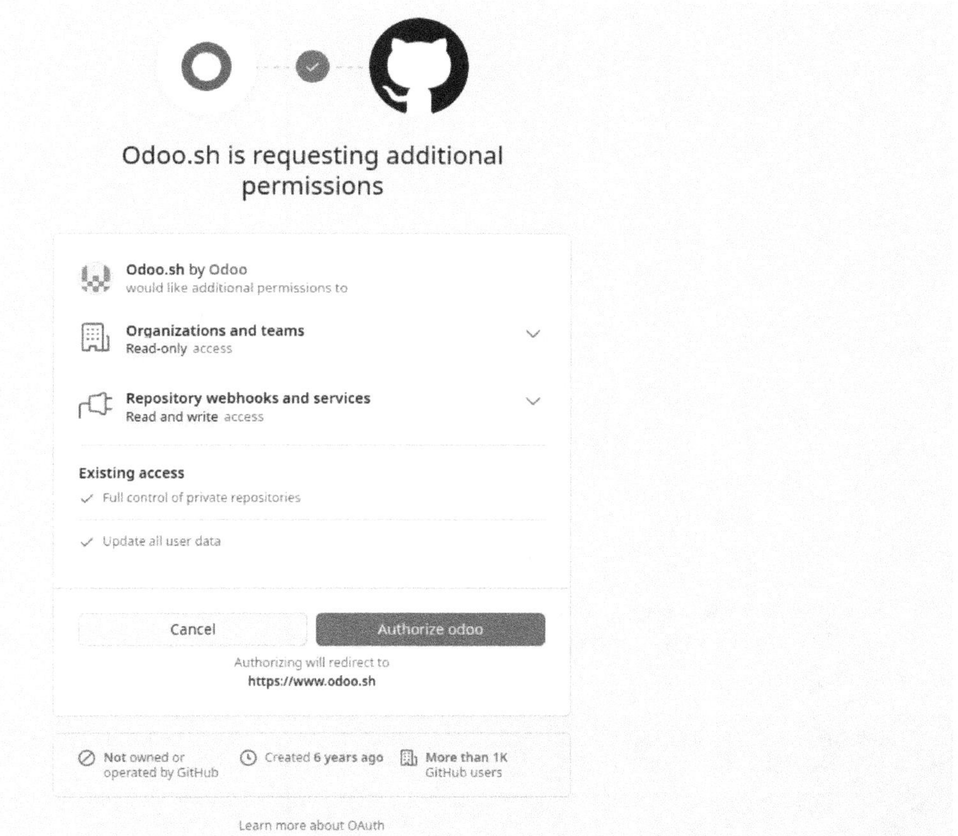

Figure 19.1 – GitHub authentication

2. Give authorization to your repositories, which will redirect you back to Odoo.sh. Fill in the form to deploy the instance:

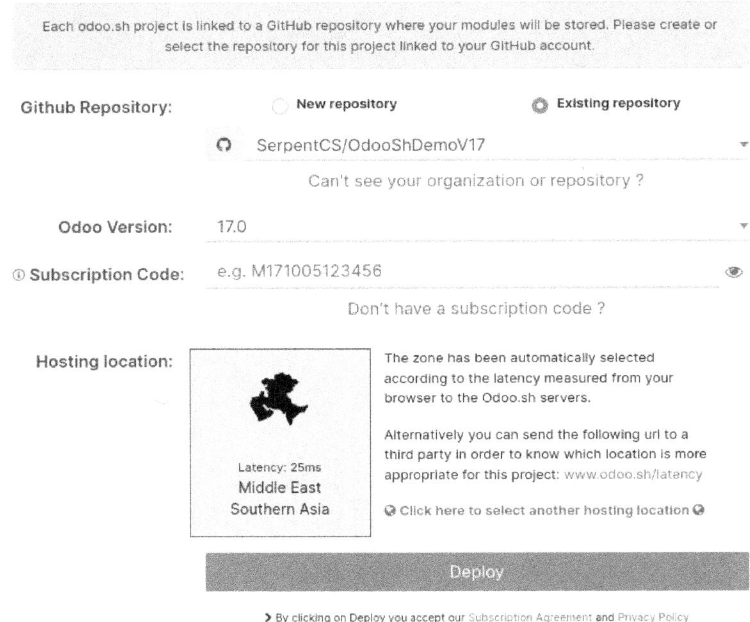

Figure 19.2 – Create an Odoo.sh instance

3. This will deploy the instance, and you will be redirected to the Odoo.sh control panel. Wait for the build status to be successful; then, you can connect to your instance with the **CONNECT** button displayed in the following screenshot:

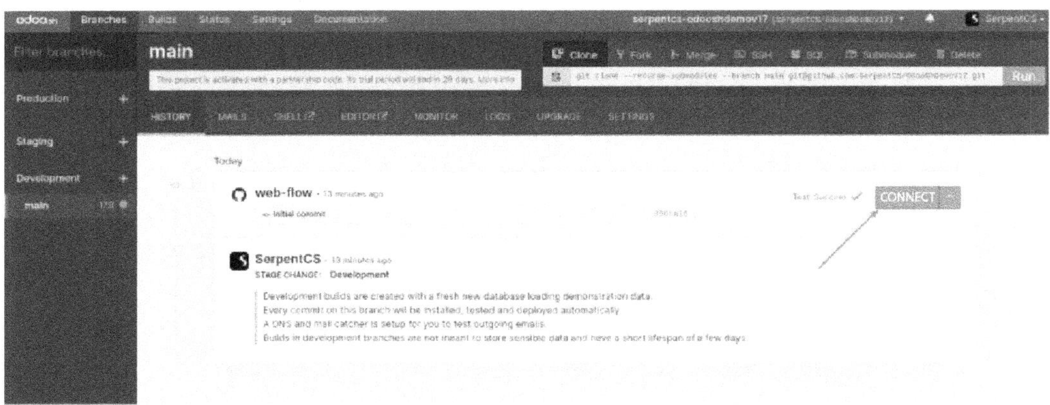

Figure 19.3 – Connect to the development instance

Upon clicking **CONNECT**, you will be automatically logged in to your instance. If you are an admin, by clicking on the arrow button at the side, you can connect as other users as well.

How it works...

The Odoo.sh platform is integrated with GitHub. You need to give full authorization to Odoo.sh so that it can access your repositories. Odoo.sh will also create the webhooks. GitHub webhooks notify the Odoo.sh platform when a new commit or branch has been added to your repository. When you sign in for the first time, Odoo.sh will redirect you to GitHub. GitHub will show a page similar to the screenshot in *step 1*, in which you will need to provide access to all of your private and public repositories. If you are not the owner of the repository, you will see the button to make an access request to the owner for the rights.

After you grant repository access to Odoo.sh, you will be redirected back to Odoo.sh, where you will see the form to deploy the Odoo instance. To create a new instance, you will need to add the following information:

- **GitHub repository**: Here, you will need to set the GitHub repository with your custom modules. The modules in this repository will be available to the Odoo instance. You will see a list of all your existing repositories. You can select one of them or create a new one.
- **Odoo version**: Choose the Odoo version you want to deploy. You can select from the currently supported Odoo LTS versions. Make sure you select the version that is compatible with the modules in the GitHub repository. For our example, we will select version 14.0.
- **Subscription code**: This is the code to activate the instance. You will receive the code via email after purchasing an Odoo.sh plan; if you are an official Odoo partner, you can ask for this code from Odoo.
- **Hosting location**: Here, you need to choose a server location based on your geographic location. The server that is nearest will give the best performance. The latency displayed under the hosting location is based on your location. So if you are creating an instance for your customer and the customer is in another country, you will need to select a server location that is near the customer's location with lower latency.
- Once you submit this form, your Odoo instances will be deployed, and you will be redirected to the Odoo.sh control panel. Here, you will see your first build. It will take a few minutes, and then you will be able to connect to your Odoo instance. If you check the left panel, you will see that there are no branches in the production and staging sections and that only one branch is in the development section. In the next few recipes, we will see how you can create staging and production branches.

There's more...

Right now, Odoo.sh only works with GitHub. Other version-control systems, such as GitLab and Bitbucket, are not supported right now. If you want to use a system other than GitHub, you can use the intermediate GitHub repository that is linked to your actual repository via the submodule. In the future, Odoo will add support for GitLab and Bitbucket, but this is not the priority at the moment, according to the Odoo officials. The method suggested here is just a workaround if you want to use GitLab or Bitbucket.

Adding and installing custom modules

As we described earlier, in the *Exploring some basic concepts of Odoo.sh* recipe, on the Odoo.sh platform, you can add custom Odoo modules. The platform is integrated with GitHub, so adding a new commit in the registered repository will create a new build in the respective branch. In this recipe, we will add a custom module in our repository and access that module in Odoo.sh.

Getting ready

For our example, we will choose the `my_hostel` module from *Chapter 18, Automated Test Cases*. You can add any valid Odoo module in this recipe, but we will use the module with test cases here, as the Odoo.sh platform will perform all the test cases automatically. For simplicity, we have added this module in the GitHub repository of this book, at `Chapter20/r0_initial_module/my_hostel`.

How to do it...

Follow these steps to add your custom modules to Odoo.sh:

1. Get your Git repository on your local machine, add the `my_hostel` module in it, and then execute the following command to push the module into the GitHub repository:

```
git add .
git commit -am"Added my_hostel module"
git push origin main
```

2. Open your project in Odoo.sh. Here, you will find a new build for this commit. It will start running test cases, and you will see the following screen:

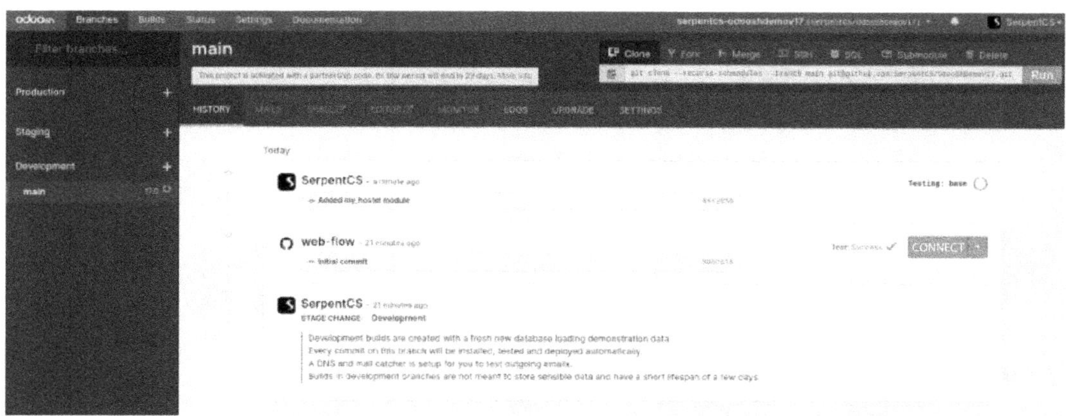

Figure 19.4 – New build for the hostel module

3. After a new commit is pulled in your Odoo.sh project, you will see the installation progress on the right side. Wait for the installation to be complete, then access your instance by clicking on the green **CONNECT** button. It will open the Odoo instance with the `my_hostel` module:

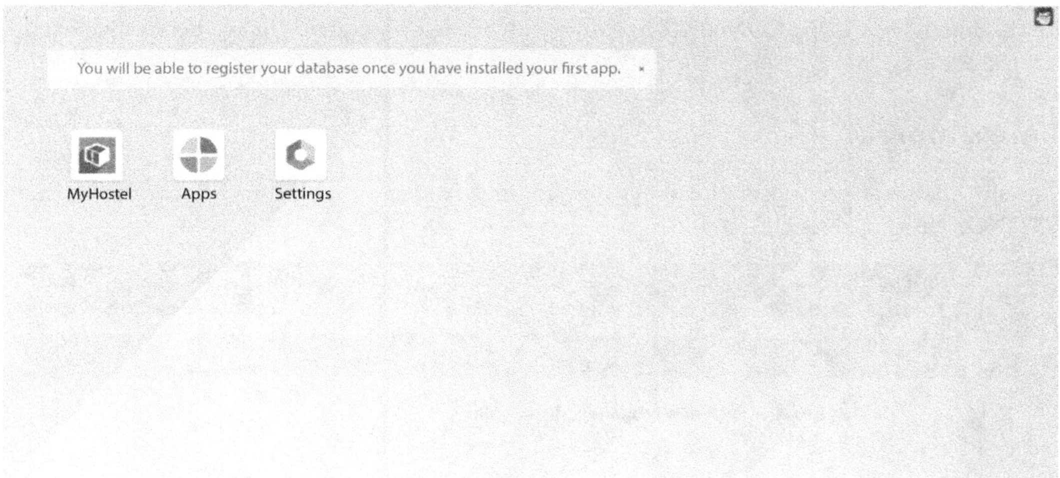

Figure 19.5 – Hostel module installed

Explore and test the `my_hostel` module. Note that this is not a production build, so you can test it however you like.

How it works...

In *step 1*, we uploaded the `my_hostel` module to the GitHub repository. Odoo.sh will be notified about these changes instantly through a webhook. Then, Odoo.sh will start building a new instance. It will install all your custom modules and their dependencies. A new build will automatically perform the test cases for the installed modules.

> **Important note**
>
> By default, Odoo.sh will only install your custom modules and their dependencies. If you want to change this behavior, you can do it from the module installation section of the global settings. We will look at these settings in detail in the next few recipes.

In the **HISTORY** tab, you will be able to see the full history of the branch. Here, you can find some basic information about the build. It will display the commit message, the author information, and the GitHub link of the commit. On the right side, you will get the live progress of the build. Note that the builds in the development section will install the modules with demonstration data. In the next few recipes, you will see the difference between the production, development, and staging branches in detail.

After a successful build, you will see a button to connect the instance. By default, you will be connected with the admin user. Using **CONNECT** as a drop-down menu, you can log in as a demo and portal user instead.

There's more...

Odoo.sh will create a new build for every new commit. You can change this behavior from the **SETTINGS** tab of the branch:

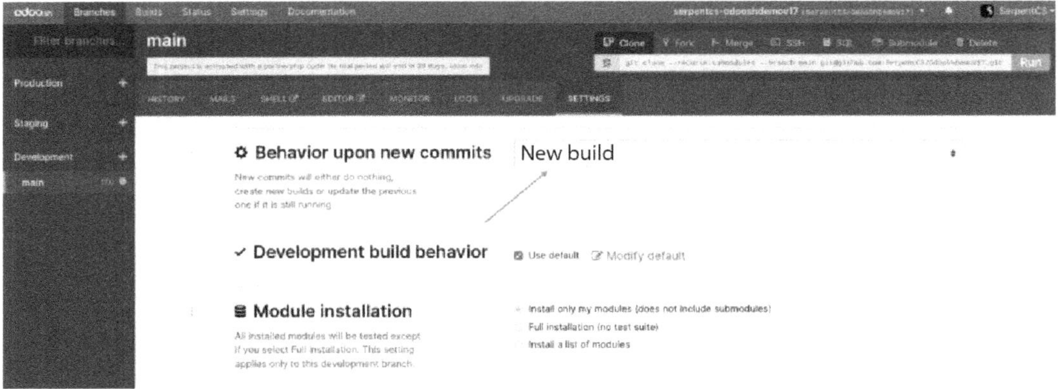

Figure 19.6 – Development branch options

Here, you will find several options. One of them is **Behavior upon new commits**. It has three possible values:

- **New Build**: This option will create a new build for each commit
- **Do Nothing**: This option will ignore the new commit and do nothing
- **Update Previous Build**: This will use an existing build for the new commit

The **Module installation** and **Test suite** options will help you control the test suites. You can disable testing and you can run specific test cases with these options.

Managing branches

In Odoo.sh, you can create multiple development and staging branches along with the production branch. In this recipe, we will create different types of branches and see the differences between them. You will see the full workflow of how you can develop, test, and deploy the new features.

Getting ready

Visit `https://www.odoo.sh/project` and open the project we created in the *Creating an Odoo.sh account* recipe. We will create a development branch for the new feature and then test it in the staging branch. Finally, we will merge the feature in the production branch.

How to do it...

In this recipe, we will create all types of branches in Odoo.sh. At the moment, we don't have any branches in production, so we will start by creating a production branch.

Creating the production branch

Right now, we only have one **main** branch in the **Development** section. The last build of the **main** branch shows a green label that reads **Test: success**, meaning that all of the automated test cases have run successfully. We can move this branch into the **Production** branch, as the test case status shows that everything is fine. In order to move your **main** branch into the **Production** branch, you just need to drag the **main** branch from the **Development** section and drop it in the **Production** section, as shown in the following screenshot:

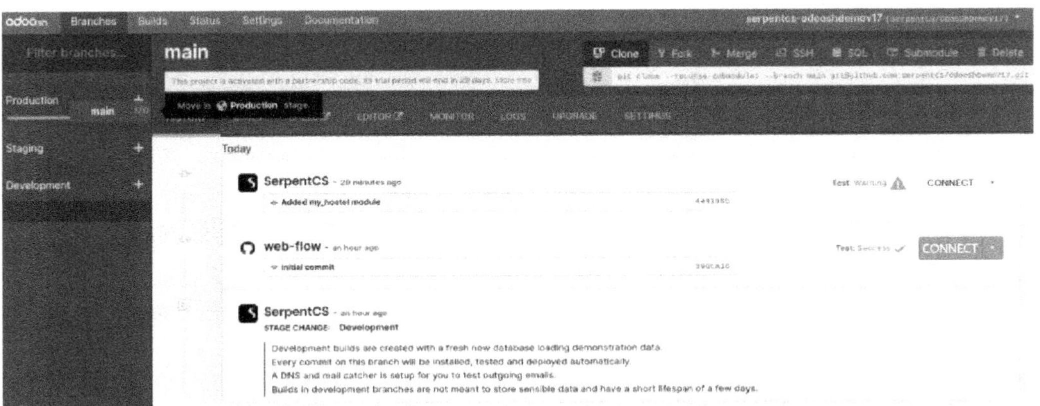

Figure 19.7 – Move the main branch to Production

This will create your **Production** branch. You can access the **Production** branch with the **Connect** button on the right side. Once you open the production instance, you will notice that there have been no applications installed in the production database. This is because the production instance requires you or your end customer to install and configure the operation according to the requirements. Note that this is a production instance, so in order to keep the instance running, you need to enter your Enterprise subscription code.

Creating a development branch

You can create development branches directly from the browser. Click on the plus (+) button next to the **Development** section. This will show two types of input. One is the branch to fork, and the other is the name of the development branch. After filling in the input, hit the *Enter* key.

This will create a new branch by forking the given branch, as shown in the following screenshot:

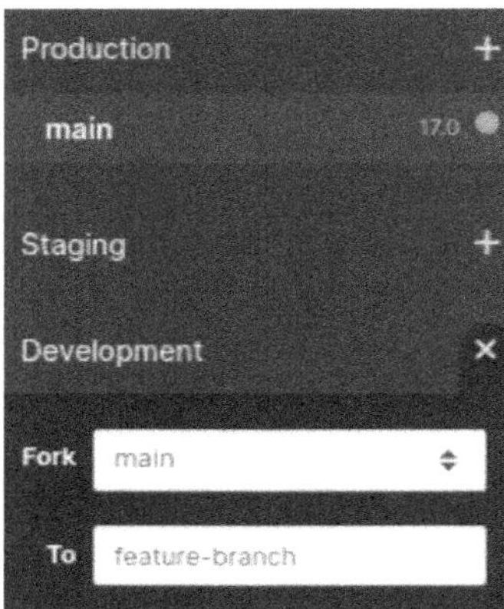

Figure 19.8 – Create a new development branch

> **Important note**
>
> If you don't want to create a development branch from the UI, you can create it directly from GitHub. If you add a new branch in the GitHub repository, Odoo.sh will create a new development branch automatically.

Branches in development are usually new feature branches. As an example, we will add a new field in the `hostel.room` model. Follow these steps to add a new HTML field in the `hostel` model:

1. Increase the module version in the `manifest` file:

```
...
'version': '17.0.1.0.1',
...
```

2. Add a new field in the model:

```
    other_info = fields.Text("Other Information",
                             help="Enter more information")
    description = fields.Html('Description')
    policy = fields.Html('Description')
```

3. Add a **policy** field in the hostel's form view:

```
<notebook>
<page string="Policy">
            <field name="policy"/>
</page>
</notebook>
```

4. Push the changes in the feature branch by executing the following command in the terminal:

```
git commit -am"Added room policy"
git push origin feature-branch
```

This will create a new build on Odoo.sh. After a successful build, you can test this new feature by accessing the instance. You will be able to see a new HTML field in the book's form view. Note that this branch is the development branch, so the new feature is only available to this branch. Your production branch has not changed.

Creating a staging branch

Once you complete the development branch and the test cases are successful, you can move the branch to the **Staging** section. This is the pre-production section. Here, the new feature will be tested with a copy of the production database. This will help us to find any issues that might be generated in the production database. To move from the development branch to the **Staging** branch, just drag and drop the branch into the **Staging** section:

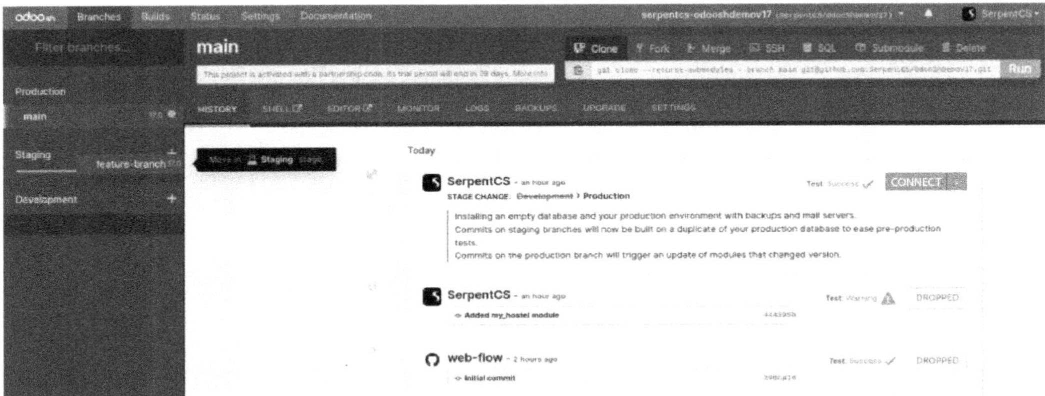

Figure 19.9 – Move the development branch to Staging

Once you move the **Development** branch to the **Staging** section, you can test your new development with production data. Just like any other build, you can access the **Staging** branch with the **CONNECT** button on the right. The only difference is that you will be able to see the data of the production database in this case. Here, your development module is only upgraded automatically if you have increased the module version from the manifest.

> **Important note**
>
> The staging branch will use a copy of the production database, so the staging instance will have real customers and their emails. For this reason, in the staging branch, real emails are disabled so that you don't send any by accident when testing a new feature in the staging branch.

If you haven't changed the module version, you will need to upgrade the modules manually to see the new features in action.

Merging new features in the production branch

After you test the new development with the production database (in the staging branch), you can deploy the new development into the **Production** branch. Like before, you just need to drag and drop the **Staging** branch into the **Production** branch. This will merge the new feature branch into the main branch. Like the **Staging** branch, your development module is only upgraded automatically if you have increased the module version from **manifest**. After this, the new module is available for the end customer:

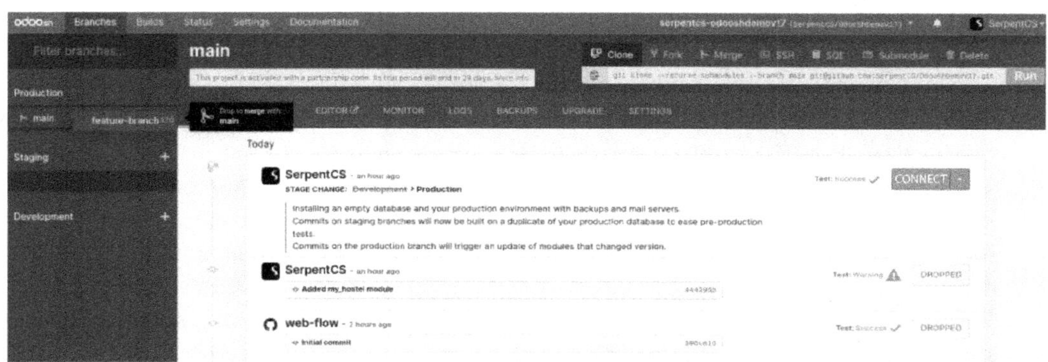

Figure 19.10 – Merge changes to production

Once you drop the staging branch to **Production**, a popup will be displayed with two options:

- **Rebase and Merge**: This will create a pull request and merge it with the rebase so you will have liner history.
- **Merge**: This will create a merge commit without fast-forwarding:

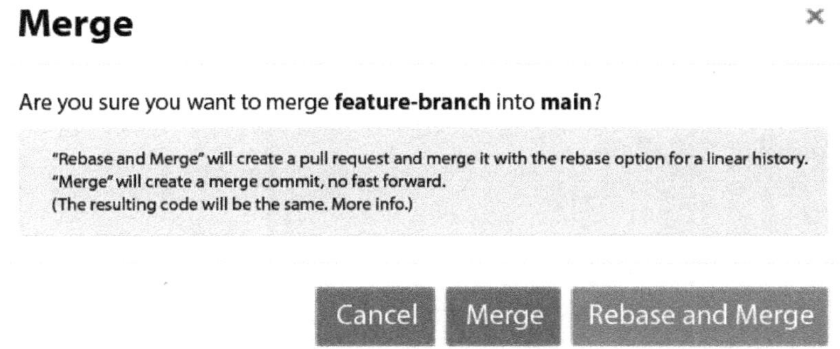

Figure 19.11 – Display popup for Merge and Rebase and Merge button

How it works...

In the previous example, we performed a full workflow to deploy a new feature into production. The following list explains the purposes of the different types of branches in Odoo.sh:

- **Production branch**: This is the actual instance that is used by the end customer. There is only one production branch, and the new features are intended to merge with this branch. In this branch, the mailing service is active, so your end customer can send and receive emails. Daily backup is also active for this branch.
- **Development branches**: This type of branch shows all the active development. You can create unlimited development branches, and every new commit in the branch will trigger a new build.

The database in this branch is loaded with the demonstration data. After the development is complete, this branch will be moved to the staging branch. The mailing service is not active in these branches.

- **Staging branches**: This is the intermediate stage in the workflow. A stable development will be moved to the staging branch to be tested with a copy of the production branch. This is a very important step in the development life cycle; it might happen that a feature that works fine in the development branch does not work as expected with the production database. The staging branches give you an opportunity to test the feature with the production database before deploying it in production. If you find any issues with the development in this branch, you can move the branch back to development. The number of staging branches is based on your Odoo.sh plan. By default, you only have one staging branch, but you can purchase more if you want to.

This is the complete workflow of how new features should be merged into production. In the next recipe, you will see some other options that we can use with these branches.

Accessing debugging options

Odoo.sh provides different features for analysis and debugging purposes. In this recipe, we will explore all of these features and options.

How to do it...

We will be using the same Odoo.sh project for this recipe. Each option will be shown in a different section with a screenshot.

Branch history

You have already seen this feature in previous recipes. The **HISTORY** tab shows the full history of the branch. You can connect to the builds from here:

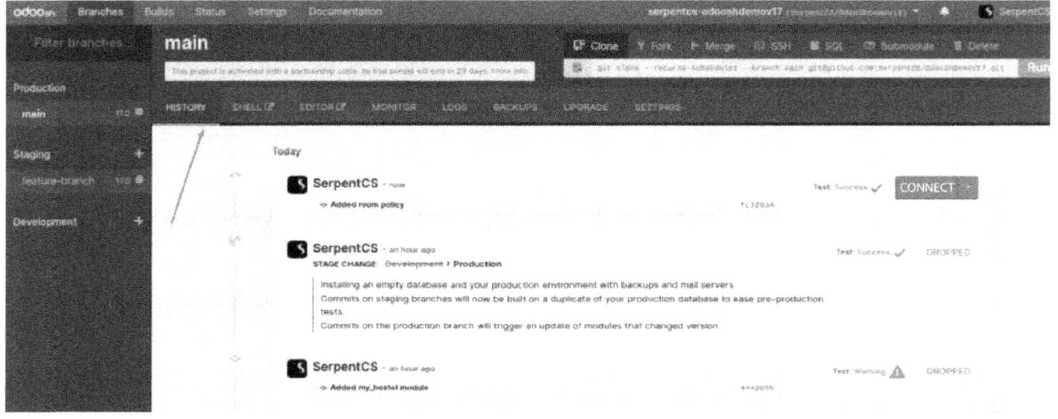

Figure 19.12 – The HISTORY tab

In the **HISTORY** tab, you can see all past actions performed on a selected branch. It will display logs, merges, new commits, and database restores.

Mail catcher

The staging branch uses a copy of your production database, so it has information about your customers. Testing the staging branch can send emails to real customers. This is why emails are only activated on production branches. The staging and development branches do not send real emails. If you want to test the email system before deploying any feature into production, you can use the mail catcher, where you can see the list of all outgoing emails. The mail catcher will be available in the staging and development branches.

The mail catcher will display an email with the source and any attachments, as shown in the following screenshot:

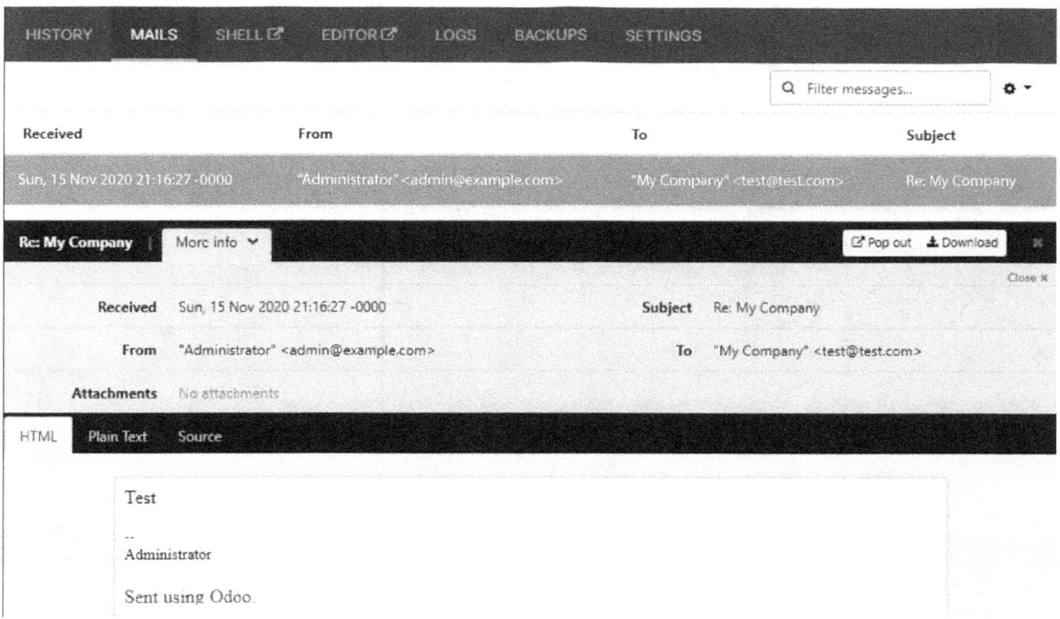

Figure 19.13 – Mail catcher

In the **MAILS** tab, you can see a list of all the captured mail with all attachments. Note that the **MAILS** tab will only be displayed in the staging and development branches.

Web shell

From the **SHELL** tab, you can access the web shell. Here, you can access the source code, the logs, the file store, and so on. It provides all of the shell features with editors such as **nano** and Vim. You can install the Python package with `pip` and maintain multiple tabs.

Take a look at the following screenshot: you can access the web shell by clicking on **SHELL**:

```
Terminal 1
serpentcs-odooshdemo-main-9700559 [production/16.0]:~$ ls
backup.daily  data  logs  src  tmp
serpentcs-odooshdemo-main-9700559 [production/16.0]:~$ cd src/odoo/
serpentcs-odooshdemo-main-9700559 [production/16.0]:~/src/odoo$ ls
addons  CONTRIBUTING.md  COPYRIGHT  debian  doc  LICENSE  MANIFEST.in  odoo  odoo-bin  README.md  requirements.txt  SECURITY.md  setup  setup.cfg  setup.py
serpentcs-odooshdemo-main-9700559 [production/16.0]:~/src/odoo$ ls -la
total 124
drwxr-xr-x   9 odoo odoo  4096 May  1 17:30 .
drwxr-xr-x   6 odoo odoo    62 Sep 14 08:58 ..
drwxr-xr-x 474 odoo odoo 16384 Sep 11 17:30 addons
-rw-r--r--   1 odoo odoo   803 Oct 22  2022 CONTRIBUTING.md
-rw-r--r--   1 odoo odoo   433 Oct  8  2022 COPYRIGHT
drwxr-xr-x   3 odoo odoo  4096 Jul  4 13:52 debian
drwxr-xr-x   3 odoo odoo    17 Oct  8  2022 doc
-rw-r--r--   1 odoo odoo    53 Oct  8  2022 .git
drwxr-xr-x   2 odoo odoo    63 Oct  8  2022 .github
-rw-rw-r--   1 odoo odoo   681 Dec  5  2022 .gitignore
-rw-r--r--   1 odoo odoo 43529 Oct  8  2022 LICENSE
-rw-r--r--   1 odoo odoo   124 Oct  8  2022 MANIFEST.in
drwxr-xr-x  11 odoo odoo  4096 Aug 29 13:22 odoo
-rwxr-xr-x   1 odoo odoo   180 Oct  8  2022 odoo-bin
-rw-r--r--   1 odoo odoo  2114 Oct 22  2022 README.md
-rw-rw-r--   1 odoo odoo  3021 May  1 17:30 requirements.txt
-rw-r--r--   1 odoo odoo  1734 Oct 22  2022 SECURITY.md
drwxr-xr-x   4 odoo odoo  4096 Nov 28  2022 setup
-rw-r--r--   1 odoo odoo   339 Oct 11  2022 setup.cfg
-rw-r--r--   1 odoo odoo  1759 Oct 11  2022 setup.py
drwxr-xr-x   2 odoo odoo    20 Jul 17 17:30 .tx
serpentcs-odooshdemo-main-9700559 [production/16.0]:~/src/odoo$
```

Figure 19.14 – Web shell

With shell access, you can traverse between different directories and perform operations. You can also use the `pip` command to install Python packages.

Here is the directory structure from the root directory:

```
.
├── data
│   ├── addons
│   ├── filestore
│   └── sessions
├── logs
├── Maildir
│   ├── cur
│   ├── new
│   └── tmp
├── repositories
│   └── git_github.com_pga-odoo_odooshdemov17.git
├── src
│   ├── enterprise
│   ├── odoo
│   ├── themes
│   └── user
└── tmp
```

These directories can be different based on the type of branch. For example, **Maildir** will only be available in the staging and development branches as it uses a mail catcher.

Sometimes, you need to restart the server or update the module from the shell. You can use the following command in the shell to restart the server:

```
odoosh-restart
```

To update the module, execute the given command in the shell:

```
odoo-bin -u my_hostel --stop-after-init

odoo-update my_hostel
```

The previous command will update the **my_hostel** module. If you want to update multiple modules, you can pass module names separated by a comma.

Code editor

If you are not comfortable with shell access, Odoo.sh provides a full-featured editor. Here, you can access the Python shell, the Odoo shell, and the Terminal. You can also edit the source code from here, as you can see in the given screenshot. After modifying the source code, you can restart the server from the **Odoo** menu at the top:

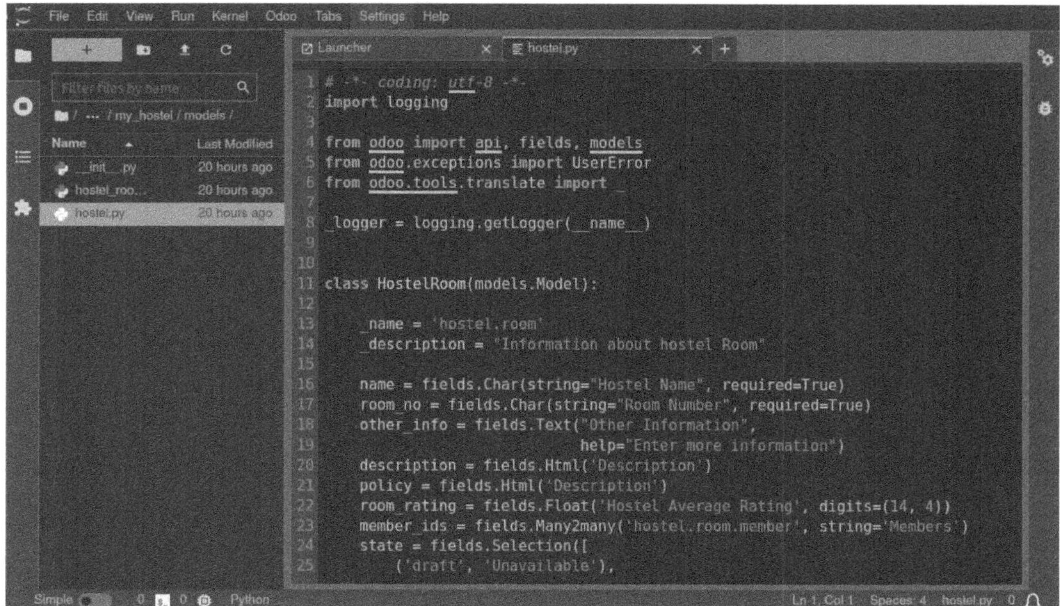

Figure 19.15 – Web code editor

As depicted in the preceding screenshot, you will be able to update files from the editor. Odoo will detect the changes automatically and restart the server. Note that if you make changes in data files, you will need to update the module.

Logs

From the **LOGS** tab, you can access all of the logs for your instance. You can see the live logs without reloading the pages. You can filter the logs from here. This allows you to find issues from the production server. Here is a list of the different log files you can find in the **LOGS** tab:

- `install.log`: This is for the logs that are generated when installing the modules. The logs of all the automated test cases will be located here.
- `pip.log`: You can add Python packages with the `requirement.txt` file. In this log file, you will find the installation log of these Python packages.
- `odoo.log`: This is the normal access log of Odoo. You will find the full access log here. You should look in this log to check production errors.
- `update.log`: When you upload a new module with a different manifest version, your module gets updated automatically. This file contains the logs of these automatic updates.

Take a look at the following screenshot. This shows the live logs for the production branch:

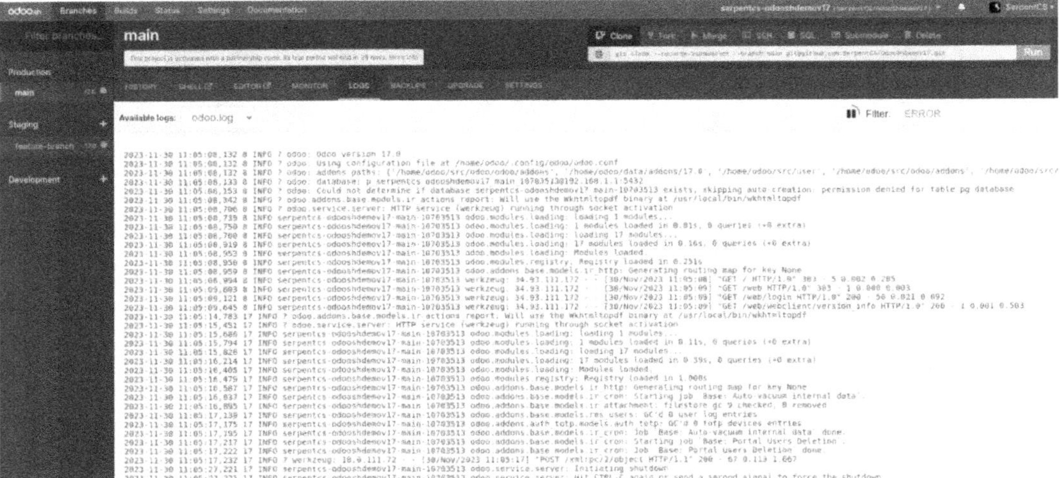

Figure 19.16 – Server log

The preceding screenshot shows that the logs are live, so you will be able to see new logs without reloading. Additionally, you can search for a particular log with the textbox in the top-right corner of the UI.

There's more...

Some commonly used `git` commands are available on top of the module, as shown in the following screenshot. You can run these by using the **Run** button on the left. These commands can't be edited, but if you want to run a modified command, you can copy it from here and then run it from the shell:

Figure 19.17 – Git commands

You can execute these `git` commands in the shell to perform various operations, as depicted in the preceding screenshot.

Getting a backup of your instance

Backups are essential for the production server. Odoo.sh provides a built-in backup facility. In this recipe, we will illustrate how you can download and restore backups from Odoo.sh.

How to do it...

In the production branch, you can access the full information about the backups from the **BACKUPS** tab at the top. This will display a list of backups:

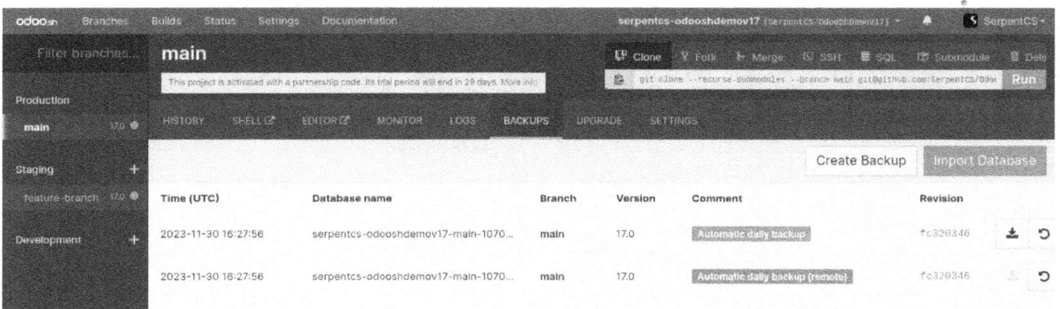

Figure 19.18 – Backups manager

From the buttons at the top, you can carry out backup operations, such as downloading the dump, performing a manual backup, or restoring from a backup. A database backup can take a long time, so it will be done in the background. You will recieve a notification on the bell icon at the top when it is completed.

How it works...

Odoo automatically takes a backup of your production instance daily. Odoo also takes an automatic backup whenever you merge a new development branch and update the module. You can also perform a manual backup using the button at the top.

Odoo.sh keeps a total of 14 full backups for the Odoo production instance for up to 3 months—1 per day for 7 days, 1 per week for 4 weeks, and 1 per month for 3 months. From the **BACKUPS** tab, you can access 1 month of backups (all 7 days of the week and 4 weekly backups).

If you are moving to Odoo.sh from the on-premises or online option, you can import your database using the **Import Database** button. If you import your database directly into production, it might cause issues. To avoid this, you should import the database into the staging branch first.

Checking the status of your builds

Whenever you make a new commit, Odoo.sh creates the new commit. It also performs automated test cases. To manage all of this, Odoo.sh has its own version of runbot. In this recipe, we will check the statuses of all the builds.

How to do it...

Click on the **Builds** menu at the top to open the list of builds. Here, you can see a full overview of all of the branches and their commits:

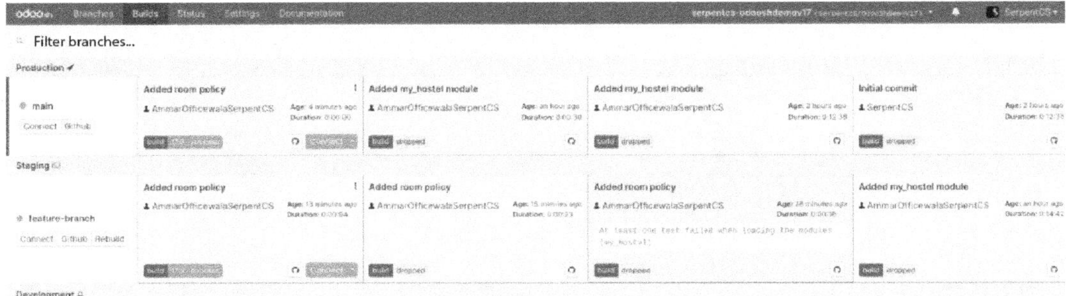

Figure 19.19 – Build status

By clicking on the **Connect** buttons, you can connect to the instances. You can see the status of the build by the background color of the branch.

How it works...

On the runbot screen, you will get extra control over the builds. You can connect to the previous builds from here. Different colors show the status of the build. Green means that everything is fine; yellow indicates a warning, which can be ignored, but it is recommended that you fix it; red means there is a critical issue that you have to fix before merging the development branch into production. The red and yellow branches show the exclamation icon, **(!)**, near the **Connect** button. When you click on this, you will get a popup with the error and warning log. Usually, you need to search the installation log files to find the error or warning logs, but this popup will filter out the other logs and only display the error and warning logs. This means that whenever a build goes red or yellow, you should come here and fix the errors and warnings before merging them into production.

Inactive development branches are destroyed after a few minutes. Normally, a new build will be created when you add a new **Commit** button. If you want to reactivate the build without a new commit; however, you can use the **Rebuild** button on the left side. The builds for the staging branches are also destroyed after a few minutes, apart from the last one, which will remain active.

There's more...

From the **Status** menu in the bar at the top, you can see the overall statistics of your instance. The platform servers are continuously monitored. On the **Status** screen, you will see the statistics of the server's availability, which will be computed automatically from the platform's monitoring system. It will show data, including the server uptime. The **Status** page will show the input and output data from the server. The **Status** page will display the following information:

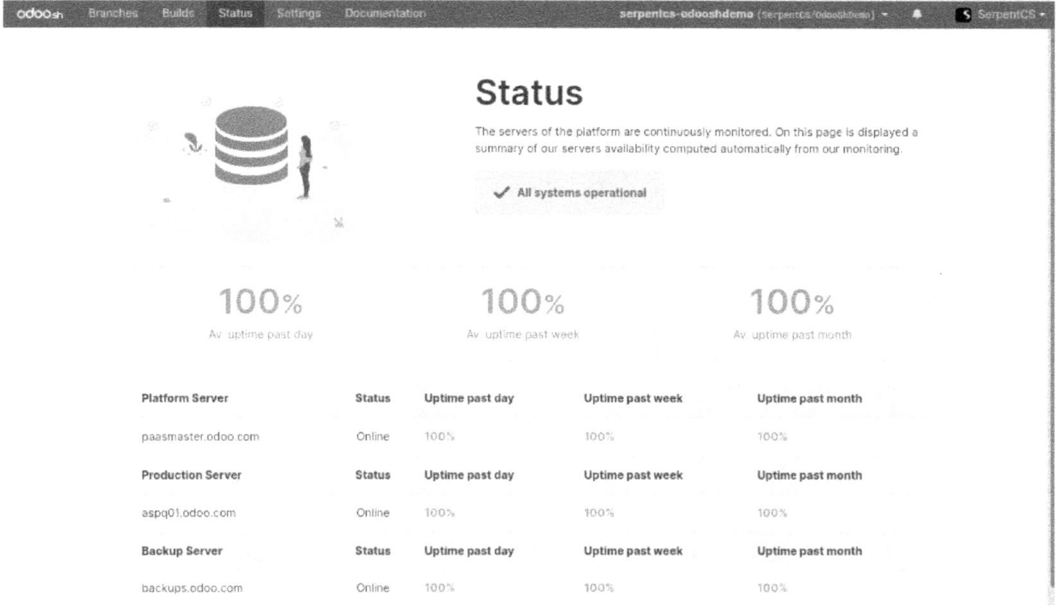

Figure 19.20 – Odoo.sh status

The data displayed in the **Status** tab is collected from the various monitoring tools used by Odoo.sh.

All Odoo.sh options

Odoo.sh provides a few further options under the **Settings** menu. In this recipe, you will see all of the important options used to modify the default behavior of certain things on the platform.

Getting ready

We will be using the same Odoo.sh project that we used in previous recipes. You can access all the Odoo.sh settings from the **Settings** menu in the top bar. If you are not able to see this menu, that means you are accessing a shared project and you don't have admin access.

How to do it...

Open the **Settings** page from the **Settings** menu in the top bar. We'll take a look at the different options in the following sections.

Project name

You can change the name of the Odoo.sh project from this option. The project name in the input will be used to generate your production URL. Development builds also use this project name as a prefix. In this case, the URL of our feature branch will be something like `https://serpentcs-odooshdemov17-feature-branch-260887.dev.odoo.com`:

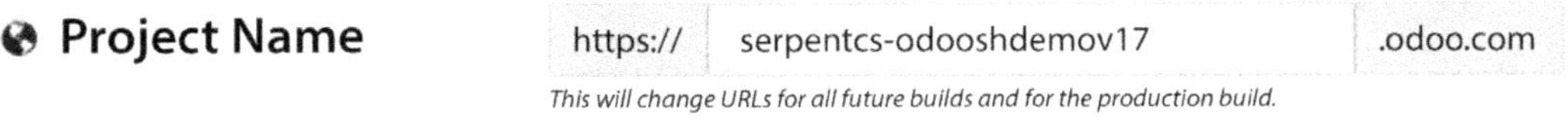

Figure 19.21 – Change the project name

> **Important note**
>
> This option will change the production URL, but you cannot get rid of `*.odoo.com`. If you want to run a production branch on a custom domain, you can add your custom domain in the **Settings** tab of the production branch. You will also need to add a CNAME entry in your DNS manager.

Collaborators

You can share the project by adding collaborators. Here, you can search for and add a new collaborator using their GitHub ID. A collaborator can have either **Admin** or **User** access rights. A collaborator with admin access rights will have full access (to the settings as well). A collaborator with user access rights, on the other hand, will have restricted access rights. They will be able to see all builds, but they will not be able to access the backups, logs, shells, or emails of the production or staging branches, although they will have full access to the development branches:

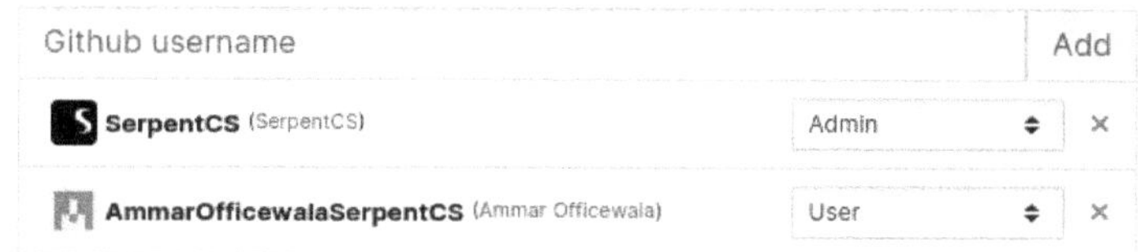
Collaborators
Grant other Github users access to this project. User rights disable access to production data and the settings. Push privileges on the repository are handled on Github.
Github username
Add
SerpentCS (SerpentCS)
Admin
AmmarOfficewalaSerpentCS (Ammar Officewala)
User

Figure 19.22 – Add collaborators

> **Important note**
>
> You will need to give these users access to the GitHub repository, too; otherwise, they won't be able to create a new repository from the browser.

Public Access

Using this option, you can share builds with your end customer. This can be used for demonstration or testing purposes. To do so, you need to enable the **Allow public access** checkbox:

Public Access
Allow public access
Expose the Builds page publicly, allowing visitors to connect to your development builds.

Figure 19.23 – Give public access to builds

Note that the staging branch will have the same password as your production branch. However, in the development branch, you will have the username and password shown in this table:

Username	Password
admin	admin
demo	demo
portal	portal

Table 19.1

Module installation

In the **Settings** tab of the development branch, you will see the **Module installation** option for the development branches. It provides three options, as shown in the following screenshot:

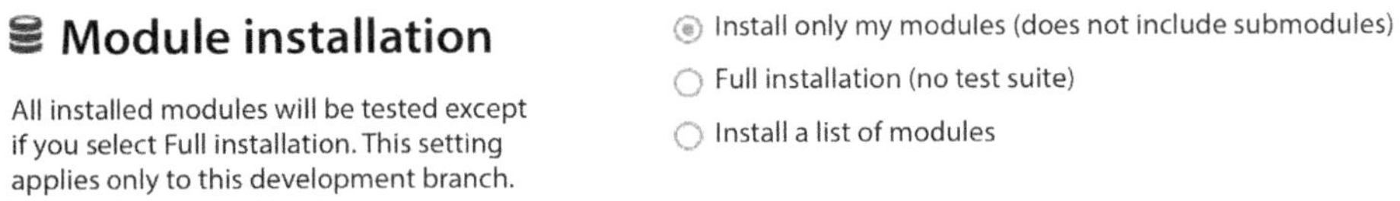

Module installation

All installed modules will be tested except if you select Full installation. This setting applies only to this development branch.

- (•) Install only my modules (does not include submodules)
- ( ) Full installation (no test suite)
- ( ) Install a list of modules

Figure 19.24 – Module installation options

By default, it is set to **Install only my modules**. This option will install all of your custom modules and their dependent modules in the new development branches. Only automated test cases are performed for these modules. The second option is **Full installation**. This option will install all of the modules and perform automated test cases for all of those modules. The final option is **Install a list of modules**. In this option, you will need to pass a list of comma-separated modules, such as **sales**, **purchases**, and **my_hostel**. This option will install the given modules and their dependencies.

This setting only applies to development builds. Staging builds duplicate the production build, so they will have the same modules installed in the production branch and perform test cases for modules that have an updated version manifest.

Submodules

The **Submodules** option is used when you are using private modules as submodules. This setting is only needed for private submodules; public submodules will work fine without any issues. It is not possible to download private repositories publicly, so you need to give repository access to Odoo.sh. Follow these steps to add access to the private submodules:

1. Copy the SSH URL of your private submodule repository in the input and click on **Add**.
2. Copy the displayed **public key**.
3. Add this **public key** as a deploy key in your private repository settings in GitHub (similar settings are also available on Bitbucket and GitLab):

Submodules

These settings are required for **private repositories** only.

If you are looking for some instructions on how to set up your submodules, please check out the documentation.

*Enter the Git URL of your private submodule hereunder and click on **Add**.*
*Then copy the **Public key** and add it as a **Deploy key** in the repository settings of your Git hosting service.*
→ You can read our documentation for more specific instructions.

git@github.com:acme/mii-theme.git Add

git@github.com

Fingerprint f5:86:09 :2b:6a:eb:6f:50

Public key ssh-rsa

Figure 19.25 – Set the private submodule

You can add multiple submodules, too, and you can remove submodules from here as well.

Database Workers

You can increase the number of workers for the production build. This is useful when you have more users; usually, a single worker can handle 25 backend users or 5,000 daily website visitors. This formula is not perfect; it can vary based on usage. This option is not free, and increasing the number of workers will increase the price of your Odoo.sh subscription:

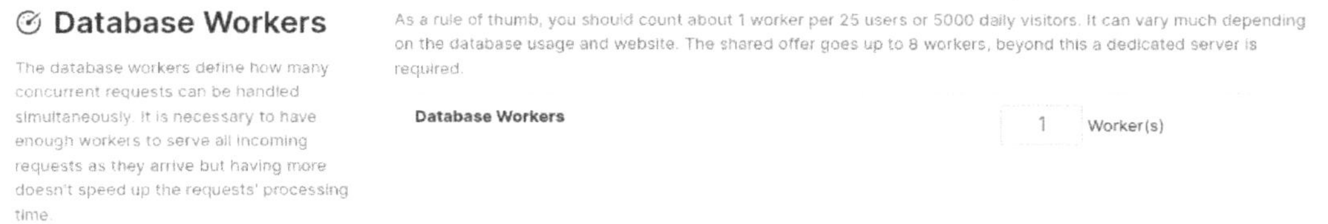

Database Workers
The database workers define how many concurrent requests can be handled simultaneously. It is necessary to have enough workers to serve all incoming requests as they arrive but having more doesn't speed up the requests' processing time.
As a rule of thumb, you should count about 1 worker per 25 users or 5000 daily visitors. It can vary much depending on the database usage and website. The shared offer goes up to 8 workers, beyond this a dedicated server is required.
Database Workers
1
Worker(s)

Figure 19.26 – Set Database Workers

These **Database Workers** are multithreaded, and each one is able to handle 15 concurrent requests. It is necessary to have enough workers to serve all incoming requests as they arrive, but increasing the number of workers does not increase the speed of the requests' processing time. It is only used to handle a large number of concurrent users.

Staging Branches

Staging branches are used to test a new development with the production database. By default, Odoo.sh gives you one staging branch. If you are working on large projects with lots of developers, this might be a bottleneck in the development process, so you can increase the number of **Staging Branches** at an extra cost:

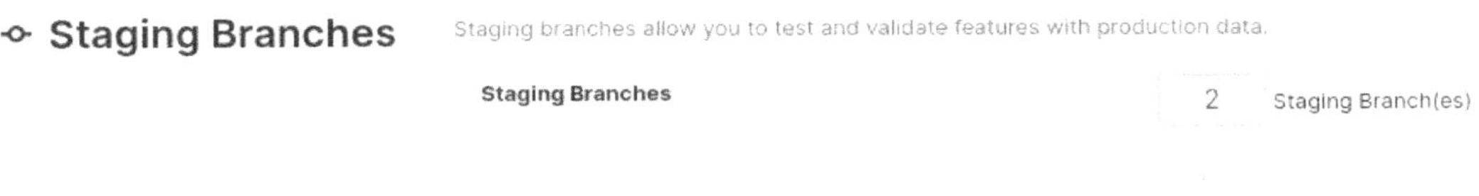

Figure 19.27 – Set staging branches

There's more...

Along with the configuration options, the **Settings** menu will also display some statistics related to the platform.

Database size

This section will display the size of your production database. The Odoo.sh platform charges the database at USD 1/GB/month. This option helps you keep track of your database. The displayed database size is only for the production database; it does not include the databases of the staging and development branches:

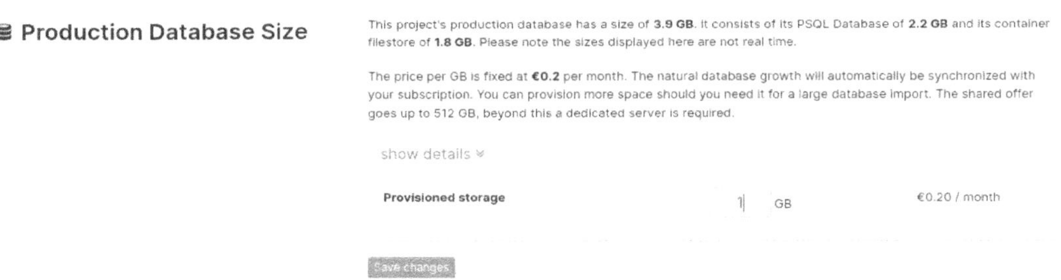

Figure 19.28 – Database size

Odoo source code revisions

This section will display the GitHub revision number of Odoo's project. It will display the revision hash for the Community, Enterprise, and theme projects that are currently being used in the platform. This source code will automatically be updated every week. This option will help you get the exact same versions on your local machine. You can also check this from the web shell, through the `git` command in the repository.

20
Remote Procedure Calls in Odoo

The Odoo server supports **remote procedure calls** (**RPCs**), which means that you can connect Odoo instances from external applications. An example is if you want to show the status of a delivery order in Fan Android application that is written in Java Here, you can fetch the delivery status from Odoo via RPC. With the Odoo RPC API, you can perform any CRUD operations on a database. Odoo RPC is not limited to CRUD operations; you can also invoke public methods of any model. Of course, you will need to have proper access to rights to perform these operations because RPC respects all of the access rights and record rules you have defined in your database. Consequently, it is very safe to use because the RPC respects all access rights and record rules. Odoo RPC is not platform-dependent, so you can use it on any platform, including Odoo.sh, online, or self-hosted platforms. Odoo RPC can be used with any programming language, so you can integrate Odoo with any external application.

Odoo provides two types of RPC API: XML-RPC and JSON-RPC. In this chapter, we will learn how to use these RPCs from an external program. Finally, you will learn how to use Odoo RPC through OCA's `odoorpc` library.

In this chapter, we will cover the following recipes:

- Logging in to/connecting Odoo with XML-RPC
- Searching/reading records using XML-RPC
- Creating/updating/deleting records using XML-RPC
- Calling methods using XML-RPC
- Logging in to/connecting Odoo with JSON-RPC
- Fetching/searching records using JSON-RPC
- Creating/updating/deleting records using JSON-RPC
- Calling methods using JSON-RPC

- The OCA odoorpc library
- Generating API keys

Technical requirements

In this chapter, we will be using the `my_hostel` module, which we created in *Chapter 19, Managing, Deploying, and Testing with Odoo.sh*. You can find the same initial `my_hostel` module in the GitHub repository: `https://github.com/PacktPublishing/Odoo-17-Development-Cookbook-Fifth-Edition/tree/main/Chapter20`.

Here, we will not Introduce a new language as you may not be familiar with it. We will continue using Python to access the RPC API. You can use another language if you want to, as the same procedure can be applied in any language to access the RPC.

To connect Odoo through the RPC, you will need a running Odoo instance to connect with. Throughout this chapter, we will assume that you have the Odoo server running on `http://localhost:8017`, that you called the `cookbook_17e` database, and that you have installed the `my_hostel` module therein. Note that you can connect any valid IP or domain through the RPC.

Logging in to/connecting Odoo with XML-RPC

In this recipe, we will carry out user authentication through RPC to check whether the credentials (server_url, db_name, username, and password) supplied are valid.

Getting ready

To connect an Odoo instance through RPC, you will need a running Odoo instance to connect with. We will assume that you have the Odoo server running on `http://localhost:8017` and that you have installed the `my_hostel` module.

How to do it...

Perform the following steps to carry out user authentication through RPC:

1. Add the `odoo_authenticate.py` file. You can place this file anywhere you want because the RPC program will work independently.
2. Add the following code to the file:

```
from xmlrpc import client

server_url = 'http://localhost:8017'
db_name = 'cookbook_17e'
```

```
username = 'admin'
password = 'admin'

common = client.ServerProxy('%s/xmlrpc/2/common' % server_url)
user_id = common.authenticate(db_name, username, password, {})

if user_id:
    print("Success: User id is", user_id)
else:
    print("Failed: wrong credentials")
```

3. Run the following Python script from the Terminal with the following command:

```
python3 odoo_authenticate.py
```

This will print a success message with the user ID if you have provided a valid login name and password.

How it works...

In this recipe, we used the Python `xmlrpc` library to access Odoo instances through XML-RPC. This is a standard Python library, and you do not have to install anything else to use it.

For authentication, Odoo provides XML-RPC on the `/xmlrpc/2/common` endpoint. This endpoint is used for meta methods, which do not require authentication. The `authentication()` method itself is a public method, so it can be called publicly. The `authentication()` method accepts four arguments—database name, username, password, and user agent environment. The user agent environment is a compulsory argument, but if you do not want to pass the user agent parameter, at least pass the empty dictionary.

When you execute the `authenticate()` method with all valid arguments, it will make a call to the Odoo server and perform authentication. It will then return the user ID, provided the given login ID and password are correct. It will return `False` if the user is not present or if the password is incorrect.

You need to use the `authenticate()` method before accessing any data through RPC. This is because accessing data with the wrong credentials will generate an error.

> **Important note**
>
> Odoo's online instances (`*.odoo.com`) use OAuth authentication, so the local password is not set on the instance. To use XML-RPC on these instances, you will need to set the user's password manually from the **Settings** | **Users** | **Users** menu of your instance.

Additionally, the methods used to access data require a user ID instead of a username, so the `authenticate()` method is needed to get the ID of the user.

There's more...

The `/xmlrpc/2/common` endpoint provides one more method: `version()`. You can call this method without credentials. It will return the version information of the Odoo instance. The following is an example of the `version()` method usage:

```
from xmlrpc import client

server_url = 'http://localhost:8017'

common = client.ServerProxy('%s/xmlrpc/2/common' % server_url)
version_info = common.version()

print(version_info)
```

The preceding program will generate the following output:

```
$ python3 version_info.py
{'server_version': '17.0+e', 'server_version_info': [17, 0, 0,
'final', 0, 'e'], 'server_serie': '17.0', 'protocol_version': 1}
```

This program will print version information based on your server.

Searching/reading records using XML-RPC

In this recipe, we will see how you can fetch the data from an Odoo instance through RPC. The user can access most data, except data that are restricted by the security access control and record rules. RPC can be used in many situations, such as collecting data for analysis, manipulating a lot of data at once, or fetching data for display in another software/system. There are endless possibilities, and you can use RPCs whenever necessary.

Getting ready

We will create a Python program to fetch the room data from the `hostel.room` model. Make sure you have installed the `my_hostel` module and that the server is running on `http://localhost:8017`.

How to do it...

Perform the following steps to fetch a room's information through RPC:

1. Add the `rooms_data.py` file. You can place this file anywhere you want because the RPC program will work independently.

2. Add the following code to the file:

```
from xmlrpc import client

# room data with search method
server_url = 'http://localhost:8017'
db_name = 'cookbook_17e'
username = 'admin'
password = 'admin'

common = client.ServerProxy('%s/xmlrpc/2/common' % server_url)
user_id = common.authenticate(db_name, username, password, {})

models = client.ServerProxy('%s/xmlrpc/2/object' % server_url)

if user_id:
    search_domain = [['name', 'ilike', 'Standard']]
    rooms_ids = models.execute_kw(db_name, user_id, password,
        'hostel.room', 'search',
        [search_domain],
        {'limit': 5})
    print('Rooms ids found:', rooms_ids)

    rooms_data = models.execute_kw(db_name, user_id, password,
        'hostel.room', 'read',
        [rooms_ids, ['name', 'room_no']])
    print("Rooms data:", rooms_data)
else:
    print('Wrong credentials')
```

3. Run the Python script from the Terminal with the following command:

```
python3 rooms_data.py
```

The preceding program will fetch the room data and give you the following output:

```
$ python3 rooms_data.py
Rooms ids found: [1, 2, 3, 4, 5]
Rooms data: [{'id': 1, 'name': '8th Standard', 'room_no': '1'}, {'id':
2, 'name': '9th Standard', 'room_no': '2'}, {'id': 3, 'name': '10th
Standard', 'room_no': '3'}, {'id': 4, 'name': '11th Standard', 'room_
no': '4'}, {'id': 5, 'name': '12th Standard', 'room_no': '5'}]
```

The output shown in the preceding screenshot is based on data in my database. The data in your Odoo instance may be different data, so the output will also be different.

How it works...

In order to access the room data, you first have to authenticate. At the beginning of the program, we did authentication in the same way as we did in the *Logging in to/connecting Odoo with XML-RPC* recipe earlier. If you provided valid credentials, the `authentication()` method will return the `id` of the user's record. We will use this user ID to fetch the room data.

The `/xmlrpc/2/object` endpoint is used for database operation. In our recipe, we used the `object` endpoint to fetch the room data. In contrast to the `/xmlrpc/2/common` endpoint, this endpoint does not work without credentials. With this endpoint, you can access the public method of any model through the `execute_kw()` method. `execute_kw()` takes the following arguments:

- Database name
- User ID (we get this from the `authenticate()` method)
- Password
- Model name, for example, `res.partner` or `hostel.room`
- Method name, for example, `search`, `read`, or `create`
- An array of positional arguments
- A dictionary for keyword arguments (optional)

In our example, we want to fetch the room's information. This can be done through a combination of `search()` and `read()`. Room information is stored in the `hostel.room` model, so in `execute_kw()`, we use `hostel.room` as the model name and `search` as the method name. This will call the ORM's `search` method and return record IDs. The only difference here is that the ORM's `search` method returns a record set, while this search method returns a list of IDs.

In `execute_kw()`, you can pass arguments and keyword arguments for the method provided. The `search()` method accepts a domain as a positional argument, so we passed a domain to filter rooms. The `search` method has other optional keyword arguments, such as `limit`, `offset`, `count`, and `order`, from which we have used the `limit` parameter to fetch only five records. This will return the list of room IDs whose names contain the `Standard` strings.

However, we need to fetch room data from the database. We will use the `read` method to do this. The `read` method accepts a list of IDs and fields to complete the task. At the end of *step 3*, we used the list of room IDs that we received from the `search` method and then used the room IDs to fetch the `name` and `room_no` of the rooms. This will return the list of the dictionary with the room's information.

> **Important note**
>
> Note that the arguments and keyword arguments passed in `execute_kw()` are based on the passed method. You can use any public ORM method via `execute_kw()`. You just need to give the method a name, the valid arguments, and the keyword arguments. These arguments are going to be passed on in the method in the ORM.

There's more...

The data fetched through a combination of the `search()` and `read()` methods is slightly time-consuming because it will make two calls. `search_read` is an alternative method for fetching data. You can search and fetch the data in a single call. Here is the alternative way to fetch a room's data with `search_read()`.

> **Important note**
>
> The `read` and `search_read` methods will return `id` fields even if the `id` field is not requested. Furthermore, for the `many2one` field, you will get an array made up of the `id` and display name. For example, the `create_uid` `many2one` field will return data like this: `[07, 'Deepak ahir']`.

It will return the same output as in the previous example:

```
from xmlrpc import client

# room data with search_read method
server_url = 'http://localhost:8017'
db_name = 'cookbook_17e'
username = 'admin'
password = 'admin'

common = client.ServerProxy('%s/xmlrpc/2/common' % server_url)
user_id = common.authenticate(db_name, username, password, {})

models = client.ServerProxy('%s/xmlrpc/2/object' % server_url)

if user_id:
    search_domain = [['name', 'ilike', 'Standard']]
    rooms_ids = models.execute_kw(db_name, user_id, password,
        'hostel.room', 'search_read',
        [search_domain, ['name', 'room_no']],
        {'limit': 5})
    print('Rooms data:', rooms_ids)
```

```
else:
    print('Wrong credentials')
```

The `search_read` methods improve performance significantly as you get your result in one RPC call, so use the `search_read` method instead of a combination of the `search` and `read` methods.

Creating/updating/deleting records using XML-RPC

In the previous recipe, we saw how to search and read data through RPC. In this recipe, we will perform the remaining **CRUD** operations through RPC, which are **create**, **update** (write), and **delete** (unlink).

Getting ready

We will create the Python program to `create`, `write`, and `unlink` data in the `hostel.room` model. Make sure you have installed the `my_hostel` module and that the server is running on `http://localhost:8017`.

How to do it...

Perform the following steps to create, write, and update a room's information through RPC:

1. Add the `rooms_operation.py` file. You can place this file anywhere you want because the RPC program will work independently.
2. Add the following code to the `rooms_operation.py` file:

```
from xmlrpc import client

server_url = 'http://localhost:8017'
db_name = 'cookbook_17e'
username = 'admin'
password = 'admin'

common = client.ServerProxy('%s/xmlrpc/2/common' % server_url)
user_id = common.authenticate(db_name, username, password, {})

models = client.ServerProxy('%s/xmlrpc/2/object' % server_url)

if user_id:
    # create new room records.
    create_data = [
        {'name': 'Room 1', 'room_no': '101'},
```

```
        {'name': 'Room 3', 'room_no': '102'},
        {'name': 'Room 5', 'room_no': '103'},
        {'name': 'Room 7', 'room_no': '104'}
    ]
    rooms_ids = models.execute_kw(db_name, user_id, password,
        'hostel.room', 'create',
        [create_data])
    print("Rooms created:", rooms_ids)

    # Write in existing room record
    room_to_write = rooms_ids[1]  # We will use ids of recently
created rooms
    write_data = {'name': 'Room 2'}
    written = models.execute_kw(db_name, user_id, password,
        'hostel.room', 'write',
        [room_to_write, write_data])
    print("Rooms written", written)

    # Delete the room record
    rooms_to_delete = rooms_ids[2:]
    deleted = models.execute_kw(db_name, user_id, password,
        'hostel.room', 'unlink',
        [rooms_to_delete])
    print('Rooms unlinked:', deleted)

else:
    print('Wrong credentials')
```

3. Run the Python script from the Terminal with the given command:

```
python3 rooms_operation.py
```

The preceding program will create four records of the rooms. Updating the data in the room records and later deleting two records gives you the following output (the IDs created may be different depending on your database):

```
$ python3 rooms_operation.py
Rooms created: [6, 7, 8, 9]
Rooms written True
Rooms unlinked: True
```

The `write` and `unlink` methods return `True` if the operation is successful. This means that if you get `True` in response, assume that a record has been updated or deleted successfully.

How it works...

In this recipe, we performed `create`, `write`, and `delete` operations through XML-RPC. This operation also uses the `/xmlrpc/2/` object endpoint and the `execute_kw()` method.

The `create()` method supports the creation of multiple records in a single call. In *step 2*, we first created a dictionary with the room's information. Then, we used the room's dictionary to create new records of the rooms through XML-RPC. The XML-RPC call needs two parameters to create new records: the `create` method name and the room data. This will create the four room records in the `hostel.room` model. In ORM, when you create the record, it returns a record set of created records, but if you create the record's RPC, this will return a list of IDs.

The `write` method works in a similar way to the `create` method. In the `write` method, you will need to pass a list of record IDs and the field values to be written. In our example, we updated the name of the room created in the first section. This will update the name of the second room from `Room 3` to `Room 2`. Here, we passed only one `id` for a room, but you can pass a list of IDs if you want to update multiple records in a single call.

In the third section of the program, we deleted two rooms that we created in the first section. You can delete records using the `unlink` method and a list of record IDs.

After the program is executed successfully, you will find two room records in the database, as indicated in *Figure 20.3*. In the program, we have created four records, but we have also deleted two of them, so you will only find two new records in the database.

There's more...

When you are performing a CRUD operation through RPC, this may generate an error if you don't have permission to do that operation. With the `check_access_rights` method, you can check whether the user has the proper access rights to perform a certain operation. The `check_access_rights` method returns `True` or `False` values based on the access rights of the user. Here is an example showing whether a user has the right to create a room record:

```
from xmlrpc import client

server_url = 'http://localhost:8017'
db_name = 'cookbook_17e'
username = 'admin'
password = 'admin'

common = client.ServerProxy('%s/xmlrpc/2/common' % server_url)
user_id = common.authenticate(db_name, username, password, {})

models = client.ServerProxy('%s/xmlrpc/2/object' % server_url)
```

```
if user_id:
    has_access = models.execute_kw(db_name, user_id, password,
        'hostel.room', 'check_access_rights',
        ['create'], {'raise_exception': False})
    print('Has create access on room:', has_access)
else:
    print('Wrong credentials')


# Output: Has create access on room: True
```

When you are doing complex operations via RPC, the `check_access_rights` method can be used prior to performing the operation to make sure you have proper access rights.

Calling methods using XML-RPC

With Odoo, the RPC API is not limited to CRUD operations; you can also invoke business methods. In this recipe, we will call the `make_available` method to change the room's state.

Getting ready

We will create the Python program to call `make_available` on the `hostel.room` model. Make sure that you have installed the `my_hostel` module and that the server is running on `http://localhost:8017`.

How to do it...

Perform the following steps to create, write, and update a room's information through RPC:

1. Add the `rooms_method.py` file. You can place this file anywhere you want because the RPC program will work independently.
2. Add the following code to the file:

```
from xmlrpc import client

server_url = 'http://localhost:8017'
db_name = 'cookbook_17e'
username = 'admin'
password = 'admin'

common = client.ServerProxy('%s/xmlrpc/2/common' % server_url)
user_id = common.authenticate(db_name, username, password, {})
```

```
models = client.ServerProxy('%s/xmlrpc/2/object' % server_url)

if user_id:
    # Create room with state draft
    room_id = models.execute_kw(db_name, user_id, password,
        'hostel.room', 'create',
        [{
            'name': 'New Room',
            'room_no': '35',
            'state': 'draft'
        }])
    # Call make_available method on new room

    models.execute_kw(db_name, user_id, password,
        'hostel.room', 'make_available',
        [[room_id]])

    # check room status after method call
    room_data = models.execute_kw(db_name, user_id, password,
        'hostel.room', 'read',
        [[room_id], ['name', 'state']])
    print('Room state after method call:', room_data[0]
['state'])
else:
    print('Wrong credentials')
```

3. Run the Python script from the Terminal with the following command:

```
python3 rooms_method.py
```

The preceding program will create one room using `draft` and then we will change the room's state by calling the `make_available` method. After that, we will fetch the room data to check the room's status, which will generate the following output:

```
$ python3 rooms_method.py
Room state after method call: available
```

The program of this recipe will create a new room record and change the state of the room by calling the `model` method. By the end of the program, we have read the room record and printed the updated state.

How it works...

You can call any modal method from RPC. This helps you to perform business logic without encountering any side effects. For example, you created the sales order from RPC and then called the `action_confirm` method of the `sale.order` method. This is equivalent to clicking on the **Confirm** button on a sales order form.

You can call any public method of the model, but you cannot call a private method from RPC. A method name that starts with _ is called a private method, such as `_get_share_url()` and `_get_data()`.

It is safe to use these methods, as they go through the ORM and follow all security rules. If the method is accessing unauthorized records, it will generate errors.

In our example, we created a room with a state of `draft`. Then, we made one more RPC call to invoke the `make_available` method, which will change the room's state to `available`. Finally, we made one more RPC call to check the state of the room. This will show that the room's state has changed to **Available**, as indicated in *Figure 20.4*.

Methods that do not return anything internally return `None` by default. Such methods cannot be used from RPC. Consequently, if you want to use your method from RPC, at least add the return `True` statement.

There's more...

If an exception is generated from a method, all of the operations performed in the transaction will be automatically rolled back to the initial state. This is only applicable to a single transaction (a single RPC call). For example, imagine you are making two RPC calls to the server, and an exception is generated during the second call. This will roll back the operation that was carried out during the second RPC call. The operation performed through the first RPC call won't be rolled back. Consequently, you want to perform a complex operation through RPC. It is recommended that this be performed in a single RPC call by creating a method in the model.

Logging in to/connecting Odoo with JSON-RPC

Odoo provides one more type of RPC API: JSON-RPC. As its name suggests, JSON-RPC works in the JSON format and uses the `jsonrpc` 2.0 specification. In this recipe, we will see how you can log in with JSON-RPC. The Odoo web client itself uses JSON-RPC to fetch data from the server.

Getting ready

In this recipe, we will perform user authentication through JSON-RPC to check whether the given credentials are valid. Make sure you have installed the `my_hostel` module and that the server is running on `http://localhost:8017`.

How to do it...

Perform the following steps to perform user authentication through RPC:

1. Add the `jsonrpc_authenticate.py` file. You can place this file anywhere you want because the RPC program will work independently.

2. Add the following code to the file:

```
import json
import random
import requests

server_url = 'http://localhost:8017'
db_name = 'cookbook_17e'
username = 'admin'
password = 'admin'

json_endpoint = "%s/jsonrpc" % server_url
headers = {"Content-Type": "application/json"}

def get_json_payload(service, method, *args):
    return json.dumps({
        "jsonrpc": "2.0",
        "method": 'call',
        "params": {
            "service": service,
            "method": method,
            "args": args
        },
        "id": random.randint(0, 1000000000),
    })

payload = get_json_payload("common", "login", db_name, username,
password)
response = requests.post(json_endpoint, data=payload,
headers=headers)
user_id = response.json()['result']

if user_id:
    print("Success: User id is", user_id)
else:
    print("Failed: wrong credentials")
```

3. Run the Python script from the Terminal with the following command:

```
python3 jsonrpc_authenticate.py
```

When you run the preceding program, and you have passed a valid login name and password, the program will print a success message with the `id` of the user, as follows:

```
$ python3 jsonrpc_authentication.py
Success: User id is 2
```

The JSON authentication works just like XML-RPC, but it returns a result in the JSON format.

How it works...

JSON-RPC uses the JSON format to communicate with the server using the `/jsonrpc` endpoint. In our example, we used the Python requests package to make `POST` requests, but if you want to, you can use other packages, such as `urllib`.

JSON-RPC only accepts a payload formatted in the **JSON-RPC 2.0** specification. You may refer to this link to learn more about the JSON-RPC format: `https://www.jsonrpc.org/specification`. In our example, we created the `get_json_payload()` method. This method will prepare the payload in the valid JSON-RPC 2.0 format. This method accepts the `service` name and the `method` to call, and the remaining arguments will be placed in `*args`. We will be using this method in all subsequent recipes. JSON-RPC accepts requests in JSON format, and these requests are only accepted if the request contains a `{"Content-Type": "application/json"}` header. The results of the requests will be in JSON format.

Like XML-RPC, all public methods, including login, come under the common service. For this reason, we passed `common` as a service and `login` as a method to prepare the JSON payload. The login method required some extra arguments, so we passed the database name, username, and password. Then, we made the `POST` request to the JSON endpoint with the payload and headers. If you passed the correct username and password, the method returns the user ID. The response will be in JSON format, and you will get the result in the result key.

> **Important note**
>
> Note that the `get_json_payload()` method created in this recipe is used to remove repetitive code from the example. It is not compulsory to use it, so feel free to apply your own adaptations.

There's more...

Like XML-RPC, the version method is also available in JSON-RPC. This version of the method comes under the common service and is accessible publicly. You can get version information without login information. See the following example showing how to fetch the version info of the Odoo server:

```
import json
import random
import requests


server_url = 'http://localhost:8017'
db_name = 'cookbook_17e'
username = 'admin'
```

```
password = 'admin'

json_endpoint = "%s/jsonrpc" % server_url
headers = {"Content-Type": "application/json"}

def get_json_payload(service, method, *args):
    return json.dumps({
        "jsonrpc": "2.0",
        "method": 'call',
        "params": {
            "service": service,
            "method": method,
            "args": args
        },
        "id": random.randint(0, 1000000000),
    })

payload = get_json_payload("common", "version")
response = requests.post(json_endpoint, data=payload, headers=headers)

print(response.json())
```

This program will display the following output:

```
$ python3 jsonrpc_version_info.py
{'jsonrpc': '2.0', 'id': 361274992, 'result': {'server_version':
'17.0+e', 'server_version_info': [17, 0, 0, 'final', 0, 'e'], 'server_
serie': '17.0', 'protocol_version': 1}}
```

This program will print version information based on your server.

Fetching/searching records using JSON-RPC

In the previous recipe, we saw how you can do authentication through JSON-RPC. In this recipe, we will see how you can fetch the data from the Odoo instance with JSON-RPC.

Getting ready

In this recipe, we will fetch room information with JSON-RPC. Make sure you have installed the `my_hostel` module and that the server is running on `http://localhost:8017`.

How to do it...

Perform the following steps to fetch room data from the `hostel.room` model:

1. Add the `jsonrpc_fetch_data.py` file. You can place this file anywhere you want because the RPC program will work independently.
2. Add the following code to the file:

```
# place authentication and get_json_payload methods (see first
jsonrpc recipe)

if user_id:
    # search for the room's ids
    search_domain = [['name', 'ilike', 'Standard']]
    payload = get_json_payload("object", "execute_kw",
        db_name, user_id, password,
        'hostel.room', 'search', [search_domain], {'limit': 5})
    res = requests.post(json_endpoint, data=payload,
headers=headers).json()
    print('Search Result:', res)  # ids will be in result keys

    # read data for rooms ids
    payload = get_json_payload("object", "execute_kw",
        db_name, user_id, password,
        'hostel.room', 'read', [res['result'], ['name', 'room_
no']])
    res = requests.post(json_endpoint, data=payload,
headers=headers).json()
    print('Rooms data:', res)
else:
    print("Failed: wrong credentials")
```

3. Run the Python script from the Terminal with the following command:

```
python3 json_fetch_data.py
```

The preceding program will give you the following output. The first RPC call will print the room's ID, and the second one will print the information for the room's ID:

```
$ python3 json_fetch_data.py
Search Result: {'jsonrpc': '2.0', 'id': 19247199, 'result': [1, 2, 3,
4, 5]}
```

```
Rooms data: {'jsonrpc': '2.0', 'id': 357582271, 'result': [{'id':
1, 'name': '8th Standard', 'room_no': '1'}, {'id': 2, 'name': '9th
Standard', 'room_no': '2'}, {'id': 3, 'name': '10th Standard', 'room_
no': '3'}, {'id': 4, 'name': '11th Standard', 'room_no': '4'}, {'id':
5, 'name': '12th Standard', 'room_no': '5'}]}
```

The output shown in the preceding screenshot is based on data in my database. The data in your Odoo instance may be different data, so the output will also be different.

How it works...

In the *Logging in to/connecting Odoo with JSON-RPC* recipe, we saw that you can validate `username` and `password`. If the login details are correct, the RPC call will return `user_id`. You can then use this `user_id` to fetch the model's data. Like XML-RPC, we need to use the `search` and `read` combination to fetch the data from the model. To fetch the data, we use `object` as a service and `execute_kw()` as the method. `execute_kw()` is the same method that we used in XML-RPC for data, so it accepts the same argument as follows:

- Database name
- User ID (we get this from the `authenticate()` method)
- Password
- Model name, for example, `res.partner` or `hostel.room`
- Method name, for example, `search`, `read`, or `create`
- An array of positional arguments (`args`)
- A dictionary for keyword arguments (`optional`) (`kwargs`)

In our example, we called the `search` method first. The `execute_kw()` method usually takes mandatory arguments as positional arguments and optional arguments as keyword arguments. In the `search` method, `domain` is a mandatory argument, so we passed it in the list and passed the `optional` argument limit as the `keyword` argument (dictionary). You will get a response in JSON format, and in this recipe, the response of the `search()` method RPC will have the room's IDs in the `result` key.

In *step 2*, we made an RPC call using the `read` method. To read the room's information, we passed two positional arguments: the list of room IDs and the list of fields to fetch. This RPC call will return the room information in JSON format, and you can access it in using the `result` key.

> **Important note**
>
> Instead of `execute_kw()`, you can use `execute` as the method. This does not support keyword arguments, so you need to pass all of the intermediate arguments if you want to pass some optional arguments.

There's more...

Similar to XML-RPC, you can use the `search_read()` method instead of the `search()` and `read()` method combination, as it is slightly time-consuming. Take a look at the following code:

```
# place authentication and get_json_payload methods (see first jsonrpc
recipe)

if user_id:
    # search for the room's ids
    search_domain = [['name', 'ilike', 'Standard']]
    payload = get_json_payload("object", "execute_kw",
        db_name, user_id, password,
        'hostel.room', 'search_read', [search_domain, ['name', 'room_
no']], {'limit': 5})
    res = requests.post(json_endpoint, data=payload, headers=headers).
json()
    print('Rooms data:', res)
else:
    print("Failed: wrong credentials")
```

The code snippet is an alternative way of fetching room data with `search_read()`. It will return the same output as in the previous example.

Creating/updating/deleting records using JSON-RPC

In the previous recipe, we looked at how to search and read data through JSON-RPC. In this recipe, we will perform the remaining **CRUD** operations through RPC: **create**, **update** (write), and **delete** (unlink).

Getting ready

We will create a Python program to `create`, `write`, and `unlink` data in the `hostel.room` model. Make sure you have installed the `my_hostel` module and that the server is running on `http://localhost:8017`.

How to do it...

Perform the following steps to create, write, and unlink a room's information through RPC:

1. Add the `jsonrpc_operation.py` file. You can place this file anywhere you want because the RPC program will work independently.
2. Add the following code to the file:

    ```
    # place authentication and get_json_payload method (see last
    recipe for more)
    ```

```
if user_id:
    # creates the room's records
    create_data = [
        {'name': 'Room 1', 'room_no': '201'},
        {'name': 'Room 3', 'room_no': '202'},
        {'name': 'Room 5', 'room_no': '205'},
        {'name': 'Room 7', 'room_no': '207'}
    ]
    payload = get_json_payload("object", "execute_kw", db_name,
user_id, password, 'hostel.room', 'create', [create_data])
    res = requests.post(json_endpoint, data=payload,
headers=headers).json()
    print("Rooms created:", res)
    rooms_ids = res['result']

    # Write in existing room record
    room_to_write = rooms_ids[1]  # We will use ids of recently
created rooms
    write_data = {'name': 'Room 2'}
    payload = get_json_payload("object", "execute_kw", db_name,
user_id, password, 'hostel.room', 'write', [room_to_write,
write_data])
    res = requests.post(json_endpoint, data=payload,
headers=headers).json()
    print("Rooms written:", res)

    # Delete in existing room record
    room_to_unlink = rooms_ids[2:]  # We will use ids of
recently created rooms
    payload = get_json_payload("object", "execute_kw", db_name,
user_id, password, 'hostel.room', 'unlink', [room_to_unlink])
    res = requests.post(json_endpoint, data=payload,
headers=headers).json()
    print("Rooms deleted:", res)

else:
    print("Failed: wrong credentials")
```

3. Run the Python script from the Terminal with the following command:

```
python3 jsonrpc_operation.py
```

The preceding program will create four rooms. Writing one room and deleting two rooms gives you the following output (the IDs created may be different based on your database):

```
$ python3 jsonrpc_operation.py
Rooms created: {'jsonrpc': '2.0', 'id': 837186761, 'result': [43, 44,
45, 46]}
Rooms written: {'jsonrpc': '2.0', 'id': 317256710, 'result': True}
Rooms deleted: {'jsonrpc': '2.0', 'id': 978974378, 'result': True}
```

The `write` and `unlink` methods return `True` if the operation is successful. This means that if you get `True` in response, assume that a record has been updated or deleted successfully.

How it works...

`execute_kw()` is used for the `create`, `update`, and `delete` operations. From Odoo version 12, the `create` method supports the creation of multiple records. So, we prepared the dictionary with information from the four rooms. Then, we made the JSON-RPC call with `hostel.room` as the model name and `create` as the method name. This will create four room records in the database and return a JSON response with the IDs of these newly created rooms. In the next RPC calls, we want to use these IDs to make an RPC call for the `update` and `delete` operations, so we assign it to the `rooms_ids` variable.

> **Important note**
>
> Both JSON-RPC and XML-RPC generate an error when you try to create the record without providing values for the required field, so make sure you have added all the required fields to the `create` values.

In the next RPC call, we used the `write` method to update the existing records. The `write` method accepts two positional arguments; the records to update and the values to write. In our example, we have updated the name of the room by using the ID of the second room from a created room's IDs. This will change the name of the second room from `Room 3` to `Room 2`.

Then, we made the last RPC call to delete two room records. To do so, we used the `unlink` method. The `unlink` method accepts only one argument, which is the ID of the records you want to delete. This RPC call will delete the last two rooms.

There's more...

Like XML-RPC, you can use the `check_access_rights` method in JSON-RPC to check whether you have access rights to perform the operation. This method requires two parameters: the model name and the operation name. In the following example, we check access rights for the `create` operation on the `hostel.room` model:

```
# place authentication and get_json_payload method (see last recipe
for more)



if user_id:
    payload = get_json_payload("object", "execute_kw",
        db_name, user_id, password,
        'hostel.room', 'check_access_rights', ['create'])
    res = requests.post(json_endpoint, data=payload, headers=headers).
json()
    print("Has create access:", res['result'])

else:
    print("Failed: wrong credentials")
```

This program will generate the following output:

```
$ python3 jsonrpc_access_rights.py
Has create access: True
```

When you are performing complex operations via RPC, the use of the `check_access_rights` method can be used before performing an operation to make sure you have proper access rights.

Calling methods using JSON-RPC

In this recipe, we will learn how to invoke a custom method of the model through JSON-RPC. We will change the status of the room by calling the `make_available()` method.

Getting ready

We will create the Python program to call `make_available` on the `hostel.room` model. Make sure you have installed the `my_hostel` module and that the server is running on `http://localhost:8017`.

How to do it...

Perform the following steps to create, write, and update a room's information through RPC:

1. Add the `jsonrpc_method.py` file. You can place this file anywhere you want because the RPC program will work independently.
2. Add the following code to the file:

```
# place authentication and get_json_payload method (see last
recipe for more)


if user_id:
    # Create the room record in draft state
    payload = get_json_payload("object", "execute_kw",
        db_name, user_id, password,
        'hostel.room', 'create', [{
            'name': 'Room 1',
            'room_no': '101',
            'state': 'draft'
        }])
    res = requests.post(json_endpoint, data=payload,
headers=headers).json()
    print("Room created with id:", res['result'])
    room_id = res['result']

    # Change the room state by calling make_available method
    payload = get_json_payload("object", "execute_kw",
        db_name, user_id, password,
        'hostel.room', 'make_available', [room_id])
    res = requests.post(json_endpoint, data=payload,
headers=headers).json()

    # Check the room status after method call
    payload = get_json_payload("object", "execute_kw",
        db_name, user_id, password,
        'hostel.room', 'read', [room_id, ['name', 'state']])
    res = requests.post(json_endpoint, data=payload,
headers=headers).json()
    print("Room state after the method call:", res['result'])

else:
    print("Failed: wrong credentials")
```

3. Run the Python script from the Terminal with the following command:

```
python3 jsonrpc_method.py
```

The preceding command will create one room using `draft` and then we will change the room state by calling the `make_available` method. After that, we will fetch the room data to check the room's status, which will generate the following output:

```
$ python3 jsonrpc_method.py
Room created with id: 53
Room state after the method call: [{'id': 53, 'name': 'Room 1',
'state': 'available'}]
```

The program of this recipe will create a new room record and change the state of the room by calling the model method. By the end of the program, we will have read the room record and printed the updated state.

How it works...

`execute_kw()` is capable of calling any public method of the model. As we saw in the *Calling methods through XML-RPC* recipe, public methods are those that have names that don't start with _ (underscore). Methods that start with _ are private, and you cannot invoke them from JSON-RPC.

In our example, we created a room with a state of `draft`. Then, we made one more RPC call to invoke the `make_available` method, which will change the room's state to `available`. Finally, we made one more RPC call to check the state of the room. This will show that the room's state has changed to **Available**, as seen in *Figure 20.10*.

Methods that do not return anything internally return `None` by default. Such methods cannot be used from RPC. Consequently, if you want to use your method from RPC, at least add the return `True` statement.

The OCA odoorpc library

The **Odoo Community Association** (**OCA**) provides a Python library called `odoorpc`. This is available at `https://github.com/OCA/odoorpc`. The `odoorpc` library provides a user-friendly syntax from which to access Odoo data through RPC. It provides a similar a syntax similar to that of the server. In this recipe, we will see how you can use the `odoorpc` library to perform operations through RPC.

Getting ready

The `odoorpc` library is registered on the Python package (`PyPI`) index. In order to use the library, you need to install it using the following command. You can use this in a separate virtual environment if you want:

```
pip install OdooRPC
```

In this recipe, we will perform some basic operations using the odoorpc library. We will use the hostel.room model to perform these operations. Make sure you have installed the my_hostel module and that the server is running on http://localhost:8017.

How to do it...

Perform the following steps to create, write, and update a room's information through RPC:

1. Add the odoorpc_hostel.py file. You can place this file anywhere you want because the RPC program will work independently.
2. Add the following code to the file:

```
import odoorpc

db_name = 'cookbook_17e'
user_name = 'admin'
password = 'admin'

# Prepare the connection to the server
odoo = odoorpc.ODOO('localhost', port=8017)
odoo.login(db_name, user_name, password)  # login

# User information
user = odoo.env.user
print(user.name)                  # name of the user connected
print(user.company_id.name)   # the name of user's company
print(user.email)                 # the email of user

RoomModel = odoo.env['hostel.room']
search_domain = [['name', 'ilike', 'Standard']]
rooms_ids = RoomModel.search(search_domain, limit=5)
for room in RoomModel.browse(rooms_ids):
    print(room.name, room.room_no)

# create the room and update the state
room_id = RoomModel.create({
    'name': 'Test Room',
    'room_no': '103',
    'state': 'draft'
})
room = RoomModel.browse(room_id)
print("Room state before make_available:", room.state)
room.make_available()
room = RoomModel.browse(room_id)
print("Room state after make_available:", room.state)
```

3. Run the Python script from the Terminal with the following command:

```
python3 odoorpc_hostel.py
```

The program will do the authentication, print user information, and perform an operation in the `hostel.room` model. It will generate the following output:

```
$ python3 odoorpc_hostel.py
Mitchell Admin
Packt publishing
admin@yourcompany.example.com
8th Standard 1
9th Standard 2
10th Standard 3
11th Standard 4
12th Standard 5
Room state before make_available: draft
Room state after make_available: available
```

The preceding output is the result of several RPC calls. We have fetched user info, some room info, and we have changed the state of the room.

How it works...

After installing the `odoorpc` library, you can start using it straight away. To do so, you will need to import the `odoorpc` package and then we will create the object of the `ODOO` class by passing the server URL and port. This will make the `/version_info` call to the server to check the connection. To log in, you need to use the `login()` method of the object. Here, you need to pass the `database name`, `username`, and `password`.

Upon successful login, you can access the user information at `odoo.env.user`. `odoorpc` provides a user-friendly version of RPC, so you can use this user object exactly like the record set in the server. In our example, we accessed the name, email, and company name from this user object.

If you want to access the model registry, you can use the `odoo.env` object. You can call any model method on the model. Under the hood, the `odoorpc` library uses `jsonrpc`, so you can't invoke any private model method name that starts with an _. In our example, we accessed the `hostel.room` model from the registry. After that, we called the `search` method with the `domain` and `limit` parameters. This will return the IDs of the rooms. By passing the room IDs to the `browse()` method, you can generate a record set for the `hostel.room` model.

By the end of the program, we will have created a new room and changed the room's state by calling the `make_available()` method. If you look closely at the syntax of the program, you will see that it uses the same syntax as the server.

There's more...

Although it provides a user-friendly syntax like the server, you can use the library just like the normal RPC syntax. To do so, you need to use the `odoo.execute` method with the model name, method name, and arguments. Here is an example of reading some room information in the raw RPC syntax:

```
import odoorpc

db_name = 'cookbook_17e'
user_name = 'admin'
password = 'admin'

# Prepare the connection to the server
odoo = odoorpc.ODOO('localhost', port=8017)
odoo.login(db_name, user_name, password)  # login

rooms_info = odoo.execute('hostel.room', 'search_read',
    [['name', 'ilike', 'Standard']])
print(rooms_info)
```

See also

There are several other implementations of RPC libraries for Odoo, as follows:

- `https://github.com/akretion/ooor`
- `https://github.com/OCA/odoorpc`
- `https://github.com/odoo/openerp-client-lib`
- `http://pythonhosted.org/OdooRPC`
- `https://github.com/abhishek-jaiswal/php-openerp-lib`

Generating API keys

Odoo v17 has built-in support for the **two-factor authentication** (**2FA**) feature. 2FA is an extra layer of security for user accounts and users need to enter a password and time-based code. If you have enabled 2FA, then you won't be able to use RPC by entering your user ID and password. To fix this, you will need to generate an API key for the user. In this recipe, we will see how you can generate API keys.

How to do it...

Perform the following steps to generate an API key for RPC:

1. Open user preferences and open the **Account Security** tab.
2. Click on the **New API Key** button:

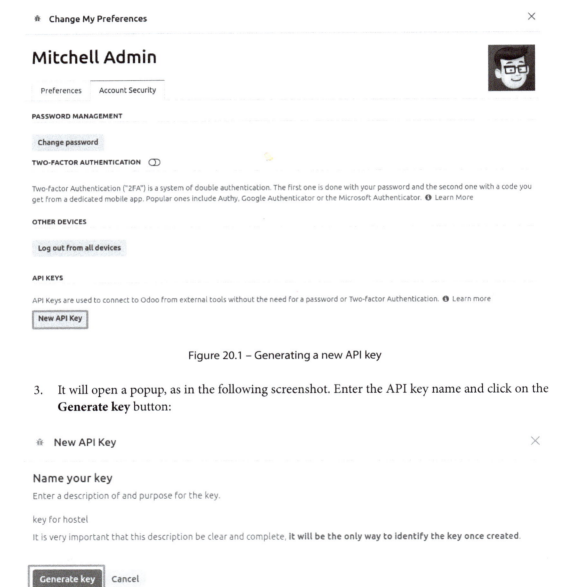

Figure 20.1 – Generating a new API key

3. It will open a popup, as in the following screenshot. Enter the API key name and click on the **Generate key** button:

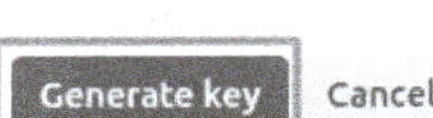

Figure 20.2 – Naming your key

4. This will generate the API key and show it in a new popup. Note down the API key because you will need this again:

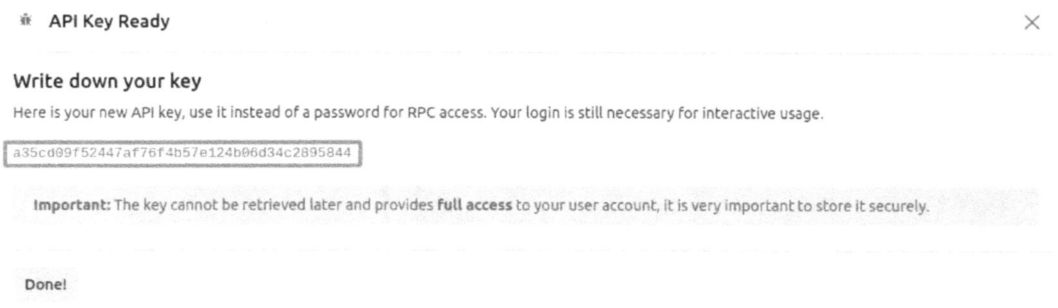

API Key Ready

Write down your key

Here is your new API key, use it instead of a password for RPC access. Your login is still necessary for interactive usage.

a35cd09f52447af76f4b57e124b06d34c2895844

Important: The key cannot be retrieved later and provides **full access** to your user account, it is very important to store it securely.

Done!

Figure 20.3 – Noting the generated API key

Once the API key is generated, you can start using the API key for RPC in the same way as the normal password.

How it works...

Using API keys is straightforward. However, there are a few things that you need to take care of. The API keys are generated per user, and if you want to utilize RPC for multiple users, you will need to generate an API key for each user. Additionally, the API key for a user will have the same access rights as the user would have, so if someone gains access to the key, they can perform all the operations that the user can. So, you need to keep the API key secret.

> **Important note**
>
> When you generate the API key, it is displayed only once. You need to note down the key. If you lose it, there is no way to get it back. In such cases, you would need to delete the API key and generate a new one.

Using the API key is very simple. During RPC calls, you just need to use the API key instead of the user password. You will be able to call RPC even if 2FA is activated.

21
Performance Optimization

With the help of the Odoo framework, you can develop large and complex applications. Good performance is key to the success of any project. In this chapter, we will explore the patterns and tools you need to optimize performance. You will also learn about the debugging techniques used to find the root cause of a performance issue.

In this chapter, we will cover the following recipes:

- The prefetching pattern for recordsets
- The in-memory cache – `ormcache`
- Generating differently sized images
- Accessing grouped data
- Creating or writing multiple records
- Accessing records through database queries
- Profiling Python code

The prefetching pattern for recordsets

When you access data from a recordset, it makes a query in the database. If you have a recordset with multiple records, fetching records on it can make a system slow because of the multiple SQL queries. In this recipe, we will explore how you can use the prefetching pattern to solve this issue. By following the prefetching pattern, you can reduce the number of queries needed, which will improve performance and make your system faster.

How to do it...

Take a look at the following code; it is a normal `compute` method. In this method, `self` is a recordset of multiple records. When you iterate directly on the recordset, prefetching works perfectly:

```
# Correct prefetching
def compute_method(self):
    for rec in self:
        print(rec.name)
```

However, in some cases, prefetching becomes more complex, such as when fetching data with the `browse` method. In the following example, we browse records one by one in the `for` loop. This will not use prefetching efficiently, and it will execute more queries than usual:

```
# Incorrect prefetching
def some_action(self):
    record_ids = []
    self.env.cr.execute("some query to fetch record id")
    for rec in self.env.cr.fetchall():
        record = self.env['res.partner'].browse(rec[0])
        print(record.name)
```

By passing a list of IDs to the `browse` method, you can create a recordset of multiple records. If you perform operations on this recordset, prefetching works perfectly fine:

```
# Correct prefetching
def some_action(self):
    record_ids = []
    self.env.cr.execute("some query to fetch record id")
    record_ids = [ rec[0] for rec in self.env.cr.fetchall() ]
    recordset = self.env['res.partner'].browse(record_ids)
    for record_id in recordset:
        print(record.name)
```

This way, you will not lose the prefetching feature, and data will be fetched in a single SQL query.

How it works...

When you work with multiple recordsets, prefetching helps reduce the number of SQL queries. It does this by fetching all of the data at once. Usually, prefetching works automatically in Odoo, but you lose this feature in certain circumstances, such as when you split records, as depicted in the following example:

```
recs = [r for r in recordset r.id not in [1,2,4,10]]
```

The preceding code given will split the recordset into parts, so you cannot take advantage of prefetching.

Using prefetching correctly can significantly improve the performance of **Object-Relational Mapping (ORM)**. Let's explore how prefetching works under the hood.

When you iterate on a recordset through a `for` loop and access the value of a field in the first iteration, the prefetching process starts its magic. Instead of fetching data for the current record in the iteration, prefetching will fetch the data for all of the records. The logic behind this is that if you access a field in a `for` loop, you are likely to fetch that data for the next record in the iteration as well. In the first iteration of the `for` loop, prefetching will fetch the data for all of the recordsets and keep it in the cache. In the next iteration of the `for` loop, data will be served from this cache, instead of making a new SQL query. This will reduce the query count from `O(n)` to `O(1)`.

Let's suppose the recordset has 10 records. When you are in the first loop and access the `name` field of the record, it will fetch the data for all 10 records. This is not only the case for the `name` field; it will also fetch all the fields for those 10 records. In the subsequent `for` loop iterations, the data will be served from the cache. This will reduce the number of queries from 10 to 1:

```
for record in recordset: # recordset with 10 records
    record.name # Prefetch data of all 10 records in the first loop
    record.email # data of email will be served from the cache.
```

Note that the prefetching will fetch the value of all of the fields (except the `*2many` fields), even if those fields are not used in the body of the `for` loop. This is because the extra columns only have a minor impact on performance compared to the extra queries for each column.

> **Note**
>
> Sometimes, prefetched fields could reduce performance. In these cases, you can disable prefetching by passing `False` into the `prefetch_fields` context, as follows: `recordset.with_context(prefetch_fields=False)`.

The prefetch mechanism uses the environment cache to store and retrieve record values. This means that once the records are fetched from the database, all subsequent calls for fields will be served from the environment cache. You can access the environment cache using the `env.cache` attribute. To invalidate the cache, you can use the `invalidate_cache()` method of the environment.

There's more...

If you split recordsets, the ORM will generate a new recordset with a new prefetch context. Performing operations on such recordsets will only prefetch the data for the respective records. If you want to prefetch all the records after `prefetch`, you can do this by passing the prefetch record IDs to the `with_prefetch()` method. In the following example, we split the recordset into two parts. Here, we passed a common prefetch context in both recordsets, so when you fetch the data from one of them, ORM will fetch the data for the other and put the data in the cache for future use:

```
recordset = ... # assume recordset has 10 records.
recordset1 = recordset[:5].with_prefetch(recordset._ids)
recordset2 = recordset[5:].with_prefetch(recordset._ids)
self.env.cr.execute("select id from sale_order limit 10")
record_ids = [rec[0] for rec in self.env.cr.fetchall()]
recordset = self.env['sale.order'].browse(record_ids)
recordset1 = recordset[:5]
for rec in recordset1:
   print(rec.name)  # Prefetch name of all 5 records in the first loop
   print(rec.attention)  # Prefetch attention of all 5 records in the
first loop
recordset2 = recordset[5:].with_prefetch(recordset._ids)
for rec in recordset1:
   print(rec.name)  # Prefetch name of all 10 records in the first
loop
   print(rec.attention)  # Prefetch attention of all 10 records in the
first loop
```

The prefetch context is not limited to splitting recordsets. You can also use the `with_prefetch()` method to have a common prefetch context between multiple recordsets. This means that when you fetch data from one record, it will fetch data for all other recordsets, too.

The in-memory cache – ormcache

The Odoo framework provides the `ormcache` decorator to manage the in-memory cache. In this recipe, we will explore how you can manage the cache for your functions.

How to do it...

The classes of this ORM cache are available at `/odoo/tools/cache.py`. In order to use these in any file, you will need to import them as follows:

```
from odoo import tools
```

After importing the classes, you can use the ORM cache decorators. Odoo provides different types of in-memory cache decorators. We'll take a look at each of these in the following subsections.

ormcache

This one is the simplest and most used cache decorator. You need to pass the parameter name upon which the method's output depends. The following is an example method with the `ormcache` decorator:

```
@tools.ormcache('mode')
def fetch_mode_data(self, mode):
```

```
    # some calculations
    return result
```

When you call this method for the first time, it will be executed, and the result will be returned. `ormcache` will store this result based on the value of the `mode` parameter. When you call the method again with the same `mode` value, the result will be served from the cache without executing the actual method.

Sometimes, your method's result depends on the environment attributes. In these cases, you can declare the method as follows:

```
@tools.ormcache('self.env.uid', 'mode')
def fetch_data(self, mode):
    # some calculations
    return result
```

The method given in this example will store the cache based on the environment user and the value of the `mode` parameter.

ormcache_context

This cache works similarly to `ormcache`, except that it depends on the parameters plus the value in the context. In this cache's decorator, you need to pass the parameter name and a list of context keys. For example, if your method's output depends on the `lang` and `website_id` keys in the context, you can use `ormcache_context`:

```
@tools.ormcache_context('mode', keys=('website_id','lang'))
def fetch_data(self, mode):
    # some calculations
    return result
```

The cache in the preceding example will depend on the `mode` argument and the values of `context`.

ormcache_multi

Some methods carry out an operation on multiple records or IDs. If you want to add a cache to these kinds of methods, you can use the `ormcache_multi` decorator. You need to pass the `multi` parameter, and during the method call, ORM will generate the cache keys by iterating on this parameter. In this method, you will need to return the result in the dictionary format with an element of the `multi` parameter as a key. Take a look at the following example:

```
@tools.ormcache_multi('mode', multi='ids')
def fetch_data(self, mode, ids):
    result = {}
    for i in ids:
        data = ... # some calculation based on ids
```

```
        result[i] = data
    return result
```

Suppose we called the preceding method with `[1,2,3]` as the IDs. The method will return a result in the `{1:... , 2:..., 3:... }` format. ORM will cache the result based on these keys. If you make another call with `[1,2,3,4,5]` as the IDs, your method will receive `[4, 5]` as the `ID` parameter, so the method will carry out the operations for the `4` and `5` IDs, and the rest of the result will be served from the cache.

How it works...

The ORM cache keeps the cache in the dictionary format (the cache lookup). The keys of this cache will be generated based on the signature of the decorated method, and the values will be the result. Put simply, when you call the method with the `x, y` parameters and the result of the method is `x+y`, the cache lookup will be `{(x, y): x+y}`. This means that the next time you call this method with the same parameters, the result will be served directly from this cache. This saves computation time and makes the response faster.

The ORM cache is an in-memory cache, so it is stored in RAM and occupies memory. Do not use `ormcache` to serve large data, such as images or files.

> **Warning**
> Methods using this decorator should never return a recordset. If they do, they will generate `psycopg2.OperationalError` because the underlying cursor of the recordset is closed.

You should use the ORM cache on pure functions. A pure function is a method that always returns the same result for the same arguments. The output of these methods only depends on the arguments, so they return the same result. If this is not the case, you need to manually clear the cache when you perform operations that make the cache's state invalid. To clear the cache, call the `clear_caches()` method:

```
self.env[model_name].clear_caches()
```

Once you have cleared the cache, the next call to the method will execute the method and store the result in the cache, and all subsequent method calls with the same parameter will be served from the cache.

There's more...

The ORM cache is the **Least Recently Used** (**LRU**) cache, meaning that if a key in the cache is not used frequently, it will be removed. If you don't use the ORM cache properly, it might do more harm than good. For instance, if the argument passed in a method is always different, then each time, Odoo will look in the cache first and then call the method to compute. If you want to learn how your cache is performing, you can pass the `SIGUSR1` signal to the Odoo process:

```
kill -SIGUSR1 <pid>
kill -SIGUSR1 496
```

Here, `496` is the process ID. After executing the command, you will see the status of the ORM cache in the logs:

```
> 2023-10-18 09:22:49,350 496 INFO odoo-book-17.0 odoo.tools.
cache:       1 entries,       31 hit,       1 miss,       0 err, 96.9%
ratio, for ir.actions.act_window._existing
> 2023-10-18 09:22:49,350 496 INFO odoo-book-17.0 odoo.tools.
cache:       1 entries,        1 hit,       1 miss,       0 err, 50.0%
ratio, for ir.actions.actions.get_bindings
> 2023-10-18 09:22:49,350 496 INFO odoo-book-17.0 odoo.tools.
cache:       4 entries,        1 hit,       9 miss,       0 err, 10.0%
ratio, for ir.config_parameter._get_param
```

The percentage in the cache is the hit-to-miss ratio. It's the success ratio of the result being found in the cache. If the cache's hit-to-miss ratio is too low, you should remove the ORM cache from the method.

Generating images in different size

Large images can be troublesome for any website. They increase the size of web pages and consequently make them slower as a result. This leads to bad SEO rankings and visitor loss. In this recipe, we will explore how you can create images of different sizes; by using the right images, you can reduce the web page size and improve the page loading time.

How to do it...

You will need to inherit `image.mixin` in your model. Here is how you can add `image.mixin` to your model:

```
class HostelStudent(models.Model):
    _name = "hostel.student"
    _description = "Hostel Student Information"
    _inherit = ["image.mixin"]
```

The mixin will automatically add five new fields to the hostel student model to store images of different sizes. See the *How it works...* section to learn about all five fields.

How it works...

The `image.mixin` instance will automatically add five new binary fields to the model. Each field stores images with a different resolution. Here is a list of the fields and their resolutions:

- `image_1920`: 1,920x1,920
- `image_1024`: 1,024x1,024
- `image_512`: 512x1,512

- `image_256`: 256x256
- `image_128`: 128x128

Of all the fields given here, only `image_1920` is editable. The other image fields are read-only and update automatically when you change the `image_1920` field. So, in the backend form view of your model, you need to use the `image_1920` field to allow the user to upload images. However, by doing so, we load large `image_1920` images in the form view. However, there is a way to improve performance by using `image_1920` images in the form view but displaying smaller images. For instance, we can utilize the `image_1920` field but display an `image_128` field. To do this, you can use the following syntax:

```
<field name="image_1920" widget="image"
          options="{'preview_image': 'image_128'}" />
```

Once you have saved the image to the field, Odoo will automatically resize the image and store it in the respective field. The form view will display the converted `image_128`, as we use it as `preview_image`.

> **Note**
>
> The `image.mixin` model is `AbstractModel`, so its table is not present in the database. You need to inherit it in your model in order to use it.

With this `image.mixin`, you can store an image with a maximum resolution of 1,920x1,920. If you save an image with a resolution higher than 1,920x1,920, Odoo will reduce it to 1,920x1,920. While doing so, Odoo will also preserve the resolution of the image, avoiding any distortion. As an example, if you upload an image with a 2,400x1,600 resolution, the `image_1920` field will have a resolution of 1,920x1,280.

There's more...

With `image.mixin`, you can get images with certain resolutions, but what if you want to use an image with another resolution? To do so, you can use a binary wrapper field image, as shown in the following example:

```
image_1500 = fields.Image("Image 1500", max_width=1500, max_
height=1500)
```

This will create a new `image_1500` field, and storing the image will resize it to 1,500x1,500 resolution. Note that this is not part of `image.mixin`. It just reduces the image to 1,500x1,500, so you need to add this field in the form view; editing it will not make changes to the other image fields in `image.mixin`. If you want to link it with an existing `image.mixin` field, add the `related="image_1920"` attribute to the field definition.

Accessing grouped data

When you want data for statistics, you often need it in a grouped form, such as a monthly sales report, or a report that shows sales per customer. It is time-consuming to search records and group them manually. In this recipe, we will explore how you can use the `read_group()` method to access grouped data.

How to do it...

Perform the following steps.

> **Note**
> The `read_group()` method is widely used for statistics and smart stat buttons.

1. Let's assume that you want to show the number of sales orders on the partner form. This can be done by searching sales orders for a customer and then counting the length:

```
# in res.partner model
so_count = fields.Integer(compute='_compute_so_count',
string='Sale order count')
def _compute_so_count(self):
    sale_orders = self.env['sale.order'].
search(domain=[('partner_id', 'in', self.ids)])
    for partner in self:
        partner.so_count = len(sale_orders.filtered(lambda so:
so.partner_id.id == partner.id))
```

   The previous example will work, but not optimally. When you display the `so_count` field on the tree view, it will fetch and filter sales orders for all the partners in a list. With this small amount of data, the `read_group()` method won›t make much difference, but as the amount of data grows, it could be a problem. To fix this issue, you can use the `read_group` method.

2. The following example will do the same as the preceding one, but it only consumes one SQL query, even for large datasets:

```
# in res.partner model
so_count = fields.Integer(compute='_compute_so_count',
string='Sale order count')
def _compute_so_count(self):
    sale_data = self.env['sale.order'].read_group(
        domain=[('partner_id', 'in', self.ids)],
        fields=['partner_id'], groupby=['partner_id'])
    mapped_data = dict([(m['partner_id'][0], m['partner_id_
count']) for m in sale_data])
    for partner in self:
        partner.so_count = mapped_data[partner.id]
```

The previous code snippet is optimized, as it obtains the sales order count directly via SQL's GROUP BY feature.

How it works...

The read_group() method internally uses the GROUP BY feature of SQL. This makes the read_group method faster, even if you have large datasets. Internally, the Odoo web client uses this method in the charts and the grouped tree view. You can tweak the behavior of the read_group method by using different arguments.

Let's explore the signature of the read_group method:

```
def read_group(self, domain, fields, groupby, offset=0, limit=None,
orderby=False, lazy=True):
```

The different parameters available for the read_group method are as follows:

- domain: This is used to filter records. This will be the search criteria for the read_group method.
- fields: This is a list of the fields to fetch with the grouping. Note that the fields mentioned here should be in the groupby parameter, unless you use some aggregate functions. The read_group method supports the SQL aggregate functions. Let›s say you want to get the average order amount per customer. If so, you can use read_group as follows:

  ```
  self.env['sale.order'].read_group([], ['partner_id', 'amount_
  total:avg'], ['partner_id'])
  ```

  If you want to access the same field twice but with a different aggregate function, the syntax is a little different. You need to pass the field name as alias:agg(field_name). This example will give you the total and average number of orders per customer:

  ```
  self.env['sale.order'].read_group([], ['partner_id',
  'total:sum(amount_total)', 'avg_total:avg(amount_total)'],
  ['partner_id'])
  ```

- groupby: This parameter will be a list of fields by which the records are grouped. It lets you group records based on multiple fields. To do this, you will need to pass a list of fields. For example, if you want to group the sales orders by customer and order state, you can pass ['partner_id ', 'state'] in this parameter.
- offset: This parameter is used for pagination. If you want to skip a few records, you can use this parameter.
- limit: This parameter is used for pagination; it indicates the maximum number of records to fetch.
- lazy: This parameter accepts Boolean values. By default, its value is True. If this parameter is True, the results are grouped only by the first field in the groupby parameter. You will

get the remaining groupby parameters and the domain in the __context and __domain keys in the result. If the value of this parameter is set to False, it will group the data by all fields in the groupby parameter.

There's more...

Grouping by date fields can be complicated because it is possible to group records based on days, weeks, quarters, months, or years. You can change the grouping behavior of the date field by passing groupby_function after : in the groupby parameter. If you want to group the monthly total of the sales orders, you can use the read_group method:

```
self.env['sale.order'].read_group([], ['total:sum(amount_total)'],
['order_date:month'])
```

The possible options for date grouping are day, week, month, quarter, and year.

See also

Refer to the documentation if you want to learn more about PostgreSQL aggregate functions: https://www.postgresql.org/docs/current/functions-aggregate.html.

Creating or writing multiple records

If you are new to Odoo development, you might execute multiple queries to write or create multiple records. In this recipe, we will look at how to create and write records in batches.

How to do it...

Creating multiple records and writing on multiple records work differently under the hood. Let's see each of these records one by one.

Creating multiple records

Odoo supports creating records in batches. If you are creating a single record, simply pass a dictionary with the field values. To create records in a batch, you just need to pass a list of these dictionaries instead of a single dictionary. The following example creates three room records in a single create call:

```
vals = [{
    'name': "Room A-101",
    'room_no': 101,
    'floor_no': 1,
    'student_per_room': 2,
}, {
    'name': "Room A-102",
```

```
    'room_no': 102,
    'floor_no': 1,
    'student_per_room': 3,
}, {
    'name': "Room B-201",
    'room_no': 201,
    'floor_no': 2,
    'student_per_room': 3,
}]
self.env['hostel.room'].create(vals)
```

This code snippet will create the records for three new books.

Writing on multiple records

When working with multiple versions of Odoo, it's important to understand how the write method behaves. In this case, it adopts a delayed approach for updates, meaning it doesn't immediately write data to the database. Instead, Odoo only writes the data to the database when necessary or when the `flush()` method is called.

Here are two examples of the `write` method:

```
# Example 1
data = {...}
for record in recordset:
    record.write(data)
# Example 2
data = {...}
recordset.write(data)
```

If you are using Odoo v13 or above, then there will not be any issues regarding performance. However, if you are using an older version, the second example will be much faster than the first one because the first example will execute a SQL query in each iteration.

How it works...

In order to create multiple records in a batch, you need to pass value dictionaries in the form of a list to create new records. This will automatically manage batch-creating the records. When you create records in a batch, doing so internally will insert a query for each record. This means that creating records in a batch is not done in a single query. However, this doesn't mean that creating records in batches does not improve performance. The performance gain is achieved through batch-calculating computing fields.

Things work differently for the `write` method. Most things are handled automatically by the framework. For instance, if you write the same data on all records, the database will be updated with only one UPDATE query. The framework will even handle it if you update the same record again and again in the same transaction, as follows:

```
recordset.name= 'Admin'
recordset.email= 'admin@example.com'
recordset.name= 'Administrator'
recordset.email= 'admin-2@example.com'
```

In the previous code snippet, only one query will be executed for `write`, with the final values of `name=Administrator` and `email=admin-2@example.com`. This does not have a bad impact on performance, as the assigned values are in the cache and written later in a single query.

Things are different if you use the `flush()` method in between, as shown in the following example:

```
recordset.name= 'Admin'
recordset.email= 'admin@example.com'
recordset.flush()
recordset.name= 'Administrator'
recordset.email= 'admin-2@example.com'
```

The `flush()` method updates the values from the cache to the database. So, in the previous example, two UPDATE queries will be executed – one with data before the flush and another with data after the flush.

There's more...

If you are using an older version, then writing a single value will execute the UPDATE query immediately. Check the following examples to explore the correct usage of the `write` operation for an older version of Odoo:

```
# incorrect usage
recordset.name= 'Admin'
recordset.email= 'admin@example.com'

# correct usage
recordset.write({'name': 'Admin', 'email'= 'admin@example.com'})
```

Here, in the first example, we have two UPDATE queries, while the second example will only take one UPDATE query.

Accessing records through database queries

Odoo ORM has limited methods, and sometimes, it is difficult to fetch certain data from ORM. In these cases, you can fetch data in the desired format, and you need to perform an operation on the data to get a certain result. Due to this, it becomes slower. To handle these special cases, you can execute SQL queries in the database. In this recipe, we will explore how you can run SQL queries from Odoo.

How to do it...

You can perform database queries using the `self._cr.execute` method:

1. Add the following code:

```
self.flush()
self._cr.execute("SELECT id, name, room_no, floor_no  FROM
hostel_room WHERE name ilike %s", ('%Room A-%',))
data = self._cr.fetchall()
print(data)
```

   Here is the output:

```
[(4, 'Room A-101', '101', 1), (5, 'Room A-103', '103', 1), (6,
'Room A-201', '201', 2)]
```

2. The result of the query will be in the form of a list of tuples. The data in the tuples will be in the same sequence as the fields in the query. If you want to fetch data in dictionary format, you can use the `dictfetchall()` method. Take a look at the following example:

```
self.flush()
self._cr.execute("SELECT id, name, room_no, floor_no  FROM
hostel_room WHERE name ilike %s", ('%Room A-%',))
data = self._cr.dictfetchall()
print(data)
```

   Here is the output:

```
[{'id': 4, 'name': 'Room A-101', 'room_no': 101, 'floor_no': 1},
{'id': 5, 'name': 'Room A-103', 'room_no': 103, 'floor_no': 1},
{'id': 6, 'name': 'Room A-201', 'room_no': 201, 'floor_no': 2}]
```

If you want to fetch only a single record, you can use the `fetchone()` and `dictfetchone()` methods. These methods work like `fetchall()` and `dictfetchall()`, but they only return a single record, and you need to call the `fetchone()` and `dictfetchone()` methods multiple times if you want to fetch multiple records.

How it works...

There are two ways to access the database cursor from the recordset – one is from the recordset itself, such as `self._cr`, and the other is from the environment (in particular, `self.env.cr`). This cursor is used to execute database queries. In the preceding example, we saw how you can fetch data through raw queries. The table name is the name of the model after replacing `.` with `_`, so the `hostel.room` model becomes `hostel_room`.

Note that we used `self.flush()` before executing a query. The reason behind this is that Odoo uses the cache excessively, and the database might not have the correct values. `self.flush()` will push all the delayed updates to the database and conduct all the dependent computations as well, and you will then get correct values from the database. The `flush()` method also supports a few parameters that help you control what is flushed in the database. The parameters are as follows:

- The `fname` parameter needs a list of fields that you want to flush to the database
- The `records` parameter needs a recordset, and it is used if you want to flush certain records only

If you are executing `INSERT` or `UPDATE` queries, you will also need to execute `flush()` after executing the query because the ORM might not be aware of the change you made, and it might have cached records.

You need to consider a few things before you execute raw queries. Only use raw queries when you have no other choice. By executing raw queries, you bypass the ORM layers. Therefore, you also bypass security rules and the ORM's performance advantages. Sometimes, wrongly built queries can introduce SQL injection vulnerabilities. Consider the following example, in which the queries could allow an attacker to perform SQL injection:

```
# very bad, SQL injection possible
self.env.cr.execute('SELECT id, name FROM hostel_room WHERE name ilike
+ search_keyword + ';')

# good
self.env.cr.execute('SELECT id, name FROM hostel_room WHERE name ilike
%s ';', (search_keyword,))
```

Don't use the string format function either; it will also allow an attacker to perform SQL injection. Using SQL queries makes your code harder to read and understand for other developers, so avoid using them wherever possible.

> **Information**
>
> A lot of Odoo developers believe that executing SQL queries makes operations faster, as it bypasses the ORM layer. This is not completely true, however; it depends on the use case. In most operations, ORM performs better and faster than `RAW` queries because data is served from the recordset cache.

There's more...

Operations made in one transaction are only committed at the end of it. If an error occurs in the ORM, the transaction is rolled back. If you have made an `INSERT` or `UPDATE` query and you want to make it permanent, you can use `self._cr.commit()` to commit the changes.

> **Note**
>
> Note that using `commit()` can be dangerous because it can put records in an inconsistent state. An error in the ORM can cause incomplete rollbacks, so only use `commit()` if you are completely sure of what you›re doing.

If you use the `commit()` method, then there›s no need to use `flush()` afterward. The `commit()` method flushes the environment internally.

Profiling

Sometimes, you will be unable to pinpoint the cause of an issue. This is especially true of performance issues. Odoo provides some built-in profiling tools that help you find the real cause of an issue.

Profiling is about analyzing the execution of a program and measuring aggregated data. These data can be the elapsed time for each function, the executed SQL queries, and so on.

While profiling does not improve the performance of a program by itself, it can prove very helpful in finding performance issues and identifying which part of the program is responsible for them.

Code profiling in Odoo can help you identify performance and optimize your code. It is a technique used to analyze the code execution time, complexity of the program, and memory usage of an application.

By using profiling techniques in Odoo, you can improve the overall performance and user experience of your application, making it faster and more efficient.

Enabling the profiler

The profiler can either be enabled from the user interface, which is the easiest way to do so but only allows you to profile web requests, or from Python code:

1. Enable developer mode.
2. The profiler must be enabled globally on the database. This can be done in two ways:

    - Open the developer mode tools, and then toggle the **Enable profiling** button. A wizard suggests a set of expiry times for the profiling. Click on **Enable profiling** to enable the profiler globally.

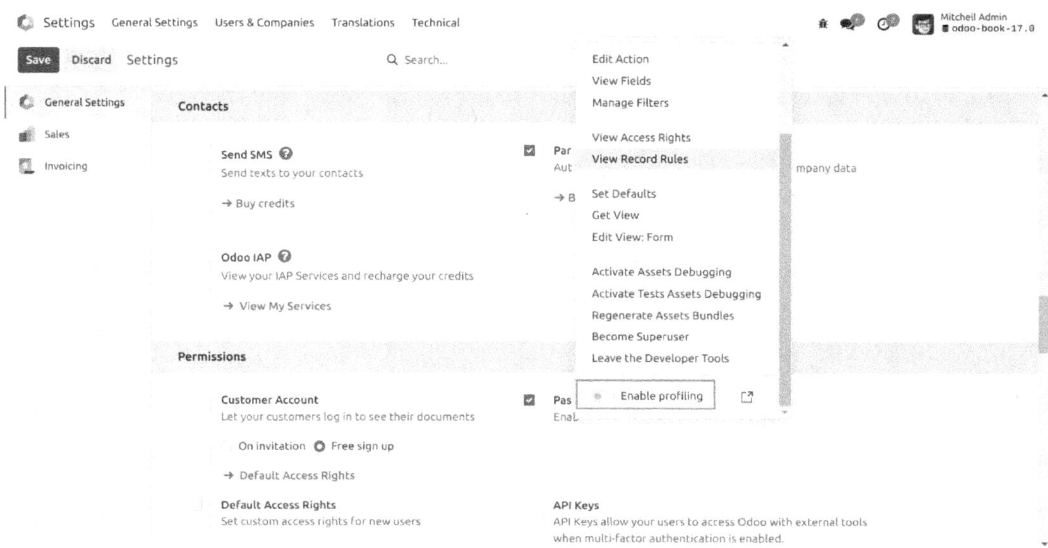

Figure 21.1 – Enabling Profiling

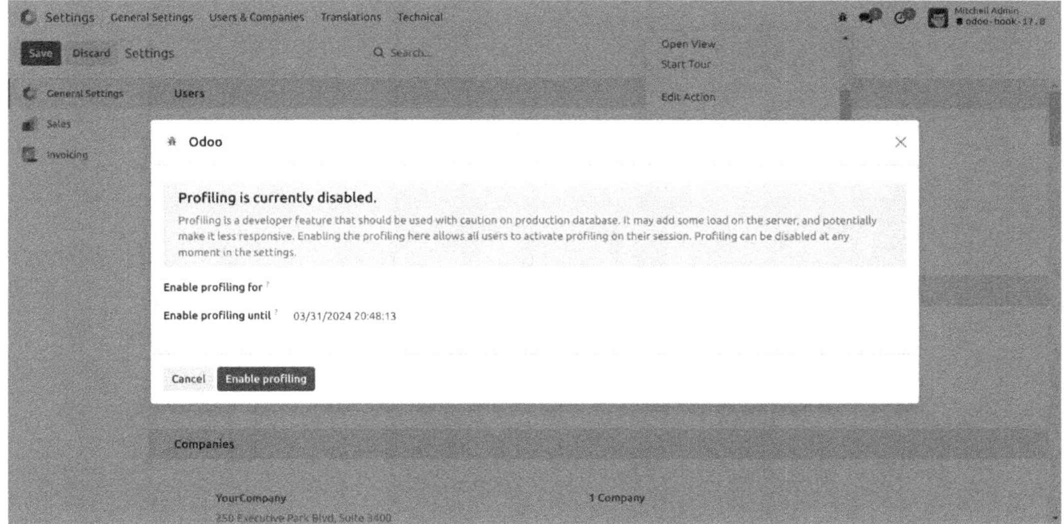

Figure 21.2 – Disabling profiling

- Go to **Settings** | **General Settings** | **Performance** and set the desired time for the field Enable profiling field.

Analyzing the results

To browse the profiling results, make sure that the profiler is enabled globally on the database, then open the developer mode tools, and click on the button in the top-right corner of the profiling section. A list view of the `ir.profile` records grouped by profiling session will open.

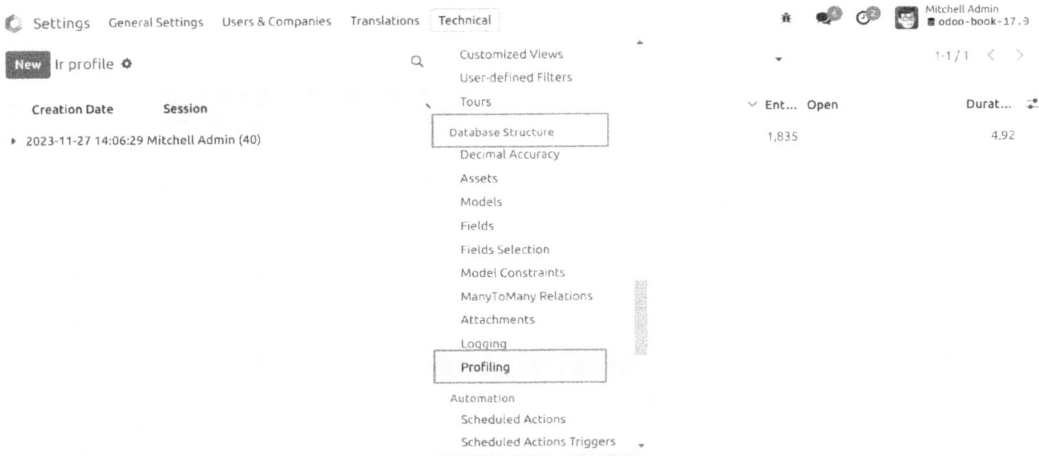

Each record has a clickable link that opens the speedscope results in a new tab.

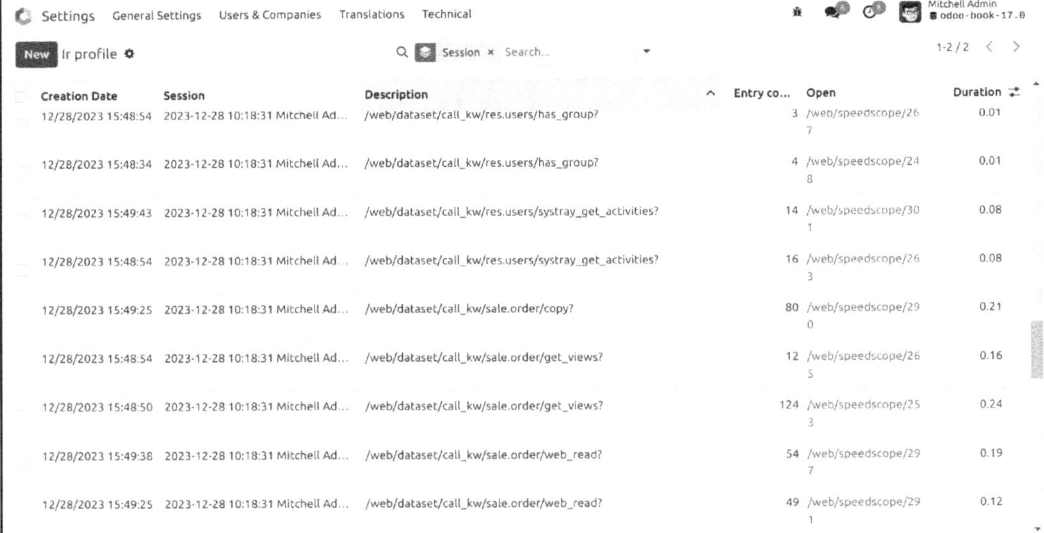

Speedscope falls out of the scope of this documentation, but there are a lot of tools to try out – search, highlight of similar frames, zoom on frame, timeline, left heavy, sandwich view, and so on.

Depending on the profiling options that were activated, Odoo generates different view modes that you can access from the top menu.

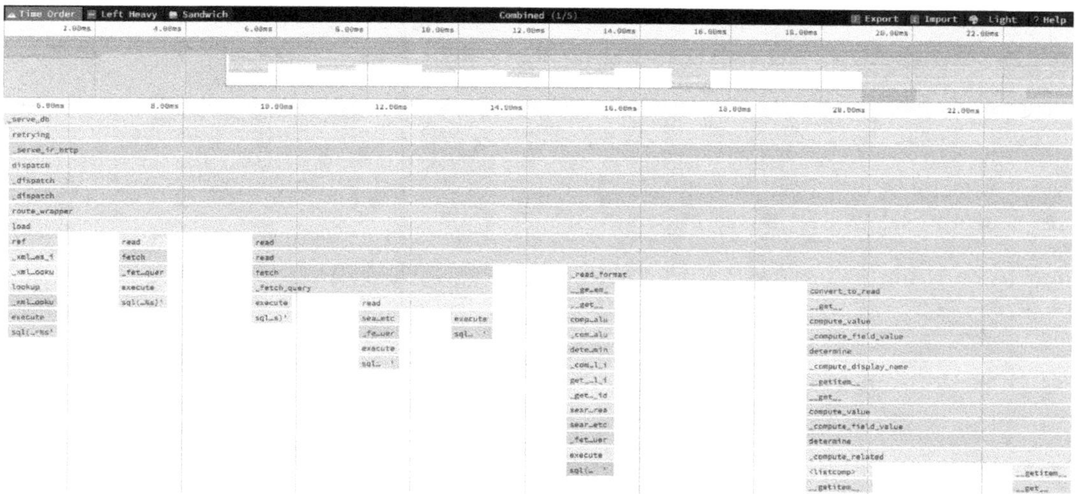

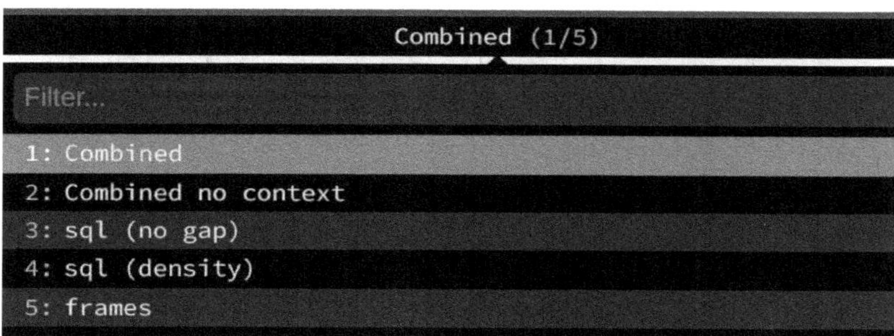

- **Combined**: The **Combined** view displays all of the SQL queries and traces that have been integrated together.
- **Combined no context**: The **Combined no context** view produces the same results but disregards the stored execution context, performance/profiling/enable>.
- **sql (no gap)**: The **sql (no gap)** view displays all SQL queries as if they were done sequentially, without any Python logic. This is solely beneficial for SQL optimization.
- **sql (density)**: Only the SQL queries are displayed in the **sql (no density)** view, with space between them. This can help you discover areas where numerous tiny queries could be batch-processed and determine whether the issue is with the Python or SQL code.
- **frames**: Only the periodic collector's results are displayed in the **frames** view.

> **Note**
>
> Despite the profiler's lightweight design, it can still affect performance, particularly when utilizing the `Sync` collector. Remember that when you examine the speedscope data.

Collectors

Every collector has a unique format and method to gather profiling data. Through their specific toggle button in the developer mode tools, or from Python code using their key or class, each can be independently enabled from the user interface.

There are currently four collectors available in Odoo:

- `SQLCollector`
- `PeriodicCollector`
- `QwebCollector`
- `SyncCollector`

SQLCollector

All SQL queries made to the database in the current thread (for all cursors) are saved by the `SQL` collector, together with the stack trace. Using the collector on a large number of tiny queries could affect execution time and other profilers, since the overhead of the collector is added to the thread that is examined for each query.

Debugging query counts and adding data to the `Periodic` collector in the combined speedscope view are two particularly helpful uses for it:

```
class SQLCollector(Collector):
    """
    Saves all executed queries in the current thread with the call
stack.
    """
    name = 'sql'
```

The Periodic collector

This collector runs in a separate thread and saves the stack trace of the analyzed thread at every interval. The interval (by default, 10 ms) can be defined through the **Interval** option in the user interface or the interval parameter in Python code.

> **Note**
>
> Memory problems will arise when profiling lengthy queries if the interval is set extremely low. The interval will lose information on brief function executions if it is set extremely high.

Because of its distinct thread, it should have relatively little effect on execution time, making it one of the finest ways to assess performance:

```
class PeriodicCollector(Collector):
    """
    Record execution frames asynchronously at most every `interval`
seconds.

    :param interval (float): time to wait in seconds between two
samples.
    """
    name = 'traces_async'
```

The Qweb collector

The Python execution time and queries for each directive are reduced by this collector. With the SQL collector, the overhead may be significant when a large number of tiny instructions are executed. In terms of data collected, the results differ from those of other collectors, and a custom widget can be used to examine them from the `ir.profile` form view.

It is most helpful when trying to maximize views:

```
class QwebCollector(Collector):
    """
    Record qweb execution with directive trace.
    """
    name = 'qweb'
```

The Sync collector

Performance is significantly impacted by this collector, since it operates on a single thread and saves the stack for each function call and return.

Debugging and comprehending intricate flows, as well as tracking their execution within the code, can be helpful. However, due to the significant overhead, performance analysis is not advised to use it:

```
class SyncCollector(Collector):
    """
    Record complete execution synchronously.
    Note that --limit-memory-hard may need to be increased when
launching Odoo.
    """
    name = 'traces_sync'
```

Performance pitfalls

- Be careful with randomness. Multiple executions may lead to different results – for example , a garbage collector being triggered during execution.
- Be careful with blocking calls. In some cases, an external `c_call` may take some time before releasing the GIL, thus leading to unexpected long frames with the Periodic collector. This should be detected by the profiler and given a warning. It is possible to trigger the profiler manually before such calls if needed.
- Pay attention to the cache. Profiling before the view/assets/... are in a cache can lead to different results.
- Be aware of the profiler's overhead. The SQL collector's overhead can be important when many small queries are executed. Profiling is practical to spot a problem, but you may want to disable the profiler to measure a code change's real impact.
- Profiling results can be memory-intensive. In some cases (e.g., profiling an install or a long request), you can reach the memory limit, especially when rendering the speedscope results, which can lead to an HTTP 500 error. In this case, you may need to start the server with a higher memory limit – `--limit-memory-hard $((8*1024**3))`.

22
Point of Sale

Point of Sale is a fully integrated application that allows you to sell products (online or offline) with any device. It also automatically registers product moves in your stock, gives you real-time statistics, and consolidations across all shops. In this chapter, we will see how to modify the Point of Sale application.

In this chapter, we will cover the following topics:

- Adding custom JavaScript/SCSS files
- Adding an action button to the keyboard
- Making RPC calls
- Modifying the Point of Sale screen UI
- Modifying existing business logic
- Modifying customer

> **Note**
> The Point of Sale application is mostly written in JavaScript. This chapter is written assuming that you have a basic knowledge of JavaScript. This chapter also uses the OWL framework, so if you are unaware of these JavaScript terms, check out *Chapter 16, The Odoo Web Library (OWL).*

Throughout this chapter, we will be using an add-on module called `point_of_sale_customization`. This `point_of_sale_customization` module will have a dependency on `point_of_sale`, as we are going to do customization in the Point of Sale application. To get started with this point quickly, we have prepared an initial `point_of_sale_customization` module, and you can grab it from the `Chapter22/00_initial_module/point_of_sale_customization` directory in the GitHub repository of this book.

Technical requirements

All the code used in this chapter can be downloaded from the following GitHub repository: https://github.com/PacktPublishing/Odoo-17-Development-Cookbook-Fifth-Edition/tree/main/Chapter22.

Adding custom JavaScript/SCSS files

The Point of Sale app uses different asset bundles to manage JavaScript and style sheet files. In this recipe, we will learn how to add **SCSS** and **JavaScript** files to the Point of Sale asset bundle.

Getting ready

First, we will load an SCSS style sheet and a JavaScript file into the Point of Sale application.

How to do it...

To load assets into the Point of Sale application, follow these steps:

1. Add a new SCSS file at `point_of_sale_customization/static/src/scss/point_of_sale_customization.scss` and insert the following code:

```
.pos .pos-content {
    .price-tag {
        background: #00abcd;
        width: 100%;
        right: 0;
        left: 0;
        top:0;
    }
}
```

2. Add a JavaScript file at `point_of_sale_customization/static/src/js/point_of_sale_customization.js` and add the following:

```
/** @odoo-module */

console.log("Point Of Sale Javascript Loaded");
```

3. Register these JavaScript and SCSS files in the `point_of_sale assets`:

```
'assets': {
      'point_of_sale._assets_pos': [
'point_of_sale_customization/static/src/scss/point_of_sale_
customization.scss',
'point_of_sale_customization/static/src/js/point_of_sale_
customization.js'
```

```
        ],
    },
```

4. Install the `point_of_sale_customization` module.

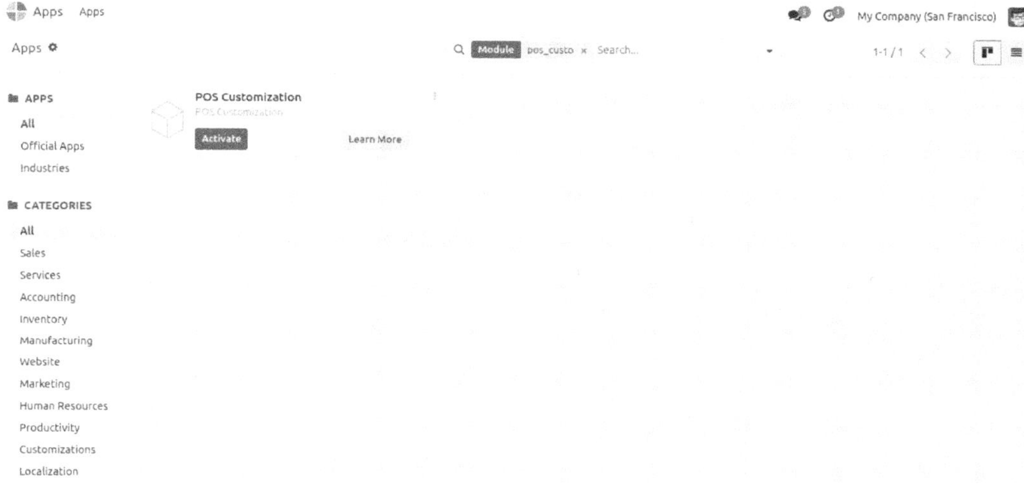

Figure 22.1 – Installing the POS Customization module

To see your changes in action, start the new session from the **Point of Sale | Dashboard** menu.

How it works...

So far, we have loaded one JavaScript file and one SCSS file into the Point of Sale application.

In *step 1*, we changed the background color of the pricing label of the product card. After installing the `point_of_sale_customization` module, you will be able to see changes to the pricing labels:

Figure 22.2 – The updated price labels

In *step 2*, we added the JavaScript file. In it, we added the log to the console. In order to see the message, you will need to open your browser's developer tools. In the **Console** tab, you will see the following log. This shows that your JavaScript file has loaded successfully. Right now, we have only added the log to the JavaScript file, but in upcoming recipes, we will add more to it:

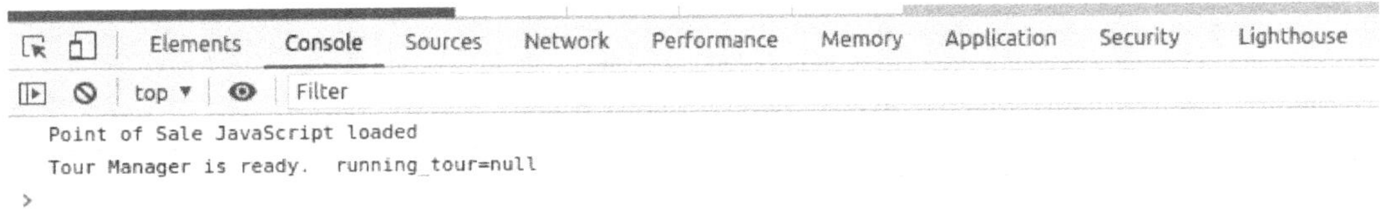

Figure 22.3 – JavaScript loaded (the log in the console)

In *step 3*, we added the JavaScript file and the SCSS file, as follows:

```
'assets': {
    'point_of_sale._assets_pos': [
         'js,scss path'
    ],
}
```

There's more...

Odoo also has an add-on module for Point of Sale solutions for restaurants. Note that this Point of Sale restaurant module is just an extension of the Point of Sale application. If you want to do customization in the restaurant module, you will need to add your JavaScript and SCSS files to the same `point_of_sale._assets_pos` asset bundle.

Adding an action button to the keyboard

As we discussed in the previous point, the Point of Sale application is designed in such a way that it works offline. Thanks to this, the code structure of the Point of Sale application is different from the remaining Odoo applications. The code base of the Point of Sale app is largely written with JavaScript and provides different utilities for customization. At this point, we will use one such utility and create an action button at the top of the keyboard panel.

Getting ready

Here, we will use the `point_of_sale_customization` module created in the *Adding custom JavaScript/SCSS files* recipe. We will add a button at the top of the keyboard panel. This button will be a shortcut to apply a discount to the order lines.

How to do it...

Follow these steps to add a 5% discount action button to the keyboard panel for the Point of Sale application:

1. Add the following code to the `/static/src/js/point_of_sale_customization.js` file, which will define the action button:

```
/** @odoo-module */

import { Component } from "@odoo/owl";
import { ProductScreen } from "@point_of_sale/app/screens/
product_screen/product_screen";
import { usePos } from "@point_of_sale/app/store/pos_hook";

export class PosDiscountButton extends Component {
    static template = "PosDiscountButton";
    setup() {
        this.pos = usePos();
    }
    async onClick() {
        const order = this.pos.get_order();
        if (order.selected_orderline) {
            order.selected_orderline.set_discount(5);
        }
    }
}
ProductScreen.addControlButton({
    component: PosDiscountButton,
    condition: function () {
        return true;
    }
});
```

2. Add the QWeb template for the button to the `/static/src/xml/point_of_sale_customization.xml` file:

```
<?xml version="1.0" encoding="UTF-8"?>
<templates id="template" xml:space="preserve">
   <t t-name="PosDiscountButton">
       <span class="control-button btn btn-light rounded-0
fw-bolder"
              t-on-click="() => this.onClick()">
           <i class="fa fa-gift"></i>
```

```
                <span>5%</span>
                <span>Discount</span>
            </span>
        </t>
    </templates>
```

3. Add a new SCSS file at `point_of_sale_customization/static/src/scss/point_of_sale_customization.scss` and insert the following code:

```
.pos .pos-content {
    .price-tag {
        background: #00abcd;
        width: 100%;
        right: 0;
        left: 0;
        top:0;
    }
}
```

4. Register the QWeb template in the manifest file as follows:

```
'assets': {
        'point_of_sale._assets_pos': [
            'point_of_sale_customization/static/src/scss/point_
of_sale_customization.scss',
            'point_of_sale_customization/static/src/xml/point_of_
sale_customization.xml',
            'point_of_sale_customization/static/src/js/point_of_
sale_customization.js'
        ],
    },
```

5. Update the `point_of_sale_customization` module to apply the changes. After that, you will be able to see a **5%Discount** button above the calculator:

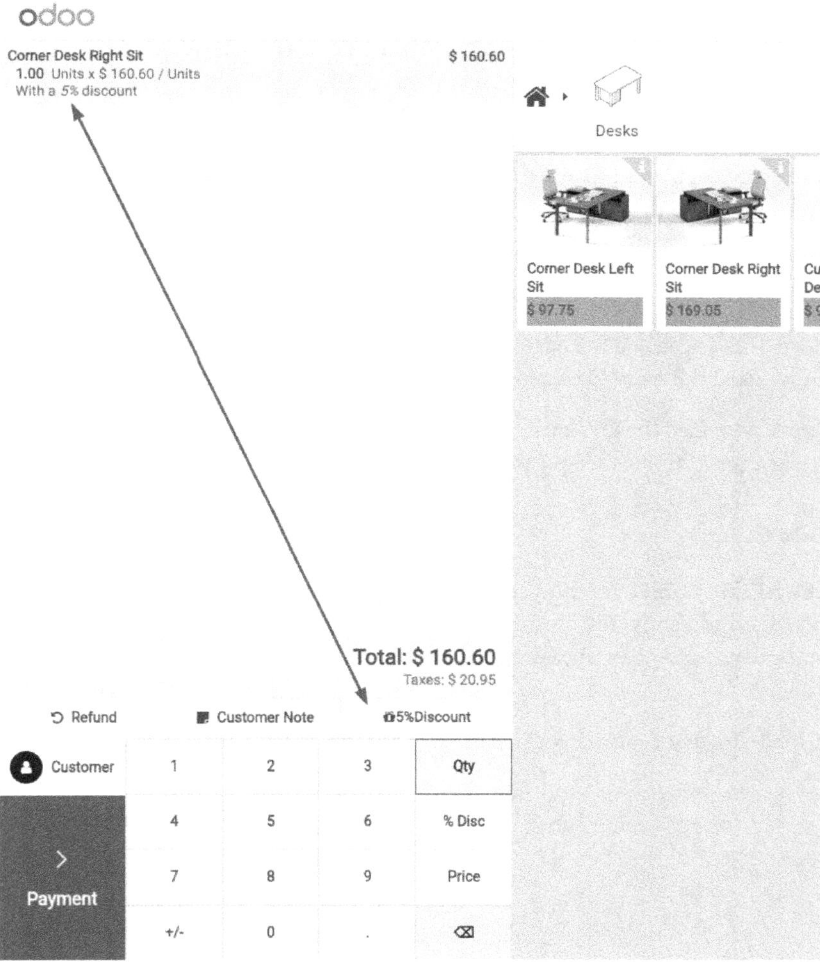

Figure 22.4 – The discount button

After clicking this, the discount will be applied to the selected order line.

How it works..

In Odoo v17, code based on the Odoo Point of Sale application is completely rewritten using the OWL framework. You can learn more about the OWL framework in *Chapter 16*, *The Odoo Web Library (OWL)*.

To create the action button in the Point of Sale application, you will need to *extend* `Component`. Now, `Component` is defined in `@odoo/owl namespace`, so to use it in your code, you will need to import it.

In *step 1*, we imported `Component` from `@odoo/owl`. Then, we created `PosDiscountButton` by extending `Component`. In *step 1*, we also imported `ProductScreen` from `@point_of_sale/app/screens/product_screen/product_screen` and `usePos` from `@point_of_sale/app/store/pos_hook`.

Now, `ProductScreen` is used to add a button to the Point of Sale screen via the `addControlButton` method.

`Component` has some built-in utilities that give access to useful information such as order details and the Point of Sale configuration. You can access it via the `this.pos = usePos()` variable.

In our example, we have accessed the current order information via the `this.pos.get_order()` method. Then, we used the `set_discount()` method to set a 5% discount.

In *step 2* and *step 3*, we added the OWL template, which will be rendered over the Point of Sale keyboard. If you wish to learn more about this, please refer to *Chapter 16, The Odoo Web Library (OWL)*.

There's more...

The `addControlButton()` method supports one more parameter, which is `condition`. This parameter is used to hide/show the button based on some condition. The value of this parameter is a function that returns a Boolean. Based on the returned value, the Point of Sale system will hide or show the button.

Take a look at the following example for more information:

```
ProductScreen.addControlButton({
    component: PosDiscountButton,
    condition: function () {
        return true;
    },
});
```

Making RPC calls

Though the Point of Sale application works offline, it is still possible to make RPC calls to the server. The RPC call can be used for any operation; you can use it for CRUD operations, or to perform an action on the server.

Now, we will make an RPC call to fetch information about a customer's last five orders.

Getting ready

Now, we will use the `point_of_sale_customization` module created for the **Adding an action** button in the keyboard recipe. We will define the action button. When the user clicks on the action button, we will make an RPC call to fetch the order information and display it on the popup.

How to do it...

Follow these steps to display the last five orders for the selected customer:

1. Add the following code to the `/static/src/js/point_of_sale_customization.js` file; this will add a new action button to fetch and display the information about the last five orders when a user clicks on the button:

```
/** @odoo-module */

import { Component } from "@odoo/owl";
import { ErrorPopup } from "@point_of_sale/app/errors/popups/
error_popup";
import { ProductScreen } from "@point_of_sale/app/screens/
product_screen/product_screen";
import { SelectionPopup } from "@point_of_sale/app/utils/input_
popups/selection_popup";
import { usePos } from "@point_of_sale/app/store/pos_hook";
import { useService } from "@web/core/utils/hooks";
import { sprintf } from "@web/core/utils/strings";

export class PosLastOrderButton extends Component {
   static template = "PosLastOrderButton";
   setup() {
        this.pos = usePos();
        this.popup = useService("popup");
   }
}
ProductScreen.addControlButton({
   component: PosLastOrderButton,
   condition: function () {
        return true;
   },
});
```

2. Add the `onClick` function to the `PosLastOrders` component to manage button clicks:

```
export class PosLastOrderButton extends Component {
    static template = "PosLastOrderButton";
```

```
    setup() {
        this.pos = usePos();
        this.popup = useService("popup");
    }
    async onClick() {
        var self = this;
        const order = this.pos.get_order();
        const client = order.get_partner();
        if (client) {
            var domain = [['partner_id', '=', client.id]];
            const orders = await this.pos.orm.call(
                "pos.order",
                "search_read",
                [],
                {
                    domain: domain,
                    fields: ['name', 'amount_total'],
                    limit:5
                }
            );
            if (orders.length > 0) {
                var order_list = orders.map((o) => {
                    return { 'label': sprintf("%s -TOTAL: %s",
o.name, o.amount_total) };
                });
                await this.popup.add(SelectionPopup, {
                    title: 'Last 5 orders',
                    list: order_list
                });
            } else {
                await this.popup.add(ErrorPopup, {
                    body: "No previous orders found"
                });
            }
        } else {
            await this.popup.add(ErrorPopup, {
                body: "No previous orders found"
            });
        }
    }
}
```

3. Add the QWeb template for the button to the `/static/src/xml/point_of_sale_customization.xml` file:

```
<t t-name="PosLastOrderButton">
        <span class="control-button btn btn-light rounded-0
fw-bolder"
                t-on-click="() => this.onClick()">
              <i class="fa fa-shopping-cart"></i>Making RPC calls
              <span></span>
              <span>Last Orders</span>
        </span>
   </t>
```

4. Update the `point_of_sale_customization` module to apply the changes. After that, you will be able to see the **Last orders** button above the keyboard panel. When this button is clicked, a popup will be displayed with the order information:

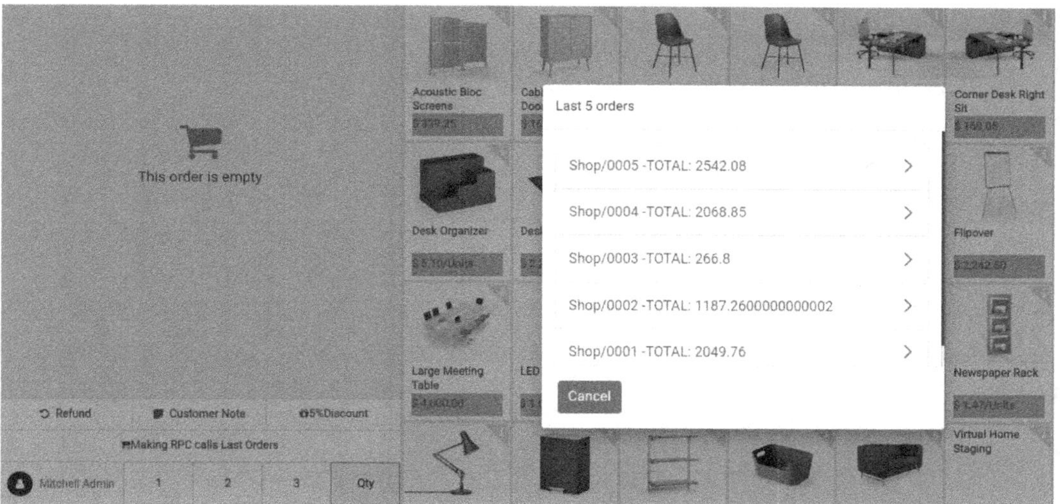

Figure 22.5 – The last five orders of a customer

If no previous orders are found, a warning will be displayed instead of an order list.

How it works...

In *step 1*, we created the action button. If you want to learn more about the action button, refer to the *Adding an action button to the keyboard* recipe in this chapter.

Before going into the technical details, let's understand what we wanted to accomplish with this action button. Once clicked, we want to display information for the last five orders for the selected customer. There will be a few cases where the customer is not selected, or customers have no previous orders. In such cases, we want to show a popup with an appropriate message. The RPC utility is available with the `this.pos.orm.call` attribute of the component.

In *step 2*, we added the click-handler function. On clicking the action button, the click-handler function will be called. This function will make the RPC call the server to fetch the order information.

We used the `this.pos.orm.call()` method to make RPC calls.

Then, we used the `search_read` method to fetch data through RPC. We passed the customer domain to filter the orders. We also passed `limit` keyword arguments to fetch only five orders. `this.pos.orm.call()` is an asynchronous method and returns a `Promise` object, so to handle the result, you can use the `await` keyword.

> **Note**
>
> The RPC call does not work in offline mode. If you have a good internet connection and you do not use offline mode frequently, you can use RPCs.
>
> Although the Odoo Point of Sale application works offline, a few operations, such as creating or updating a customer, require an internet connection, as those features use RPC to call internally.

We displayed the previous order information in the popup. We used `SelectionPopup`, which is used to display a selectable list; we used it to show the last five orders. We also used `ErrorPopup` to display a warning message when a customer is not selected or no previous orders were found.

In *step 3*, we added the QWeb template for the action button. The Point of Sale application will render this template to display the action button.

There's more...

There are plenty of other pop-up utilities. For example, `NumberPopup` is used to take a number input from the user. Refer to the files in the `@point_of_sale/app/utils/input_popups/number_popup` directory to see all these utilities. The `NumberPopup` module is probably a custom component or utility function to handle number input popups within a POS application. Depending on the context, this module could be responsible for displaying a pop-up dialog to input numerical data in a user-friendly way, such as for entering quantities or prices in a retail system. Use the following code to open a number popup:

```
import { NumberPopup } from "@point_of_sale/app/utils/input_popups/
number_popup";
this.popup.add(NumberPopup, { title: ("Set the new quantity")});
```

Modifying the Point of Sale screen UI

The UI of the Point of Sale application is written with the OWL QWeb template. In this recipe, we will learn how you can modify UI elements in the Point of Sale application.

Getting ready

In this recipe, we will use the `point_of_sale_customization` module created in the *Making RPC calls* recipe. We will modify the UI of the product card and display the profit margin per product.

How to do it...

Follow these steps to display the profit margin on the product card:

1. Add the following code to the `/models/pos_session.py` file to fetch the extra field for the product's actual price:

```
from odoo import models

class PosSession(models.Model):
    _inherit = 'pos.session'

    def _loader_params_product_product(self):
        result = super()._loader_params_product_product()
  result['search_params']['fields'].append('standard_price')
        return result
```

2. Add the following code to `/static/src/xml/point_of_sale_customization.xml` in order to display a profit margin product card:

```
<t t-name="ProductsWidget"
t-inherit="point_of_sale.ProductsWidget"
       t-inherit-mode="extension">
       <xpath expr="//ProductCard" position="attributes">
           <attribute name="standard_price">
               pos.env.utils.formatCurrency(product.get_display_
price() - product.standard_price)
           </attribute>
       </xpath>
   </t>
   <t t-name="ProductCard" t-inherit="point_of_sale.ProductCard"
       t-inherit-mode="extension">
       <xpath expr="//span[hasclass('price-tag')]"
position="after">
           <span t-if="props.standard_price"
```

```
                    class="sale_margin py-1 fw-bolder">
                  <t t-esc="props.standard_price"/>
              </span>
          </xpath>
      </t>
```

3. Add the following style sheet to style the margin text:

```
    .sale_margin {
    line-height: 21px;
    background: #CDDC39;
    padding: 0px 5px;
}
```

4. Update the `point_of_sale_customization` module to apply the changes. After that, you will be able to see the profit margin on the product card:

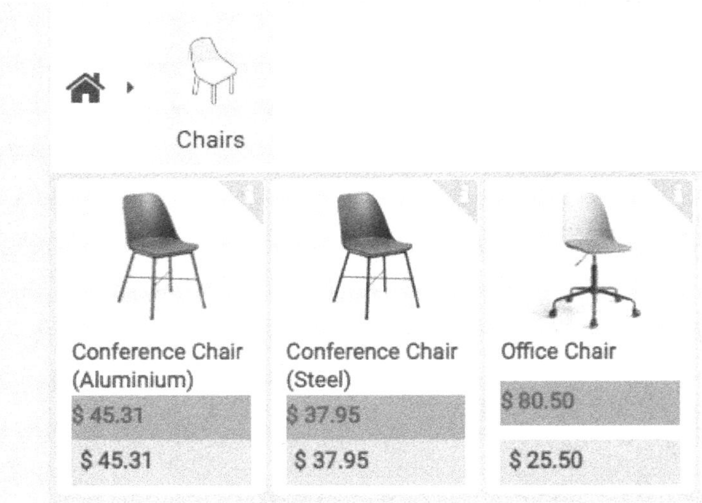

Figure 22.6 – The profit margins for products

If the product cost is not set on a product, then the product card will not display a profit margin, so make sure you set the product cost.

How it works...

In this recipe, we want to use the `standard_price` field as the purchase cost of the product. This field is not loaded by default in Point of Sale applications.

In *step 1*, we added the `standard_price` field for the `product.product` model. After this, the product data will have one more field – `standard_price`.

In *step 2*, we extended the default product card template. You will need to use the `t- inherit` attribute to extend the existing **QWeb** template.

Then, you need to use XPath to select the element on which you want to perform the operation. If you want to learn more about XPaths, refer to the *Changing existing views – view inheritance* recipe in *Chapter 9, Backend Views*.

To fetch the product sale price, we used the `product` properties sent from the parent OWL component. Then, we calculated the margin by using the product price and product cost. If you want to learn more about this, please refer to *Chapter 16, The Odoo Web Library (OWL)*.

In *step 3*, we added the style sheet to modify the position of the margin element. This will add a background color to the margin element and place it under the price pill.

Modifying existing business logic

In the previous recipes, we saw how to fetch data through an RPC and how to modify the UI of the Point of Sale application. In this recipe, we will see how you can modify or extend the existing business logic.

Getting ready

In this recipe, we will use the `point_of_sale_customization` module created in the *Modifying the Point of Sale screen UI* recipe, which is where we fetched the purchase price of a product and displayed the product margin. Now, in this recipe, we will show a warning to the user if they sell the product below the product margin.

How to do it...

Most of the business logic of the Point of Sale application is written in JavaScript, so we just need to make changes to it to achieve the goal of this recipe. Add the following code to `/static/src/js/point_of_sale_customization.js` to show a warning when the user sells a product below the purchase price:

```
/** @odoo-module */

import { ErrorPopup } from "@point_of_sale/app/errors/popups/error_
popup";
import { ProductScreen } from "@point_of_sale/app/screens/product_
screen/product_screen";
import { patch } from "@web/core/utils/patch";
patch(ProductScreen.prototype, {
    _setValue(val) {
```

```
            super._setValue(val);
            const orderline = this.currentOrder.get_selected_orderline();
            if (orderline && orderline.product.standard_price) {
                var price_unit = orderline.get_unit_price() * (1.0 -
    (orderline.get_discount() / 100.0));
                if (orderline.product.standard_price > price_unit) {
                    this.popup.add(ErrorPopup, {
                        title: 'Warning',
                        body: 'Product price set below cost of product.'
                    });
                }
            }
        }
    });
```

Update the `point_of_sale_customization` module to apply the changes. After the update, add the discount on the order line in such a way that the product price becomes less than the purchase price. A popup will appear with the following warning:

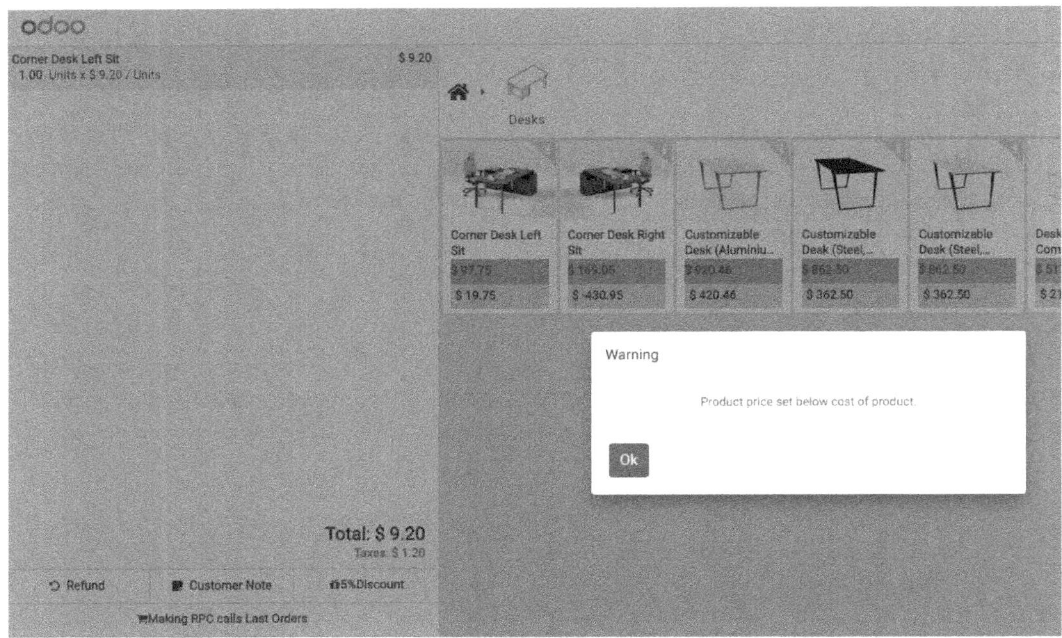

Figure 22.7– A warning on a big discount

Note that when you set the product price below the actual cost, a warning will be displayed, and it will continue to pop up every time you take an action, such as when you change the quantity for the product order.

How it works...

The Point of Sale component register provides an `extend` method to make changes to an existing function. Internally, it monkey-patches the actual component definition.

In our example, we modified the `_setValue()` method. The `_setValue()` method of `ProductScreen` is called whenever the user makes a change to the order line. We wanted to show a warning if the user sets the product price below the product cost. So, we defined a new `_setValue()` method and called the `super` method; this will make sure that whatever actions the user performs are applied. After the call to the `super` method, we wrote our logic, which checks whether the product sale price is higher than the actual cost of the product. If not, we then show a warning to the user.

> **Note**
>
> Using `super` can break things if it's not used carefully. If the method is inherited from several files, you must call the `super` method; otherwise, it will skip the logic in the subsequent inheritance. This sometimes leads to a broken internal data state.

We placed our business logic after the default implementation (`super`) is called. If you want to write business logic before the default implementation, you can do so by moving the `super` call to the end of the function.

Modifying customer receipts

When you customize a Point of Sale application, a common request you get from customers is to modify customer receipts. In this recipe, you will learn how to modify customer receipts.

Getting ready

In this recipe, we will use the `point_of_sale_customization` module created in the *Modifying existing business logic* recipe. We will add one line to the Point of Sale receipt to show how much money the customer saved in the order.

How to do it...

Follow these steps to modify a customer receipt in the Point of Sale application:

1. Add the following code to the `/static/src/js/point_of_sale_customization.js` file. This will add extra data to the receipt environment:

```
/** @odoo-module */

import { Order } from "@point_of_sale/app/store/models";
import { patch } from "@web/core/utils/patch";
patch(Order.prototype, {
   saved_amount(){
        const order = this;
        return order.orderlines.reduce((rem, line) => {
             var diffrence = (line.product.lst_price * line.
quantity) - line.get_base_price();
             return rem + diffrence;
        }, 0);
   },
   export_for_printing() {
        const json = super.export_for_printing(...arguments);
        var savedAmount = this.saved_amount();
        if (savedAmount > 0) {
             json.saved_amount = this.env.utils.
formatCurrency(savedAmount);
        }
        return json;
   }
})
```

2. Add the following code to `/static/src/xml/point_of_sale_customization.xml`. This will extend the default receipt template and add our customization:

```
<t t-name="OrderReceipt"
      t-inherit="point_of_sale.OrderReceipt"
      t-inherit-mode="extension">
        <xpath expr="//div[hasclass('pos-receipt')]//
div[hasclass('before-footer')]" position="before">
             <div style="text-align:center;" t-if="props.data.
saved_amount">
                  <br/>
                  <div >
                       You saved
                       <t t-esc="props.data.saved_amount"/>
```

```
                    on this order.
                </div>
            </div>
        </xpath>
    </t>
```

Update the `point_of_sale_customization` module to apply the changes. After that, add a product with the discount and check the receipt; you will see one extra line in the receipt:

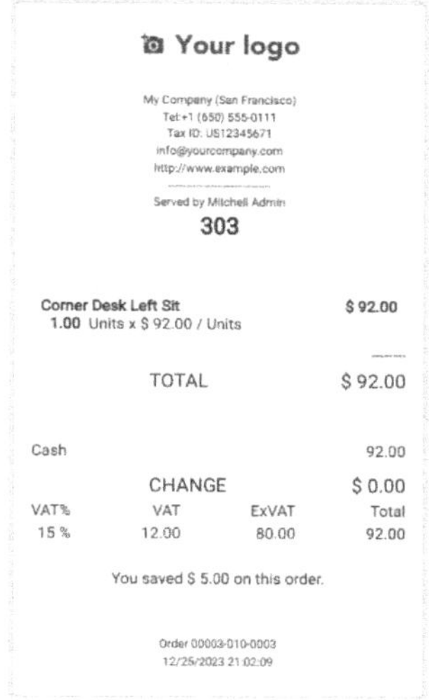

Figure 22.8 – The updated receipt

The receipt will not display the **Amount saved** screen if it is zero or negative.

How it works...

There is nothing new in this recipe. We just updated the receipt by using the previous recipes.

In *step 1*, we overrode the `export_for_printing()` function to send more data to the receipt environment. Whatever you are sending from the `export_for_printing()` method will be available in the QWeb template of the receipt. We compared the product's base price with the product price in the receipt to calculate how much money the customer saved. We sent this data to the receipt environment via the `saved_amount` key.

In *step 2*, we modified the default QWeb template of the receipt. The template name of the actual receipt is `OrderReceipt`, so we used it as a value in the `t-inherit` attribute. In *step 1*, we'd already sent the information needed to modify the receipt. In the QWeb template, we get the saved amount in the `props.data.saved_amount` key, so we just add one more `<div>` element before the footer. This will print the saved amount in the receipt. If you want to learn more about overriding, refer to the *Modifying the Point of Sale screen UI* recipe.

23

Managing Emails in Odoo

Email integration is the most prominent feature of Odoo. You can send and receive emails directly from the Odoo user interface. You can even manage email threads on business documents, such as leads, sales orders, and projects. In this chapter, we will explore a few important ways to deal with emails in Odoo.

Here, we'll cover the following recipes:

- Configuring incoming and outgoing email servers
- Managing chatter on documents
- Managing activities on documents
- Sending emails using the Jinja template
- Sending emails using the QWeb template
- Managing the email alias
- Logging user changes in a chatter
- Sending periodic digest emails

Technical requirements

All the code used in this chapter can be downloaded from `https://github.com/PacktPublishing/Odoo-17-Development-Cookbook-Fifth-Edition/tree/main/Chapter23`.

Configuring incoming and outgoing email servers

Before you start sending and receiving emails in Odoo, you will need to configure the incoming and outgoing email servers. In this recipe, you will learn how to configure email servers in Odoo.

Getting ready

There is no development needed for this recipe, but you will require email server information, such as the server URL, port, server type, username, and password. We will use this information to configure the email servers.

> **Note**
>
> If you are using **Odoo Online** or **Odoo.sh**, you do not need to configure the email servers. You can send and receive emails without any complex configurations on those platforms. This recipe is for on-premises Odoo instances.

How to do it...

Configuring incoming and outgoing email servers involves a few steps that are common to the processes for incoming and outgoing servers and a few steps that are unique to each kind of server. So, first, we will see the common configuration steps, and then we will configure the incoming and outgoing email servers individually. The following are the steps required for both incoming and outgoing email servers:

1. Open the **General Settings** form menu, at **Settings** | **General Settings**.
2. Go to the **Discuss** section and inside **Alias Domain**. This will display the following options:

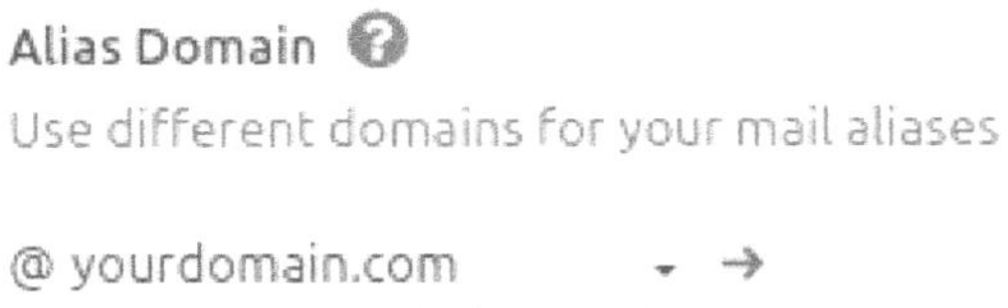

Figure 23.1 – Setting an alias domain

3. In the **Alias Domain** field, enter the domain name on which your email server is running. Then, save the configuration.

Configuring the incoming email server

Perform the following steps to configure the incoming email server:

1. Open **General Settings** and click on the **Incoming Email Servers** link under Technical | Email. This will redirect you to a list view of incoming email servers.

2. Click on the **Create** button, which will open the following form view. Enter the details of your incoming email server (see the *How it works…* section for an explanation of each field):

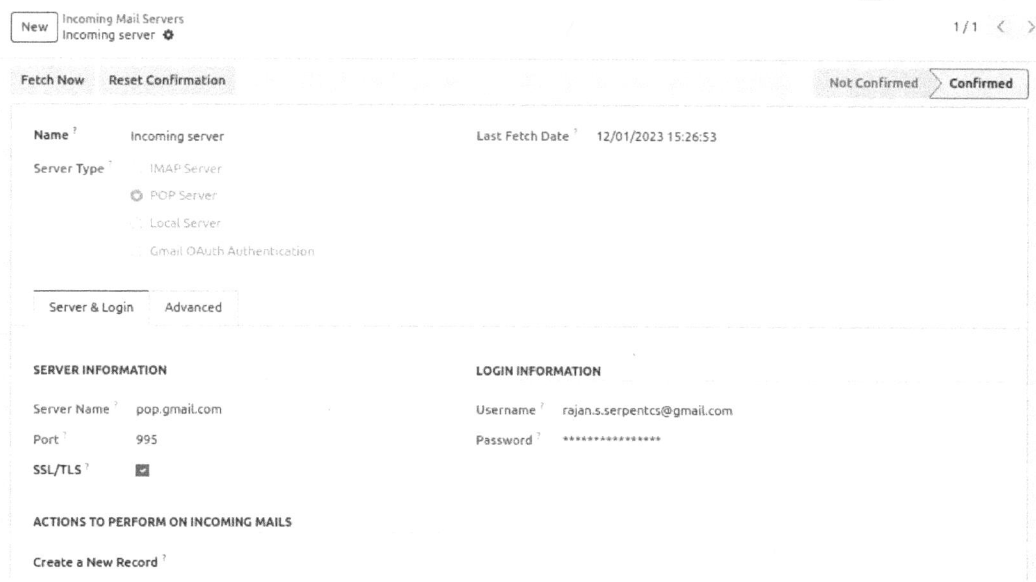

Figure 23.2 – Configuring the incoming email server

3. Click on the **Test & Confirm** button to verify your configuration. It will show an error message if you have wrongly configured the incoming email server.

Configuring the outgoing email server

Follow these steps to configure the outgoing email server:

1. Open **General Settings** and enable the **Custom Email Servers** option, then click on the **Outgoing Email Servers** link. This will redirect you to the list view of outgoing email servers.
2. Click on **Create**, which will open the following form view. Enter the details of your outgoing email server (see the *How it works…* section for an explanation of each field):

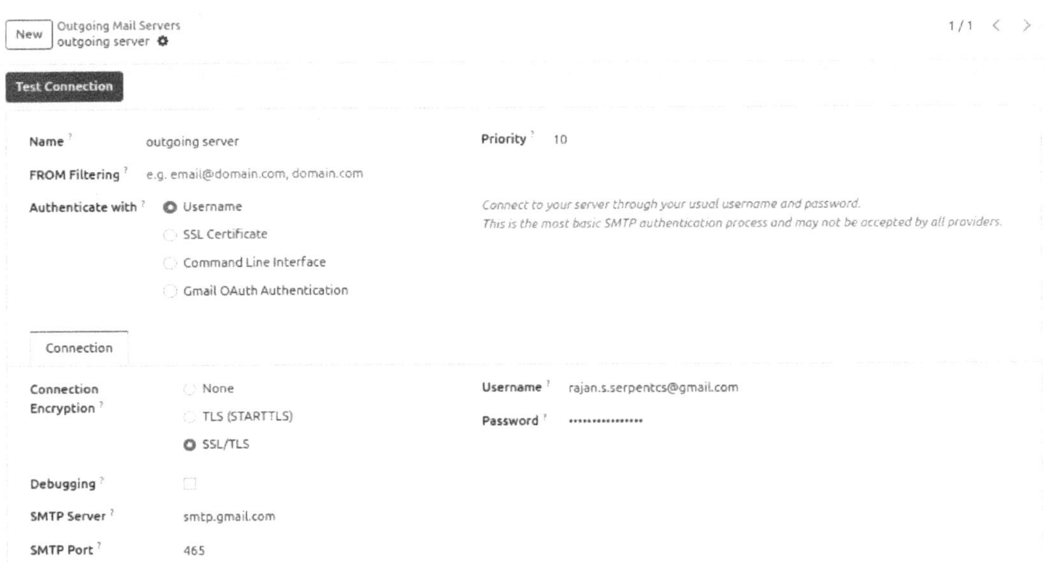

Figure 23.3 – Configuring the outgoing email server

3. Click on **Test Connection** at the bottom of the screen to verify your configuration. It will show an error message if you have wrongly configured the outgoing email server.

The outgoing email server will display the error dialog even if you have configured it properly. Look for a **Connection Test Successful!** message in the error dialog body. It means your outgoing server is configured correctly.

How it works...

The steps given in this recipe are self-explanatory and do not require further explanation. But the outgoing email and incoming email records have several fields, so let's see their purpose.

Here is a list of fields used to configure the incoming email server:

- **Name**: The name of the server, which helps you identify a specific incoming email server when you have configured multiple incoming email servers.
- **Server Type**: Here, you need to choose from three options: **POP Server**, **IMAP Server**, and **Local Server**. The value of this field will be based on your email service provider.
- **Server Name**: The domain of the server on which the service is running.
- **Port**: The number of the port on which the server is running.
- **SSL/TLS**: Check this field if you are using SSL/TLS encryption.
- **Username**: The email address for which you are fetching emails.
- **Password**: The password for the email address provided.

- **Active**: This field is used to enable or disable the incoming email server.
- **Keep Attachment**: Turn off this option if you do not want to manage attachments from incoming emails.
- **Keep Original**: Turn on this option if you want to keep the original email along with the preceding one.

The following is a list of fields used for configuring the outgoing email server:

- **Name**: The name of the server, which helps you identify a specific incoming email server when you have configured multiple incoming email servers.
- **Priority**: This field is used to define the priority of the outgoing email server. Lower numbers get higher priority, so email servers with a lower priority number will be used most.
- **SMTP Server**: The domain of the server on which the service is running.
- **SMTP Port**: The number of the port on which the server is running.
- **Connection Encryption**: The type of security used to send emails.
- **Username**: The email account used for sending emails.
- **Password**: The password for the email account provided.
- **Active**: This field is used to enable or disable the outgoing email server.

There's more...

By default, incoming emails are fetched every 5 minutes. If you want to change this interval, follow these steps:

1. Activate developer mode.
2. Open **Scheduled Actions** at **Settings** | **Technical** | **Automation** | **Scheduled Actions**.
3. Search for and open the scheduled action named **Mail: Fetchmail Service**.
4. Change the interval using the field labeled **Execute Every**.

Managing chatter on documents

In this recipe, you will learn how to manage chatter on your documents and add a communication thread to a record.

Getting ready

For this recipe, we will reuse the `my_hostel` module from *Chapter 8, Advanced Server-Side Development Techniques*. You can grab an initial copy of the module from the `Chapter23/ 00_initial_module` directory of the GitHub repository for this hostel room. In this recipe, we will add chatter to the `hostel.student` model.

How to do it...

Follow these steps to add chatter on the records of the `hostel.student` model:

1. Add the `mail` module dependency in the `__manifest__.py` file:

```
...
'depends': ['mail'],
...
```

2. Inherit `mail.thread` in the Python definition of the `hostel.student` model:

```
class HostelStudent(models.Model):
    _name = "hostel.student"
    _description = "Hostel Student Information"
    _inherit = ['mail.thread']
...
```

3. Add chatter widgets on the form view of the `hostel.student` model:

```
...
</sheet>
                    <div class="oe_chatter">
                        <field name="message_follower_ids" 
widget="mail_followers"/>
                        <field name="message_ids" widget="mail_
thread"/>
                    </div>
                </form>
...
```

4. Install the `my_hostel` module to see the changes in action:

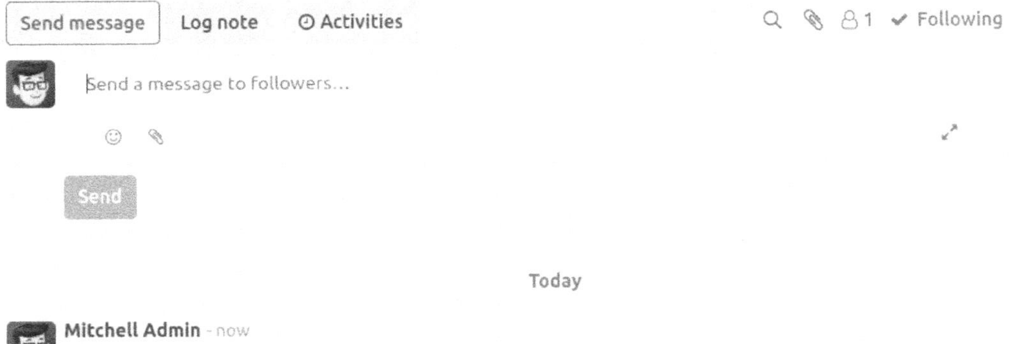

Figure 23.4 – Chatter on the hostel student form view

As shown in the preceding screenshot, after installing the module, you will be able to see chatter in the form view.

How it works...

In order to enable chatter on any model, you will need to install the `mail` module first. This is because all the code required to enable chatter or mailing capabilities is part of the `mail` module. That's why, in *step 1*, we added the `mail` module dependency in the manifest file of the `my_hostel` module. This will automatically install the `mail` module whenever you install the `my_hostel` module.

The fields and methods required to operate chatter are part of the `mail.thread` model. The `mail.thread` model is an abstract model and is just used for inheritance purposes. In *step 2*, we inherited the `mail.thread` model in the `hostel.student` model. This will add all the necessary fields and methods required for chatter in the `hostel.student` model. If you don't know how model inheritance works, refer to the *Using abstract models for reusable model features* recipe in *Chapter 4, Application Models*.

In *steps 1* and *2*, we added all the fields and methods required for chatter. The only remaining thing for chatter is adding a user interface in the form view. In *step 3*, we added a message thread and follower widget. You might be wondering about the `message_follower_ids` and `message_ids` fields. These fields are not added in the `hostel.student` model definition but they are added from the `mail.thread` model through inheritance.

There's more...

When you post messages in a chatter, emails will be sent to the followers. If you noticed in the example of this recipe, the room of the student is not the follower of the records, so they will not receive the messages. If you want to send an email notification to the student, you will need to add them to the student list. You can add the follower manually from the user interface, but if you want to add them automatically, you can use the `message_subscribe()` method. Take a look at the following code—when we assign a hostel room, the given code will automatically add the student to the list of followers:

```
@api.model
def create(self, values):
    result = super().create(values)
    partner_id = self.env['res.partner'].create({
        'name': result.name,
        'email': result.email
    })
    result.message_subscribe(partner_ids=[partner_id.id])
    return result
```

Similarly, if you want to remove followers from the list, you can use the `message_unsubscribe()` method.

Managing activities on documents

When using chatter, you can also add activities. These are used to plan your actions on the record. It is kind of a to-do list for each record. In this recipe, you will learn how to enable activities on any model.

Getting ready

For this recipe, we will be using the `my_hostel` module from the previous recipe, *Managing chatter on documents*. We will add activities to the `hostel.student` model.

How to do it...

Follow these steps to add activities to the `hostel.student` model:

1. Inherit `mail.activity.mixin` in the Python definition of the `hostel.student` model:

```
class HostelStudent(models.Model):
    _name = "hostel.student"
    _description = "Hostel Student Information"
    _inherit = ['mail.thread', 'mail.activity.mixin']
...
```

2. Add the `mail_activity` widget in the chatter of the `hostel.student` model:

```
...
<div class="oe_chatter">
                        <field name="message_follower_ids"
                        widget="mail_followers"/>
                        <field name="activity_ids" widget="mail_
                        activity"/>
                        <field name="message_ids" widget="mail_
                        thread"/>
                    </div>
...
```

3. Update the `my_hostel` module to apply the changes. This will display chatter activities:

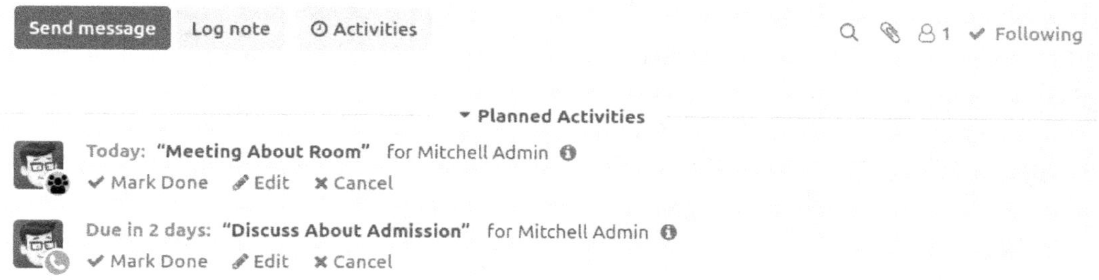

Figure 23.5 – Activity manager on the hostel student form view

This is how the user will be able to manage different chatter activities. Note that an activity scheduled by one user is visible to all other users too.

How it works...

Activities are part of the `mail` module, and you can optionally enable them in chatter. In order to enable activities on records, you need to inherit `mail.activity.mixin`. Similar to the `mail.thread` model, `mail.activity.mixin` is also an abstract model. Inheriting `mail.activity.mixin` will add all the necessary fields and methods in the module. These methods and fields are used to manage activities on records. In *step 1*, we added `mail.activity.mixin` into the `hostel.student` model. Because of this, the inheritance of `hostel.student` will get all the methods and fields required to manage activities.

In *step 2*, we added the `mail_activity` widget in the form view. This will display the UI for managing activities. The `activity_ids` field is added in the `hostel.student` model through inheritance.

Activities can be of different types. By default, you can create activities with types such as `Email`, `Call`, `Meeting`, and `To-Do`. If you want to add your own activity type, you can do it by going to **Settings** | **Technical** | **Discuss** | **Activity Types** in developer mode.

There's more...

If you want to schedule an activity automatically, you can use the `activity_schedule()` method of the `mail.activity.mixin` model. This will create an activity on a given discharge date. You can schedule the activity manually with the `activity_schedule()` method, as follows:

```
@api.model
    def create(self, values):
        result = super(HostelStudent, self).create(values)
```

```
            if result.discharge_date:
                result.activity_schedule('mail.mail_activity_data_call',
                                         date_deadline=result.discharge_date)
            return result
        return res
```

This example will schedule a call activity for the student whenever someone discharges a hostel. The deadline for the activity will be set as the discharge date of the hostel so that the rector can make a call to the student on that date.

Sending emails using the Jinja template

Odoo supports creating dynamic emails through Jinja templates. Jinja is a text-based templating engine used to generate dynamic HTML content. In this recipe, we will create a Jinja email template and then send emails with its help.

Getting ready

For this recipe, we will be using the `my_hostel` module from the previous recipe, *Managing activities on documents*. We will add the Jinja template to send an email to the student to tell them about the admission to the hostel.

How to do it...

Follow these steps to send a reminder email to the student:

1. Create a new file called `my_hostel/data/mail_template.xml` and add the email template:

```
<?xml version="1.0" encoding="utf-8"?>
<odoo noupdate="1">
   <record id="assign_room_to_student" model="mail.template">
       <field name="name">Assign Room To Student</field>
       <field name="model_id" ref="my_hostel.model_hostel_
student"/>
       <field name="email_from">{{ (object.room_id.create_uid.
email) }}</field>
       <field name="email_to">{{ (object.email) }}</field>
       <field name="subject">Assign Room</field>
       <field name="body_html" type="html">
           <div style="margin: 0px; padding: 0px;">
               <p style="margin: 0px; padding: 0px; font-size:
13px;">
                   Dear <t t-out="object.name"></t>,
                   <br/><br/>
```

```
                    <p>You have been assigned hostel
                        <b><t t-out="object.hostel_id.name"></
t></b> and room no <t t-out="object.room_id.room_no"></t>.
                    <br/>
                    Your admission date in a hostel is <b
style="color:red;"><t t-out="format_date(object.admission_
date)"></t>.</b>
                    </p>
                    <br/>

                    <p>Best regards,
                    <br/><t t-out="object.hostel_id.name"></t></
p>
                </p>
            </div>
        </field>
    </record>
</odoo>
```

2. Register the template file in the manifest file:

```
...
"data": [
        "security/hostel_security.xml",
        "security/ir.model.access.csv",
        "data/categ_data.xml",
        "data/mail_template.xml",
        "views/hostel.xml",
        "views/hostel_room.xml",
        "views/hostel_amenities.xml",
        "views/hostel_student.xml",
        "views/hostel_categ.xml",
        "views/hostel_room_category_view.xml",
   ],
...
```

3. Add a **Send Email For Assign Room** button in the form view of the `hostel.student` model to send the email:

```
...
<header>
    <button name="send_mail_assign_room"
            string="Send Email For Assign Room"
            type="object"/>
    <button name="action_assign_room"
            string="Assign Room"
```

```
            type="object"
            class="btn-primary"/>
    <field name="status" widget="statusbar"
options="{'clickable':           '1'}"/>
</header>
...
```

4. Add the `send_mail_assign_room()` method to the `hostel.student` model:

```
...
def send_mail_assign_room(self):
     self.message_post_with_source('my_hostel.assign_room_to_
student')
```

Update the `my_hostel` module to apply the changes. This will add a **Send Email For Assign Room** button in the form view of the `hostel.student` model. When they click on the button, followers will get this message:

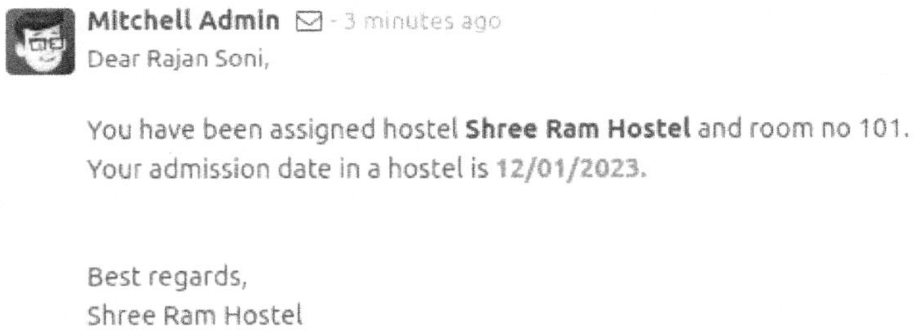

Mitchell Admin - 3 minutes ago
Dear Rajan Soni,

You have been assigned hostel **Shree Ram Hostel** and room no 101.
Your admission date in a hostel is **12/01/2023.**

Best regards,
Shree Ram Hostel

Figure 23.6 – Email sent via a Jinja template

The procedure shown in this recipe is useful when you want to send updates to your customers through emails. Because of the Jinja template, you can send emails dynamically based on individual records.

How it works...

In *step 1*, we created an email template using Jinja. Jinja templates help us generate a dynamic email based on record data. The email template is stored in the `mail.template` model. Let's see a list of fields you will need to pass in order to create a Jinja email template:

- `name`: The name of the template that is used to identify a specific template.
- `email_from`: The value of this field will be the email address from which this email is sent.
- `email_to`: The value of this field will be the email address of the recipient.
- `email_cc`: The value of this field will be used for the email address to send a copy of the email.
- `subject`: This field contains the subject of the email.
- `model_id`: This field contains the reference of the model. The email template will be rendered with the data of this model.

- `body_html`: This field will contain the body of the email template. It is a Jinja template, so you can use variables, loops, conditions, and so on. If you want to learn more about Jinja templates, go to `http://jinja.pocoo.org/docs/2.10/`. Usually, we wrap the content in the `CDATA` tag so that the content in the body is considered as character data and not as markup.
- `auto_delete`: This is a Boolean field that deletes an email once the email is sent. The default value of this field is `False`.
- `lang`: This field is used to translate the email template into another language.
- `scheduled_date`: This field is used to schedule emails in the future.

> **Information**
>
> You can use `${}` in the `email_form`, `email_to`, `email_cc`, `subject`, `scheduled_date`, and `lang` fields. This helps you to set values dynamically. Take a look at *step 1* in our recipe—we used `{{ (object.email) }}` to set the `email_to` field dynamically.

If you look closely at the content of the `body_html` field, you will notice we used `<t t-out="object.name">`. Here, the object is the recordset of the `hostel.student` model. During the rendering,`<t t-out="object.hostel_id.name"></t>` will be replaced with the hostel name. As well as `object`, some other helper functions and variables are passed in the rendering context. Here is a list of helpers passed to the renderer context:

- `object`: This variable will contain the recordset of the model, which is set in the template by the `model_id` field
- `format_date`: This is a reference to the method used to format date-time objects
- `format_datetime`: This is a reference to the method used to convert the UTC date and time into the date and time for another time zone
- `format_amount`: This is a reference to the method used to convert `float` into `string` with the currency symbol
- `format_duration`: This method is used to convert `float` into `time`—for instance, to convert 1.5 to 01:30
- `user`: This will be the recordset of the current user
- `ctx`: This will contain the dictionary of the environment context

> **Note**
>
> If you want to see the list of templates, activate developer mode, and open the **Settings | Technical | Email | Templates** menu. The form view of the template also provides a button to preview the rendered template.

In *step 2*, we registered the template file in the manifest file.

In *step 3*, we added a button in the form view to invoke the `send_mail_assign_room()` method, which will send the email to the followers.

In *step 4*, we added the `send_mail_assign_room()` method, which will be invoked by clicking the button. The `message_post_with_source()` method is used to send the email. The `message_post_with_source()` method is inherited in the model through `mail.thread` inheritance. To send the email, you just need to pass the template ID as the parameter.

There's more...

The `message_post_with_source()` method is used to send emails with the Jinja template. If you just want to send an email with plain text, you can use the `message_post()` method:

```
self.message_post(body="Your hostel admission process is completed.")
```

The preceding code will add a **Your hostel admission process is completed** message in the chatter. All followers will be notified with this message. If you just want to log the message, call the method with the `subtype_id` parameter.

Sending emails using the QWeb template

In the previous recipe, we learned how to send emails using the Jinja template. In this recipe, we will see another way to send dynamic emails. We will send emails with the help of the QWeb template.

Getting ready

For this recipe, we will use the `my_hostel` module from the previous recipe, *Sending emails using the Jinja template*. We will use the QWeb template to send an email to the student informing them that their admission was completed in the hostel.

How to do it...

Follow these steps to send a reminder email to the student:

1. Add the QWeb template into the `my_hostel/data/mail_template.xml` file:

```
<template id="assign_room_to_student_qweb">
    <p>Dear <span t-field="object.name"/>,</p>
    <br/>
    <p>You have been assigned hostel
        <b>
            <span t-field="object.hostel_id.name"/>
        </b> and room no <span t-field="object.room_id.room_
no"/>.
```

```
            <br/>
            Your admission date in a hostel is
            <b style="color:red;">
                <span t-field="object.admission_date"/>.
            </b>
        </p>
        <br/>

        <p>Best regards,
            <br/>
            <span t-field="object.hostel_id.name"/>

        </p>
    </template>
```

2. Add a **Send Email For Assign Room (QWeb)** button in the form view of the `hostel.student` model to send the email:

```
...
<header>
     <button name="send_mail_assign_room"
             string="Send Email For Assign Room"
             type="object"/>
     <button name="send_mail_assign_room_qweb"
             string="Send Email For Assign Room (QWeb)"
             type="object"/>
     <button name="action_assign_room"
             string="Assign Room"
             type="object"
             class="btn-primary"/>
     <field name="status" widget="statusbar"
     options="{'clickable': '1'}"/>
</header>
...
```

3. Add the `send_mail_assign_room_qweb()` method in the `hostel.student` model:

```
...
def send_mail_assign_room_qweb(self):
   self.message_post_with_source('my_hostel.assign_room_to_
student_qweb')
```

4. Update the `my_hostel` module to apply the changes. This will add a **Send Email For Assign Room (QWeb)** button in the form view of the `hostel.student` model. When the button is clicked, followers will get a message like this:

Mitchell Admin - now
Dear Rajan Soni,

You have been assigned hostel **Shree Ram Hostel** and room no 101.
Your admission date in a hostel is **12/01/2023**.

Best regards,
Shree Ram Hostel

Figure 23.7 – Email sent via the QWeb template

The procedure shown in this recipe works exactly like the previous recipe, *Sending emails using the Jinja template*. The only difference is the template type, as this recipe uses QWeb templates.

How it works...

In *step 1*, we created a QWeb template with the `send_mail_assign_room_qweb` ID. If you look in the template, you'll see we are not using the `format_date()` data field method anymore. This is because the QWeb rendering engine handles this automatically and displays the date based on the user's language. For the same reason, you are not required to use the `format_amount()` method to display currency symbols. The QWeb rendering engine will manage this automatically. If you want to learn more about QWeb templates, refer to the *Creating or modifying templates* recipe in *Chapter 14, CMS Website Development*.

In *step 2*, we added a button in the form view to invoke the `send_mail_assign_room_qweb()` method, which sends the email to the followers.

In *step 3*, we added the `send_mail_assign_room_qweb()` method, which will be invoked by a button click. The `message_post_with_source()` method is used to send the email. The `message_post_with_source()` method is inherited in the model through `mail.thread` inheritance. To send the email, you just need to pass the web template's XML ID as the parameter.

Sending emails with the QWeb template works exactly the same as in the previous recipe, but there are some subtle differences between the QWeb email template and the Jinja email template. Here is a quick comparison between both templates:

- There is no simple way to send extra parameters in the email templates. You have to use a recordset in the `object` variable to fetch dynamic data. On the other hand, with QWeb email templates, you can pass extra values in the renderer context through the `values` parameter:

```
self.message_post_with_source('my_hostel.assign_room_to_student_
qweb',
                                        values={'extra_data':
'test'})
```

- To manage the date format, time zone, and amount with currency symbols, in the Jinja template, you have to use the `format_date`, `format_tz`, and `format_amount` functions, while in QWeb templates, it is managed automatically.
- It is not possible to modify an existing template for other modules in Jinja, whereas in QWeb templates, you can modify the email template through inheritance. If you want to learn more about QWeb inheritance, refer to the *Creating or modifying templates* recipe in *Chapter 14, CMS Website Development*.
- You can select and use a Jinja template directly from the message composer. In the following screenshot, the drop-down menu in the bottom-right corner is used to select a Jinja template:

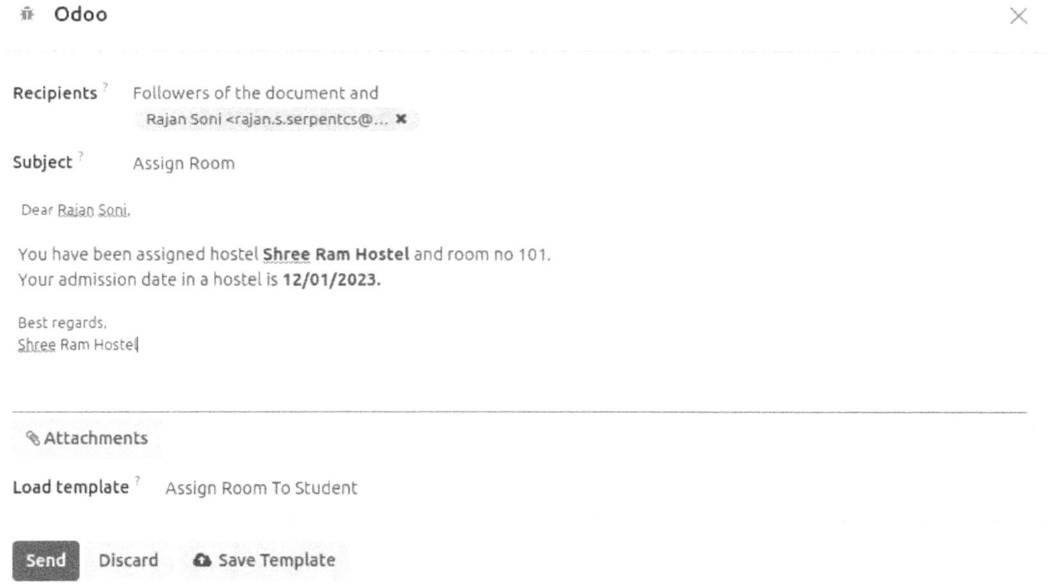

Figure 23.8 – Template selection option

- Using QWeb, selecting a template directly from the message composer is not an option.

There's more...

All methods (`message_post` and `message_post_with_source`) respect the user's preference. If the user changes the notification-management option from the user preferences, the user will not receive emails; instead, they will receive notifications in Odoo's UI. This is the same for customers; if a customer opts out of emails, they will not receive any updates through email.

Additionally, the Odoo message thread follows a concept called **subtypes**. Subtypes are used to receive emails only for information you are interested in. You can pass an extra parameter, `subtype_id`, in `message_post_*` methods to send emails based on the subtype. Usually, the user will manage their subtypes from the dropdown of the **Follow** button. Let's suppose the user has set their subtypes as follows:

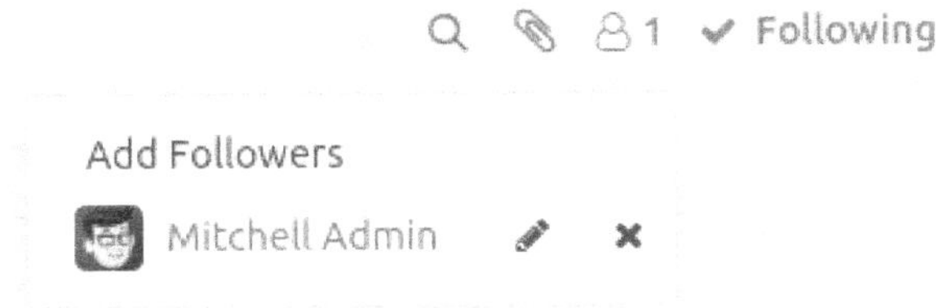

Figure 23.9 – Option to edit subtype

Based on the user's preference, the user will only get emails for **Discussions** messages.

Managing the email alias

Email aliasing is a feature in Odoo that is used to create a record through incoming emails. The simplest example of an email alias is sales teams. You just need to send an email to `sale@yourdomain.com`, and Odoo will create a new record for `crm.lead` in the sales team. In this recipe, we will create one email alias to create a hostel student record.

Getting ready

For this recipe, we will be using the `my_hostel` module from the previous recipe, *Sending emails using the QWeb template*. We will create our email alias with the `hostelstudent@yourdomain.com` email address. If you send an email to this email address with the book's name in the subject, a record is created in the `hostel.student` model.

How to do it...

Follow these steps to add an email alias for the `hostel.student` model:

1. Add the email alias data in the `my_hostel/data/mail_template.xml` file:

```
<record id="mail_alias_room_assign" model="mail.alias">
        <field name="alias_name">room</field>
        <field name="alias_model_id" ref="model_hostel_student"/>
        <field name="alias_contact">partners</field>
    </record>
```

2. Add the following imports in the my_hostel/models/hostel_student.py file:

```
import re
from odoo.tools import email_split, email_escape_char
```

3. Override the message_new() method in the hostel.student model:

```
    @api.model
    def message_new(self, msg_dict, custom_values=None):
        self = self.with_context(default_user_id=False)
        if custom_values is None:
            custom_values = {}
        custom_values['name'] = re.match(r"(.+?)\s*<(.+?)>", msg_
dict.get('from')).group(1)
        custom_values['email'] = email_escape_char(email_
split(msg_dict.get('from'))[0])
        return super(HostelStudent, self).message_new(msg_dict,
custom_values)
```

Update the my_hostel module to apply the changes. Then, send an email to hostelstudent@yourdomain.com. This will create a new hostel.student record, and it will be displayed as follows:

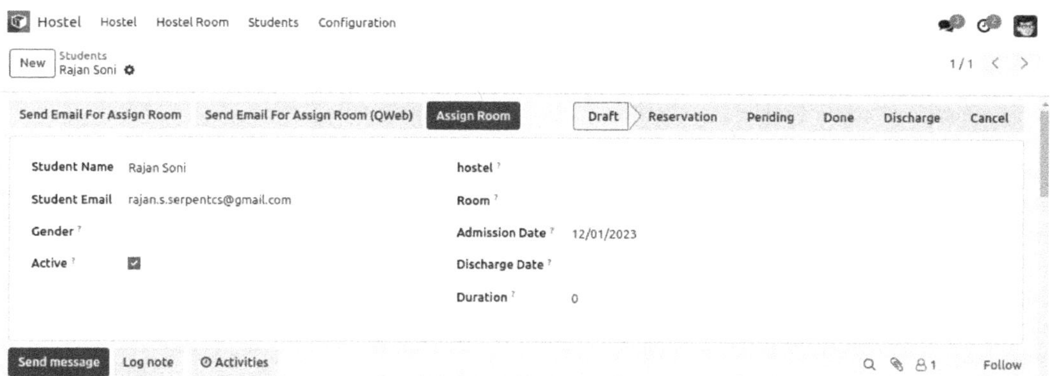

Figure 23.10 – Record generated via email

Whenever you send an email to hostelstudent@yourdomain.com, Odoo will generate a new student record.

How it works...

In *step 1*, we created a mail.alias record. This alias will handle the hostelstudent@yourdomain.com email address. When you send an email to this address, Odoo will create a new record in the `hostel.student` model. If you want to see the list of active aliases in the system, open **Settings** | **Technical** | **Email** | **Aliases**. Here is a list of fields available to configure the alias:

- `alias_name`: This field holds the local part of the email address; for example, the `hostelstudent` part in `hostelstudent@yourdomain.com` is the local part of the email address.
- `alias_model_id`: The model reference on which a record should be created for the incoming email.
- `alias_contact`: This field holds the security preferences for the alias. Possible options are `everyone`, `partners`, `followers`, and `employees`.
- `alias_defaults`: When an incoming email is received, its record is created in the model specified on the alias. If you want to set default values in the record, give the values in the form of a dictionary in this field.

In *step 2*, we added the necessary imports. In *step 3*, we overrode the `message_new()` method. This method is invoked automatically when a new email is received on the alias email address. This method will take two parameters:

- `msg_dict`: This parameter will be the dictionary that contains information about the received email. It contains email information such as the sender's email address, the receiver's email address, the email subject, and the email body.
- `custom_values`: This is a custom value used to create a new record. This is the same value you set on the alias record using the `alias_defaults` field.

In our recipe, we overrode the `message_new()` method and fetched the name from the email through a regular expression. Then, we fetched the email address of the sender with the help of the tools we imported in *step 2*. We used the sender's email address to create a student record. Then, we updated `custom_values` with these two values: `name` and `email`. We passed this updated `custom_values` data to the `super()` method, which created a new `hostel.student` record with the given `name` and `email` values. This is how a record is created when you send an email to the alias.

There's more...

Some business models have a requirement that means you need a separate alias per record. For example, the sales team model has separate aliases for each team, such as `sale-in@example.com` for Team India and `sale-be@example.com` for Team Belgium. If you want to manage such aliases in your model, you can use `mail.alias.mixin`. In order to use it in your model, you will need to inherit the mixin:

```
class Team(models.Model):
    _name = 'crm.team'
    _inherit = ['mail.alias.mixin', 'mail.thread']
```

After inheriting the mixin, you will need to add the `alias_name` field into the form view so that end users can add aliases by themselves.

Logging user changes in a chatter

The Odoo framework provides a built-in facility to log field changes in a chatter. In this recipe, we will enable logging on some of the fields so that when changes are made in them, Odoo will add logs in the chatter.

Getting ready

For this recipe, we will be using the `my_hostel` module from the previous recipe, *Managing the email alias*. In this recipe, we will log changes from a few fields in the `hostel.student` model.

How to do it...

Modify the definitions of the fields, to enable logs for the fields when you change them. This is shown in the following code snippet:

```
class HostelStudent(models.Model):
    _name = "hostel.student"
    _description = "Hostel Student Information"
    _inherit = ['mail.thread', 'mail.activity.mixin']

    name = fields.Char("Student Name")
    email = fields.Char("Student Email")
    gender = fields.Selection([("male", "Male"),
        ("female", "Female"), ("other", "Other")],
        string="Gender", help="Student gender")
    active = fields.Boolean("Active", default=True,
        help="Activate/Deactivate hostel record")
    hostel_id = fields.Many2one("hostel.hostel", "hostel", help="Name
of hostel")
    room_id = fields.Many2one("hostel.room", "Room",
        help="Select hostel room")
    status = fields.Selection([("draft", "Draft"),
        ("reservation", "Reservation"), ("pending", "Pending"),
        ("paid", "Done"),("discharge", "Discharge"), ("cancel",
"Cancel")],
        string="Status", copy=False, default="draft",
        help="State of the student hostel")
    admission_date = fields.Date("Admission Date",
        help="Date of admission in hostel",
        default=fields.Datetime.today,
```

```
        tracking=True)
    discharge_date = fields.Date("Discharge Date",
        help="Date on which student discharge",
        tracking=True)
    duration = fields.Integer("Duration", compute="_compute_check_
duration", inverse="_inverse_duration",
                              help="Enter duration of living")
```

Update the `my_hostel` module to apply the changes. Create a new record in the `hostel.student` model, make some changes in the fields, and then admission and discharge the hostel. If you check the chatter, you will see the following logs:

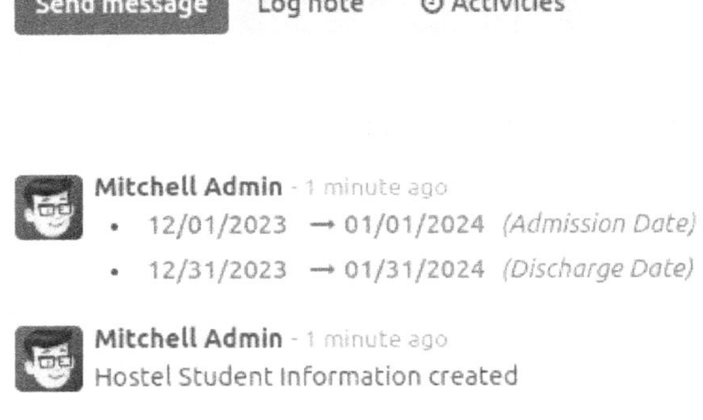

Figure 23.11 – Changelogs in the chatter

Whenever you make changes to `state`, `admission_date`, or `discharge_date`, you will see a new log in the chatter. This will help you to see the full history of the record.

How it works...

By adding the `tracking=True` attribute on the field, you can enable logging for that field. When you set the `tracking=True` attribute, Odoo will add a log that changes in the chatter whenever you update the field value. If you enable tracking on multiple records and you want to provide a sequence in the tracking values, you can also pass a number in the tracking parameter like this: `tracking=20`. When you pass `tracking=True`, then the default sequence is used, which is `100`.

In our recipe, we added `tracking=True` on the `state`, `admission_date`, and `discharge_date` fields. This means Odoo will log changes when you update the values of the `admission_date`, `discharge_date`, or `state` fields. Take a look at the screenshot in the *How to do it...* section; we have only changed the `admission_date` and `discharge_date` fields.

Note that the `track_visibility` feature only works if your model inherits the `mail.thread` model because the code-related chatter and logs are part of the `mail.thread` model.

Sending periodic digest emails

The Odoo framework has built-in support for sending out periodic digest emails. With digest emails, you can send an email with information about business KPIs. In this recipe, we will send data about the hostel room to the rector (or any other authorized person).

Getting ready

For this recipe, we will be using the `my_hostel` module from the previous recipe, *Logging user changes in a chatter*.

How to do it...

Follow these steps to generate digest emails for room rent records:

1. Inherit the `digest.digest` model and add fields for the KPIs:

```
class Digest(models.Model):
    _inherit = 'digest.digest'

    kpi_room_rent = fields.Boolean('Room Rent')
    kpi_room_rent_value = fields.Integer(compute='_compute_kpi_
room_rent_value')

    def _compute_kpi_room_rent_value(self):
        for record in self:
            start, end, company = record._get_kpi_compute_
parameters()
            record.kpi_room_rent_value = self.env['hostel.room'].
search_count([
                ('create_date', '>=', start),
                ('create_date', '<', end)
            ])
```

2. Inherit the `digest.digest` model's form view and add the KPI fields:

```
<?xml version='1.0' encoding='utf-8'?>
<odoo>
    <record id="digest_digest_view_form" model="ir.ui.view">
        <field name="name">digest.digest.view.form.inherit.
hostel</field>
        <field name="model">digest.digest</field>
        <field name="inherit_id" ref="digest.digest_digest_view_
form"/>
        <field name="arch" type="xml">
            <xpath expr="//group[@name='kpi_general']"
```

```
position="after">
                <group name="kpi_hostel" string="Hostel">
                    <field name="kpi_room_rent"/>
                </group>
            </xpath>
        </field>
    </record>
</odoo>
```

Update the module to apply the changes. Once you update the module, enable developer mode and open **Settings** | **Technical** | **Emails** | **Digest Emails**, as seen in the following screenshot:

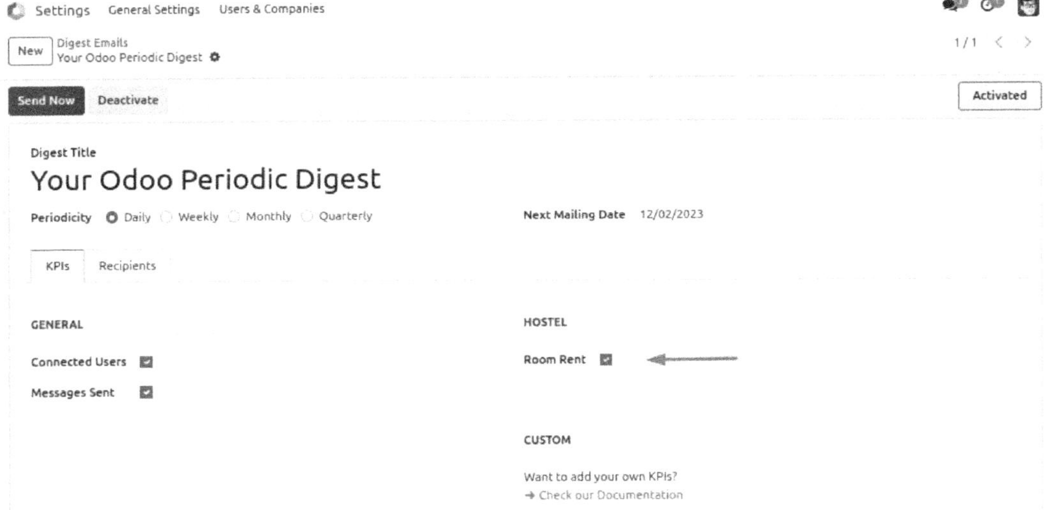

Figure 23.12 – Enabling the digest email for room rent data

Once you enable this and if you have subscribed to digest emails, you will start receiving digest emails.

How it works...

In order to build a customized digest email, you need two fields. The first field will be a `Boolean` field, used to enable and disable the KPI, while the second field will be a `compute` field and will be called to acquire the KPI value. We created both of the fields in *step 1*. If you check the definition of the `compute` field, it uses the `_get_kpi_compute_parameters` method. This method returns three parameters: a start date, an end date, and the company record. You can use these parameters to generate a value for your KPI. We have returned the number of rooms rented during a particular period of time. If your KPI is multi-website compatible, then you can use a `company` parameter.

In *step 2*, we added a field to the digest form view. This field is used to enable/disable digest emails. When you enable it, you will start receiving digest emails:

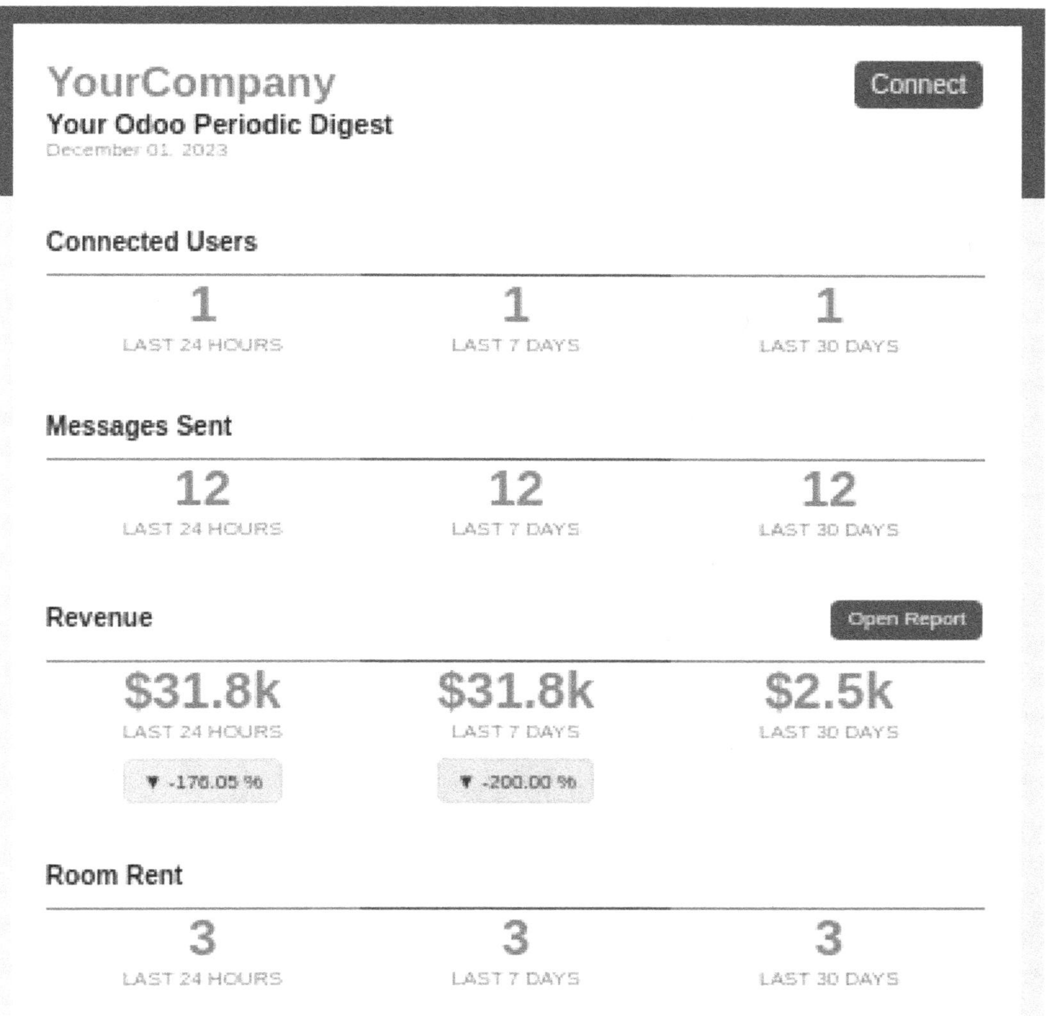

Figure 23.13 – Digest email for room rent records

Enable developer mode, then open **Settings** | **Technical** | **Emails** | **Digest Emails.** Here, you can configure the recipients of digest emails and set the periodicity for digest emails. You can also enable/disable digest emails from here.

24
Managing the IoT Box

Odoo provides support for the **Internet of Things** (**IoT**). The IoT is a network of devices/sensors that exchange data over the internet. By connecting such devices with a system, you can use them. For instance, by connecting a printer with Odoo, you can send PDF reports directly to the printer. Odoo uses a piece of hardware called the **IoT Box**, which is used to connect devices such as printers, calipers, payment devices, footswitches, and more. In this chapter, you will learn how to set up and configure the IoT Box. Here, we'll cover the following recipes:

- Flashing the IoT Box image for Raspberry Pi
- Connecting the IoT Box with a network
- Adding the IoT Box to Odoo
- Loading drivers and listing connected devices
- Taking input from devices
- Accessing the IoT Box through SSH
- Configuring a **point of sale** (**POS**)
- Sending PDF reports directly to a printer

Note that the goal of this chapter is to install and configure the IoT Box. Developing hardware drivers is outside the scope of this book. If you want to learn about the IoT Box in more depth, explore the `iot` module in the Enterprise Edition.

Technical requirements

The IoT Box is a **Raspberry Pi**-based device. The recipes in this chapter are based on the **Raspberry Pi 3 Model B+**, available at `https://www.raspberrypi.org/products/raspberry-pi-3-model-b-plus/`. The IoT Box is part of the Enterprise Edition, so you will need to use the Enterprise Edition to follow the recipes in this chapter.

All code used in this chapter can be downloaded from the following GitHub repository: `https://github.com/PacktPublishing/Odoo-17-Development-Cookbook-Fifth-Edition/tree/main/Chapter24`.

Flashing the IoT Box image for Raspberry Pi

In this recipe, you will learn how to flash a **microSD** card with an image of the IoT Box. Note that this recipe is only for those who have purchased the **blank Raspberry Pi**. If you have purchased the official IoT Box from Odoo, you can skip this recipe as it is preloaded with the IoT Box image.

Getting ready

Raspberry Pi 3 Model B+ uses a microSD card, so we have used a microSD card for this recipe. You will need to connect a microSD card to your computer.

How to do it...

Perform the following steps to install an IoT Box image onto your SD card:

1. Insert a microSD card into your computer (use an adapter if your computer doesn't have a dedicated slot).
2. Download the IoT Box image from Odoo's nightly builds. The image is available at `https://nightly.odoo.com/master/iotbox/`.
3. Download and install **balenaEtcher** on your computer. You can download this from `https://www.balena.io/etcher/`.
4. Open balenaEtcher, select the IoT Box image (we are using version 23.09 of the IoT Box image), and choose to flash your microSD card. You'll see the following screen:

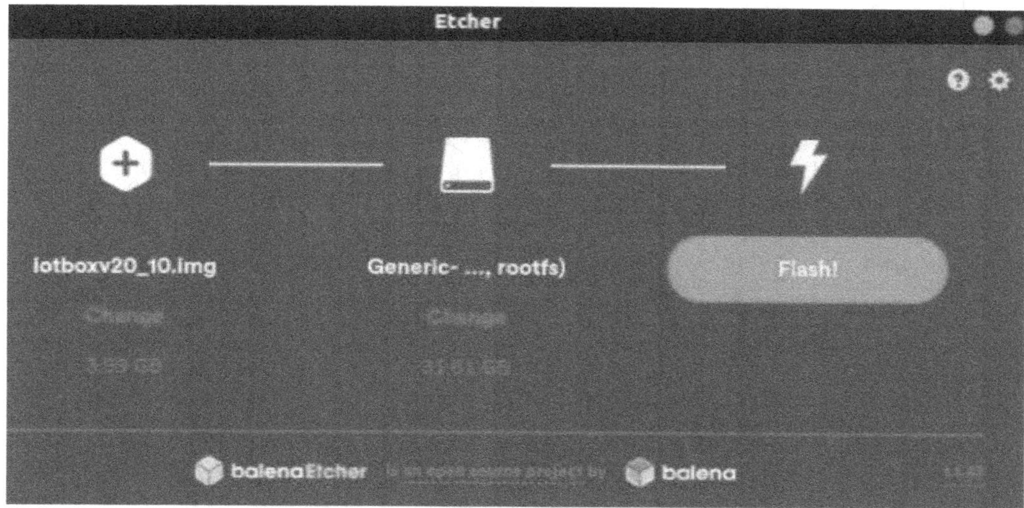

Figure 24.1 – Flashing the SD card with the IoT Box image

5. Click on the **Flash!** button and wait until the process completes.
6. Remove the microSD card and place it in the Raspberry Pi.

After these steps, your microSD card should be loaded with the IoT Box image and ready to be used in the IoT Box.

How it works...

In this recipe, we have installed the IoT Box image on a microSD card. In the second step, we downloaded the IoT Box image from the Odoo nightly builds. On the nightly page, you can find different images for the IoT Box. You need to choose the latest image from the Odoo nightly builds. When writing this book, we used the latest image, which was `iotboxv23_11.zip`. The Odoo IoT Box image is based on the Raspbian Stretch Lite OS, and the image is loaded with the libraries and modules required to integrate the IoT Box with the Odoo instance.

In *step 3*, we downloaded the balenaEtcher utility tool to flash the microSD card.

> **Note**
> In this recipe, we used balenaEtcher to flash the microSD card, but you can use any other tools to flash the microSD card.

In *step 4*, we flashed the microSD card with the IoT Box image. Note that this process can take several minutes. On completion of the process, the microSD card will be ready to be used.

Perform the following steps if you want to verify whether the image was flashed successfully:

1. Mount the microSD card into Raspberry Pi.
2. Connect it to the power supply and attach the external display through an HDMI cable (in practical usage, an external display is not compulsory; we have used it here just for verification purposes).
3. The OS will boot up and show the following page:

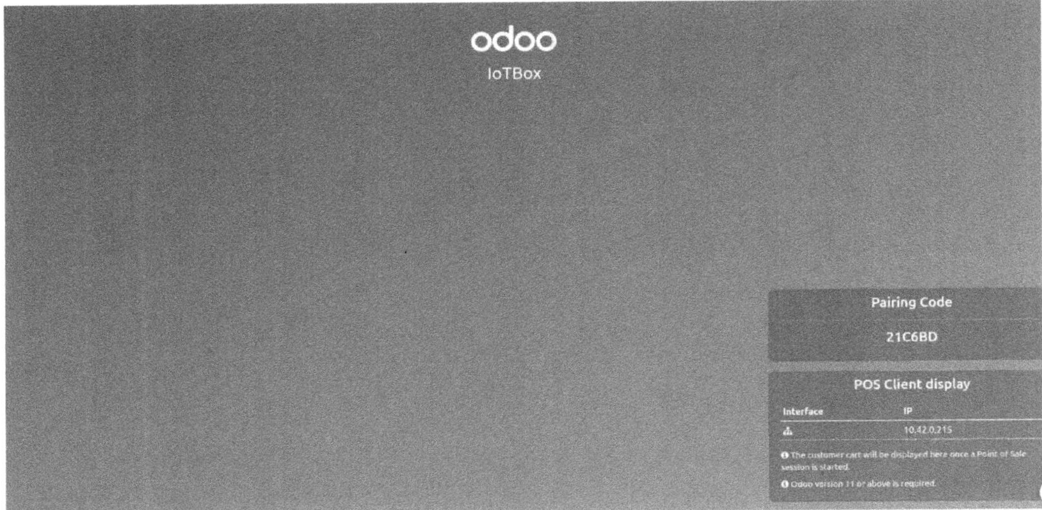

Figure 24.2 – The IoT Box screen

If you are not using a display, you can just connect the IoT Box to a power supply, and after some time, you will see the Wi-Fi network of the IoT Box.

There's more...

In previous versions of Odoo, the PosBox was used in POS applications. The IoT Box supports all the features of the PosBox, so if you are using the Community Edition of Odoo and you want to integrate devices, you can use the same IoT Box image to connect Odoo instances with different devices. See the *Configuring a POS* recipe for more information.

Connecting the IoT Box with a network

The IoT Box communicates with an Odoo instance through the network. Connecting the IoT Box is a crucial step, and if you make a mistake here, you might encounter errors when connecting the IoT Box with Odoo.

Getting ready

Mount the microSD card with the IoT Box image into the Raspberry Pi and then connect the Raspberry Pi to the power supply.

How to do it...

Raspberry Pi 3 Model B+ supports two types of network connection—**Ethernet and Wi-Fi**.

Connecting the IoT Box through Ethernet is simple; you just need to connect your IoT Box with the **RJ45 Ethernet cable**, and the IoT Box is then ready to be used. Connecting the IoT Box through Wi-Fi is complicated as you might not have a display attached to it. Perform the following steps to connect the IoT Box through Wi-Fi:

1. Connect the IoT Box to the power supply (if the Ethernet cable is plugged into the IoT Box, remove it and restart the IoT Box).
2. Open your computer and connect to the Wi-Fi network, named `IoTBox`, as shown in the following screenshot (no password is needed):

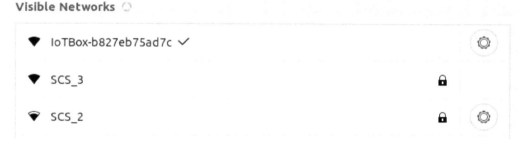

Figure 24.3 – IoT Box Wi-Fi network

3. After connecting to the Wi-Fi network, you'll see a popup with the IoT Box home page, as shown in the following screenshot (if this does not work, open the IP address of the box in the browser):

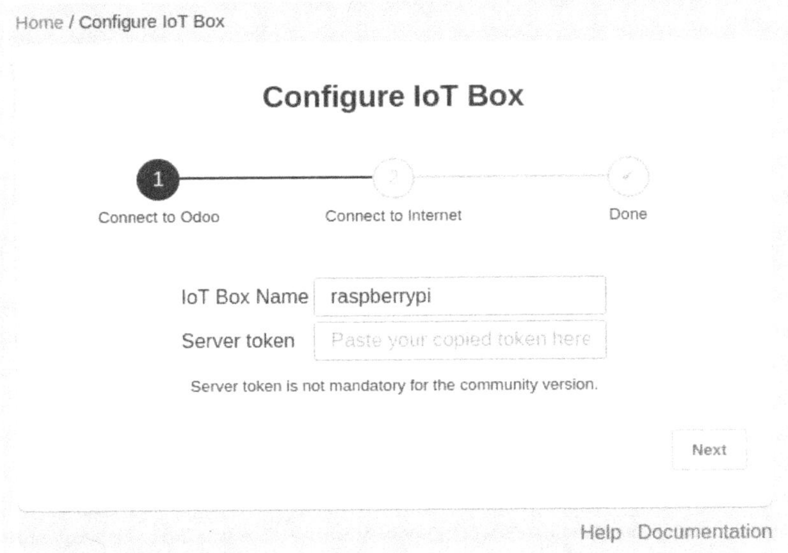

Figure 24.4 – Connecting to the IoT Box

4. Set **IoT Box Name** and keep **Server token** empty, then click on **Next**. This will redirect you to a page where you can see a list of Wi-Fi networks:

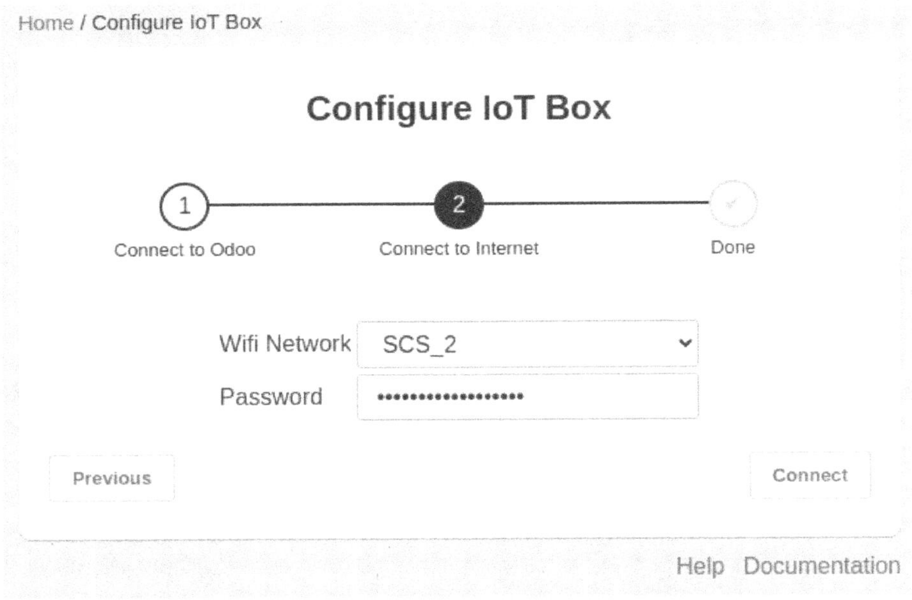

Figure 24.5 – Connecting to Wi-Fi

> **Note**
>
> You can use a server token if you are using the Enterprise Edition and you want to connect the IoT Box with Odoo right away. You can get a server token from your Odoo instance; refer to the next recipe to learn more about it.

5. Select the Wi-Fi network that you want to connect to and fill in the **Password** field. After doing this, click on the **Connect** button. If you entered the correct information, you will be redirected to the final page:

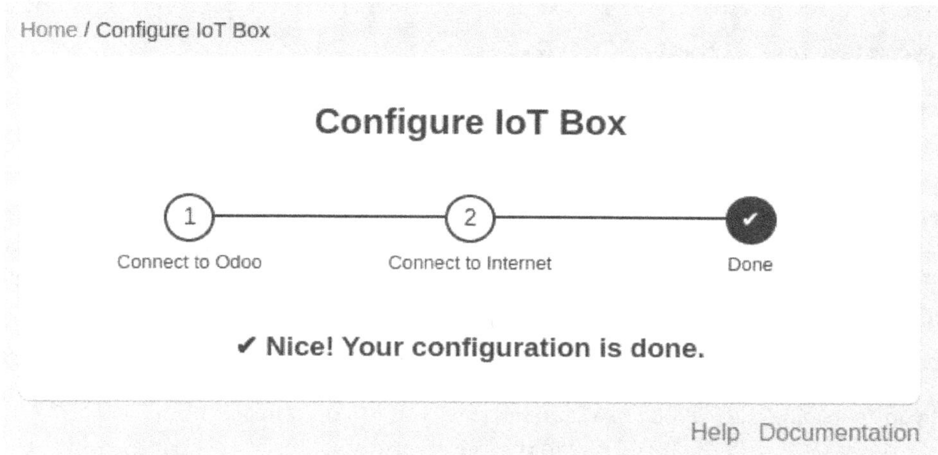

Figure 24.6 – Confirmation page

After performing these steps, your IoT Box is connected to the network and ready to be integrated with the Odoo instance.

How it works...

Connecting the Odoo instance to the IoT Box through Ethernet is simple; just connect your IoT Box with the RJ45 Ethernet cable, and the IoT Box is ready to be used. It's different when you want to connect the IoT Box with Wi-Fi; this is difficult because the IoT Box doesn't have a display or GUI. You do not have an interface to enter your Wi-Fi network password. Consequently, the solution to this problem is to disconnect your IoT Box from the Ethernet cable (if it is connected) and restart it. In such cases, the IoT Box will create its own Wi-Fi hotspot, named `IoTBox` or similar, as shown in *step 2*. You need to connect the Wi-Fi with the name `IoTBox`; luckily, it does not require a password. Once you connect to the `IoTBox` Wi-Fi, you'll get a popup, as shown in *step 3*. Here, you can name your IoT Box something like `Assembly-line IoT Box`. Keep the server token empty for now; we will learn more about it in the *Adding the IoT Box to Odoo* recipe. Then, click on the **Next** button.

Upon clicking the **Next** button, you will be shown a list of Wi-Fi networks, as shown in *step 4*. Here, you can connect the IoT Box to your Wi-Fi network. Make sure you choose the right network. You need to connect the IoT Box with the same Wi-Fi network as the computer on which the Odoo instance is going to be used. The IoT Box and the Odoo instance communicate within a **local area network** (**LAN**). This means that if both are connected to different networks, they cannot communicate, and so IoT will not work.

After choosing the right Wi-Fi network, click on **Connect**. Then, the IoT Box will turn off its hotspot and reconnect to your configured Wi-Fi network. That's it—the IoT Box is ready to be used.

Adding the IoT Box to Odoo

Our IoT Box is connected to the local network and ready to be used with Odoo. In this recipe, we will connect the IoT Box with the Odoo instance.

Getting ready

Make sure the IoT Box is on and that you have connected the IoT Box to the same Wi-Fi network as the computer with the Odoo instance.

There are a few things you need to take care of; otherwise, the IoT Box will not be added to Odoo:

- If you are testing the IoT Box in a local instance, you will need to use `http://192.168.*.*:8069` (your local IP) instead of `http://localhost:8069`. If you use localhost, the IoT Box will not be added to your Odoo instance.
- You need to connect the IoT Box with the same Wi-Fi/Ethernet network as the computer on which the Odoo instance is being used. Otherwise, the IoT Box will not be added to your Odoo instance.
- If your Odoo instance is running with multiple databases, IoT Box will not auto-connect with the Odoo instance. Use the `--db-filter` option to avoid this issue.

How to do it...

In order to connect the IoT Box with Odoo, first you will need to install the `iot` module on your Odoo instance:

1. To do so, go to the **Apps** menu and search for the **Internet of Things** module. The module will look like this. Activate the module, and we are good to go:

Figure 24.7 – Installing the iot module

2. After installing the `iot` module, you can connect your instance with the IoT Box. Then, connect your IoT Box manually with the Odoo instance by clicking on the **IoT** menu.
3. Click on the **Connect** button on the control panel. This will show the following popup. Copy the **Token** value:

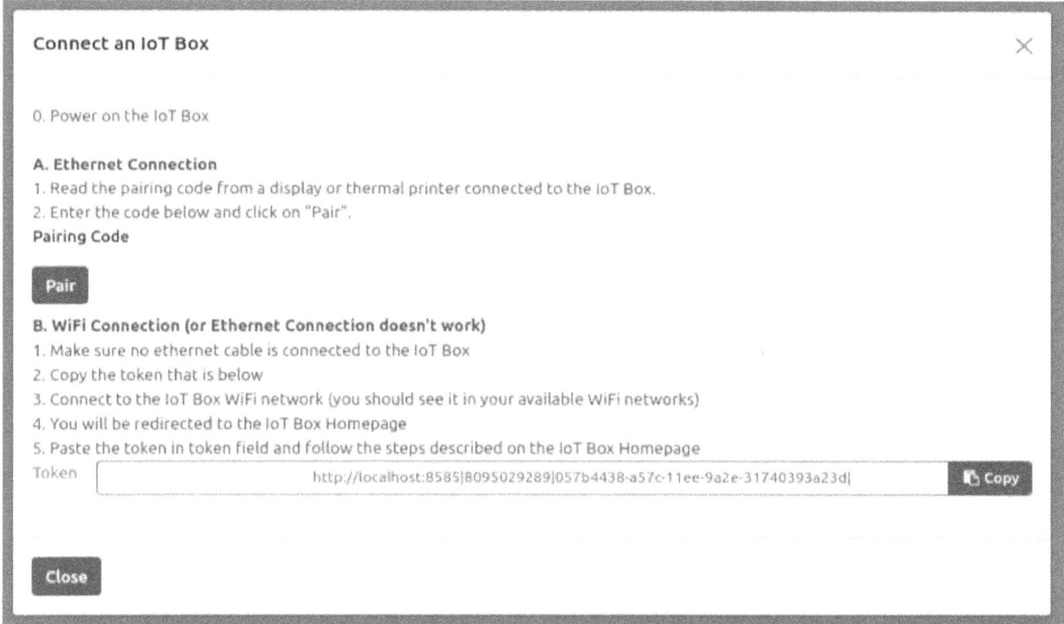

Figure 24.8 – Dialog to connect the IoT Box with Odoo

4. Open the IP of the IoT Box with port `8069`. This will display the home page of the IoT Box. Click on the **configure** button in the **Name** section:

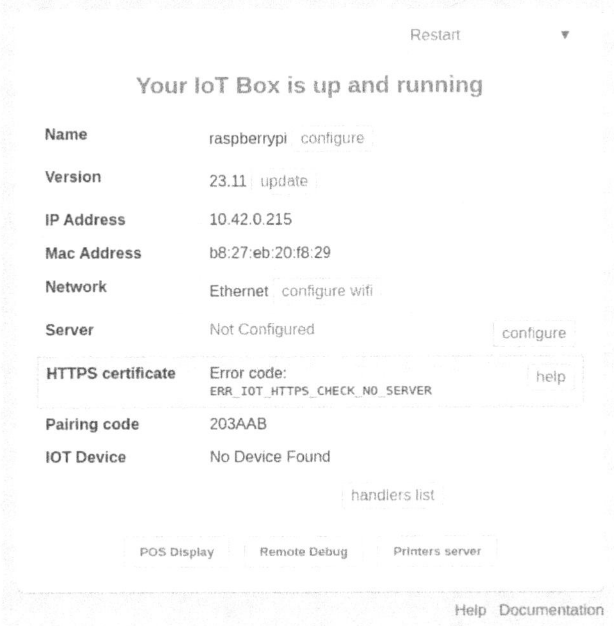

Figure 24.9 – The IoT Box home page

5. Set the **IoT Box Name** setting and paste in the server token. Then, click on the **Connect** button. This will start configuring the IoT Box. Wait for the process to complete:

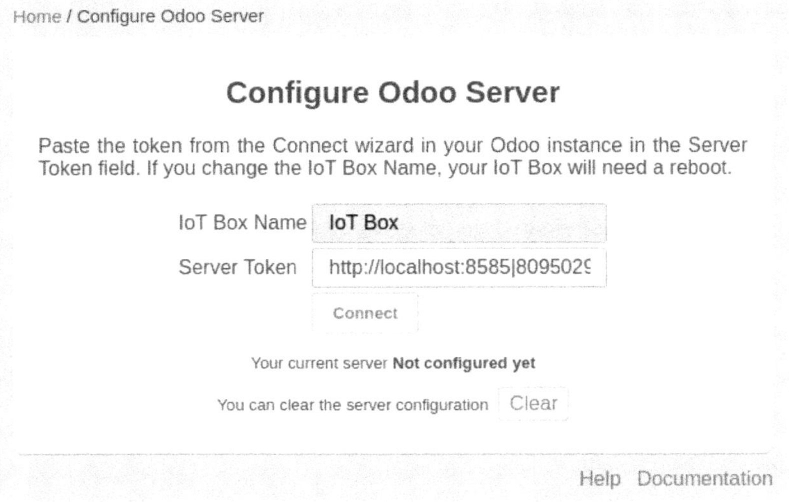

Figure 24.10 – The IoT Box home page

6. Check the **IoT** menu in your Odoo instance. You will find a new IoT Box:

IoT IoT Boxes Devices Clear Selected Devices My Company (San Francisco)
Connect IoT Boxes Search... 1-3 / 3
Shop http://0.0.0.0:8069 5
Workshop http://1.1.1.1:8069 3
IoTBox http://10.42.0.215:8069 0

Figure 24.11 – Successfully connected IoT Box

How it works...

Connecting the IoT Box with Odoo is important. This way, Odoo will know the IP of the IoT Box. The IP will be used by Odoo to communicate with devices connected to that device. This will also make sure, in the case of multiple IoT Boxes, that Odoo communicates with the right one. The rest is straightforward.

If you want to add an IoT Box to an Odoo instance during Wi-Fi configuration, that can be done. In the *Connecting the IoT Box with a network* recipe, we kept the **Server token** field empty. You just need to add the server token in this step:

Figure 24.12 – Adding the server token during Wi-Fi configuration

> **Note**
>
> Avoid using the DHCP network when using the IoT Box. This is because the IoT Box network configuration is added based on the IP address. If you use the DHCP network, then the IP address is assigned dynamically. So, there is a chance that your IoT Box will stop responding due to the new IP address. To avoid this issue, you can map the MAC address of the IoT Box to the fixed IP address.

Connecting an IoT Box with a pairing code

There is one more alternative way to connect an IoT Box, which is through a **pairing code**. The pairing code can be found on the POS display page of the IoT Box. There are two ways to open a POS client display. The first is by connecting the IoT Box with an external display. When you start your IoT Box with a display connected, it will open the POS client display by default. The second way is to open the POS client via the IoT Box IP. The URL for the POS client display is as follows: `<IoTBOX IP>:8069/point_of_sale/display`. Once you open the POS client display, you will be able to see the pairing code as follows:

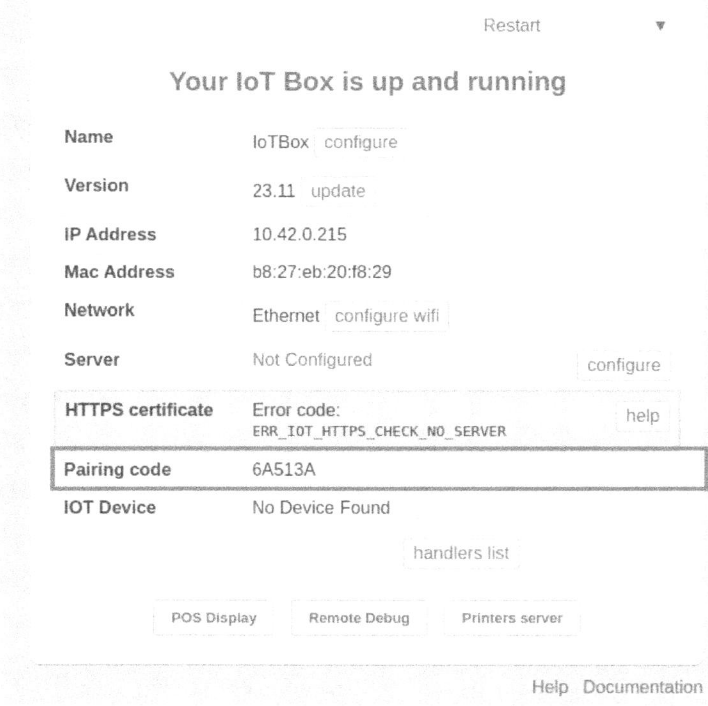

Figure 24.13 – The pairing code for the IoT Box

Then, you just need to use the pairing code in the IoT Box connection dialog in your Odoo instance.

> **Note**
>
> The pairing code will not be displayed if you are not connected to the internet.

In the preceding screenshot, we have seen how you can get the pairing code for the POS client display. But if you have an Ethernet connection and a printer, you can get the pairing code without a display. You just need to connect the IoT Box with the Ethernet and the printer. Once the IoT Box is booted, it will print a receipt with the pairing code. Then, you just need to use the pairing code in the IoT Box connection dialog in your Odoo instance.

There's more...

If you want to connect an existing IoT Box with any other Odoo instance, you will need to clear the configuration. You can clear the IoT Box configuration with the **Clear** button on the Odoo server configuration page of the IoT Box:

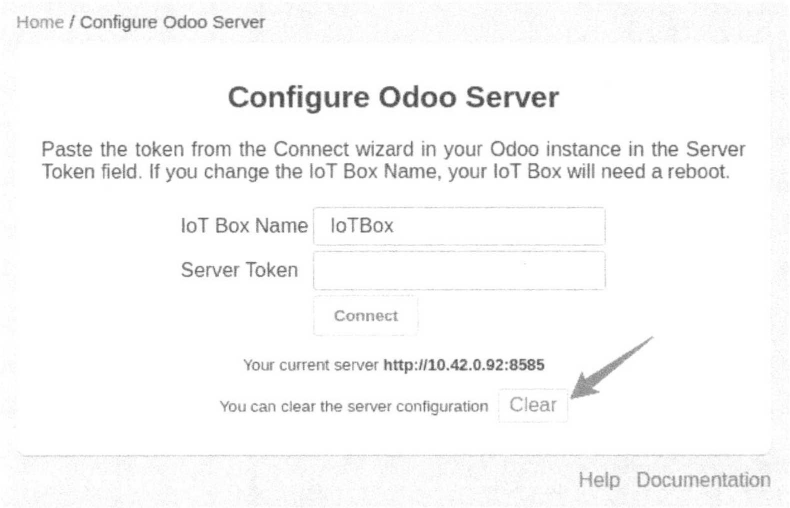

Figure 24.14 – Clearing the IoT Box configuration

Loading drivers and listing connected devices

The IoT Box is not just limited to the Enterprise Edition. You can use it like the PosBox in the Community Edition. The device's integration is part of the Enterprise Edition, so the IoT Box image does not come with device drivers; you need to load them manually. Usually, if you connect the IoT Box with an Enterprise Odoo instance, the IoT Box loads the device driver interfaces automatically.

But sometimes, you might have custom drivers or drivers that are not loaded correctly. In that case, you can manually load the drivers. In this recipe, we will see how you can load drivers and get a list of connected devices.

Getting ready

Make sure the IoT Box is on and that you have connected it to the same Wi-Fi network as the computer with the Odoo instance.

How to do it...

Perform the following steps to load device drivers into the IoT Box:

1. Open the IoT Box home page and click on the **handlers list** button at the bottom:

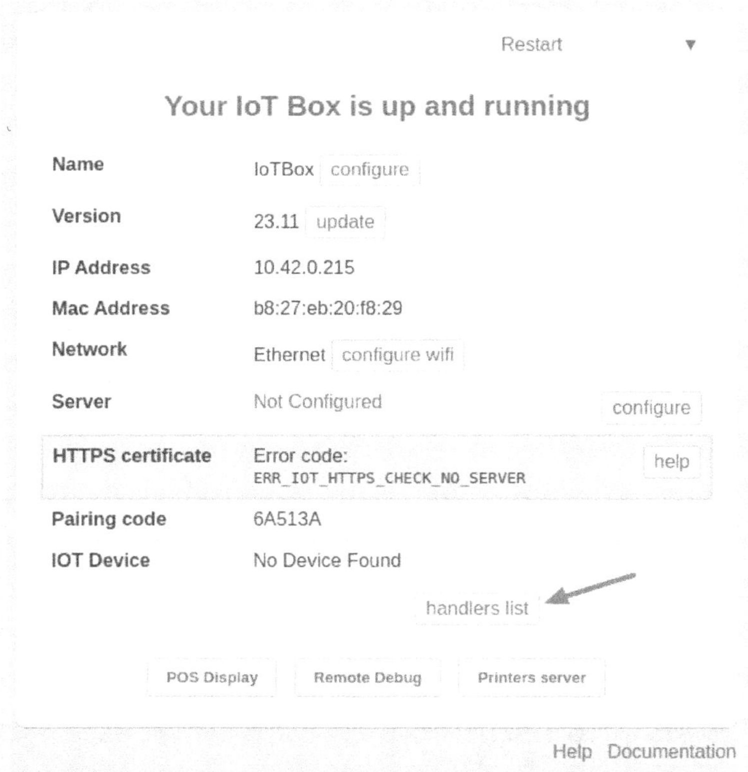

Figure 24.15 – handlers list button

2. The **handlers list** button will redirect you to the **Handlers list** page, where you will find the **Load handlers** button. Click on the button to load the drivers:

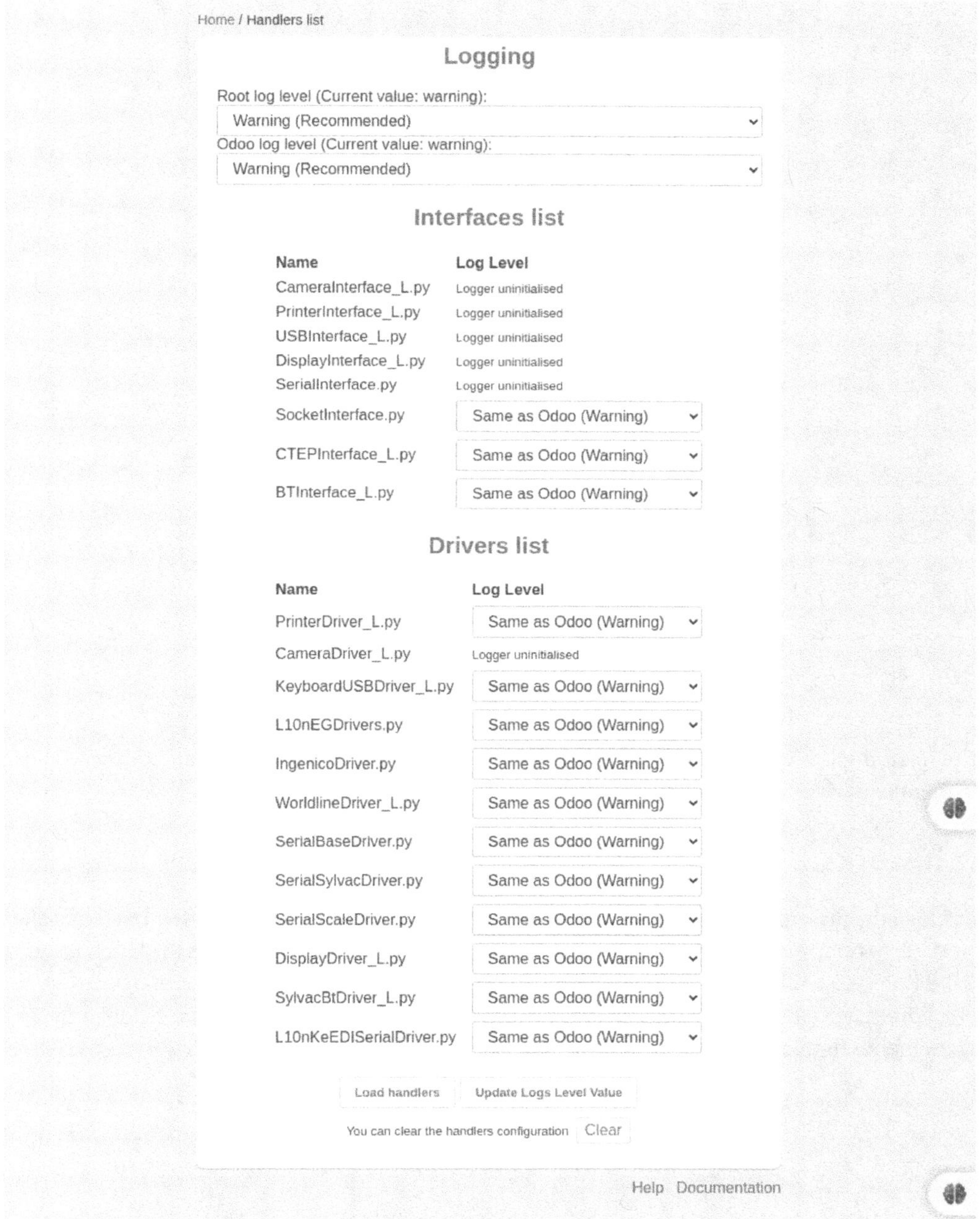

Figure 24.16 – Drivers list

3. Go back to the **IoT Box** home page. Here, you will see a list of connected devices:

IOT Device

Cameras

HD Webcam C615 :
usb-3f980000.usb-1.5

Displays

Distant Display :
distant_display

Printers

Generic Generic Text-Only Printer :
f998fdb8b5973e83709c19f22fb7027c

Hewlett-Packard HP LaserJet Pro MFP M126nw :
dnssdHP20LaserJet20Pro20MFP20M126nw5BA8BF205D_pdld

HP LaserJet M1005 :
7d0eccc980213e4e483f11f75788819d

HP LaserJet Pro MFP m126nw :
cba1bd1fec3839f26954d519590cf081

HP LaserJet m1005 Multifunction Printer :
2ae3325b1c3f38795e88bb7bb7464c2e

HP Deskjet Ink Advantage Ultra 2529 All-in-one Printer :
91058fc7097c34ce647207c770ea71fc

HP LaserJet Pro MFP M126nw :
socket1042096

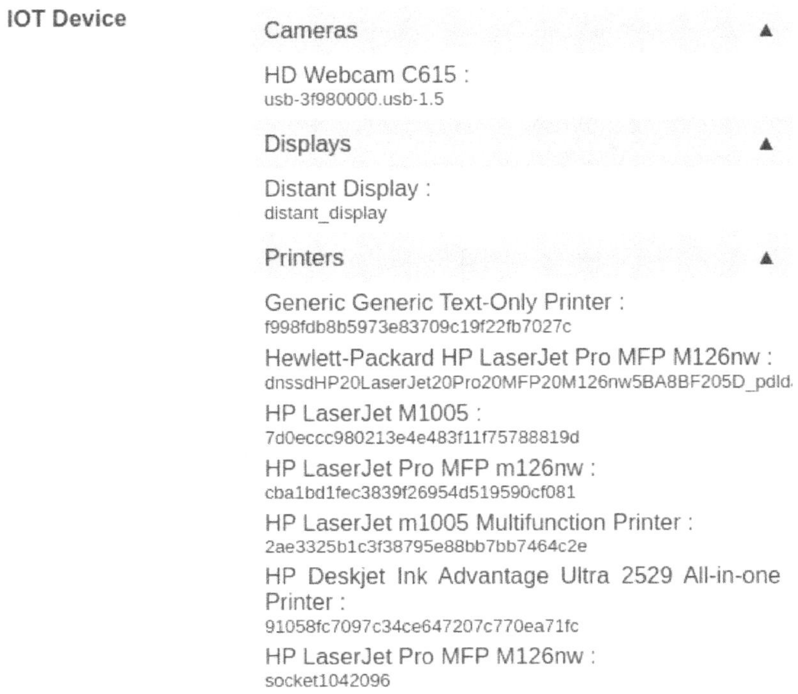

Figure 24.17 – Connected devices

After performing these steps, the IoT Box will be ready with the devices you specified, and you can start using the devices in your applications.

How it works...

You can load the drivers from the home page of the IoT Box. You can do this using the **Load handlers** button at the bottom. Note that this will only work if your IoT Box is connected with the Odoo instance using the Enterprise Edition. After loading the drivers, you will be able to see a list of devices on the IoT Box home page. You can also see a list of connected devices in the Odoo instance through the **IoT | Devices** menu. In this menu, you will see a list of connected devices for each IoT Box:

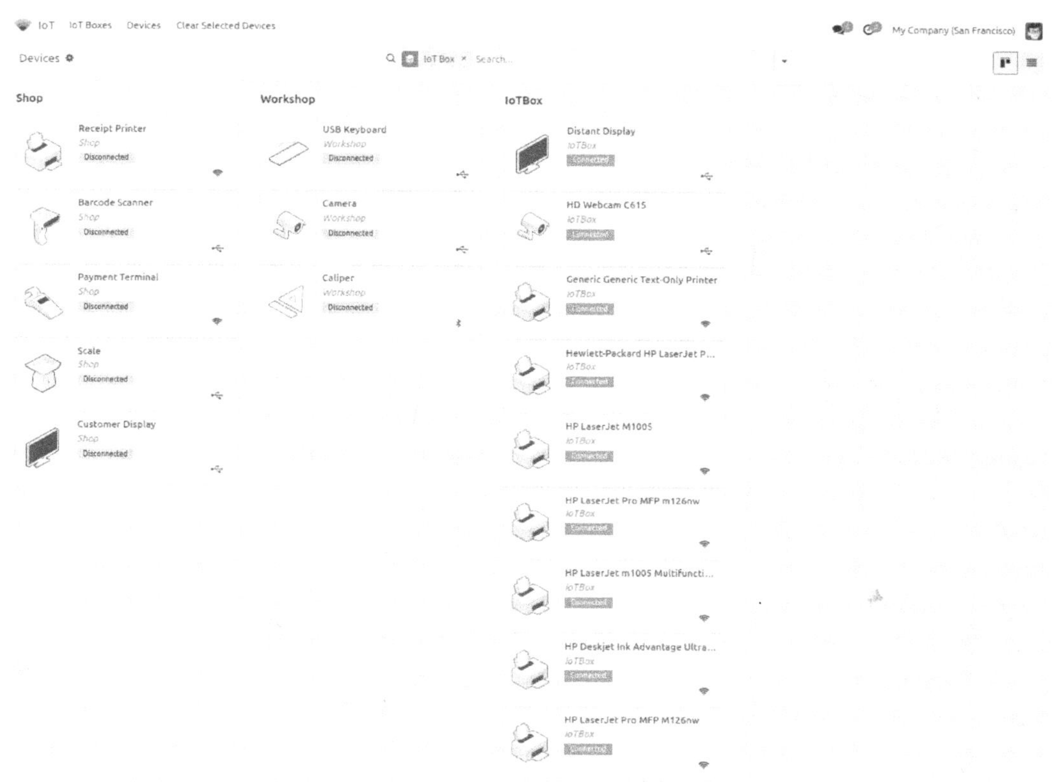

Figure 24.18 – Connected devices list

Right now, the IoT Box supports a few hardware devices, such as cameras, footswitches, printers, and calipers. A list of devices that are recommended by Odoo can be found here: `https://www.odoo.com/page/iot-hardware`. If your device is not supported, you can pay for driver development.

Taking input from devices

The IoT Box only supports limited devices. Right now, these hardware devices are integrated with the manufacturing application. But if you want, you can integrate supported devices with your module. In this recipe, we will capture a picture from a camera through our IoT Box.

Getting ready

We will be using the `my_hostel` module from the *Logging user changes in a chatter* recipe of *Chapter 23*, *Managing Emails in Odoo*. In this recipe, we will add a new field to capture and store images when a borrower returns a book. Make sure the IoT Box is on and that you have connected a supported camera device with it.

How to do it...

Perform the following steps to capture a picture using a camera with the IoT Box:

1. Add a dependency in the manifest file:

```
'depends': ['base', 'quality_iot'],
```

2. Add new fields in the `hostel.student` model:

```
test_type_id = fields.Many2one('quality.point.test_type', 'Test
Type',help="Defines the type of the quality control point.",
required=True, default=_get_default_test_type_id)
test_type = fields.Char(related='test_type_id.technical_name',
readonly=True)
device_id = fields.Many2one('iot.device', string='IoT Device',
domain="[('type', '=', 'camera')]")
ip = fields.Char(related="device_id.iot_id.ip")
identifier = fields.Char(related='device_id.identifier')
picture = fields.Binary()
```

3. Add these fields into the form view of the `hostel.student` model:

```
<group>
    <field name="test_type_id" invisible="1"/>
    <field name="test_type" invisible="1"/>

    <field name="ip" invisible="0"/>
    <field name="identifier" invisible="0"/>
    <field name="device_id" required="1"/>
    <field name="picture" widget="iot_picture"
          options="{'ip_field': 'ip', 'identifier':
'identifier'}"/>
    <field name="name"/>
    <field name="gender"/>
    <field name="active"/>
</group>
```

4. Update the `my_hostel` module to apply the changes. After the update, you will have a button to capture images:

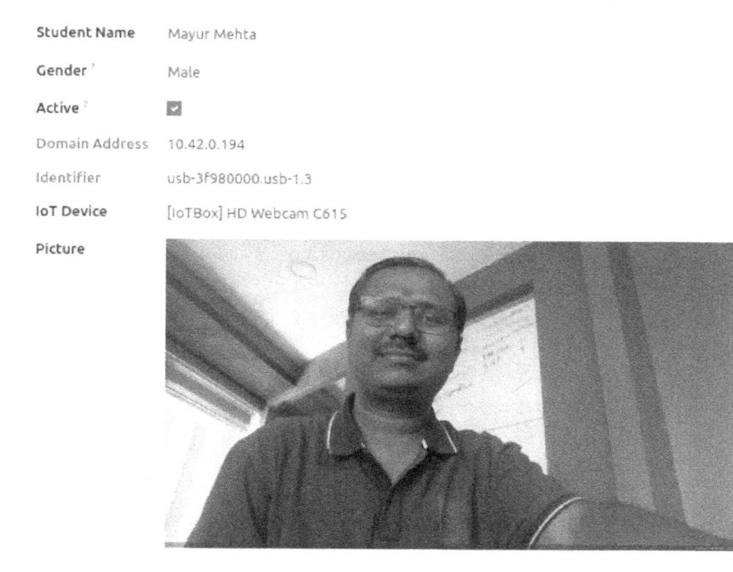

Figure 24.19 – Capturing an image via IoT

Note that the button will not capture images if the webcam is not connected to the IoT Box or drivers are not loaded in the IoT Box.

How it works...

In *step 1*, we added a dependency to the `quality_iot` module in the manifest file. The `quality_iot` module is part of the Enterprise Edition and contains a widget that allows you to request an image from a camera through the IoT Box. This will install `stock` modules, but for the sake of simplicity, we will use `quality_iot` as a dependency. If you do not want to use this dependency, you can create your own field widget. Refer to the *Creating custom widgets* recipe in *Chapter 15, Web Client Development*, to learn more about widgets.

In *step 2*, we added fields required to capture an image from the camera. To capture the image, we need two things: the device identifier and the IP address of the IoT Box. We want to give the user the option to select the camera, so we added a `device_id` field. The user will choose a camera to capture the image, and based on the selected camera device, we extracted IP and device identifier information from related fields. Based on these fields, Odoo will know where to capture the image, if you have multiple IoT Boxes. We have also added a binary field, `picture`, to save the image.

In *step 3*, we added fields in the form view. Note that we used the `iot_picture` widget on the `picture` field. We added `ip` and `identifier` fields as invisible fields because we do not want to show them to the user; rather, we want to use them in the `picture` field options. This widget will add the button in the form view; upon clicking the button, Odoo will make a request to the IoT Box to capture the image. The IoT Box will return image data as the response. This response will be saved in the `picture` binary field.

There's more...

The IoT Box supports Bluetooth calipers. If you want to take measurements in your module, you can use the `iot_measure` widget to fetch them in Odoo. Note that as with `iot_picture`, here, you will also need to add `ip` and `identifier` invisible fields in the form view:

```
<field name="measure" widget="iot_measure"
          options="{'ip_field': 'ip', 'identifier': 'identifier'}"/>
```

This will fill the `measure` field with the data captured from the IoT caliper.

Accessing the IoT Box through SSH

The IoT Box is running on the Raspbian OS, and it is possible to access the IoT Box through SSH. In this recipe, we will learn how to access the IoT Box through SSH.

Getting ready

Make sure the IoT Box is on and you have connected the IoT Box to the same Wi-Fi network as the computer with the Odoo instance.

How to do it...

In order to connect the IoT Box through SSH, you will need the IP address of the IoT Box. You can see this IP address in its form view. As an example, in this recipe, we will use `192.168.43.6` as the IoT Box IP address, so replace this with your IP address. Perform the following steps to access the IoT Box through SSH:

1. Open the terminal and execute the following command:

   ```
   $ ssh pi@192.168.43.6
   pi@192.168.43.6's password:
   ```

2. The Terminal will ask you for a password; enter `raspberry` as the password.

3. If you add the right password, you can access the shell. Execute the following command to see the directory:

```
total 24
-rw-r--r-- 1 root root    6 Oct 26 08:12 iotbox_version
drwxr-xr-x 5 pi   pi   4096 Oct 23 09:05 odoo
-rw-r--r-- 1 pi   pi     36 Nov 15 13:10 odoo-db-uuid.conf
-rw-r--r-- 1 pi   pi      0 Nov 15 13:10 odoo-enterprise-code.
conf
-rw-r--r-- 1 pi   pi     26 Nov 15 13:10 odoo-remote-server.conf
-rw-r--r-- 1 pi   pi     11 Nov 15 13:10 token
-rw-r--r-- 1 pi   pi     26 Aug 20 12:03 wifi_network.txt
```

As you have SSH access, you can explore the full filesystem of the IoT Box.

How it works...

We used the Pi user with the password `raspberry` to access the IoT Box through SSH. SSH connection is used when you want to debug a problem in the IoT Box. SSH doesn't need any explanation, but let's see how Odoo works in the IoT Box.

Here is some information that might help you debug the issue:

- The IoT Box is internally running some Odoo modules. The name of these modules usually starts with hw_, and they are available in the Community Edition. You can find all the modules in the `/home/pi/odoo/addon` directory.
- If you want to see the Odoo server log, you can access it from the `/var/log/odoo/odoo-server.log` file.
- Odoo is running through a service named odoo; you can use the following command to start, stop, or restart the service:
  - `sudo service odoo start/restart/stop`
- Customers mostly turn the IoT Box off by disconnecting the power. This means that the IoT Box OS does not shut down properly in such cases. To avoid corruption of the system, the IoT Box filesystem is read-only.

There's more...

Note that the IoT Box is only connected to the local machine. Consequently, you cannot access the shell directly from a remote location (through the internet). If you want to access the IoT Box remotely, you can paste the `ngrok` authentication token key into the IoT Box's remote debug page, as shown in the following screenshot. This will enable the TCP tunnel from the IoT Box so that you can connect the IoT Box through SSH from anywhere. Learn more about `ngrok` at `https://ngrok.com/`:

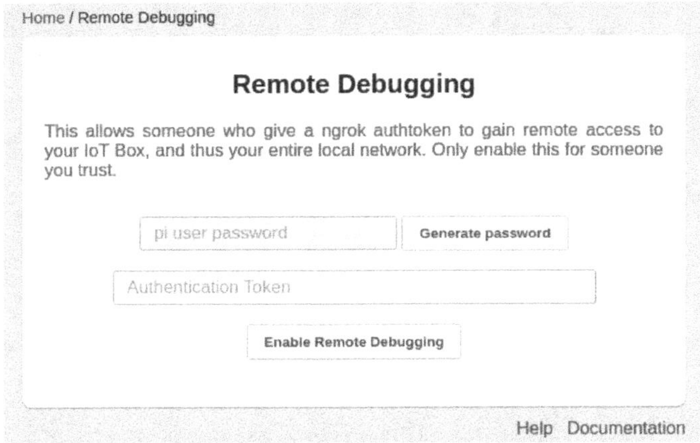

Figure 24.20 – Debugging with a ngrok token

Once you add your token, you will be able to access the IoT Box from remote locations.

Configuring a POS

The IoT Box works with POS applications. In this recipe, we will learn how to configure the IoT Box for POS applications.

Getting ready

Make sure the IoT Box is on and you have connected the IoT Box to the same Wi-Fi network as the computer with the Odoo instance. Also, install the POS application if it is not already installed.

How to do it...

Perform the following steps to configure the IoT Box for the POS application:

1. Open the POS application, and open **Settings** from the POS session dropdown:

Figure 24.21 – POS session settings

2. Click on the **Edit** button and click on the **IoT Box** checkbox. This will enable more options:

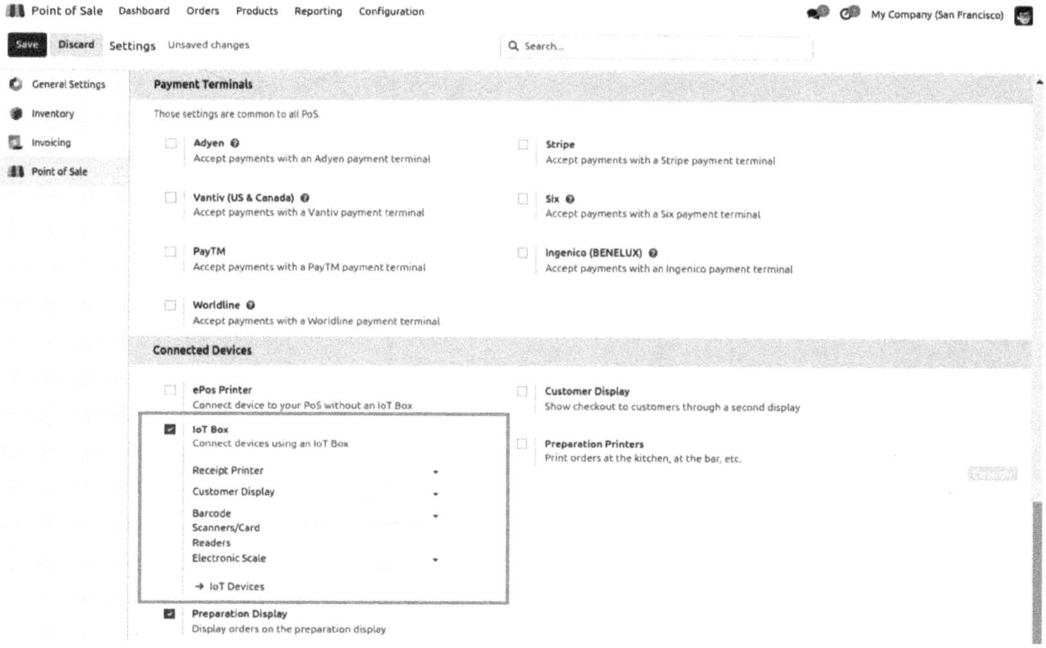

Figure 24.22 – Selecting IoT devices

3. Select the devices that you want to use in a POS session. If you are going to use hardware, such as a barcode scanner, select the relevant devices.
4. Save the changes by clicking the **Save** button in the control panel.

After the configuration, you will be able to use the IoT Box in the POS application.

How it works...

The IoT Box can be used with POS applications such as the PosBox. In order to use the IoT Box in a POS application, you have to connect the IoT Box to the Odoo instance. If you don't know how to connect the IoT Box, follow the *Adding the IoT Box to Odoo* recipe. Once you have connected the IoT Box to Odoo, you will be able to select the IoT Box in the POS application, as shown in *step 2*.

Here, you can select the hardware you want to use in the POS session. After saving the changes, if you open the POS session, you will be able to use the enabled hardware at the POS. If you enabled specific hardware from the settings but the hardware is not connected to the IoT Box, you will see the following warning in the top bar:

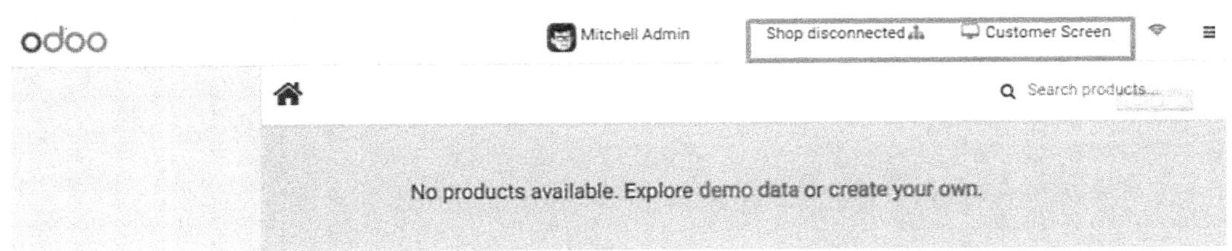

Figure 24.23 – IoT Box connection issues

You can click on these warnings to try to connect again.

There's more...

The POS application is part of the Community Edition. If you are using the Community Edition, instead of the **IoT Box** selection, you will see the **IoT Box IP Address** field in the POS settings:

Connected Devices

ePos Printer
Connect device to your PoS without an IoT Box

Customer Display
Show checkout to customers through a second display

IoT Box
Connect devices using an IoT Box

IoT Box IP Address
Barcode Scanner/Card Reader
Electronic Scale
Receipt Printer
Customer Display

Preparation Printers
Print orders at the kitchen, at the bar, etc.

Preparation Display Enterprise
Display orders on the preparation display

Figure 24.24 – IoT Box settings in the Community Edition

If you want to integrate hardware in the Community Edition, you will need to use the IP address of the IoT Box in the field.

Sending PDF reports directly to a printer

The IoT Box runs the **Common UNIX Printing System** (**CUPS**) server by default. CUPS is a printing system that allows a computer to act as a printing server. You can learn more about it at `https://www.cups.org/`. So, as the IoT Box runs CUPS internally, you can connect network printers with the IoT Box. In this recipe, we will see how you can print PDF reports directly from Odoo.

Getting ready

Make sure the IoT Box is on and you have connected the IoT Box with Odoo.

How to do it...

Follow these steps to print reports directly from Odoo:

1. Open the IoT Box home page via IP.
2. Click on the **Printer Server** button at the bottom.
3. This will open the CUPS configuration home page. Configure your printer here.
4. Once you have configured the printer, you will be able to see the printer in the IoT device list. Activate developer mode and open **Settings**| **Technical** | **Actions** | **Report**.
5. Search for the report that you want to print, open the form view, and select the printer in the **IoT Device** field, as shown in the following screenshot:

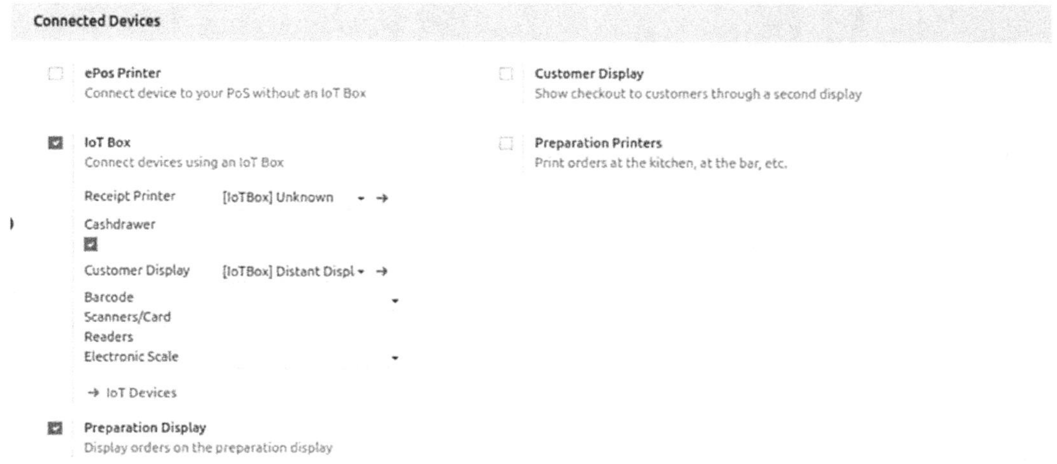

Figure 24.25 – Options to select an IoT device

Once this configuration is done, report PDFs will be sent directly to the printer.

How it works...

This recipe is straightforward in terms of configuration, but there are a few things that you should know. The IoT Box uses the CUPS server to print reports. You can access the CUPS home page at `http://<IoT Box IP>:631`.

With CUPS, you can add/remove your printer. On the home page of CUPS, you will be able to see all the documentation that you need to help you connect different types of printers. Once you have configured the printer, you will find your printer in the IoT device list. Then, you can select this IoT device (printer) in the report record. Usually, when you print a report in Odoo, it will download a PDF of the report. But when this configuration is done, instead of downloading the report, Odoo will send the PDF report directly to the selected printer. Note that only reports whose record has the printer set in the IoT device field will be sent to the printer.

25
Web Studio

Odoo Web Studio is a feature exclusive to the Odoo Enterprise edition. It's a toolbox that lets you customize the Odoo user interface and its reports directly from the user interface without any code, such as by dragging and dropping components onto the view directly. Users can create or customize reports from the user interface itself.

Odoo Web Studio is a visual development tool that allows users to customize and create applications within the Odoo **Enterprise Resource Planning** (**ERP**) platform. With Odoo Web Studio, users can design, modify, and extend various aspects of their Odoo applications without the need for extensive programming or coding skills. It offers a drag-and-drop interface, making it accessible to users with varying levels of technical expertise.

Odoo Web Studio empowers users to take full control of their Odoo ERP system by providing a user-friendly environment for module creation, report customization, automation, and more. It's a valuable tool for businesses looking to adapt and optimize their Odoo applications to meet their unique requirements and preferences. So, Odoo Web Studio is a powerful tool that empowers users to create and customize applications within the Odoo ERP system with ease. Whether you're building new modules, customizing existing ones, or designing reports, Odoo Web Studio provides a user-friendly and visual interface to streamline these processes.

Here are some key features and capabilities of Odoo Web Studio:

- **Visual customization**: Odoo Web Studio provides a visual interface that allows users to customize the layout, fields, and forms of their applications. You can modify existing modules or create entirely new ones.
- **Data model editor**: Users can define new data models, fields, and relationships between objects in their applications. This helps tailor the database structure to specific business needs.
- **Workflow configuration**: Workflow automation is a critical aspect of ERP systems. With Web Studio, users can design and configure workflows, automation rules, and triggers to streamline business processes.
- **Reports and dashboards**: Users can design custom reports and dashboards to visualize data and gain insights into their business operations.

- **Mobile responsiveness**: Odoo Web Studio applications are designed to be responsive, meaning they can adapt to different screen sizes and devices, including smartphones and tablets.
- **No-code or low-code**: While some level of technical knowledge can be helpful, Odoo Web Studio is designed to be user-friendly and accessible to those without extensive coding skills. This makes it possible for business users to make changes and adapt Odoo to their specific needs.
- **Real-time collaboration**: Multiple users can collaborate on designing and modifying applications simultaneously.
- **Integration**: Odoo Web Studio applications can be integrated with other Odoo modules and external systems to ensure seamless data flow and connectivity.

In this chapter, we will cover the following recipes:

- Installing Odoo Web Studio
- Starting with a new app
- Suggested features
- Components
- Field properties
- Views
- Build a new app
- Customizing an existing app
- Built-in functions
- Reports

Installing Odoo Web Studio

In this recipe, you'll learn how to install Odoo Web Studio.

Log into your Odoo instance with administrative or superuser credentials. In the Odoo interface, go to the **Apps** module. This is where you can install or activate new modules and features:

1. Go to **Apps**.
2. Search for `Web Studio`.
3. Click **INSTALL**.

After installation, you should see a new menu item or section called **Studio** in your Odoo instance. Click on it to access Odoo Web Studio:

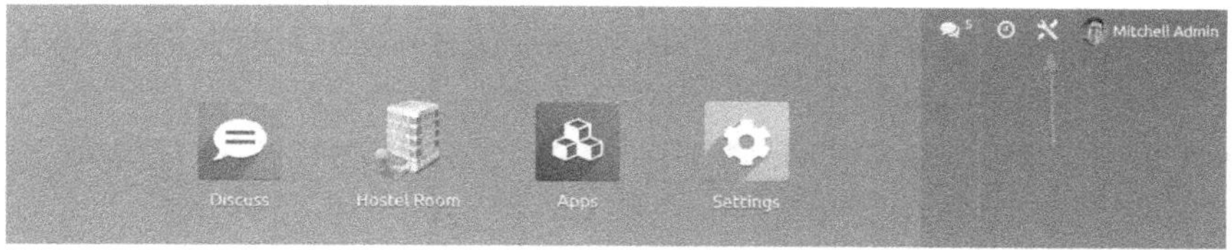

Figure 25.1 – Screenshot of the Studio button

Once you're in Odoo Web Studio, you can start customizing your Odoo applications, designing workflows, creating reports, and making other modifications using the visual tools and interface provided. By clicking on the icon, the studio customization mode is activated.

Starting with a new app

In Odoo Web Studio, you typically start by creating a new application. An application can be thought of as a module or a part of your ERP system. Click on the **Create** or **New** button to begin:

1. Go to the **App menu** screen.
2. Click on the **Customize** icon.
3. Click the **New App** button to start creating a new app:

Figure 25.2 – Screenshot of the New App creation screen

4. After clicking on **New App**, you'll see the following:

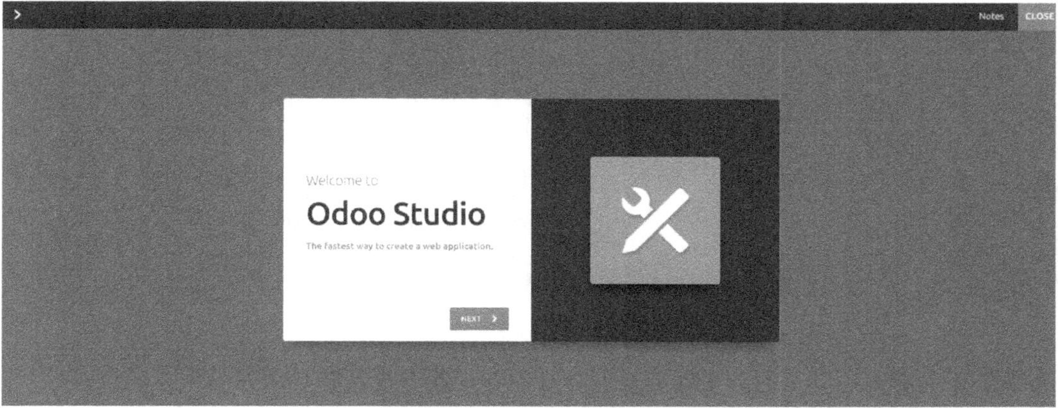

Figure 25.3 – What you'll see after clicking New App

Click **Next**. From here, you can define your module's name and change the logo of your module. You can upload a custom logo or customize the logo's icon, icon color, and logo background:

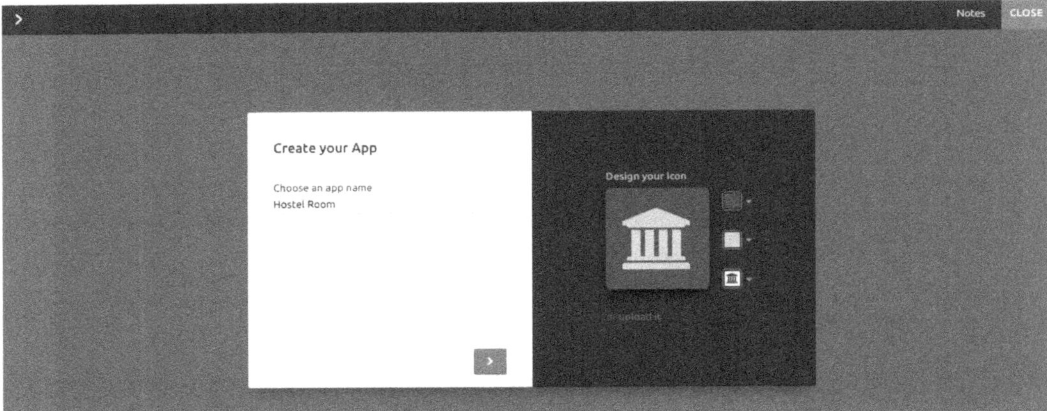

Figure 25.4 – Creating a new app

Choose a name for your application. You can customize the icon by choosing any of the in-built icons. You'll also have the chance to modify the background color and icon color as per your corporate branding. After adding the module's name, click the > button. At this point, you can add your first menu's name. Here, you have to build a new menu, so name it as you wish. Once you've done this, you can choose the type of model you wish to create. If you're creating an app from scratch, choose **New Model**. Otherwise, choose **Existing Model**:

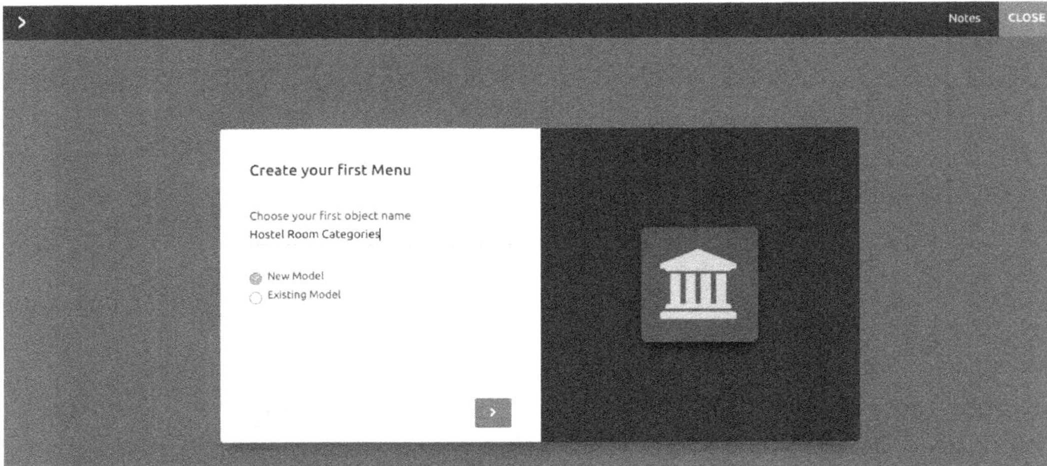

Figure 25.5 – Creating your first menu

Once you've done this, click the > button. Your app will be ready for the next level of customization.

Suggested features

Odoo Web Studio is a powerful tool that allows users to customize and extend their Odoo applications without the need for extensive coding. Depending on your business needs, there are several suggested features and capabilities that you can leverage when using Odoo Web Studio:

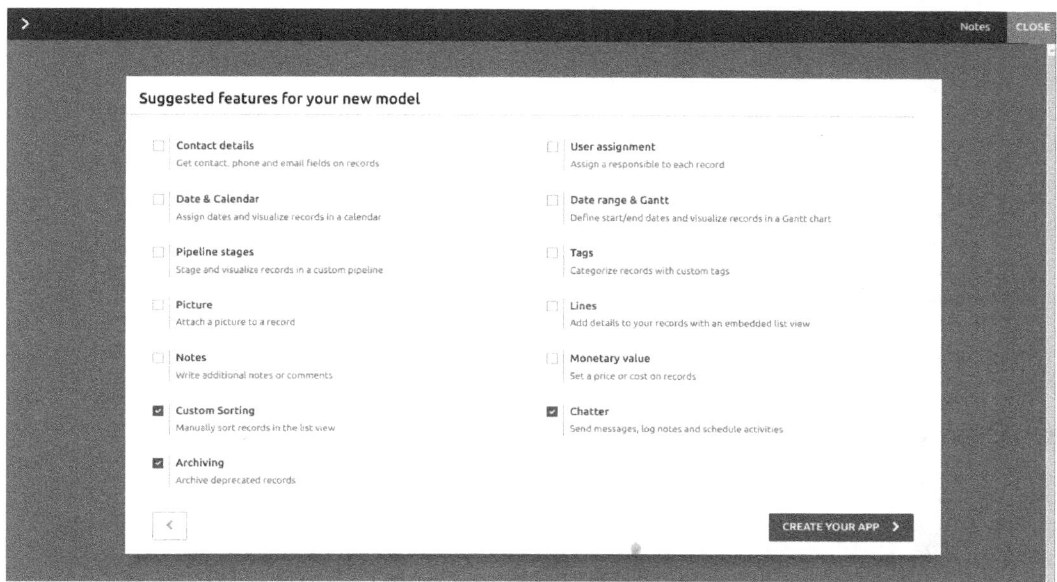

Figure 25.6 – Suggested features

Once you click on the **CREATE YOUR APP** button, you'll see the following screen. Here, you can add components and new fields, as well as modify or reuse existing model fields. You just need to drag and drop the fields:

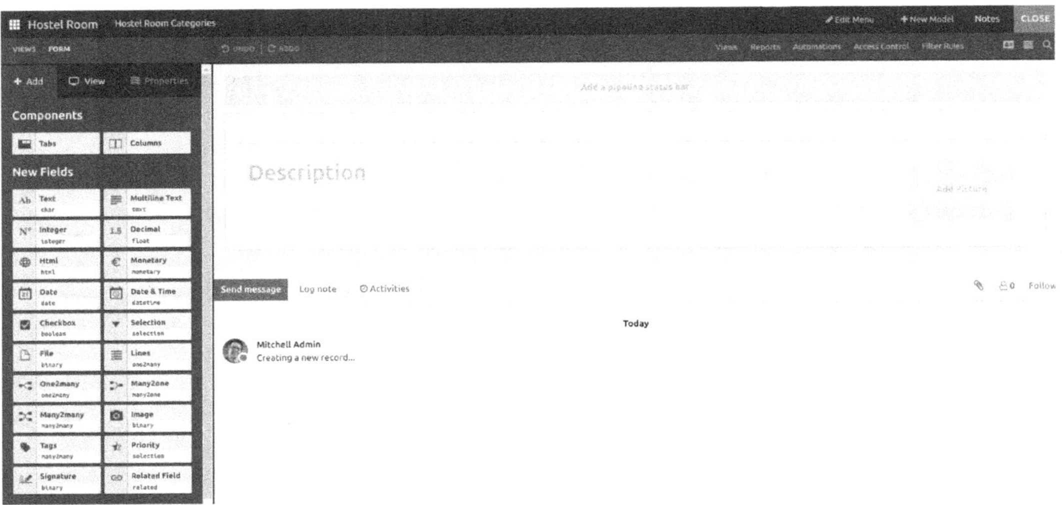

Figure 25.7 – Model components

Components

Odoo Web Studio provides a set of components that you can use to create and customize modules within the Odoo ERP system. These components enable you to design data models, user interfaces, workflows, and reports without the need for extensive coding. Here are some of the key components and features available in Odoo Web Studio:

- **Data Model Designer**: This component allows you to create and modify data models, define fields, specify data types, set default values, and establish relationships between objects. You can create custom objects to store data relevant to your business processes.
- **Form Builder**: The Form Builder component lets you design and customize forms for data entry and display. You can drag and drop fields onto forms, arrange them, and set field properties such as labels, help text, and validation rules.
- **Workflow Editor**: With the Workflow Editor component, you can design custom workflows to automate business processes. You can define triggers, actions, and transitions, allowing you to model how data moves through your application and what should happen at each stage.
- **Report Designer**: The Report Designer component enables you to create custom reports and dashboards. You can design templates for reports, add charts, tables, and graphs to visualize data, and generate printable or digital reports.
- **Menu Editor**: The Menu Editor component lets you create and modify menus and navigation structures within your Odoo modules. You can define menus for different user roles and organize them to provide easy access to various parts of your application.

- **Views and widgets**: You can customize the way data is displayed using views and widgets. Odoo Web Studio provides various view types, such as list views, form views, and kanban views, which you can configure to suit your needs.
- **Actions and triggers**: Actions and triggers allow you to define what should happen in response to certain events or user actions. For example, you can set up actions to send email notifications, update records, or trigger specific workflows.
- **Access control**: Odoo Web Studio allows you to set permissions and access rights for different user roles. You can control who can view, edit, or delete records and access specific features within your modules.
- **Localization support**: Customize your modules to accommodate regional or industry-specific requirements, including tax rules, languages, and accounting standards.
- **Data import and export**: Enable data import and export functionality to facilitate data migration and integration with external systems.
- **Scheduled actions**: You can automate tasks and actions on a scheduled basis, such as data backups or automated email notifications.
- **Integration tools**: Odoo Web Studio provides tools so that you can integrate your custom modules with other Odoo modules or external systems, ensuring seamless data exchange and synchronization.

Here is a default list view with a **Description** field:

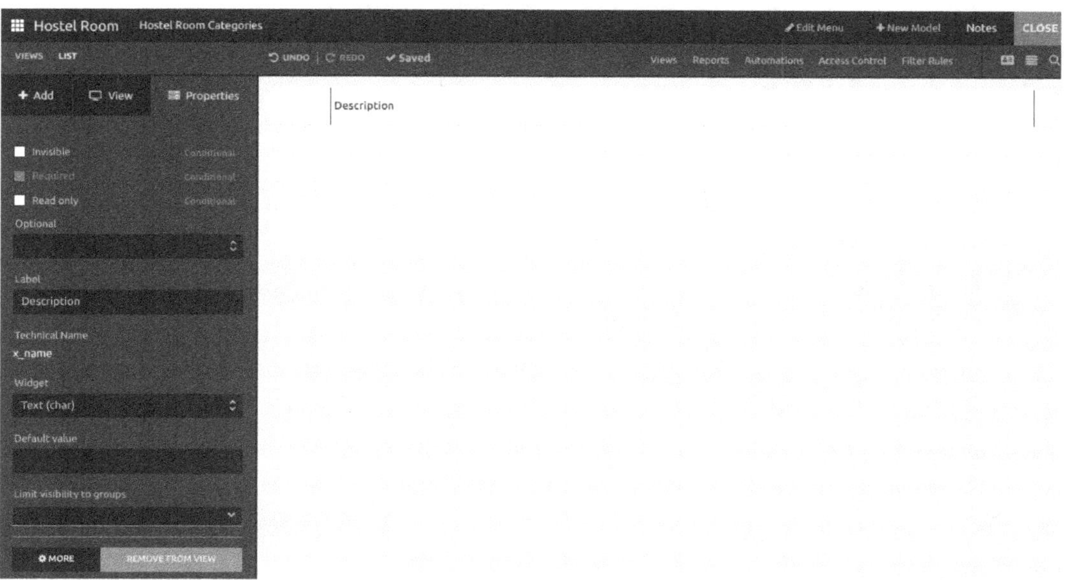

Figure 25.8 – The default list view

Field properties

In Odoo Web Studio, when creating or customizing fields in your data models, you have various options to configure and customize these fields to suit your business needs. Here are some of the common options that are available when creating new fields using Odoo Web Studio:

- **Field Name**: Give your field a descriptive name that reflects the type of data it will store.
- **Field Type**: Select the appropriate data type for your field. Odoo provides a wide range of field types, including text, integer, float, date, datetime, selection, many2one (for relationships with other records), and more.
- **Required Field**: You can make the field required, meaning that users must provide a value for this field when creating or editing records.
- **Default Value**: Set a default value for the field. This value will be pre-filled when creating a new record.
- **Read-Only**: Here, you can make the field read-only so that it cannot be edited by users. This is useful for fields that should not be modified once set.
- **Help Text**: Add some help text or a description to provide additional information about the field or instructions for users.
- **Placeholder Text**: For text or char fields, you can specify a placeholder text that appears in the input field to guide users.
- **Validation Constraints**: Here, you can set validation constraints, such as character limits, numeric ranges, or patterns for text fields.
- **Compute and Default Functions**: You can define compute functions to calculate the value of the field based on other fields or conditions. Default functions allow you to set dynamic default values.
- **Dependencies**: Here, you can define field dependencies, which determine when a field is visible or required based on the values of other fields.
- **Selection Values**: For selection fields, specify the list of values that users can choose from. This is often used for fields such as drop-down menus.
- **Domain Filters**: Apply domain filters to restrict the available choices for many2one or many2many fields based on certain conditions.
- **Advanced Options**: Odoo Web Studio also offers advanced options, such as setting a related field, specifying on-change actions, or setting access rights.
- **Groups and Access Rights**: Configure which user groups have access to view or edit this field. You can define different access rights based on user roles.
- **Computed Fields**: Create computed fields that display calculated values based on other fields in the record. These fields do not store data and instead dynamically calculate values.

- **Widgets**: Choose different widgets to control how the field is displayed, such as text, selection, date, or color picker widgets.
- **Depends On**: Define field dependencies, indicating which other fields affect the visibility or behavior of this field.
- **Related Fields**: Create related fields to display information from related records. For example, you can display a customer's name on an invoice by creating a related field.
- **Invisible or Hidden Fields**: Make fields invisible or hidden to control their visibility on forms.
- **Attachment Fields**: Configure fields to allow attachments to be added or documents or files to be uploaded.

The following screenshot shows the new fields:

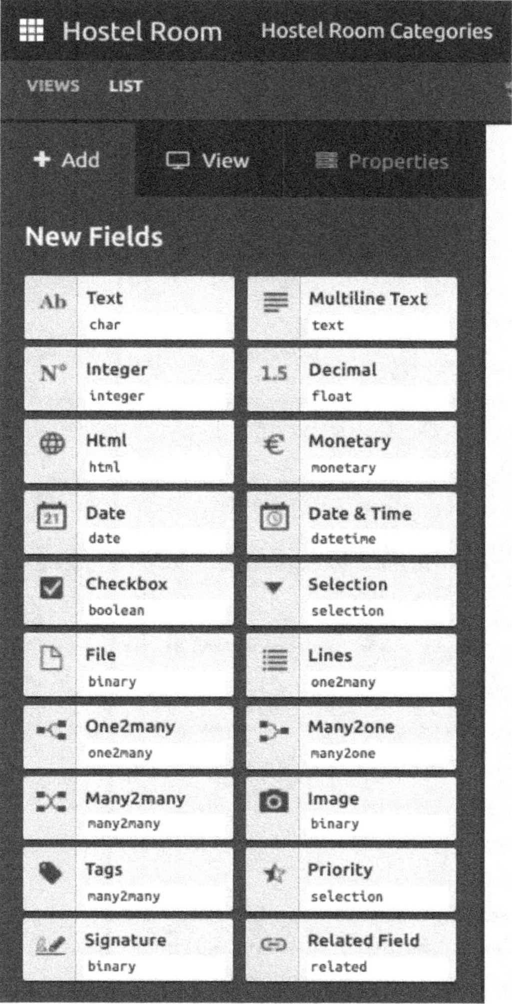

Figure 25.9 – The new fields

Here's a screenshot of the existing fields:

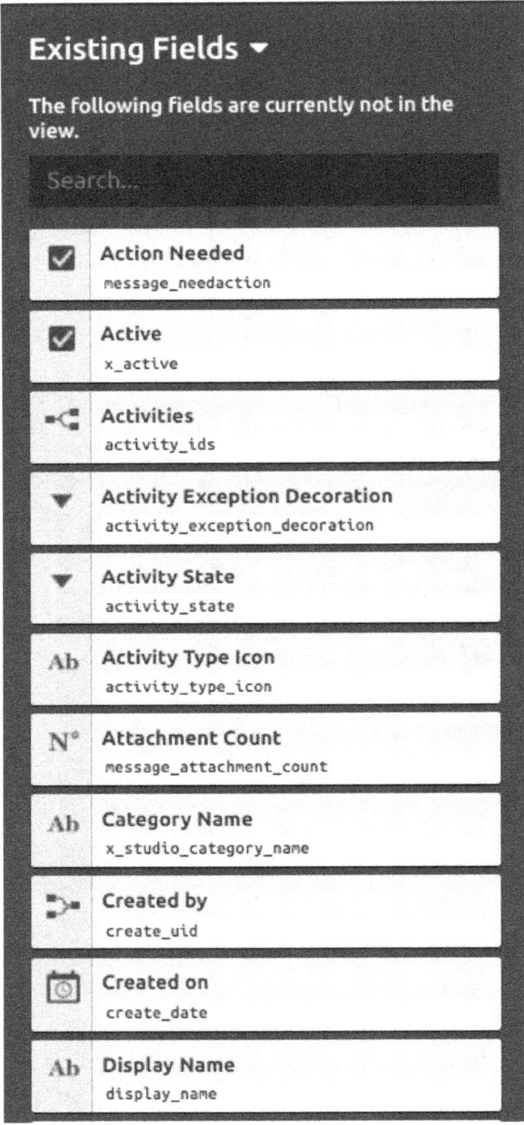

Figure 25.10 – The existing fields

Views

In Odoo Web Studio, views are fundamental components for designing the user interface of your custom modules. Views determine how data is displayed and interacted with in your Odoo applications.

There are several types of views you can work with in Odoo Web Studio to create and customize the user interface of your modules:

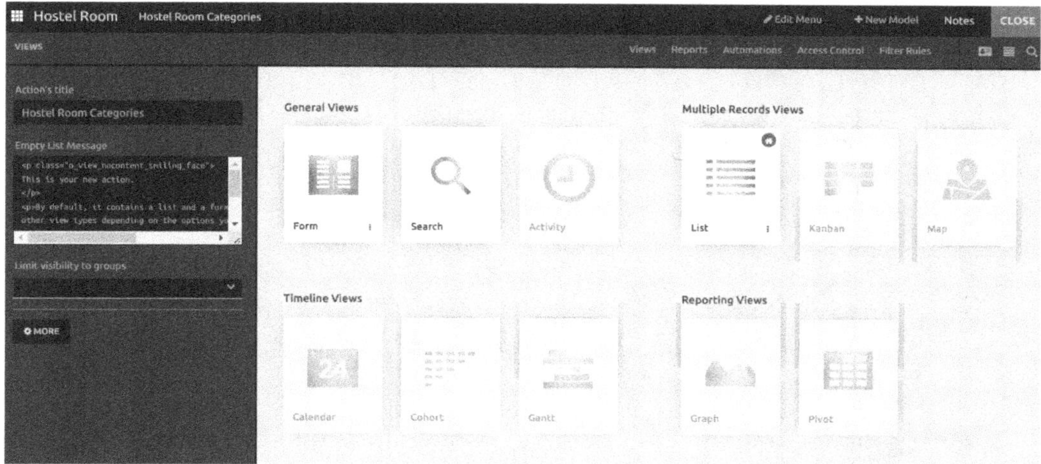

Figure 25.11 – Views

Let's look at some of the commonly used view types.

Form views

Form views allow users to view and edit individual records. You can customize the layout of form views by adding, removing, or rearranging fields. This view is commonly used for detailed record editing:

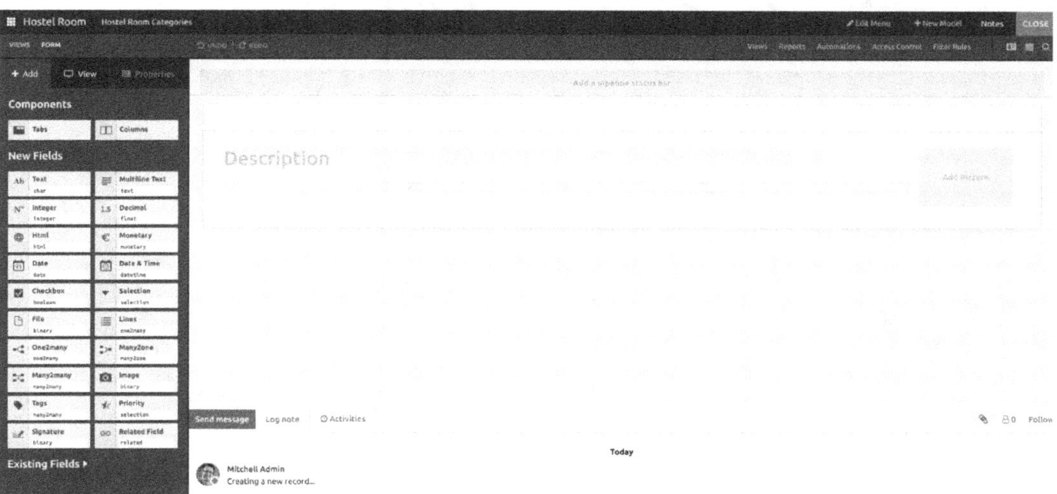

Figure 25.12 – Form view

In Odoo Web Studio, the form view is a crucial component for designing the user interface of your custom modules. Form views allow users to view and edit individual records within your application:

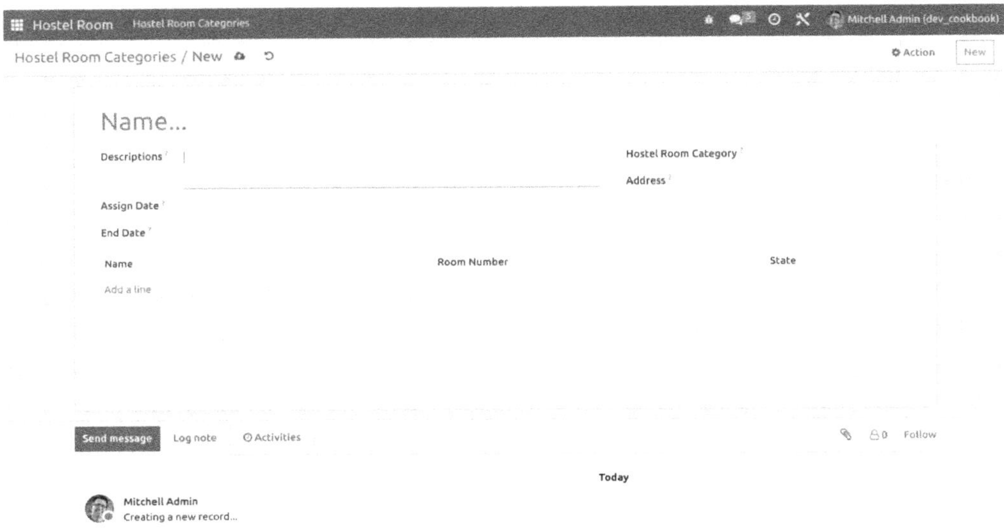

Figure 25.13 – Form view fields

We can create a form view using the fields shown in the preceding screenshot. Just drag and drop the field to create a new field in the form view where we want to display it:

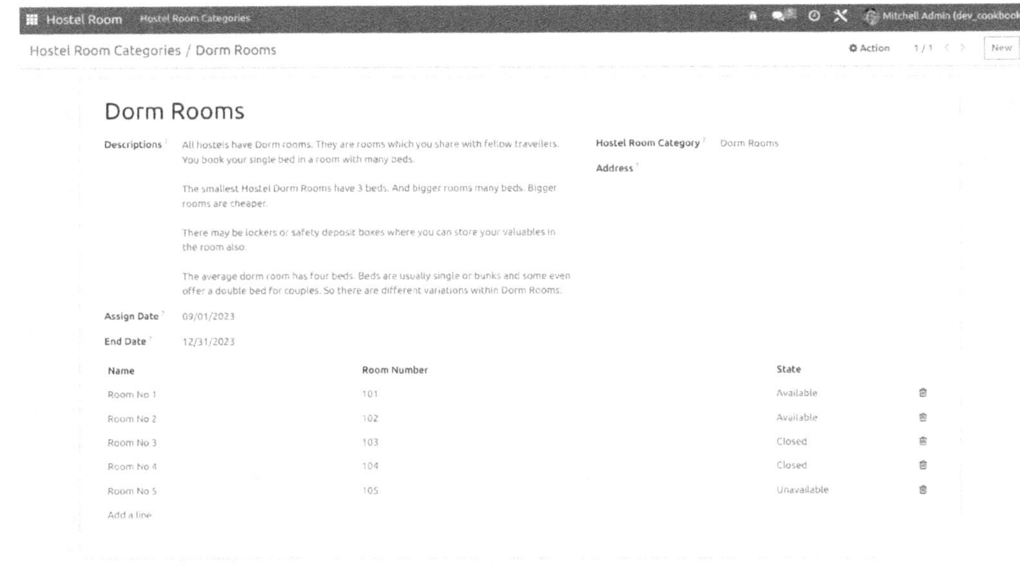

Figure 25.14 – The form view's field details

These are the field properties:

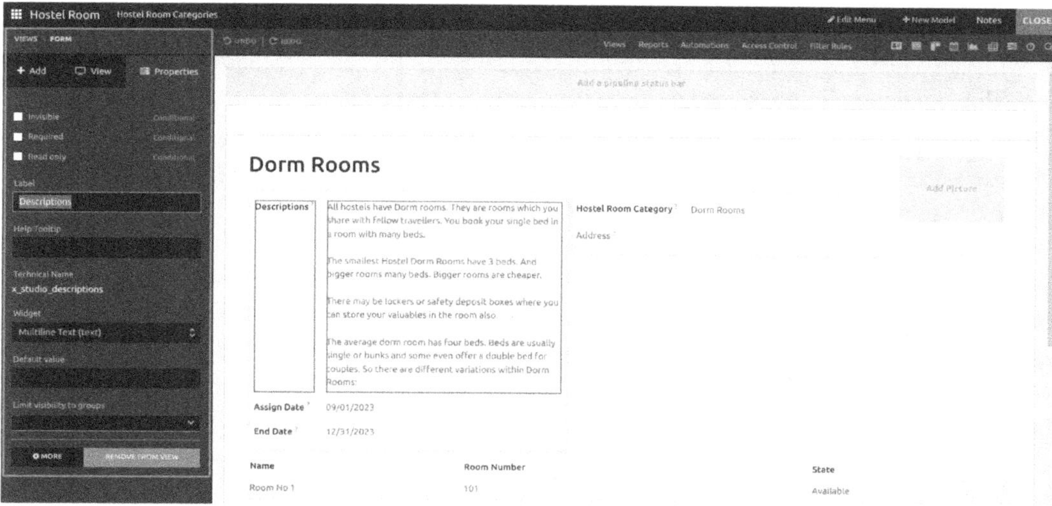

Figure 25.15 – The form view's field properties

Here, we can see the view options for particular fields:

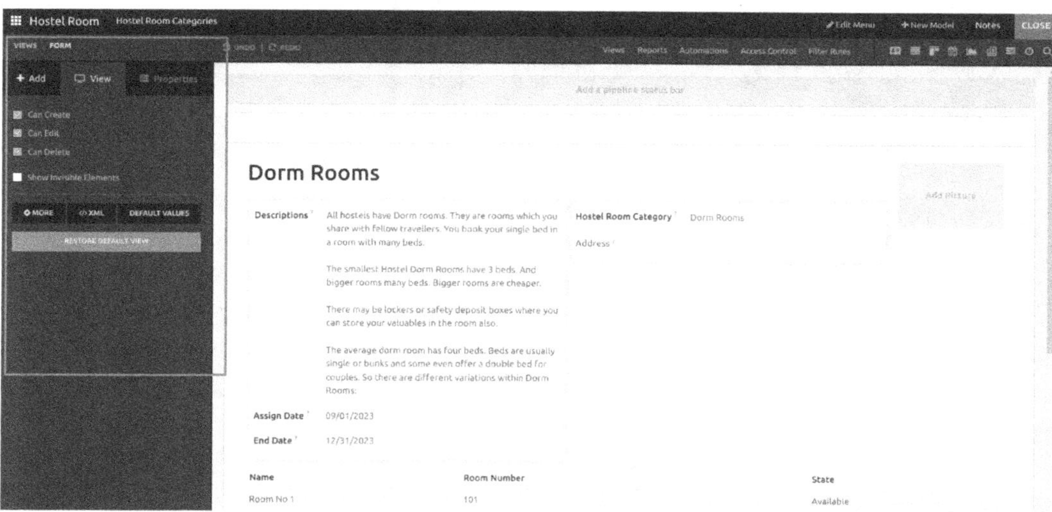

Figure 25.16 – Various form view field view options

Let's take a look at some of the important properties of the form view:

- **View inheritance**: In Odoo Web Studio, you can work with view inheritance. This allows you to base a new form view on an existing one and make specific modifications or additions to it. This can save you time when you're creating similar views.
- **Dependencies**: You can configure field dependencies within the form view. For example, you can make certain fields visible or required based on the values entered in other fields.
- **Validation rules**: Form views can have validation rules to ensure data accuracy. You can define constraints on fields to control the input data.
- **Saving and testing**: Save your changes when you're satisfied with the form view's design. To test the form view, go to the application or module where it's used, create or edit a record, and observe how your form view is displayed and functions.
- **Custom actions**: You can also link custom actions to buttons within the form view, allowing users to perform specific actions when they're interacting with records.

List views

List views display records in a tabular format, making it easy to browse and search multiple records. You can customize list views by selecting which fields to display, setting sorting options, and adding filters:

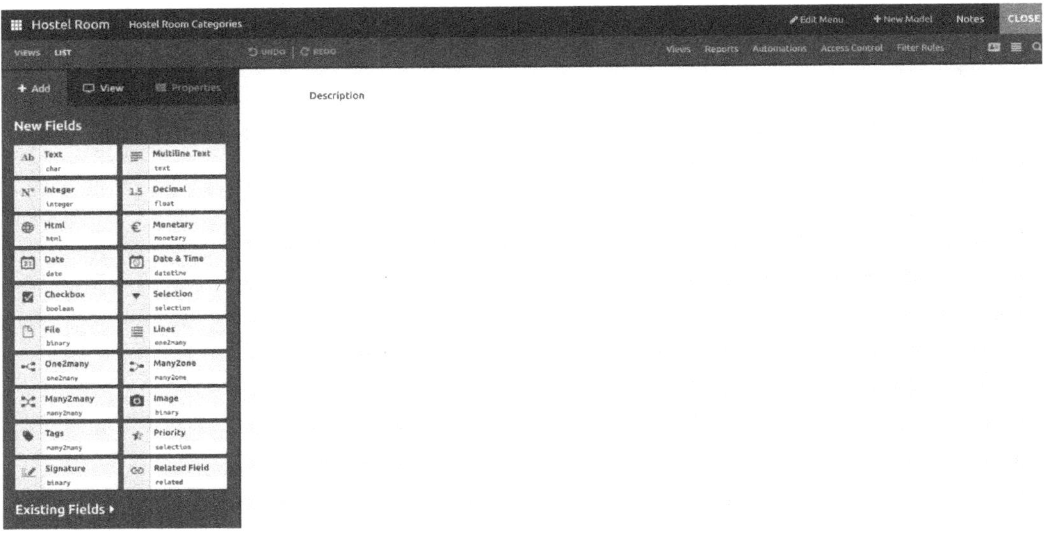

Figure 25.17 – List view

When you click a column in the list view, you can edit the properties of that field. Users can set the following properties of the field:

- **Invisible**
- **Required**
- **Read only**
- **Optional**
- **Label**
- **Widget**
- **Default value**
- **Limit visibility to the groups**:

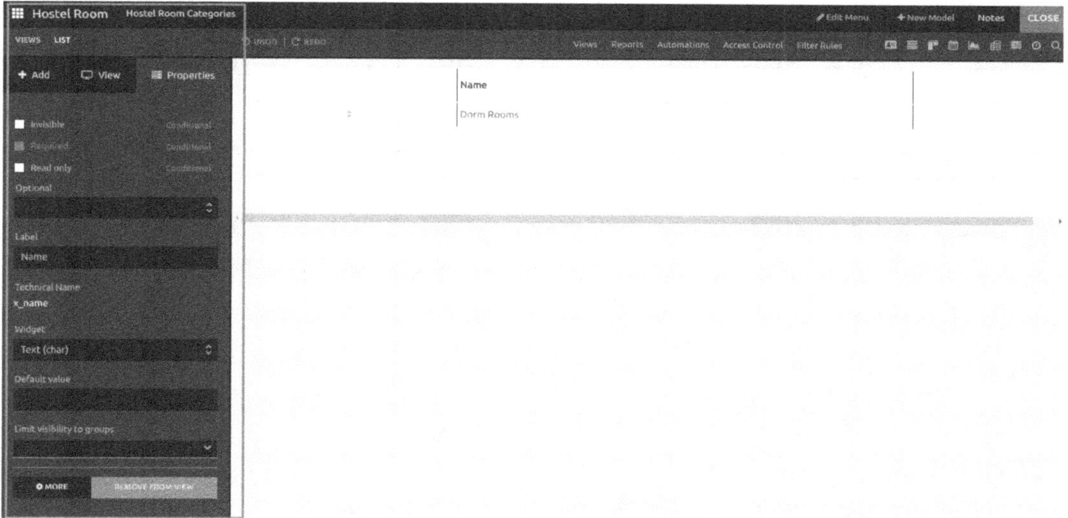

Figure 25.18 – List view properties

In Odoo Web Studio, the list view is an essential component for designing the user interface of your custom modules. Here's how you can work with list views in Odoo Web Studio:

1. **Create a new list view**: To create a new list view, click the **Create** button. Give it a name that reflects its purpose or function within your module.
2. **Design the list view**: Once you've created the list view, you can start designing it.
3. **Select the necessary fields**: Choose which fields you want to display in the list view by dragging and dropping them from the **Fields** section onto the list view canvas. You can arrange these fields as columns.

4. **Column properties**: Click on each column to access its properties. You can set labels, formatting options, and sorting behavior for each column.
5. **Sorting and grouping**: Configure how the records should be sorted and grouped in the list view.
6. **Filter criteria**: Add filter criteria to limit the records that are displayed in the list view based on specific conditions.

List view settings

Click on the list view itself to access its settings. You can configure various aspects, including the following:

- **Access Rights**: Define which user roles can view or access this list view.
- **Advanced Options**: Specify whether the list view should be visible, invisible, or read-only in specific situations.
- **Groups**: Set permissions and access rights for different user groups.
- **View Inheritance**: Similar to form views, you can also work with view inheritance for list views. This allows you to create new list views based on existing ones and make specific modifications or additions.
- **Search and Filter**: List views typically include a search and filter functionality, allowing users to quickly find records based on various criteria.
- **Group By and Totals**: You can enable grouping of records in the list view based on specific fields. Additionally, you can display totals and subtotals for numerical fields.
- **Batch Actions**: List views often include batch actions that allow users to perform actions on multiple selected records simultaneously, such as deleting, archiving, or updating records.
- **Column Visibility**: Users can often customize the visibility of columns in the list view, showing or hiding specific columns based on their preferences.
- **Sorting and Pagination**: Configure how records are sorted and displayed on the list view, including options for ascending or descending order and pagination.

Kanban views

Kanban views visualize records as cards or tiles, so they're often used for managing tasks or workflows. You can customize Kanban views by defining columns and cards' content and appearance:

Figure 25.19 – Kanban view

In Odoo Web Studio, Kanban views are useful components for designing user interfaces that visualize records as cards or tiles. These are often used for managing tasks, workflows, or project stages. Kanban views allow users to easily track the progress of records as they move through different stages. Let's learn how to work with Kanban views in Odoo Web Studio.

Accessing Kanban Views

Follow these steps:

1. To create or customize a Kanban view, go to the **Studio** module in your Odoo instance.
2. Click on the application or module for which you want to create or modify the Kanban view.
3. In the left sidebar, you will find a **Views** section, which includes **Kanban Views**. Click **Kanban Views** to see the existing Kanban views or create a new one.

Creating a new Kanban view

To create a new Kanban view, click the **Create** button. Provide a name for the Kanban view that reflects its purpose or function within your module.

Designing the Kanban view

Once you've created the Kanban view, you can start designing it:

- **Define columns**: Kanban views are organized into columns, representing different stages or categories. Define the columns you need for your workflow.
- **Add cards**: Drag and drop fields from the **Fields** section onto the Kanban view to define what information should be displayed on each card.
- **Configure card properties**: Click on each card to access its properties. You can set labels, formatting options, and sorting behavior for each card.

Kanban view settings

Click on the Kanban view itself to access its settings. You can configure various aspects, including the following:

- **Access Rights**: Define which user roles can view or access this Kanban view
- **Advanced Options**: Specify whether the Kanban view should be visible, invisible, or read-only in specific situations
- **Groups**: Set permissions and access rights for different user groups

View inheritance

As with other view types, you can work with view inheritance for Kanban views. This allows you to create new Kanban views based on existing ones and make specific modifications or additions.

Record movement and actions

In Kanban views, records can typically be moved from one column to another to indicate progress. You can configure actions or triggers that occur when records are moved or when specific actions are taken on cards.

Filter and search

Kanban views often include filtering and search capabilities to help users find and organize cards based on various criteria.

Card colors

You can use color-coding to highlight cards or records that require attention or have specific attributes.

Custom actions

Like other views, you can link custom actions to buttons or card interactions within the Kanban View.

Calendar views

Calendar views display records with date fields in a calendar format, making it suitable for scheduling and event management applications:

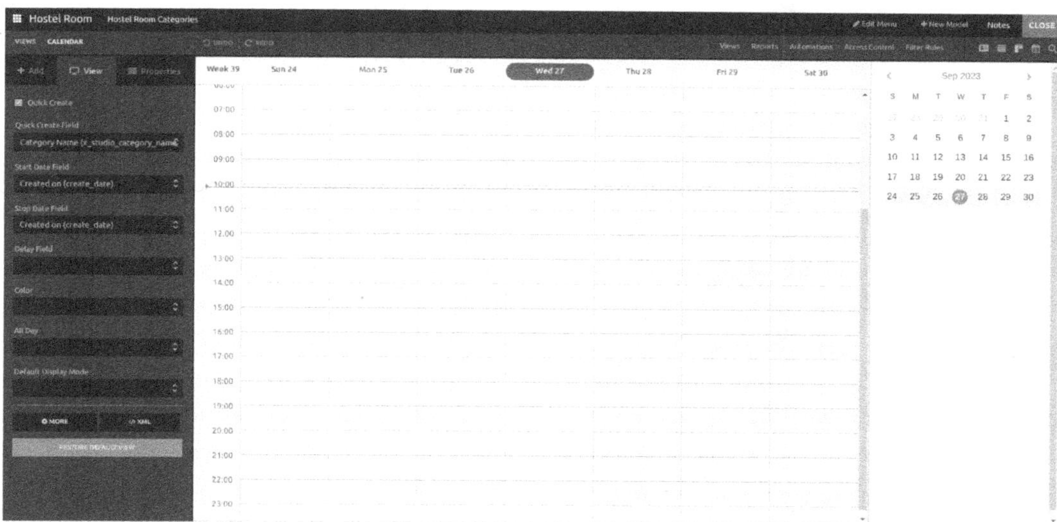

Figure 25.20 – Calendar view

In Odoo Web Studio, the calendar view is a component that allows you to present records with date-related information in a calendar format. This view is particularly useful for applications that involve scheduling, events, appointments, or any data that can be associated with dates and times. Let's learn how to work with calendar views in Odoo Web Studio.

Accessing calendar views

Follow these steps:

1. To create or customize a calendar view, go to the **Studio** module in your Odoo instance.
2. Click on the application or module for which you want to create or modify the calendar view.
3. In the left sidebar, you will find a **Views** section, which includes **Calendar Views**. Click **Calendar Views** to see the existing calendar views or create a new one.

Creating a new calendar view

To create a new calendar view, click the **Create** button. Provide a name for the calendar view that reflects its purpose or function within your module.

Designing the calendar view

Once you've created the calendar view, you can start designing it:

- **Define events**: Calendar views typically represent events or records associated with specific dates and times. You can define which fields from your data model will be displayed in the calendar, such as event titles, start and end dates, descriptions, and more.
- **Customize the events' appearance**: You can configure how events are displayed in the calendar, including the colors, text labels, and tooltips.

Calendar view settings

Click on the calendar view itself to access its settings. You can configure various aspects, including the following:

- **Access Rights**: Define which user roles can view or access this calendar view
- **Advanced Options**: Specify whether the calendar view should be visible, invisible, or read-only in specific situations
- **Groups**: Set permissions and access rights for different user groups

View inheritance

Similar to other view types, you can work with view inheritance for calendar views. This allows you to create new calendar views based on existing ones and make specific modifications or additions.

Drag-and-drop interaction

Users can typically interact with the calendar by dragging and dropping events to reschedule or modify them.

Filter and search

Calendar views often include filtering and search capabilities to help users find and organize events based on various criteria, such as date ranges or event types.

Event Details

Clicking on an event in the calendar view typically displays detailed information about the event, allowing users to view or edit event details.

Custom actions

As with other views, you can link custom actions to buttons or event interactions within the calendar view.

Graph views

Graph views allow you to create bar charts, line charts, and pie charts to visualize data based on selected fields. This is useful for data analysis and reporting:

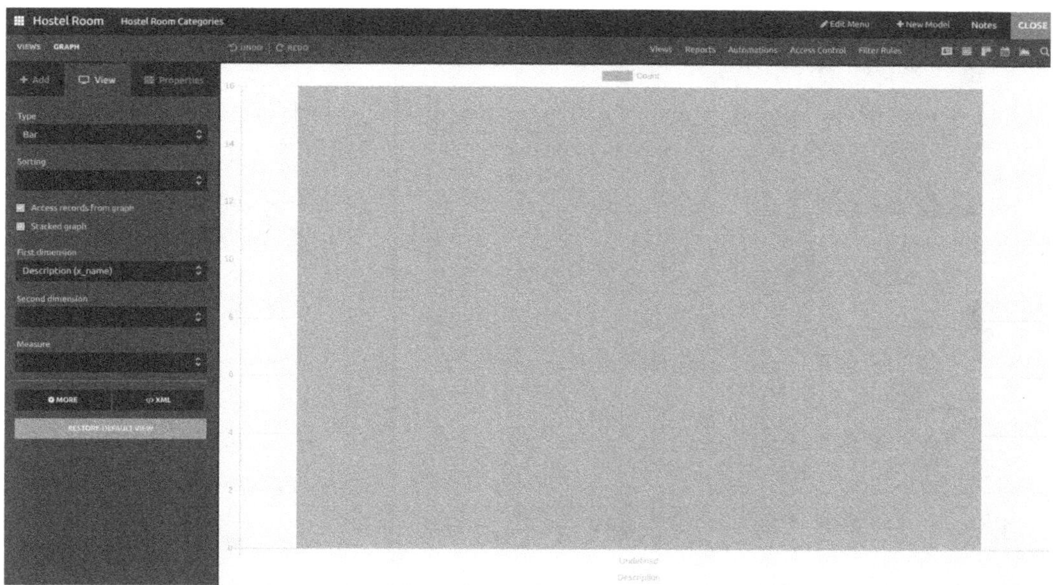

Figure 25.21 – Graph view

Pivot views

Pivot views provide an interactive way to analyze data by aggregating and summarizing records based on selected fields. Users can create custom reports and perform ad hoc analysis:

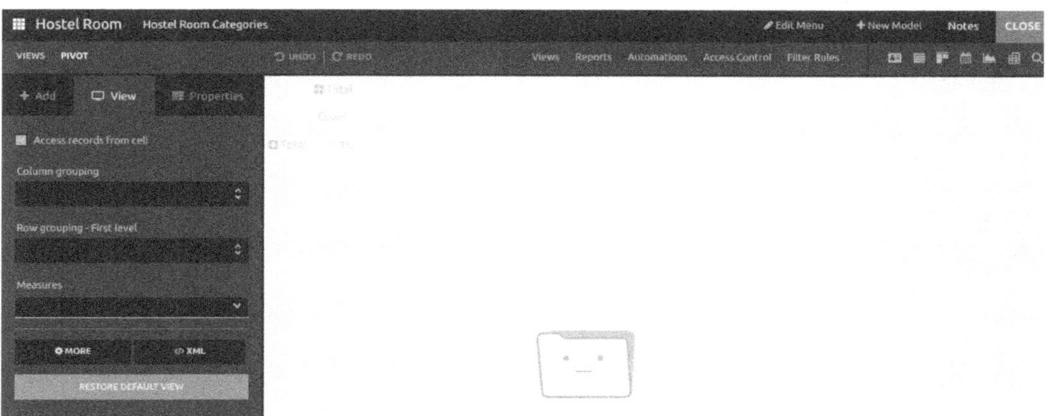

Figure 25.22 – Pivot view

Search views

Search views enable users to filter records based on specified criteria. You can customize search views by defining search filters and filter groups:

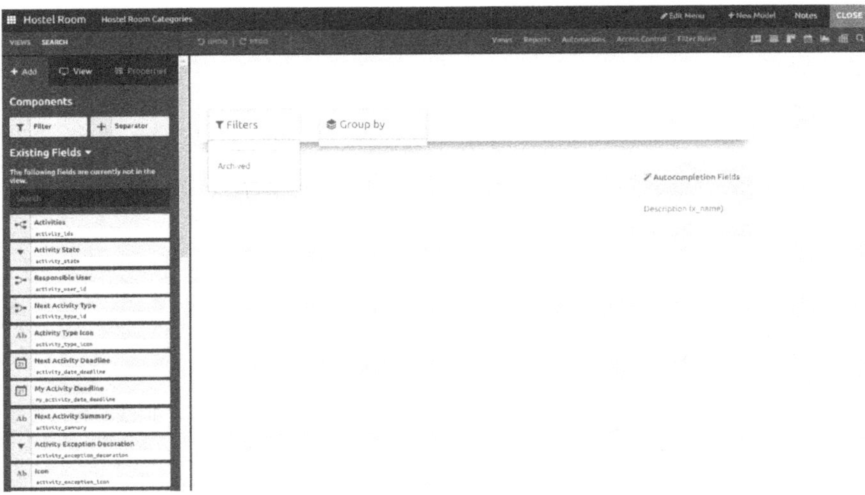

Figure 25.23 – Search view

Gantt views

Gantt views are used for project management and to display tasks or events along a timeline. Users can view and manage project schedules using this view:

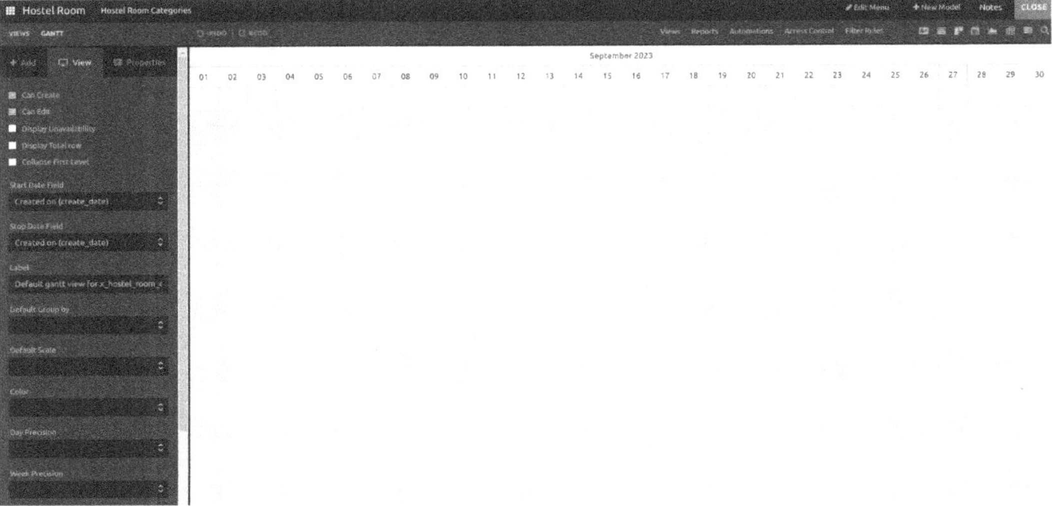

Figure 25.24 – Gantt view

Resource views

Resource views are used for resource management and to display resources (for example, employees and machines) and their availability over time.

Map views

Map views display records with geographic information on a map, making them suitable for location-based applications.

Activity views

Activity views show a timeline of activities related to a record, helping users track interactions and history:

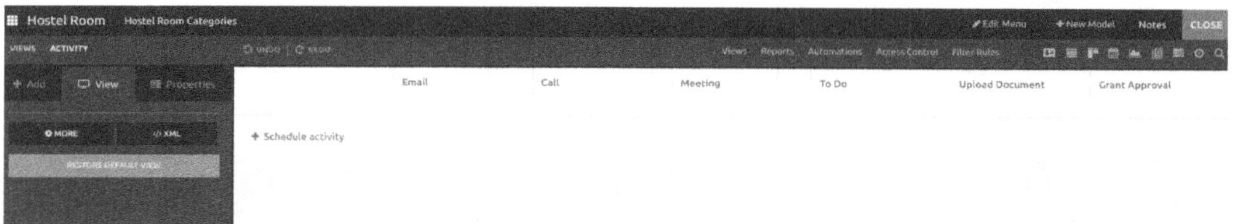

Figure 25.25 – Activity view

Building a new app

Creating a new app in Odoo Web Studio involves a series of steps to design and configure the data model, user interface, and functionality according to your specific business requirements. Here, we'll cover the general steps you'll need to follow to build a new app using Odoo Web Studio.

Defining the data model

In Odoo Web Studio, you can define the data model for your application. This includes creating custom objects (database tables) to store your data.

Use the visual interface to add fields, specify data types, set default values, and create relationships between objects:

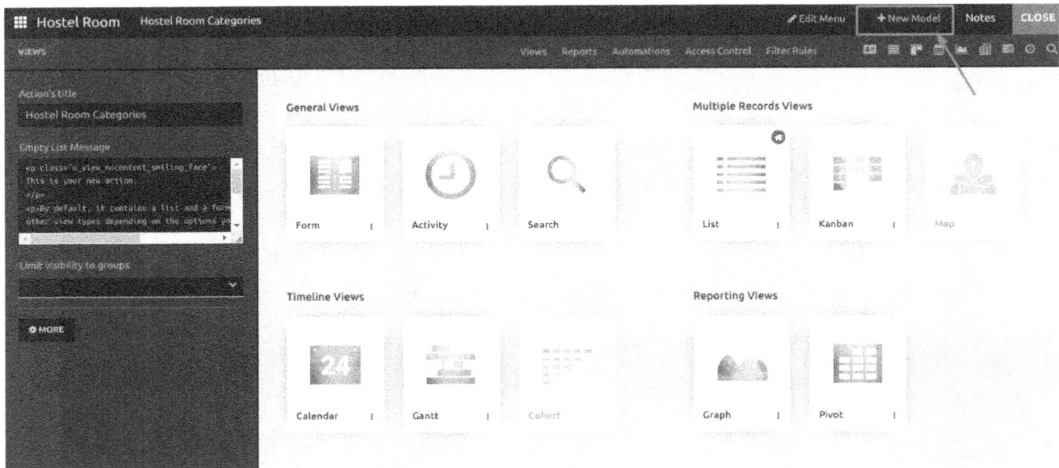

Figure 25.26 – Defining the data model

Once you click the **New Model** button, the next step is to specify the name of the model:

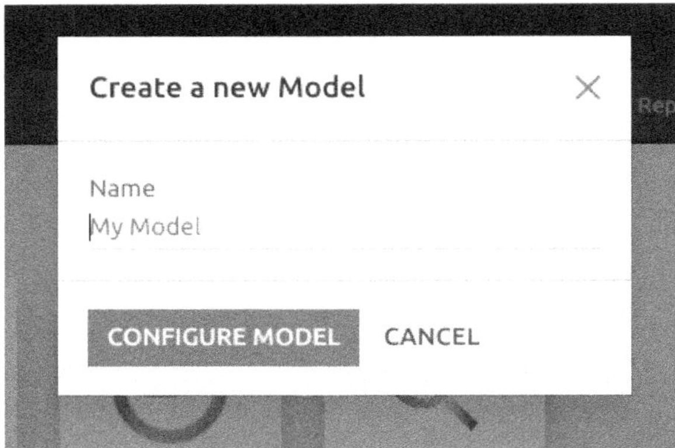

Figure 25.27 – Specifying the model's name

Once you've done this, you must choose the features of that model and then click **CREATE YOUR APP**:

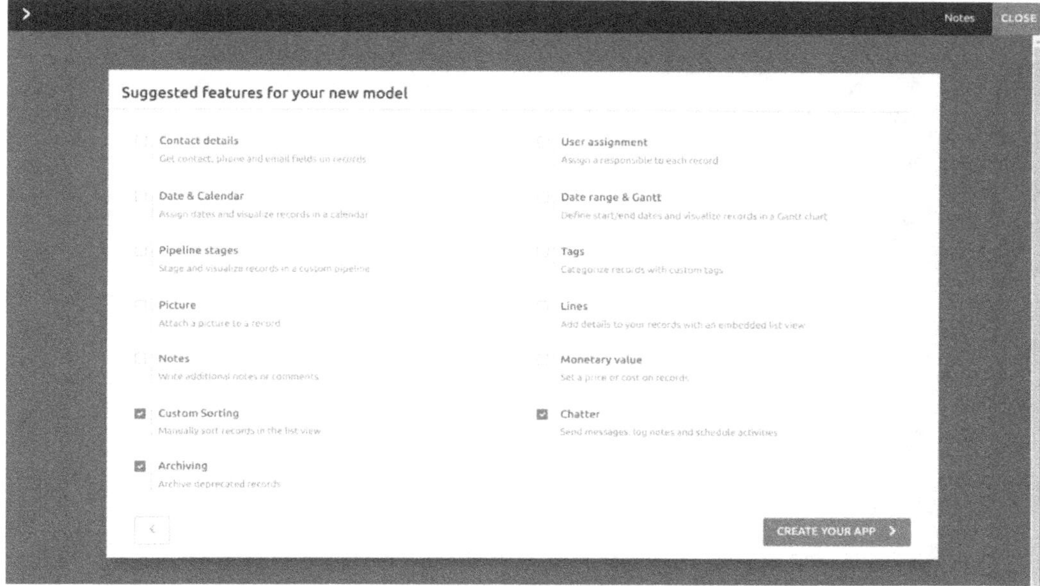

Figure 25.28 – Choosing model features

At this point, we'll have different options to customize the app:

Figure 25.29 – Various model options

Defining the general views

As explained in the *Views* recipe, we must choose the views of the model by clicking the **VIEWS** button:

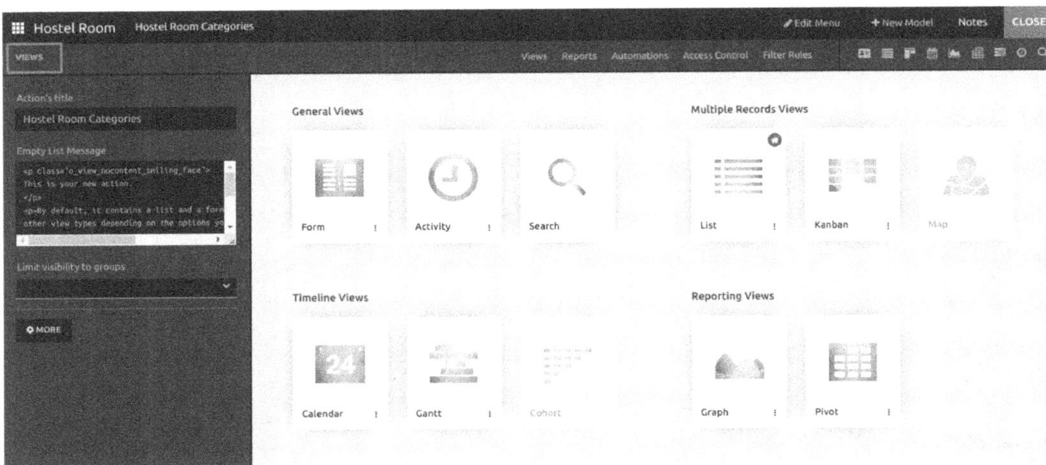

Figure 25.30 – View options

Choose the views you wish to use as per your requirements and add fields as per your needs and functions. These can be chosen from the left sidebar.

Defining the fields and components

In the form view, we can add **Tabs** and **Columns** from the **Components** section:

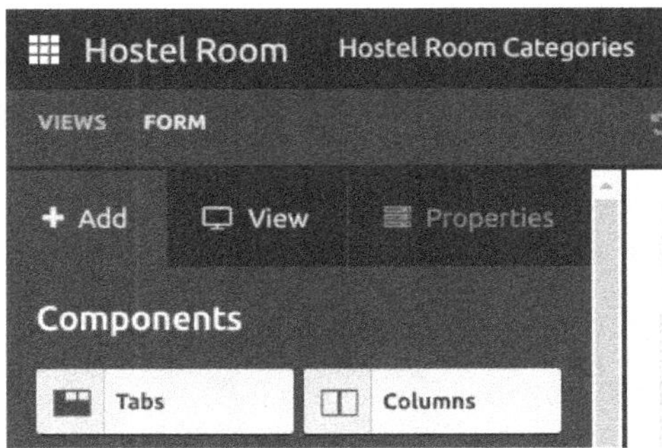

Figure 25.31 – Components options

Once you add tabs to the form view as **one2many** fields, you can edit the list and form view, as well as the **one2many** field itself:

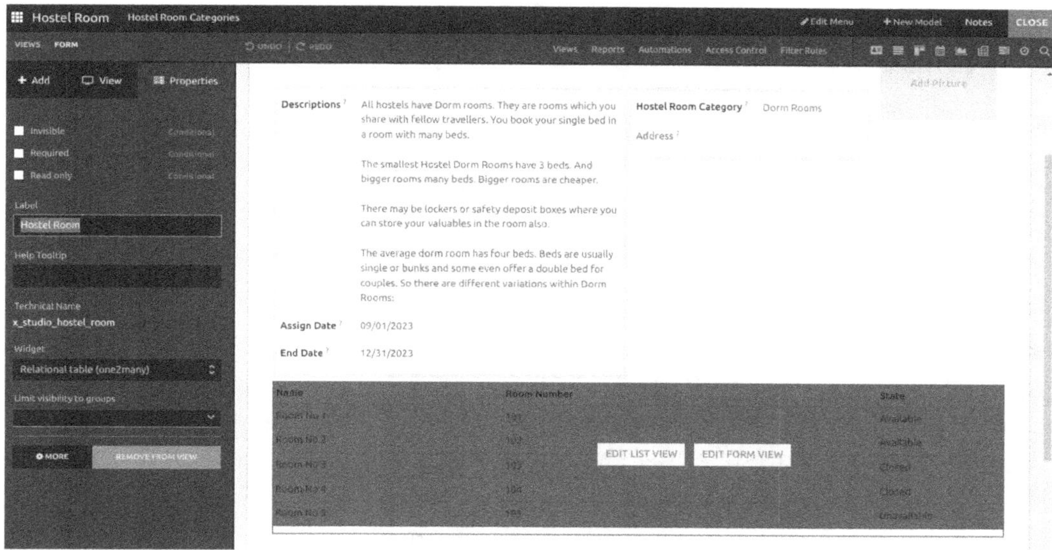

Figure 25.32 – Tabs options

We can also set the details for **Widget**, **Domain**, and **Limit visibility to groups**, **Context**, and more, as per the field types:

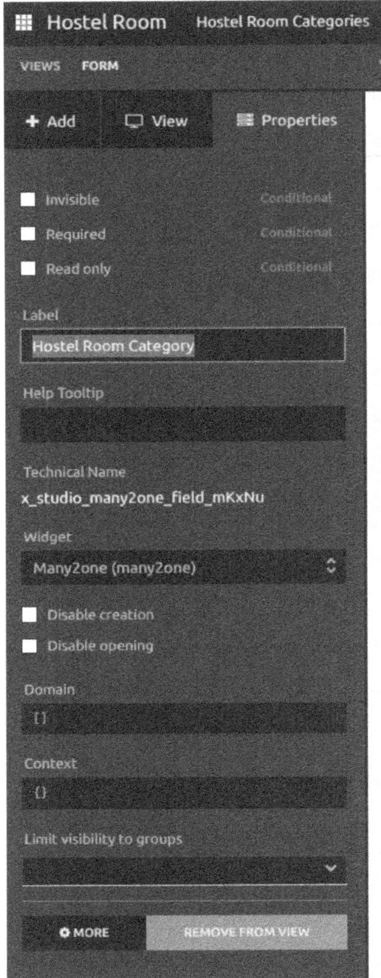

Figure 25.33 – Field properties

Text (char)

In Odoo Web Studio, Text fields are a common field type that's used to store and display textual information. Text fields are versatile and can be used to capture various types of textual data, such as names, descriptions, comments, and notes.

Multiline text (text)

In Odoo Web Studio, a Multiline text field allows users to input and display text that spans multiple lines or paragraphs. This type of field is useful when you need to capture longer descriptions, comments, notes, or any other form of text that extends beyond a single line.

Integer (integer)

In Odoo Web Studio, an Integer field is used to store and display integer (whole number) values. Integer fields are commonly used for various purposes, such as counting, quantifying, or capturing numeric data that does not require decimal points.

Decimal (float)

In Odoo Web Studio, a Decimal field is used to store and display numeric values with decimal points or fractions. Decimal fields are versatile and can be used to capture and store data that requires precision in terms of decimal places.

HTML (html)

In Odoo Web Studio, an HTML field allows you to store and display HTML-formatted content within your records. This field type is especially useful when you need to include rich text, formatted descriptions, or multimedia content within your application.

Monetary (monetary)

In Odoo Web Studio, a Monetary field is used to store and display monetary values, such as currency amounts. Monetary fields are essential for applications that involve financial transactions, accounting, or any scenario where you need to handle currency-related data.

Date (date)

In Odoo Web Studio, a Date field is used to store and display date values. Date fields are essential for applications that involve tracking events, scheduling, and recording dates associated with various records.

Date & Time (datetime)

In Odoo Web Studio, a Date & Time field is used to store and display both date and time values. This field is especially useful for applications that must record events, appointments, or transactions with precise timestamps.

Checkbox (Boolean)

In Odoo Web Studio, a Checkbox field is used to capture binary or Boolean values, which represent two states: checked (true) or unchecked (false). Checkbox fields are commonly used to record yes/no, on/off, or true/false responses to questions or conditions.

Selection (selection)

In Odoo Web Studio, a Selection field is used to provide users with a predefined list of options from which they can choose a single value. This type of field is commonly used when you want to capture categorical or discrete data with a limited set of choices.

File (binary)

In Odoo Web Studio, a File field is used to allow users to upload and store files, such as documents, images, spreadsheets, or any other type of digital files, within records. File fields are commonly used when you need to associate files with specific records, such as invoices, contracts, or product images.

Lines (one2many)

In Odoo Web Studio, a Lines field, also known as a one2many field, is used to create a relationship between two models (database tables) by establishing a one-to-many relationship. It allows you to associate multiple records from one model with a single record in another model. Lines fields are commonly used for scenarios where you need to link related records, such as order lines in an invoice or tasks in a project.

One2many (one2many)

In Odoo Web Studio, a One2many field is used to establish a one-to-many relationship between two models (database tables), allowing you to associate multiple records from one model with a single record in another model. One2many fields are commonly used in scenarios where you need to link related records, such as order lines in an invoice, tasks in a project, or products in a sales order.

Many2one (many2one)

In Odoo Web Studio, a Many2one field is used to establish a many-to-one relationship between two models (database tables), allowing you to associate a single record from one model with multiple records in another model. Many2one fields are commonly used in scenarios where you need to link records to a parent or reference record, such as linking a product to a category or a task to a project.

Many2many (many2many)

In Odoo Web Studio, a Many2many field is used to establish a many-to-many relationship between two models (database tables), allowing you to associate multiple records from one model with multiple records in another model. Many2many fields are commonly used in scenarios where you need to link multiple records to each other, such as tagging products with multiple categories or associating employees with multiple skills.

Image (binary)

In Odoo Web Studio, an Image field is used to allow users to upload and display images within records. Image fields are commonly used when you need to associate images with specific records, such as product images, profile pictures, or images related to marketing materials.

Tags (many2many)

In Odoo Web Studio, a Tags field is used to allow users to assign one or more tags or labels to records. Tags are short descriptive labels that help categorize and organize records based on specific criteria or attributes. Tags fields are commonly used in scenarios where you want to implement a flexible and user-driven categorization system, such as tagging products with product categories or labeling tasks with project stages.

Priority (selection)

In Odoo Web Studio, a Priority field is typically a selection field that's used to indicate the priority or importance level of a record or a task. Priority fields are commonly used in applications such as project management, task tracking, and issue tracking to help users and teams prioritize their work.

Signature (binary)

In Odoo Web Studio, a Signature field allows users to capture and store digital signatures within records. Signature fields are commonly used in scenarios where you need to collect and verify signatures as part of a workflow or approval process, such as signing off on documents, contracts, or delivery confirmations.

Related Field (related)

In Odoo Web Studio, Related Field is a powerful field type that allows you to display data from a related record on the current record's form view without creating a physical database link between the two records. It's commonly used when you want to display information from another model (database table) related to the current record.

Defining the compute method for a field

In Odoo Web Studio, you can define a computed field using a Python method to calculate its value dynamically based on other fields or data. Computed fields are useful when you want to display calculated or derived values in your records.

To define the compute method for the field, you'll need to write Python code in the Odoo model class associated with your custom module. Let's look at a basic example of how to define a compute method.

Defining a compute method using code

In this example, we've created a computed field called `computed_field` that calculates its value based on `field1` and `field2`. The `@api.depends` decorator specifies the fields that trigger the computation when their values change:

```
from odoo import models, fields, api

class YourModelName(models.Model):
    _name = 'your.module.name'
```

```
    _description = 'Your Module Description'

    # Define the fields used in the computation
    field1 = fields.Float('Field 1')
    field2 = fields.Float('Field 2')

    # Define the computed field
    computed_field = fields.Float('Computed Field', compute='_compute_
computed_field')

    # Define the compute method
    @api.depends('field1', 'field2')
    def _compute_computed_field(self):
        for record in self:
            # Perform the computation and assign the result to the
computed field
            record.computed_field = record.field1 + record.field2
```

Let's see how we can add the compute field using Odoo Web Studio.

Here, we can take a look at a small Compute field for calculating the total of a sale order line.

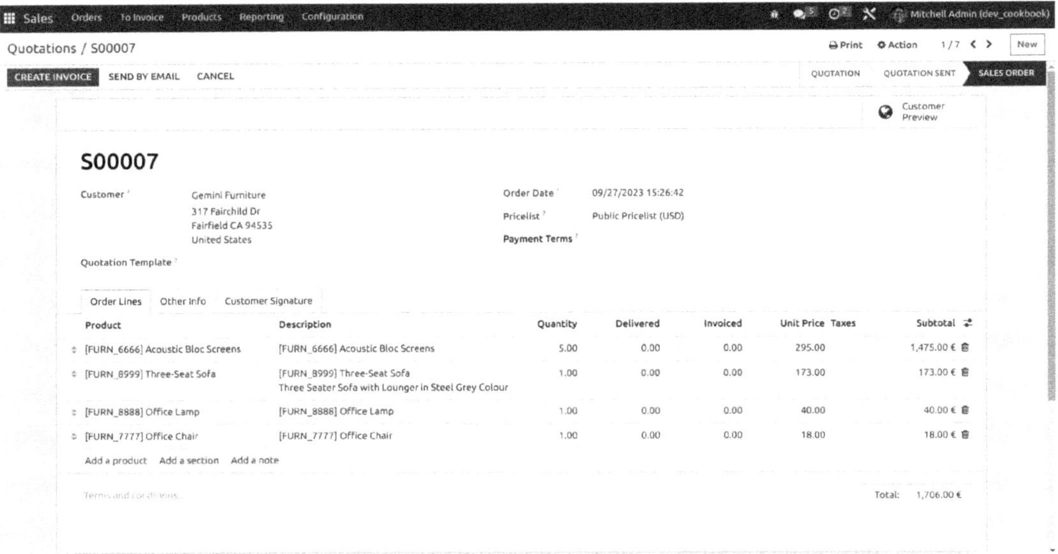

Figure 25.34 – Sales order

As shown in the following screenshot, we added one Compute field (**Total**) using Odoo Web Studio:

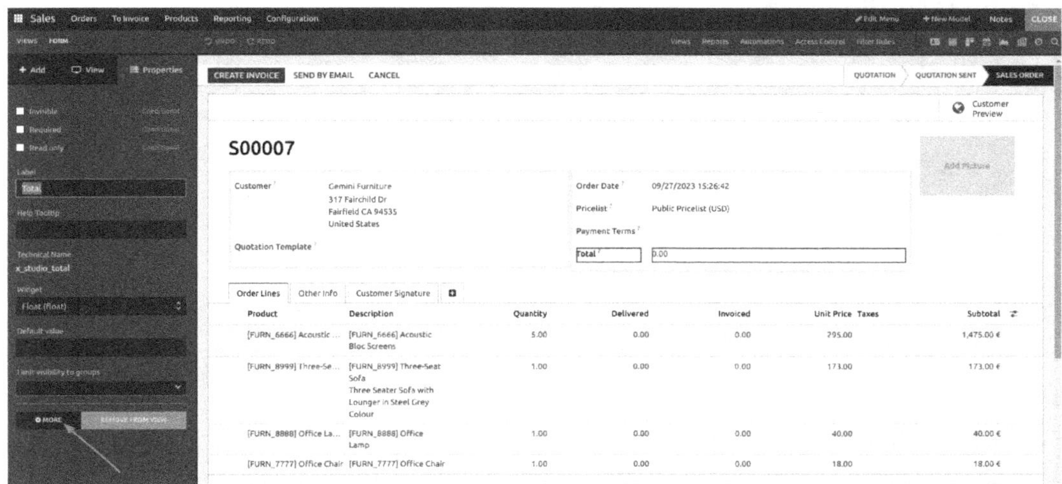

Figure 25.35 – Screenshot of the Added Total float field to write a compute method

Click the **MORE** button to see all the properties of the field:

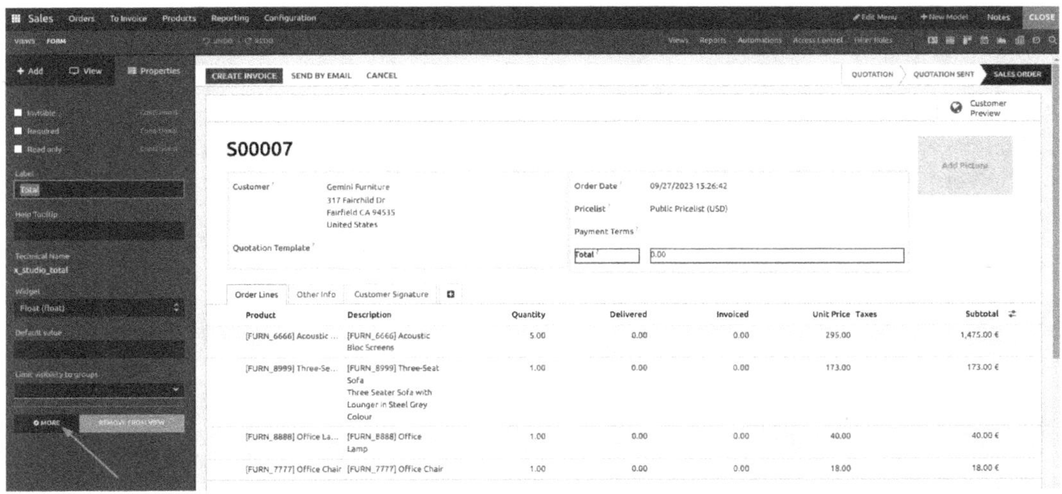

Figure 25.36 – MORE

Once you've done this, you'll see the **Dependencies** and **Compute** options under **ADVANCED PROPERTIES**:

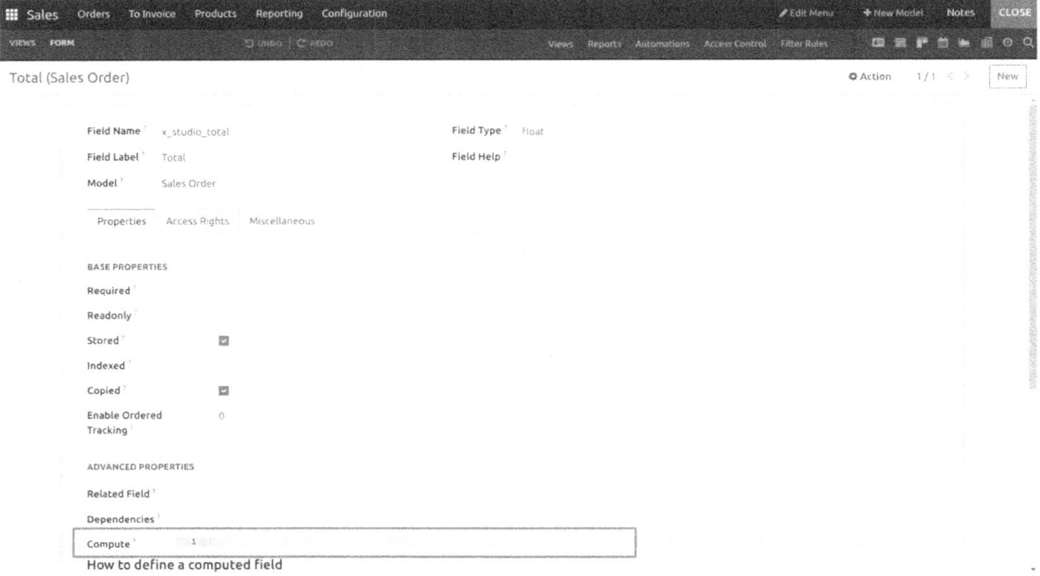

Figure 25.37 – The ADVANCED PROPERTIES area

In Odoo, a Computed field is a field that is not stored in the database but is dynamically calculated based on the values of other fields or data. Computed fields are used to display calculated or derived values in your records. They are especially useful when you need to perform calculations or apply business logic to fields in your database records.

The only predefined variables are as follows:

- self (the set of records to compute)
- datetime (Python module)
- dateutil (Python module)
- time (Python module)

Other features are accessible through self, such as **self.env**.

So, add some field dependencies and write the Python code in the **Compute** box:

ADVANCED PROPERTIES

Related Field ?

Dependencies ? order_line,order_line.price_subtotal

Compute ?

```
total = 0
for rec in self:
  for line in self.order_line:
    total = total+ line.price_subtotal
    rec['x_studio_total'] = total

```

Figure 25.38 – The compute method's code

Now the compute method calculates the field values and stores them in the **Total** field:

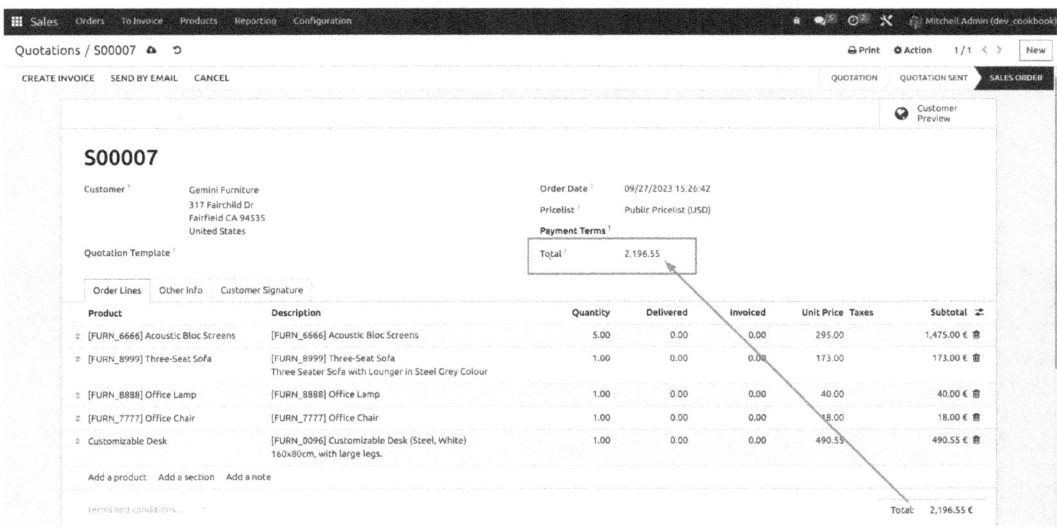

Figure 25.39 – The compute method's calculation in the Total field

Adding a button

In Odoo Web Studio, you can add a button to your custom views to trigger specific actions or functions within your Odoo application. Buttons are commonly used to initiate processes, validate data, or perform custom actions.

To add a button to your views, click on the XML section and add a new button through the code. Note that the button must be an action type button:

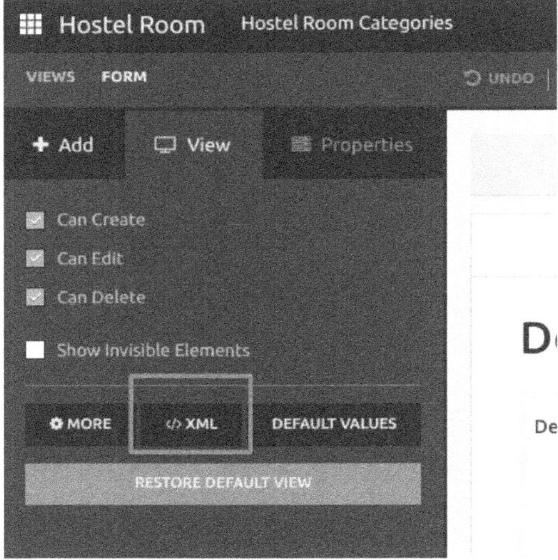

Figure 25.40 – The XML section, where you can add/modify anything through code

Once we click **XML**, the editor will open so that we can modify or add anything through the code:

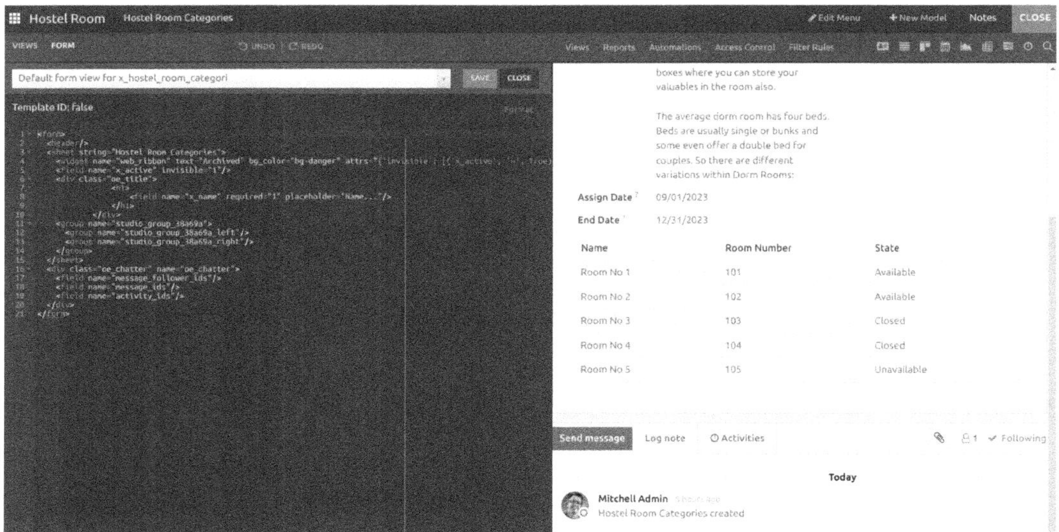

Figure 25.41 – Using the XML editor to add/modify anything through code

Adding a smart button

In Odoo Web Studio, a smart button is a dynamic UI element that displays summarized information and provides quick access to related records. Smart buttons are commonly used to display counts of related records, such as the number of orders, tasks, or leads associated with a specific record, and they allow users to navigate to those related records with a single click.

Hover your mouse cursor over the top-right corner; a + sign will become visible. You can use this to add a smart button:

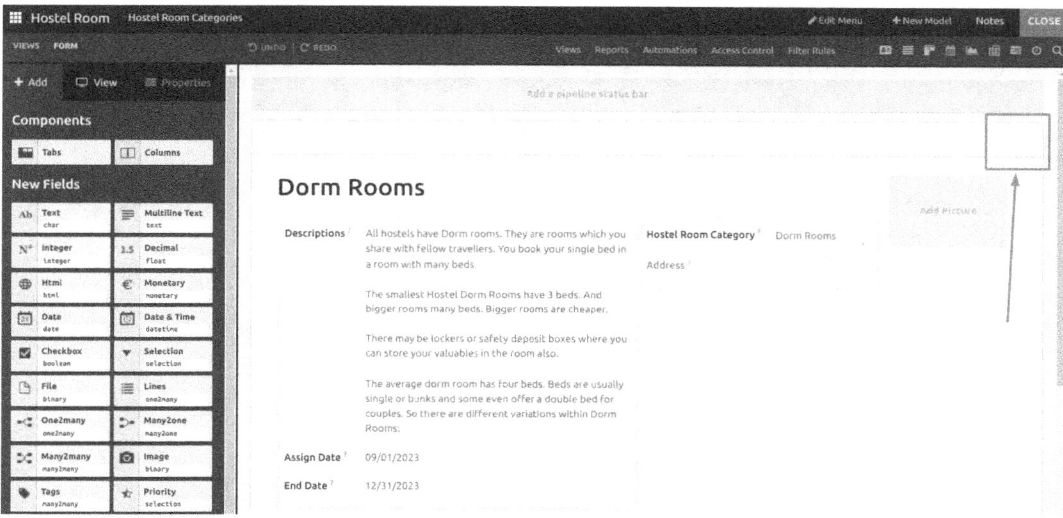

Figure 25.42 – Adding a smart button

Once you click on the + sign, a new window will open called **Add a Button**. Here, you can add a label and choose the icon of the smart button:

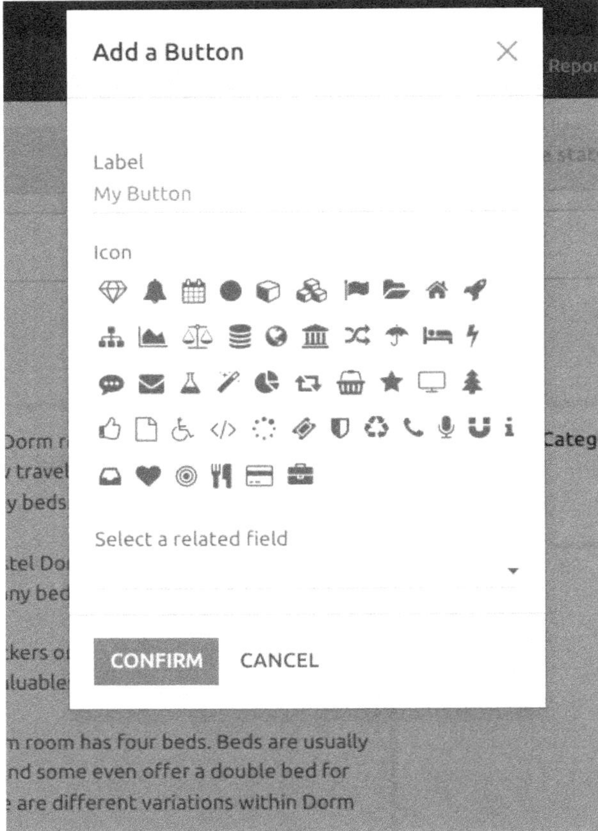

Figure 25.43 – The Add a Button options

Adding a status bar and filters

In Odoo Web Studio, you can create and customize a status bar and filters to enhance the user experience and improve the navigation of your custom views. Status bars typically display key information about the current record or context, while filters allow users to refine the records that are displayed in a list or search view:

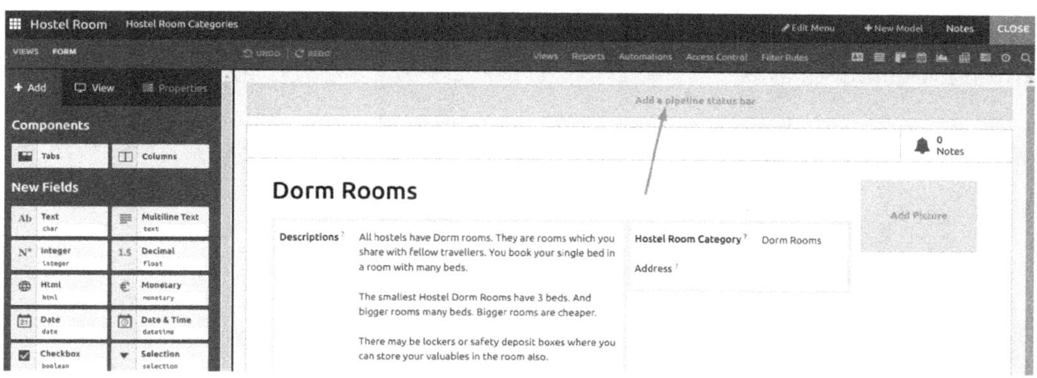

Figure 25.44 – Add a pipeline status bar

Once you click on the **Add a pipeline status bar** button, a window will open where you can add status bar options:

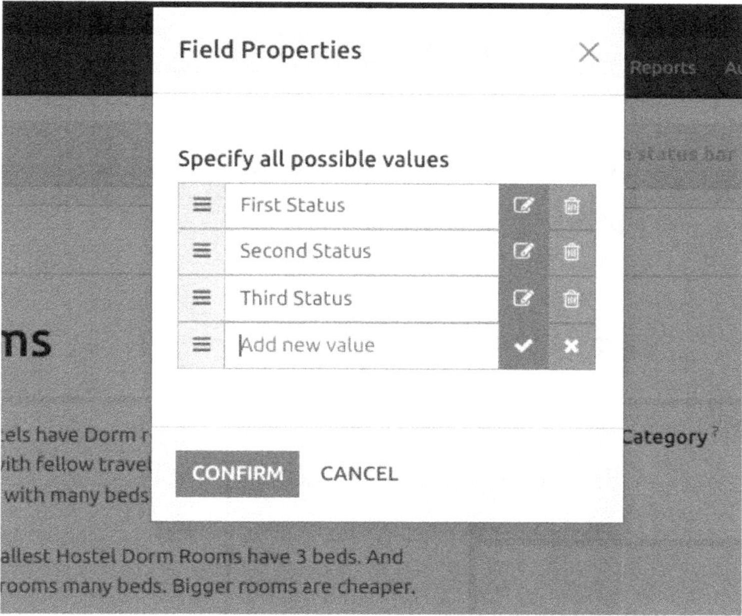

Figure 25.45 – Status bar properties

Once you've edited and added the status bar's field properties, click **CONFIRM**. The status bar will now be visible in your views:

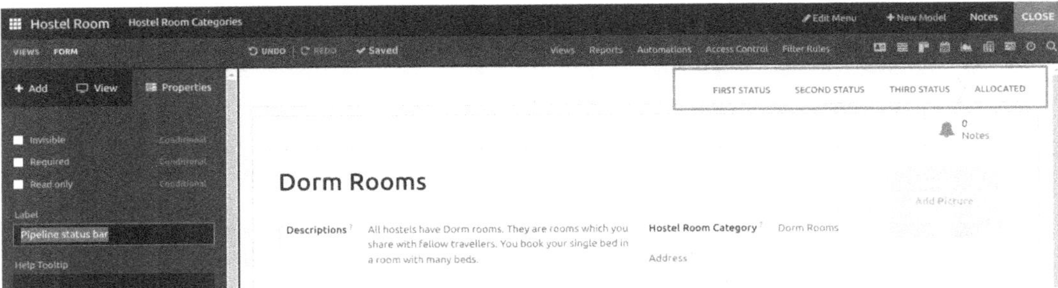

Figure 25.46 – The added status bar

Filters

In Odoo Web Studio, you can create and customize filters to allow users to refine and filter records in list views and search views. Filters are valuable in helping users find specific information within a large dataset:

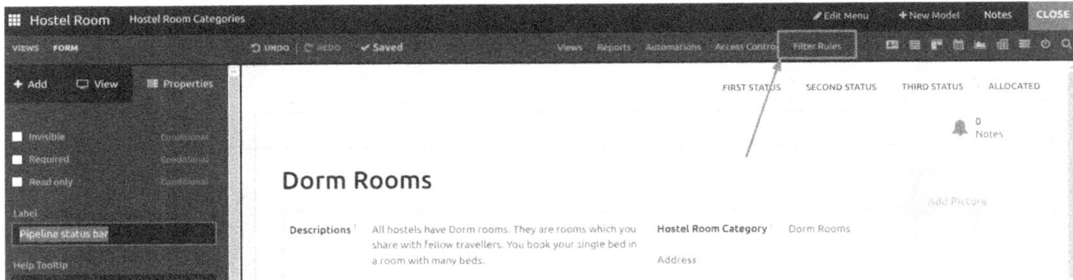

Figure 25.47 – Filter Rules

The following screenshot shows the common filter rules you must configure, as per your needs. You can also customize the domain as per the filter rules:

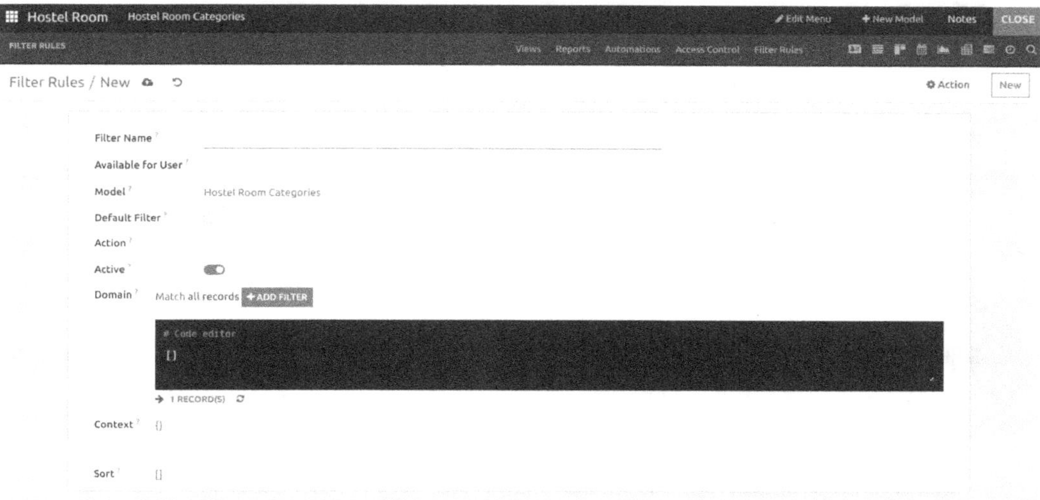

Figure 25.48 – Adding new filter rules

Edit Menu

In Odoo Web Studio, you can customize the menu structure of your Odoo application by adding, editing, or removing menu items. These menu items allow users to access different parts of the application, such as modules, views, and actions:

Figure 25.49 – Edit Menu

Once you click the **Edit Menu** button, a window will open where you can edit menu items:

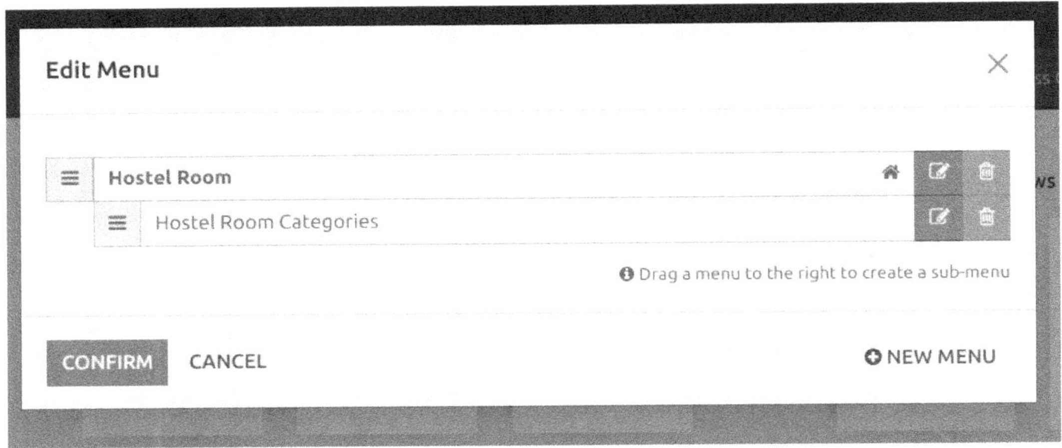

Figure 25.50 – Editing menu items

Click on the **Edit** icon of the menu item you wish to edit. Here, you can edit the following:

- **Name**: Change the display name of the menu item.
- **Action**: Modify the action associated with the menu item. Actions define what happens when users click on the menu item. You can associate a specific view, action, or function with the menu item.
- **Parent Menu**: Specify the parent menu under which the menu item should appear. This controls the hierarchy and organization of the menu's structure.
- **Visibility**: Define the visibility of the menu item based on user roles, groups, or conditions.
- **Icon**: Optionally, add an icon to represent the menu item.
- **Sequence**: Adjust the order in which the menu item appears within its parent menu.
- **Access Rights**: Configure access rights and permissions for the menu item, specifying which user roles can see or access it.

Customizing an existing app

Customizing an existing app in Odoo Web Studio involves making modifications to the app's functionality, views, and data structures so that it can be aligned with your specific business requirements. This recipe will cover the general steps you must follow to customize an existing app in Odoo Web Studio:

> **Note**
>
> Customizing an existing app typically requires developer-level access or the use of Odoo Web Studio for simpler customizations. Ensure you have the necessary permissions and access.

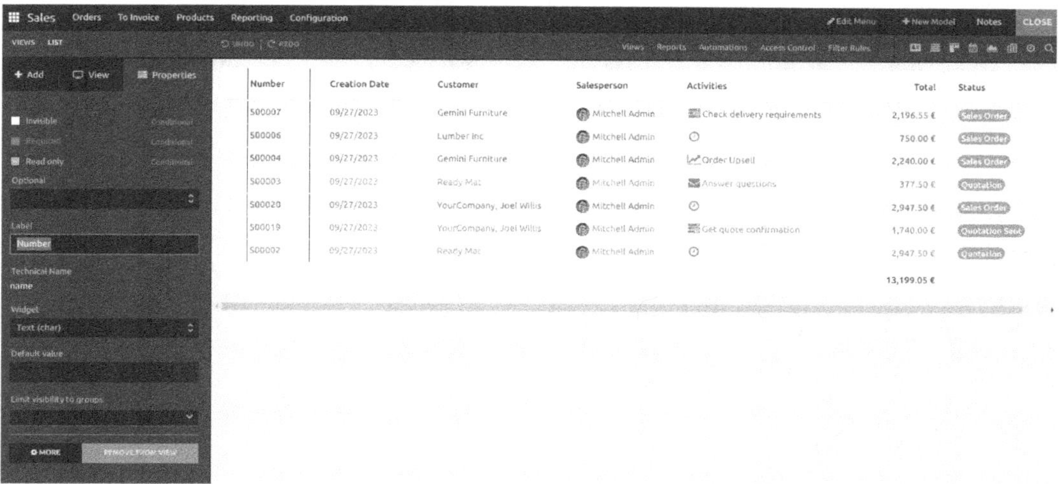

Figure 25.51 – Customizing an existing app

Choosing an existing app to customize

From the main dashboard or menu, select the app or module you want to customize. This could be any existing app in Odoo. Here, you have various options:

- **Customize views**: Use Odoo Web Studio to customize the views of the app. You can modify existing views or create new ones to display data in the way you want. You can add or remove fields, change their labels, and adjust their positions on form views, list views, Kanban views, and more.
- **Add or modify fields**: Add new fields to the app's data model or modify existing fields. You can define field types, labels, default values, and other field properties using Odoo Web Studio.
- **Create or edit actions**: Define actions and workflows for the app. Actions determine what happens when users perform specific actions, such as clicking buttons or menu items. You can create custom actions or modify existing ones.
- **Add buttons and menu items**: Customize the app's menu structure by adding buttons, menu items, and links to various parts of the app. This allows users to navigate easily between different views and functionalities.

- **Configure access rights**: Set access rights and permissions for the app's views, models, and actions. Define who can view, edit, or delete data and access specific features within the app.
- **Implement business logic**: Use Odoo Web Studio to implement custom business logic by defining computed fields, server actions, and other rules that automate processes and calculations within the app.
- **Add custom reports**: If needed, create custom reports and documents using Odoo's reporting tools. Define report templates and layouts to generate documents such as invoices, purchase orders, and sales quotes.

Built-in functions

Odoo Web Studio provides a set of built-in features and tools that allow users to customize, extend, and enhance their Odoo applications without the need for extensive programming or development skills. These built-in features are designed to streamline app customization and empower users to adapt their Odoo instance to their specific business needs.

Importing modules

Once you click on the **Import** link, a pop-up window will appear where you can upload the module's ZIP file:

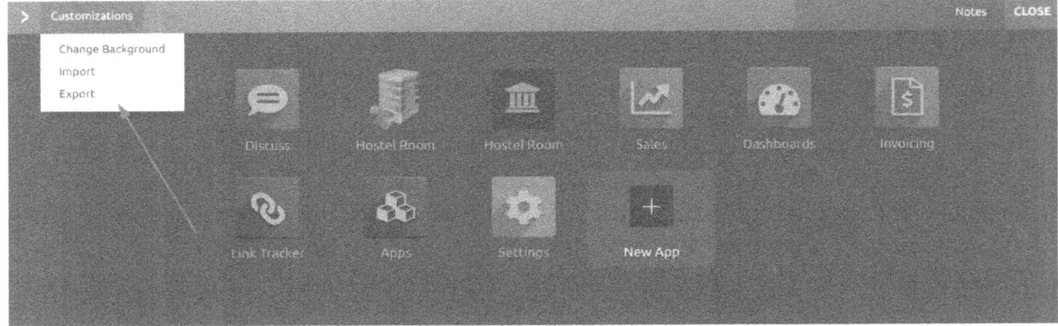

Figure 25.52 – Screenshot of the Import Export options

Here's what the **Import modules** option looks like:

Figure 25.53 – Import modules

Upload your **Module file (.zip)** and check **Force init**. Force init mode, even if installed, will update 'noupdate == 1' records:

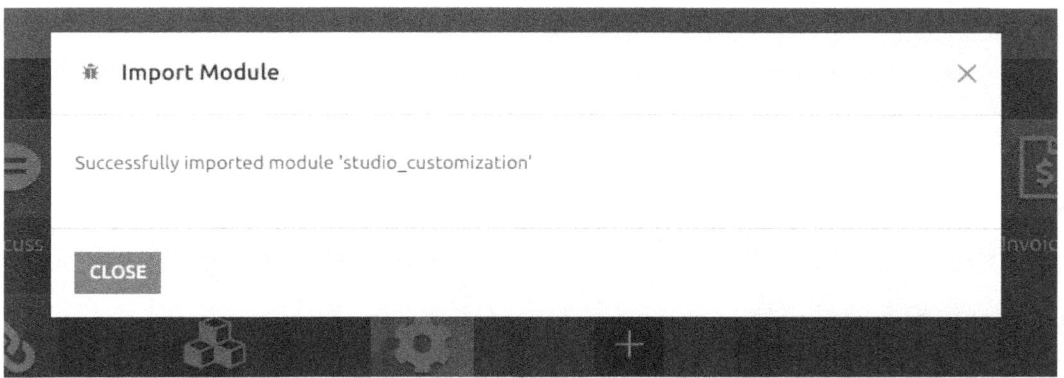

Figure 25.54 – Successfully importing a module

Once you've uploaded the module file and clicked **Import**, you will get a message stating that the module was imported successfully:

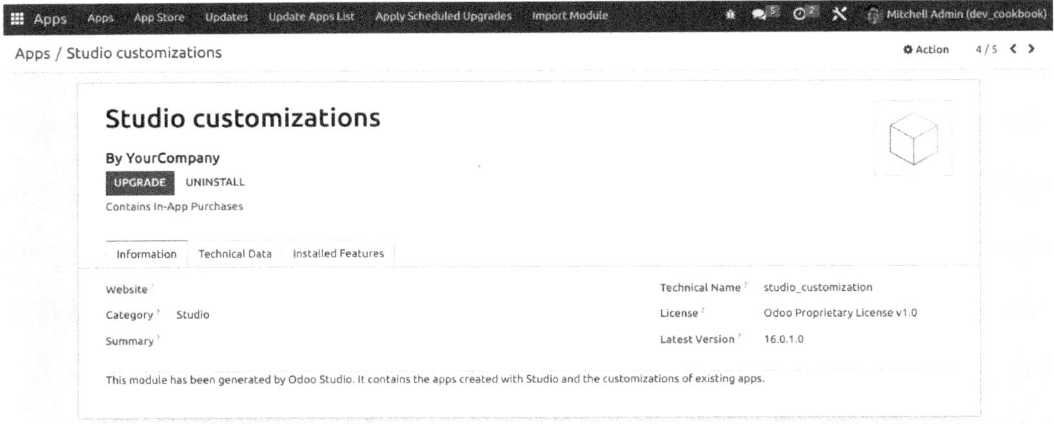

Figure 25.55 – The imported module

Go to the **App** list and search for the **Studio** customizations module.

Exporting modules

To export modules from the Odoo database, you have to install the Odoo Studio module and then customize it in the database. If we do any customization from Odoo Web Studio, a new module will be created in the database:

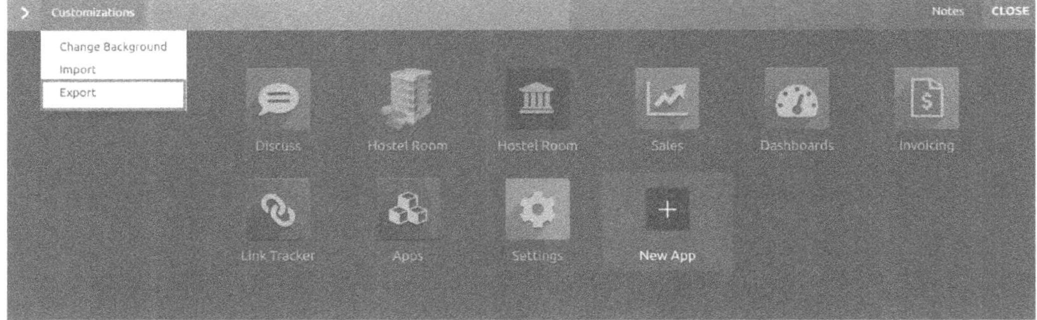

Figure 25.56 – The Export option

Once you click on the **Export** link it will download the Studio customize module. You can also import this module into other Odoo databases.

Search view

In Odoo Web Studio, you can customize search views to tailor the way users search for and filter records within a specific module or application. Search views allow users to refine their search criteria, making it easier to find specific records:

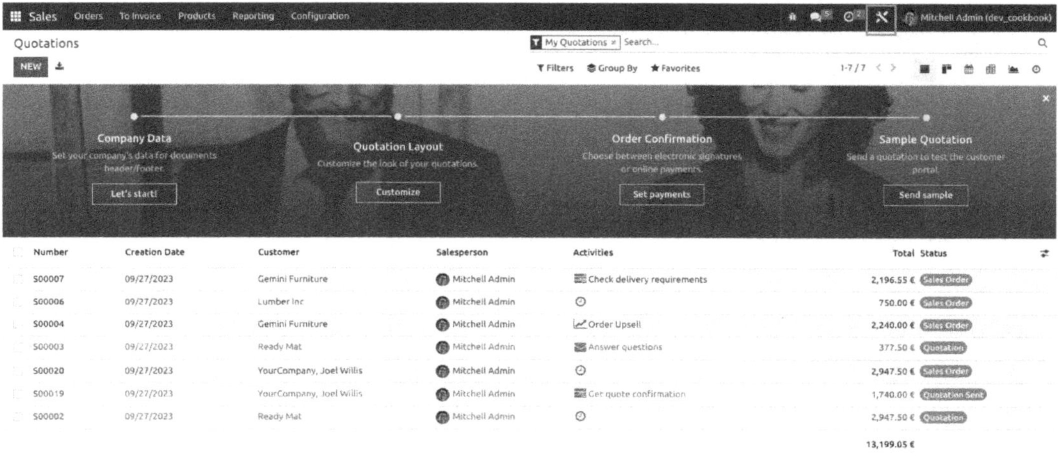

Figure 25.57 – Search filters

Click on the Studio icon, then **VIEWS**. Select the **Search** view:

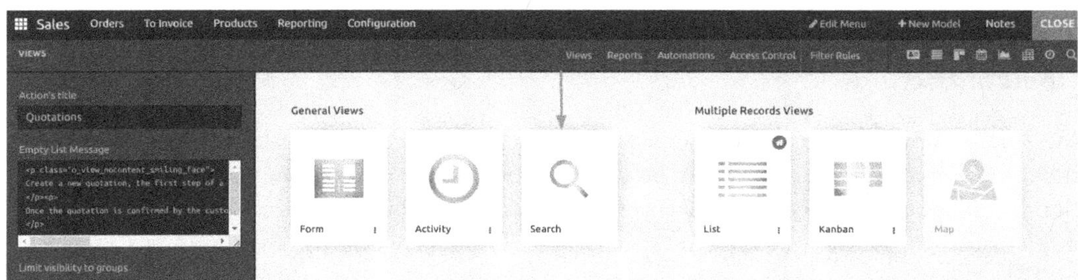

Figure 25.58 – The Search view

After clicking the **Search** view, the following screen will open. Here, you can modify or add search filters:

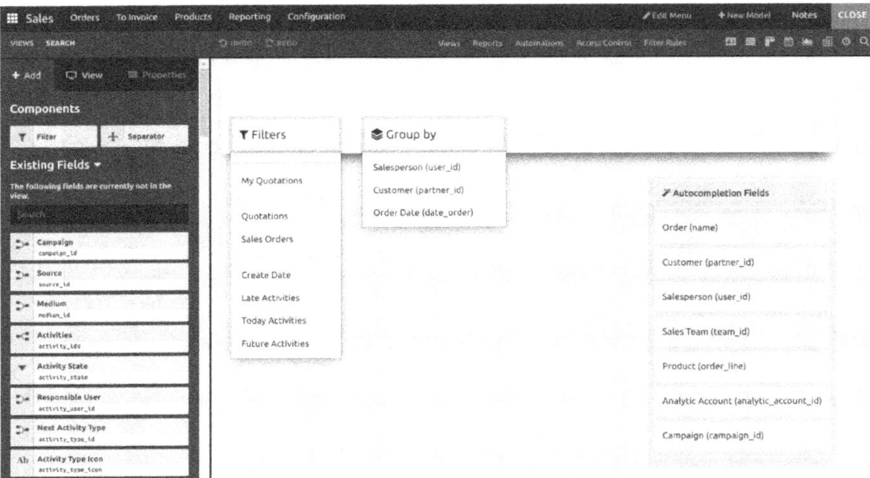

Figure 25.59 – Search view filters

Automations

Odoo provides automation capabilities to help streamline business processes and reduce manual tasks. These automation features are designed to make it easier for users to configure and customize automated actions within their Odoo applications.

Automated actions

Odoo Studio allows users to create automated actions that trigger specific tasks based on predefined conditions or events. These actions can be associated with various Odoo modules and can include actions such as creating records, sending emails, updating fields, and more. Users can define the conditions that trigger these actions and specify what should happen when the conditions are met.

Scheduled actions

Users can schedule automated actions to run at specific times or intervals. This is useful for tasks such as sending automated reminders, generating reports, or performing data maintenance tasks. Scheduled actions can be configured to execute daily, weekly, monthly, or on a custom schedule.

Email automation

Odoo Studio enables users to automate email notifications and communications. Users can set up automated email triggers for events such as order confirmation, invoice generation, or when specific conditions are met. Email templates can be customized to include dynamic data from Odoo records.

Server actions

Server actions allow users to define custom Python code or server-side logic that can be executed as part of an automated action. This provides advanced customization options for complex automation tasks that require custom programming:

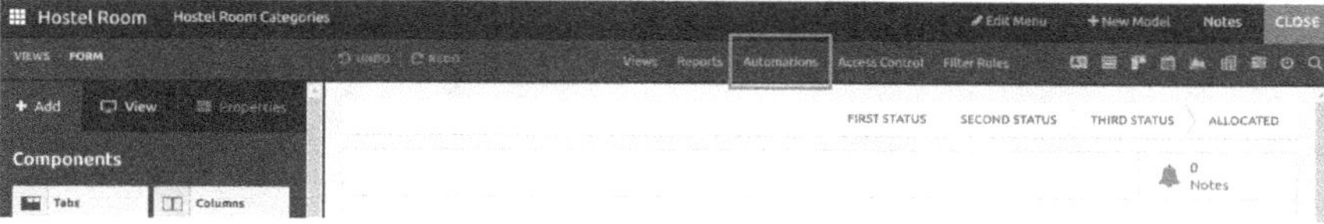

Figure 25.60 – Automations

Once you click on the **Automation** link, the following screen will appear. Here, you can add a new automation action:

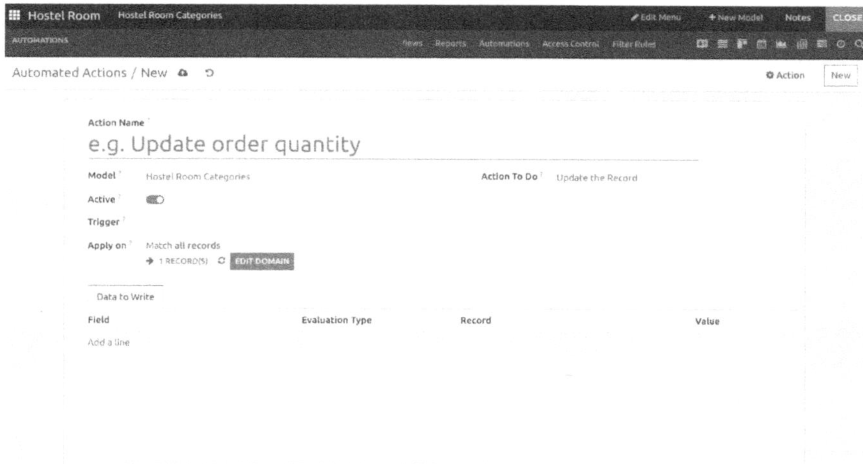

Figure 25.61 – Adding a new automation

There are multiple options you can choose from when performing an automation action:

- **Execute Python Code**
- **Create a new Record**
- **Update the Record**
- **Execute several actions**
- **Send Email**
- **Add Followers**
- **Create Next Activity**
- **Send SMS Text Message**

Here, choose **Execute Python Code** from the **Action To Do** dropdown to run the Python code:

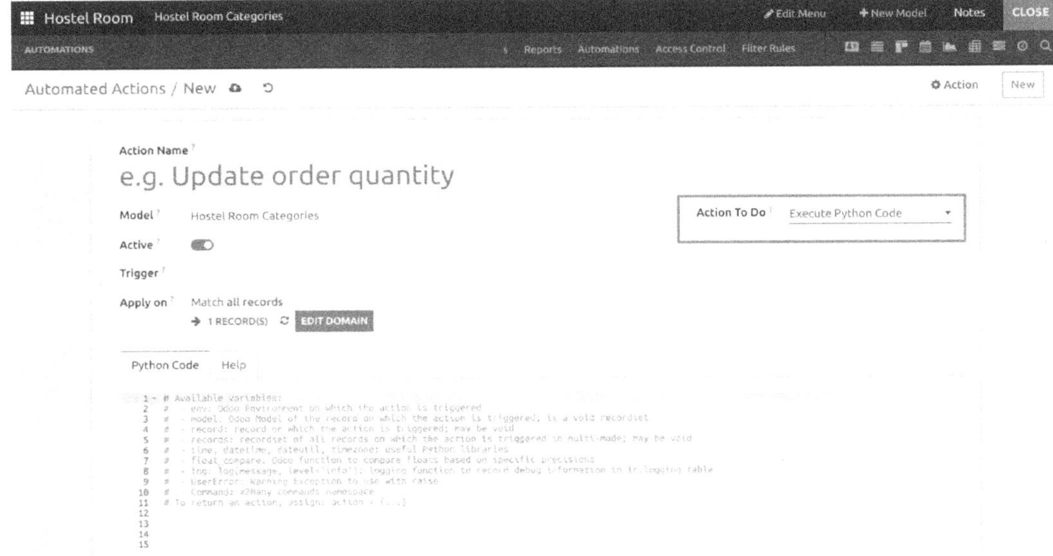

Figure 25.62 – Action To Do

Reports

In Odoo Web Studio, you can customize reports in terms of their layout, content, and appearance so that they meet your business needs. Report customization allows you to create professional and branded documents, such as invoices, purchase orders, quotations, and more.

Navigate to the **Studio** module within Odoo to access the report customization features.

Choose the app or module for which you want to customize reports. Typically, reports are associated with specific modules, such as **Sales**, **Purchase**, **Inventory**, or **Accounting**:

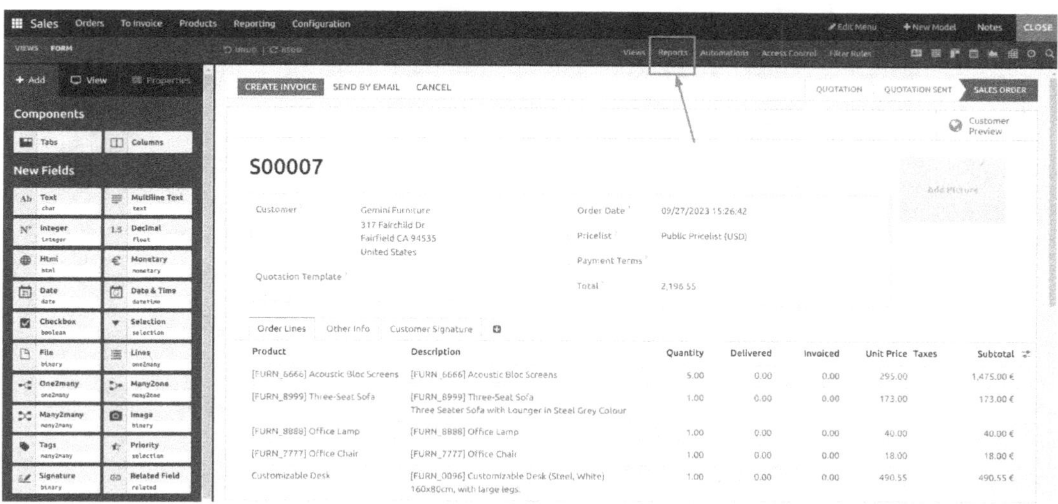

Figure 25.63 – The Reports menu

Once you click on the **Reports** menu, a screen will appear where you can choose existing model reports or create new reports:

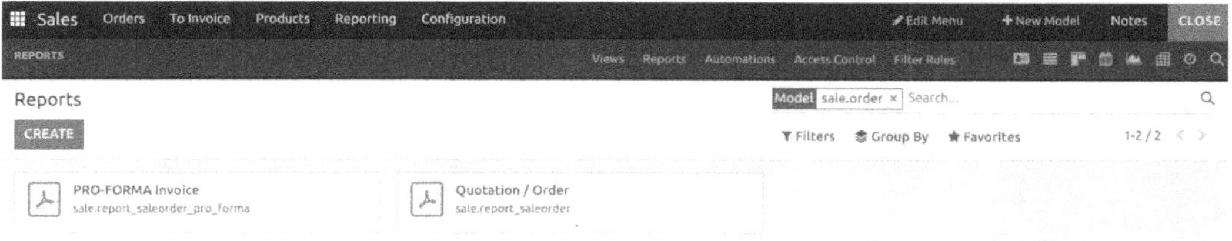

Figure 25.64 – Reports

Click **CREATE** to create a new report for the model:

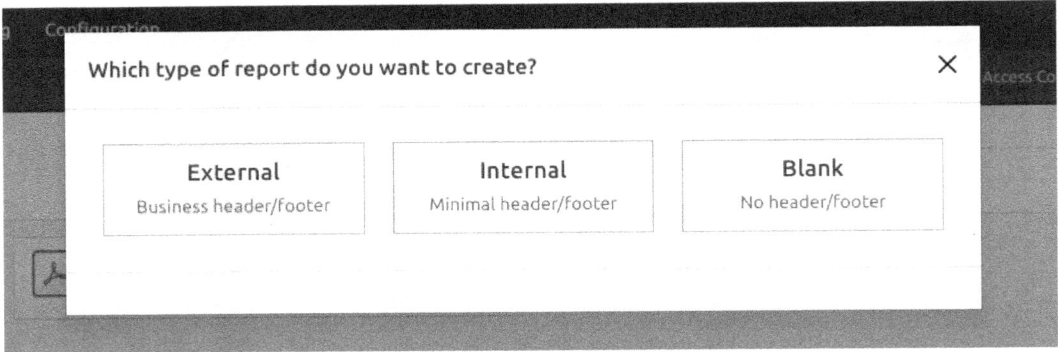

Figure 25.65 – Choosing a report type

Odoo Web Studio allows you to create external reports, also known as custom reports, so that you can generate documents and reports outside of the standard built-in reports provided by Odoo. External reports can be highly customized to meet specific business requirements.

External reports

Here's what the **External** report template looks like:

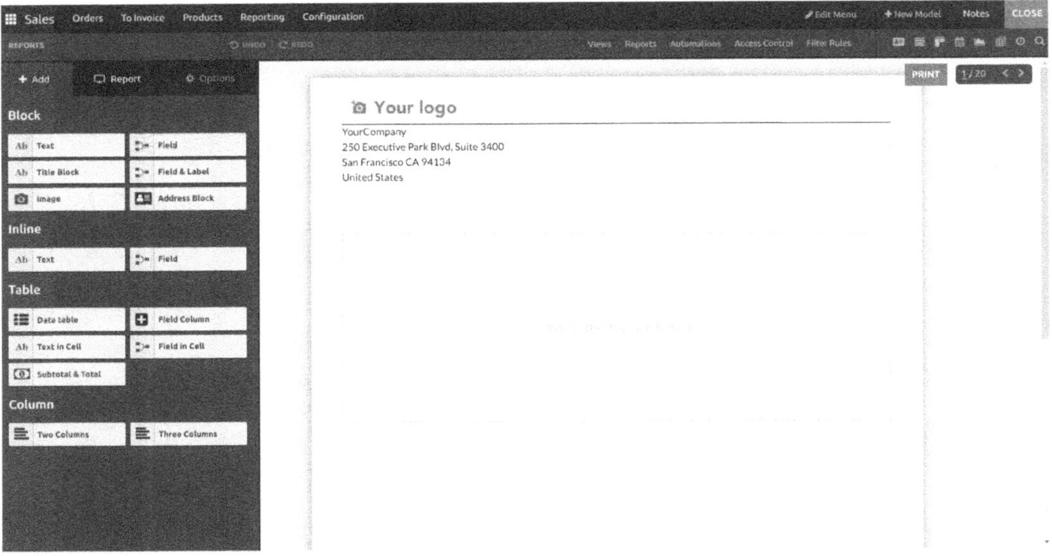

Figure 25.66 – An external report template

Internal reports

To create a PDF report in Odoo without a header and footer, choose the **Internal** report template:

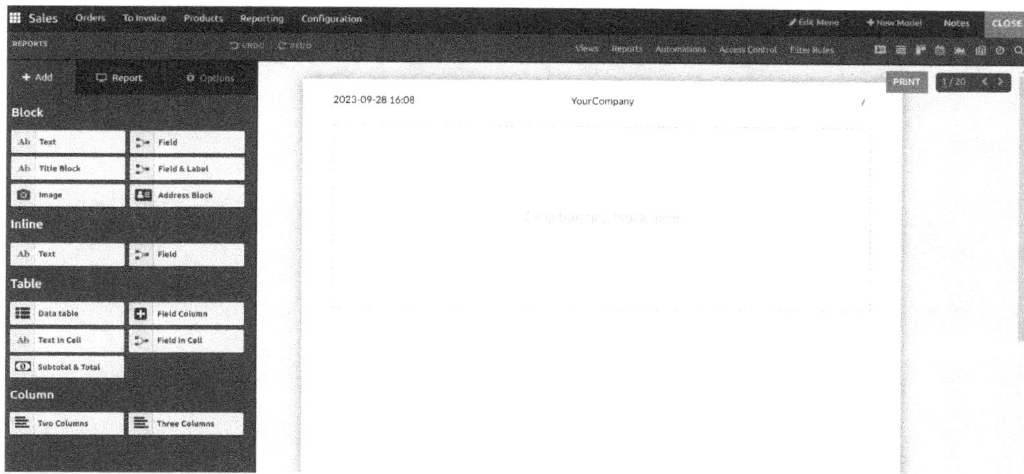

Figure 25.67 – An internal report template

Blank reports

To create a PDF report in Odoo without any predefined structure, choose the **Blank** report template:

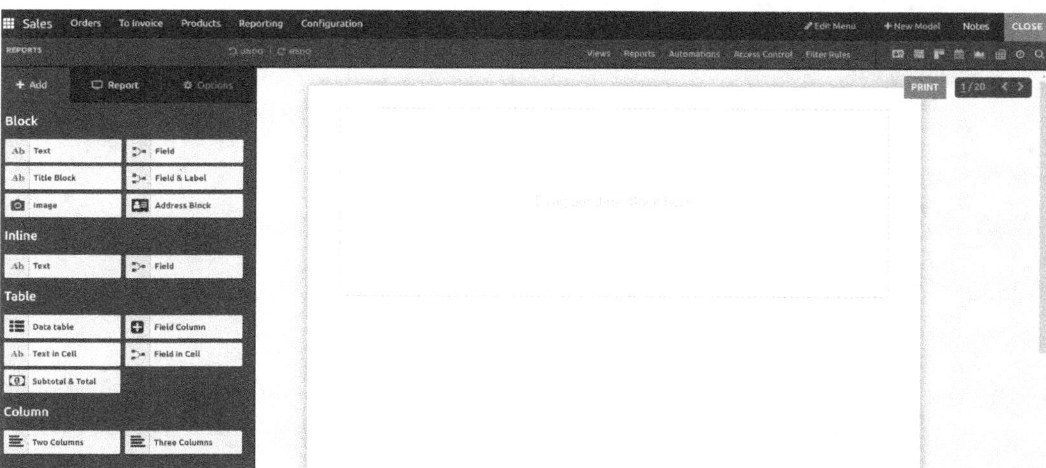

Figure 25.68 – A blank report template

In Odoo Web Studio, you can customize existing reports in terms of their layout, content, and appearance to meet your specific business needs. This customization allows you to make adjustments to standard reports provided by Odoo so that they match your company's branding and presentation requirements:

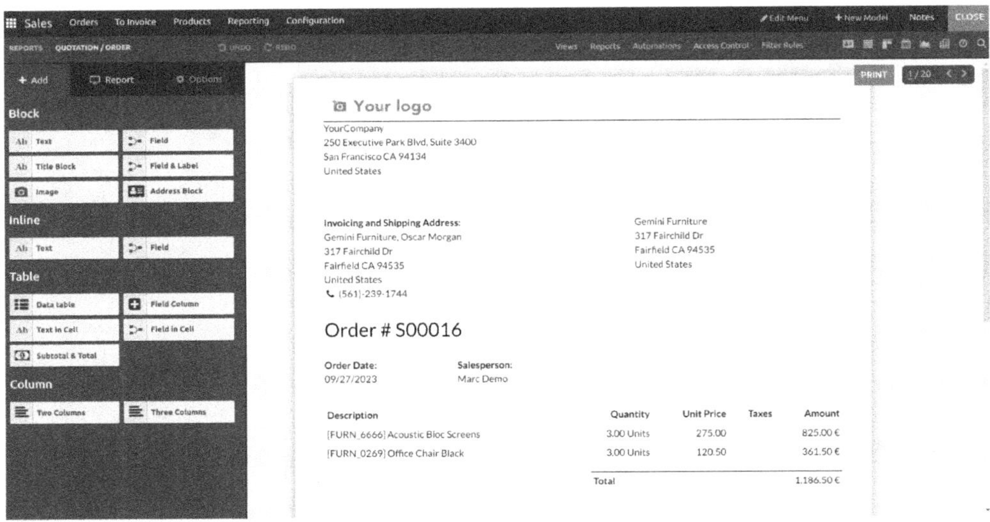

Figure 25.69 – The QUOTATION/ORDER sales report

There is also an **XML** editor option so that you can design complex parts of the report via code:

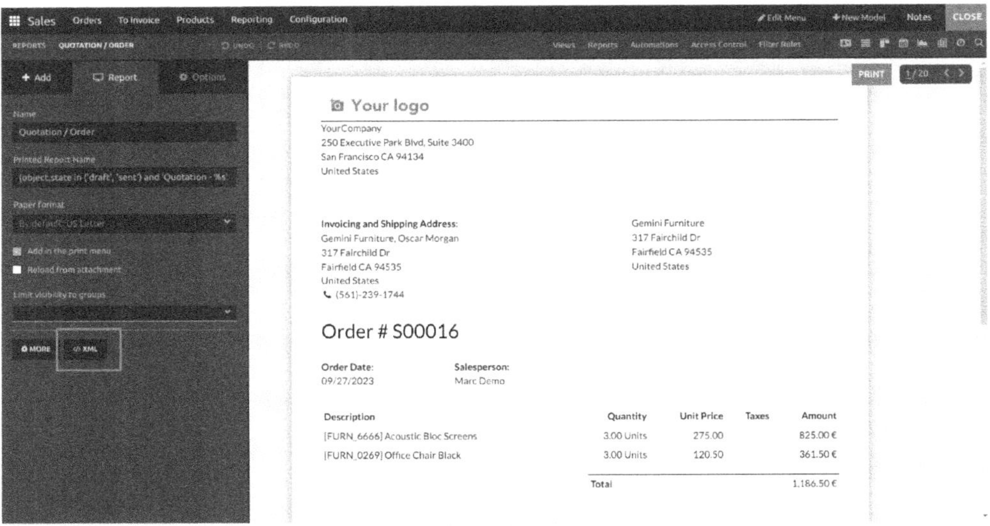

Figure 25.70 – The XML option

So, click **XML** to customize the report's design through code:

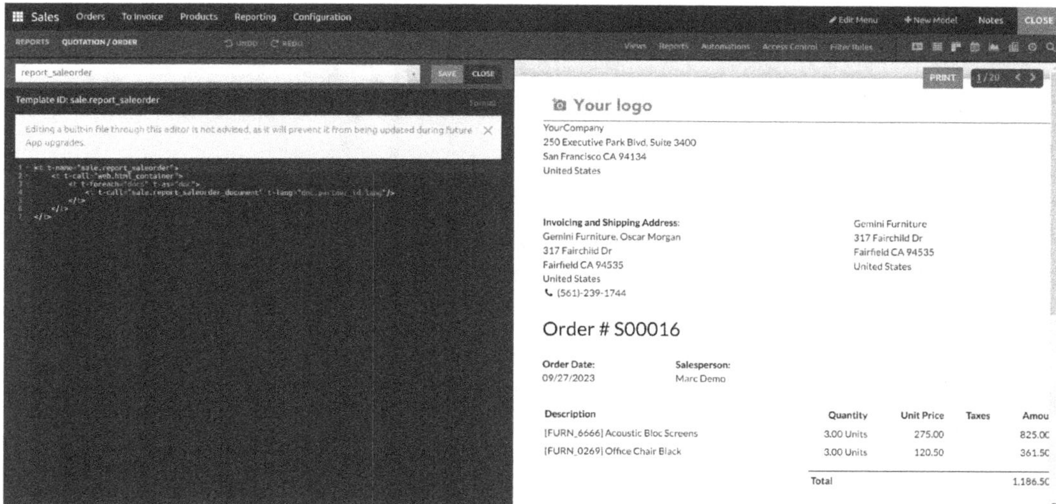

Figure 25.71 – The XML editor

Choose a different report template from the selection:

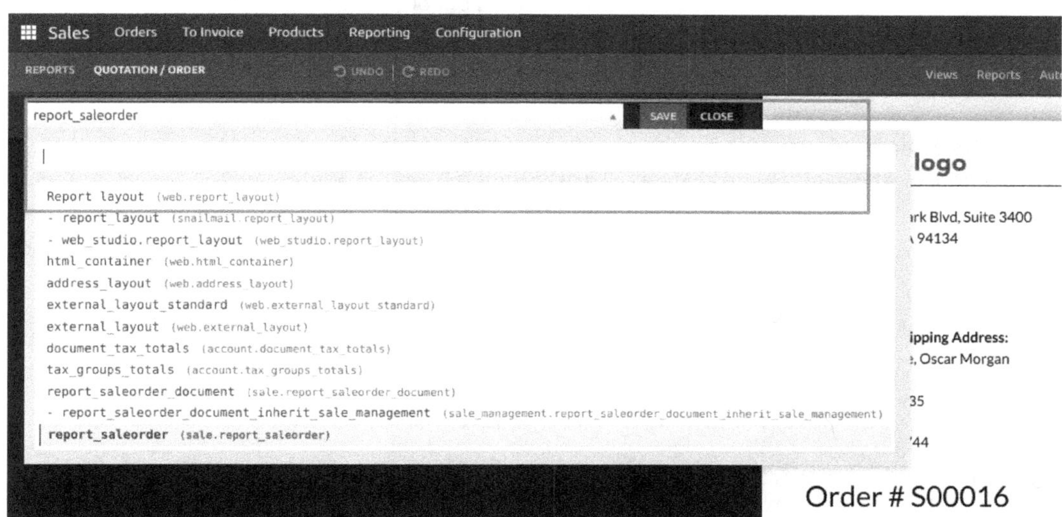

Figure 25.72 – The XML editor – choosing a different report

All newly created reports will be displayed under **Reports**:

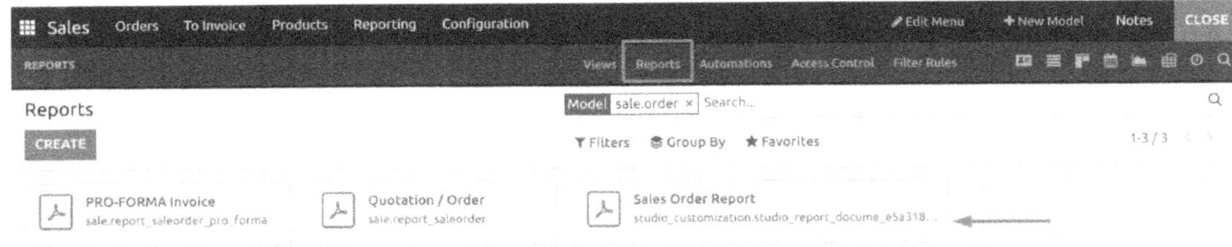

Figure 25.73 – Screenshot of the created report

Note that you now have the option to **Print** the report:

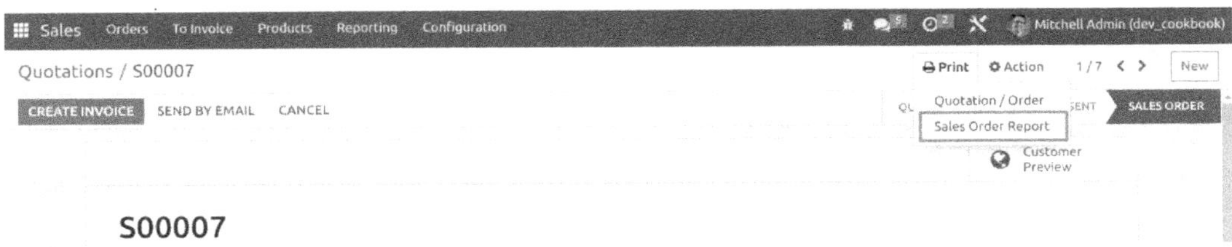

Figure 25.74 – The Print option

Modules

To export modules from the Odoo database, you have to install the Odoo Studio module and then start customizing the database. If we do any customization from the Studio module, a new module will be created in the database:

Figure 25.75 – The Export module

Once you click on the **Export** link, the Studio customize module will be downloaded. You can also import this module into other Odoo databases.

Search views

In Odoo Web Studio, you can customize search views to tailor the way users search for and filter records within a specific module or application. Search views allow users to refine their search criteria, making it easier to find specific records:

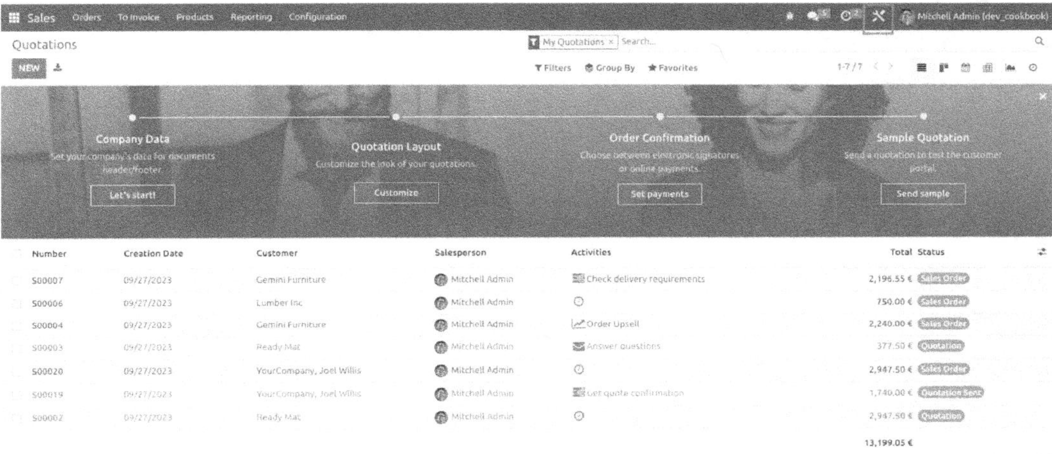

Figure 25.76 – Search filters

Click the **Studio** icon, then **VIEWS**. Once you've done this, select the **Search** view:

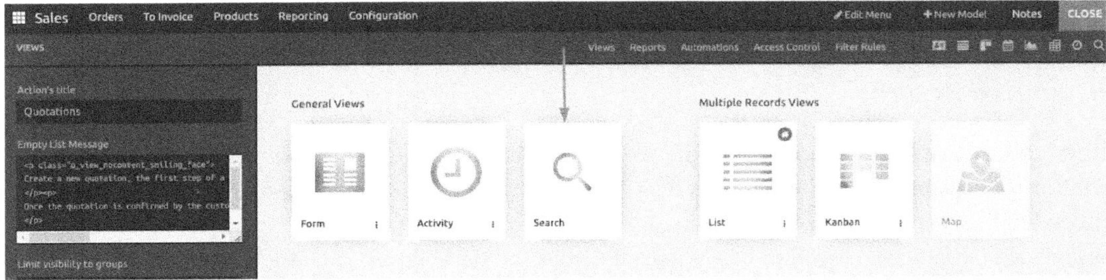

Figure 25.77 – The Search view

After clicking on the **Search** view, the following screen will open. Here, you can modify or add search filters:

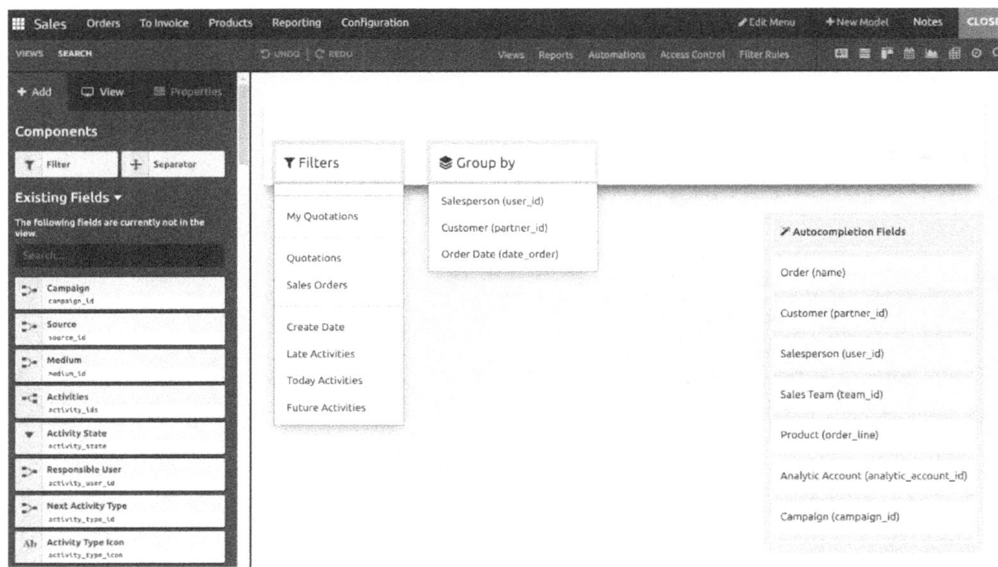

Figure 25.78 – Search view filters

Automations

Odoo provides automation capabilities to help streamline business processes and reduce manual tasks. These automation features are designed to make it easier for users to configure and customize automated actions within their Odoo applications.

Automated Actions

Odoo Studio allows users to create automated actions that trigger specific tasks based on predefined conditions or events. These actions can be associated with various Odoo modules and can include actions such as creating records, sending emails, updating fields, and more. Users can define the conditions that trigger these actions and specify what should happen when the conditions are met.

Scheduled Actions

Users can schedule automated actions to run at specific times or intervals. This is useful for tasks like sending automated reminders, generating reports, or performing data maintenance tasks. Scheduled actions can be configured to execute daily, weekly, monthly, or on a custom schedule.

Email Automation

Odoo Studio enables users to automate email notifications and communications. Users can set up automated email triggers for events such as order confirmation, invoice generation, or when specific conditions are met. Email templates can be customized to include dynamic data from Odoo records.

Server Actions

Server actions allow users to define custom Python code or server-side logic that can be executed as part of an automated action. This provides advanced customization options for complex automation tasks that require custom programming.

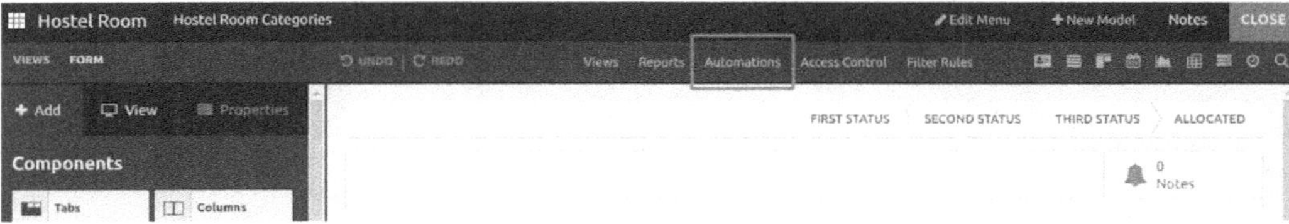

Figure 25.79 – Screenshot of the Automations

Once you click on the Automation link, it will open the screen to add a new automation action.

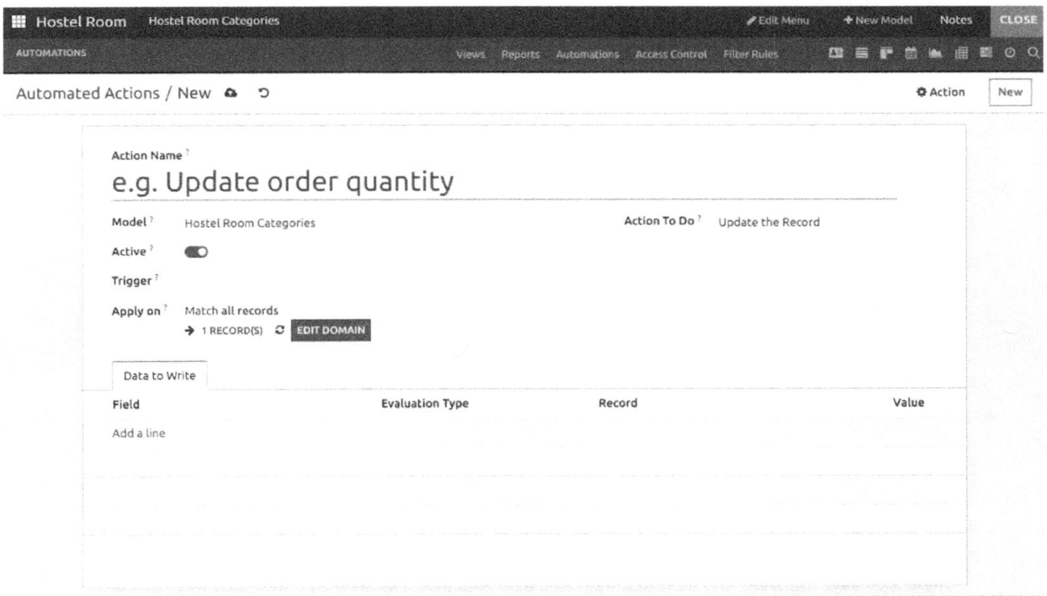

Figure 25.80 – Screenshot of the Add New Automation

There are multiple options to do an automation action.

- Execute Python Code
- Create a new Record
- Update the Record
- Execute several actions

- Send Email
- Add Followers
- Create Next Activity
- Send SMS Text Message

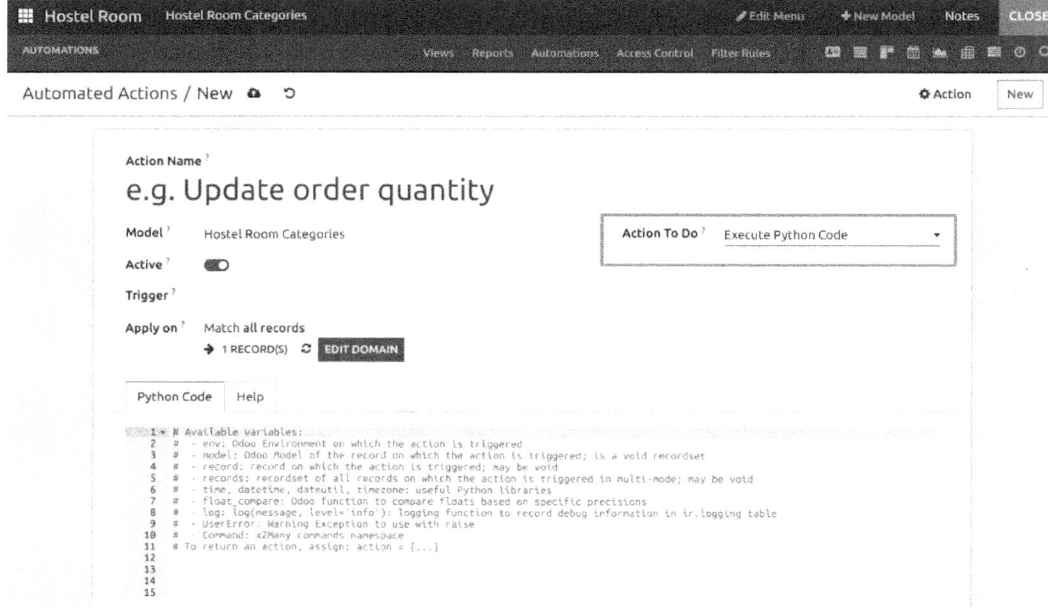

Figure 25.81 – Screenshot of the Action To Do

Index

Symbols

A

E

F

J

K

L

M

N

O

P

W

www.packtpub.com

Subscribe to our online digital library for full access to over 7,000 books and videos, as well as industry leading tools to help you plan your personal development and advance your career. For more information, please visit our website.

Why subscribe?

- Spend less time learning and more time coding with practical eBooks and Videos from over 4,000 industry professionals
- Improve your learning with Skill Plans built especially for you
- Get a free eBook or video every month
- Fully searchable for easy access to vital information
- Copy and paste, print, and bookmark content

Did you know that Packt offers eBook versions of every book published, with PDF and ePub files available? You can upgrade to the eBook version at packtpub.com and as a print book customer, you are entitled to a discount on the eBook copy. Get in touch with us at customercare@packtpub.com for more details.

At www.packtpub.com, you can also read a collection of free technical articles, sign up for a range of free newsletters, and receive exclusive discounts and offers on Packt books and eBooks.

Other Books You May Enjoy

If you enjoyed this book, you may be interested in these other books by Packt:

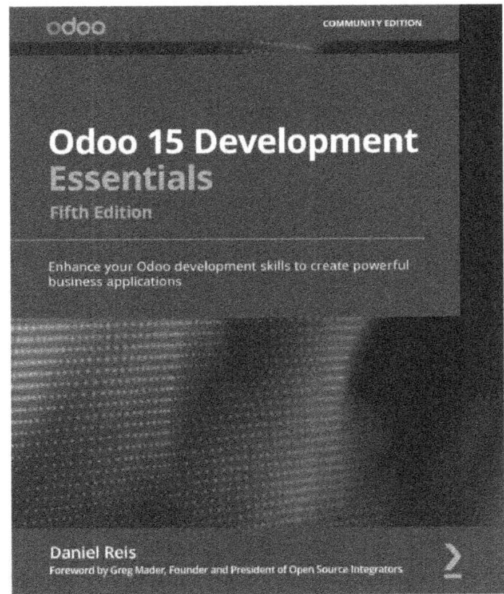

Odoo 15 Development Essentials

Daniel Reis

ISBN: 978-1-80020-006-7

- Install Odoo from source and organize the development environment
- Create your first Odoo app from scratch
- Understand the application components available in Odoo
- Structure the application's data model using ORM features
- Use the ORM API to implement the business logic layer
- Design a graphical user interface (GUI) for the web client and website
- Use the Odoo External API to interface with external systems
- Deploy and maintain your application in production environments

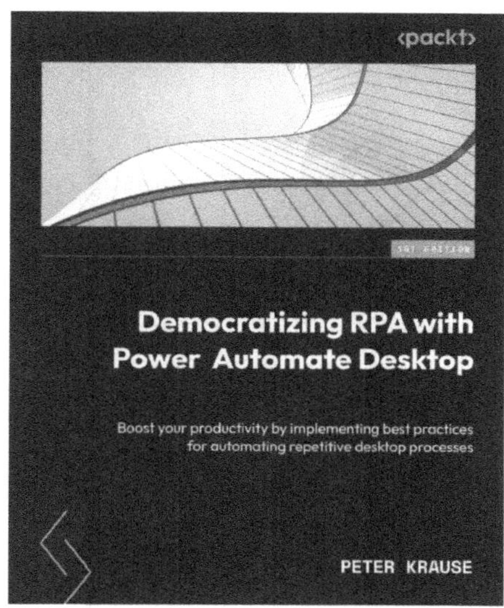

Democratizing RPA with Power Automate Desktop

Peter Krause

ISBN: 978-1-80324-594-2

- Master RPA with Power Automate Desktop to commence your debut flow
- Grasp all essential product concepts such as UI flow creation and modification, debugging, and error handling
- Use PAD to automate tasks in conjunction with the frequently used systems on your desktop
- Attain proficiency in configuring flows that run unattended to achieve seamless automation
- Discover how to use AI to enrich your flows with insights from different AI models
- Explore how to integrate a flow in a broader cloud context

Packt is searching for authors like you

If you're interested in becoming an author for Packt, please visit `authors.packtpub.com` and apply today. We have worked with thousands of developers and tech professionals, just like you, to help them share their insight with the global tech community. You can make a general application, apply for a specific hot topic that we are recruiting an author for, or submit your own idea.

Share your thoughts

Now you've finished *Odoo Development Cookbook*, we'd love to hear your thoughts! Scan the QR code below to go straight to the Amazon review page for this book and share your feedback or leave a review on the site that you purchased it from.

`https://packt.link/r/1805124277`

Your review is important to us and the tech community and will help us make sure we're delivering excellent quality content.

Download a free PDF copy of this book

Thanks for purchasing this book!

Do you like to read on the go but are unable to carry your print books everywhere?

Is your eBook purchase not compatible with the device of your choice?

Don't worry, now with every Packt book you get a DRM-free PDF version of that book at no cost.

Read anywhere, any place, on any device. Search, copy, and paste code from your favorite technical books directly into your application.

The perks don't stop there, you can get exclusive access to discounts, newsletters, and great free content in your inbox daily

Follow these simple steps to get the benefits:

1. Scan the QR code or visit the link below

`https://packt.link/free-ebook/9781805124276`

2. Submit your proof of purchase
3. That's it! We'll send your free PDF and other benefits to your email directly

www.ingramcontent.com/pod-product-compliance
Lightning Source LLC
LaVergne TN
LVHW081504050326
832903LV00025B/1389

* 9 7 8 1 8 0 5 1 2 4 2 7 6 *